Studies in Computational Intelligence

Volume 708

Series editor

Janusz Kacprzyk, Polish Academy of Sciences, Warsaw, Poland
e-mail: kacprzyk@ibspan.waw.pl

About this Series

The series "Studies in Computational Intelligence" (SCI) publishes new developments and advances in the various areas of computational intelligence—quickly and with a high quality. The intent is to cover the theory, applications, and design methods of computational intelligence, as embedded in the fields of engineering, computer science, physics and life sciences, as well as the methodologies behind them. The series contains monographs, lecture notes and edited volumes in computational intelligence spanning the areas of neural networks, connectionist systems, genetic algorithms, evolutionary computation, artificial intelligence, cellular automata, self-organizing systems, soft computing, fuzzy systems, and hybrid intelligent systems. Of particular value to both the contributors and the readership are the short publication timeframe and the worldwide distribution, which enable both wide and rapid dissemination of research output.

More information about this series at http://www.springer.com/series/7092

Guoyin Wang · Andrzej Skowron
Yiyu Yao · Dominik Ślęzak
Lech Polkowski
Editors

Thriving Rough Sets

10th Anniversary—Honoring Professor Zdzisław Pawlak's Life and Legacy & 35 Years of Rough Sets

Editors
Guoyin Wang
Chongqing Key Laboratory of Computational Intelligence
Chongqing University of Posts and Telecommunications
Chongqing
China

Andrzej Skowron
Institute of Mathematics
University of Warsaw
Warsaw
Poland

and

Systems Research Institute
Polish Academy of Sciences
Warsaw
Poland

Yiyu Yao
Department of Computer Science
University of Regina
Regina, SK
Canada

Dominik Ślęzak
Institute of Informatics
University of Warsaw
Warsaw
Poland

and

Infobright
Warsaw
Poland

Lech Polkowski
Polish-Japanese Academy of Information Technology
Warsaw
Poland

and

Department of Mathematics and Computer Science
University of Warmia and Mazury
Olsztyn
Poland

ISSN 1860-949X ISSN 1860-9503 (electronic)
Studies in Computational Intelligence
ISBN 978-3-319-85533-2 ISBN 978-3-319-54966-8 (eBook)
DOI 10.1007/978-3-319-54966-8

Printed on acid-free paper

This Springer imprint is published by Springer Nature
The registered company is Springer International Publishing AG
The registered company address is: Gewerbestrasse 11, 6330 Cham, Switzerland

Preface

It is the 10th anniversary of the death of Prof. Zdzisław Pawlak, the father of rough set theory, in 2016. He set up the rough set theory in 1982, which has become one of the major theories for processing uncertain information.

As Prof. Lotfi A. Zadeh said, Prof. Pawlak was a great scientist and a great human being. He entered science history as not only the father of rough set theory, but also as the designer of the first grammar of DNA, the precursor of mosaic and picture grammars, the inventor of the digital computer based on -2 system and a random numbers generator. Besides these great scientific achievements, he was a very warm, cordial man, demanding but fair, and a man of many talents and interests, a painter, a poet, an ardent tourist, a craftsman, a truly renaissance man.

In observance of the 10th anniversary of his departure, the International Rough Set Society (IRSS) organized two special memorial sessions commemorating him in 2016, that is, the plenary panel on the legacy of Prof. Zdzisław Pawlak at FedCSIS'16, September 11–14, 2016, and the special memorial session for Prof. Pawlak at IJCRS2016, October 7–11, 2016. Polish Information Processing Society under the auspices of the Institute of Computer Science at Warsaw University of Technology organized a special session celebrating the 90th anniversary of birth of Prof. Zdzisław Pawlak on December 6, 2016. Moreover, a special plenary session celebrating the 35th anniversary of the pioneering work on rough sets by Prof. Zdzisław Pawlak will be held at IJCRS'17, in Olsztyn, Poland, July 3–7, 2017.

In addition to these special sessions, IRSS is going to publish this special memorial book entitled "Thriving Rough Sets: 10th Anniversary—Honoring Professor Zdzisław Pawlak's Life and Legacy & 35 years of Rough Sets", in 2017.

This book includes 20 chapters. They are divided into four sections, that is, historical review of Prof. Zdzisław Pawlak and rough set, review of rough set research, rough set theory, and rough set-based data mining.

The first part of this book is about the historical review of Prof. Zdzisław Pawlak and rough set. In Chapter "The Born and Growing of Chinese Rough Set Community with Help of Professor Zdzisław Pawlak", Prof. Guoyin Wang introduces the history of the born and quick growing of Chinese rough set community with the help of Prof. Pawlak. China is becoming a very active country in the field of rough set theory. In Chapter "Zdzisław Pawlak as I Saw Him and Remember Him Now", Prof. Lech Polkowski sums up his experiences of working and living with Prof. Pawlak, shares what he knows about him, and introduces some less known achievements of him. In Chapter "Recent Development of Rough Computing: A Scientometrics View" by JingTao Yao and Adeniyi Onasanya, the authors use scientometrics approach to quantitatively analyze the contents and citations of rough set publications. They find some interesting results in key indicators between 2013 and 2016 results. Their study results indicate that rough sets as a research domain is attracting more researchers and growing healthily in recent years.

The second part of this book is about the review of rough set research. In Chapter "Rough Sets, Rough Mereology and Uncertainty", Prof. Lech Polkowski reviews the rough set research in many realms like morphology, intelligent agents, linguistics, behavioral robotics, mereology, and granular computing. He sums up his personal experience and results, and in a sense to unify them into a coherent conceptual scheme following the main themes of rough set theory: to understand uncertainty and to cope with it in data. In Chapter "Rough Sets in Machine Learning: A Review" by Rafael Bello and Rafael Falcon, the authors survey the existing literature and report the most relevant theoretical developments and applications of rough set theory in a broad field of machine learning. Chapter "Application of Tolerance Rough Sets in Structured and Unstructured Text Categorization: A Survey" by Sheela Ramanna, James Francis Peters, and Cenker Sengoz, presents a survey of literature, where tolerance rough set model is used as a text categorization and learning model. It demonstrates the versatility of the tolerance form of rough sets and its successful application in text categorization and labeling. In Chapter "Medical Diagnosis: Rough Set View", Prof. Shusaku Tsumoto discusses the formalization of medical diagnosis from the viewpoint of rule reasoning based on rough sets. In Chapter "Rough Set Analysis of Imprecise Classes", Prof. Masahiro Inuiguchi proposes to use the lower approximations of unions of k decision classes to enrich the applicability of rough set approaches instead of the lower approximations of single classes in the classical rough set approaches. In Chapter "Pawlak's Many Valued Information System, Non-deterministic Information System, and a Proposal of New Topics on Information Incompleteness Toward the Actual Application" by Hiroshi Sakai, Michinori Nakata, and Yiyu Yao, the authors discuss Pawlak's many valued information systems (MVISs), non-deterministic information systems (NISs), and related new topics on information incompleteness toward the actual application. They survey their previous research and propose new topics toward the actual application of NIS, namely data mining under various types of uncertainty, rough set-based estimation of an actual value, machine learning by rule generation, information dilution, and privacy-preserving issue, in NISs.

The third part is about recent achievements of the study of rough set theory. In Chapter "From Information Systems to Interactive Information Systems" by Andrzej Skowron and Soma Dutta, the authors propose a departure from classical notion of information systems, and bring in the background of agent's interaction with physical reality in arriving at a specific information system. They propose to generalize the notion of information systems from two aspects. In Chapter "Back to the Beginnings: Pawlak's Definitions of the Terms Information System and Rough Set", Prof. Davide Ciucci discusses two basic notions and terms, rough set and information system, which have no crystal clear definitions in rough set theory. In Chapter "Knowledge and Consequence in AC Semantics for General Rough Sets", Prof. A. Mani introduces an antichain based semantics for general rough sets. She develops two different semantics, one for general rough sets and another for general approximation spaces over quasi-equivalence relations, and studies the epistemological aspects of the semantics. Chapter "Measuring Soft Roughness of Soft Rough Sets Induced by Covering" by Amr Zakaria studies some important properties of soft rough sets induced by soft covering. A measure of soft roughness is introduced via soft covering approximation. A new approach of soft rough approximation space is presented via a measure of soft roughness. Chapter "Rough Search of Vague Knowledge" by Edward Bryniarski and Anna Bryniarska discusses the theoretical basis of the vague knowledge search, introduces some data granulation methods in semantic networks. In Chapter "Vagueness and Uncertainty: An F-Rough Set Perspective" by Dayong Deng and Houkuan Huang, the authors investigate vagueness and uncertainty from the viewpoints of F-rough sets. Some indexes, including two types of F-roughness, two types of F-membership-degree and F-dependence degree etc., are defined. In Chapter "Directions of Use of the Pawlak's Approach to Conflict Analysis", Prof. Malgorzata Przybyla-Kasperek applies the Pawlak's model to analyze the conflicts that arise between classifiers in decision making. Chapter "Lattice Structure of Variable Precision Rough Sets" by Sumita Basu studies the algebraic properties of set of variable precision rough sets for a particular imprecise set.

The fourth part of this book is about the application of rough set in data mining. In Chapter "Mining for Actionable Knowledge in Tinnitus Datasets" by Katarzyna A. Tarnowska, Zbigniew W. Ras, and Pawel J. Jastreboff, the authors verify the possibility of applying theory of traditional machine learning techniques, such as classification and association rules, as well as novel data mining methods, including action rules and meta actions, to a practical decision problem in the area of medicine. Knowledge discovery approaches with an ultimate goal of building rule-based recommender system for tinnitus treatment and diagnosis are investigated. Chapter "Rough-Granular Computing for Relational Data" by Piotr Honko introduces three rough-granular approaches dedicated to handle complex data such as relational one. The three approaches are also compared in terms of construction of information systems, information granules, and approximation spaces. In Chapter "The Boosting and Bootstrap Ensembles for the Pair Classifier Based on the Dual Indiscernibility Matrix" by Piotr Artiemjew, Lech Polkowski, Bartosz Nowak, and Przemysław Gorecki, the authors examine selected methods for stabilization of the pair classifier like bootstrap ensemble, arcing based bootstrap, Ada-Boost with Monte Carlo split.

Many distinguished researchers helped to review papers for this book. We express our gratefulness to Profs. Jerzy W. Grzymała-Busse, Mihir K. Chakraborty,

Davide Ciucci, Chris Cornelis, Salvatore Greco, Qinghua Hu, Huaxiong Li, Pawan Lingras, Dun Liu, Jusheng Mi, Duoqian Miao, Yuhua Qian, Sheela Ramanna, Hiroshi Sakai, Shusaku Tsumoto, Jingtao Yao, Zhiwen Yu, Xianzhong Zhou, and Wojciech Ziarko, for serving as reviewers. We are thankful to Dr Xin Deng and Dr Zhixing Li for their help to edit the book.

We are also thankful to Prof. Janusz Kacprzyk, Series Editor of "Studies in Computational Intelligence" for Springer, Dr. Thomas Ditzinger, Executive Editor of Springer, Interdisciplinary and Applied Sciences & Engineering, and Mr. Ramamoorthy Rajangam, Project Coordinator of Books Production at Springer, for their support and cooperation to publish the book.

Chongqing, China	Guoyin Wang
Warsaw, Poland	Andrzej Skowron
Regina, Canada	Yiyu Yao
Warsaw, Poland	Dominik Ślęzak
Warsaw/Olsztyn, Poland	Lech Polkowski
December 2016	

Contents

Part I
Historical Review of Professor Zdzisław Pawlak and Rough Set

The Born and Growing of Chinese Rough Set Community with Help of Professor Zdzisław Pawlak

Guoyin Wang

Abstract Rough Set is a mathematical theory for processing uncertain data and information, which was first described by Professor Pawlak in 1982. The rough set research in China is growing very quickly in recent years. Professor Pawlak helped the development of Chinese rough set community a lot. It is the 10th anniversary of Professor Pawlak's death this year. This paper is a record of the history of the born and growing of Chinese rough set community and a commemoration of Professor Pawlak, our old friend.

Keywords Rough Set · China · Professor Zdzisław Pawlak

1 The Born of Chinese Rough Set Community

Rough Set is a mathematical theory for processing uncertain data and information, which was first described by Professor Zdzisław Pawlak in 1982 [1]. As the father of rough set theory, Professor Pawlak had great contribution to its growing and developing in the whole world [2, 3], including China.

The rough set study in China was started in 1990s. The Chinese Rough Set and Soft Computing Society (CRSSC) became a branch of the Chinese Association of Artificial Intelligence (CAAI) in 2003. It has about 800 members now. It becomes the largest national rough set society in the whole world.

I am writing this article to record the history of the born and growing of rough set community in China in the past 20 years, and introduce the great help and support of Professor Pawlak for the born and quick growing of the rough set in China to commemorate the 10 anniversary of his death.

G. Wang (✉)
Chongqing Key Laboratory of Computational Intelligence, Chongqing University of Posts and Telecommunications, 400065 Nan'an District, Chongqing, People's Republic of China
e-mail: wanggy@ieee.org

G. Wang et al. (eds.), *Thriving Rough Sets*, Studies in Computational Intelligence 708, DOI 10.1007/978-3-319-54966-8_1

After graduating from the Xi'an Jiaotong University at the end of 1996, I joined the Chongqing University of Posts and Telecommunications. I read some rough set papers and start to learn and study rough set based knowledge acquisition from 1997. This is my first time to know the name of rough set theory and Professor Pawlak. I got my 1st NSFC research grant "Rough Set Based Automatic Knowledge Acquisition Technology and Its Application" in 1998, and visited Professor Yiyu Yao at University of Regina for about 1 month in 1999. Professor Yiyu Yao introduced a lot about rough set theory and Professor Pawlak to me when I was in Regina. After coming back at Chongqing, I started to set up a Chinese rough set society and initialize a Chinese rough set workshop/conference. Professor Qing Liu, a key senior rough set researcher in China, planed it together with me. We sent an email to Professor Pawlak to invite him to visit Chongqing, attend the 1st Chinese Conference on Rough Sets and Soft Computing (CRSSC2001) and give a keynote talk. We received his positive response immediately.

The First Chinese Conference on Rough Set and Soft Computing (CRSSC2001) [4] was held in Chongqing University of Posts and Telecommunications May 25–27, 2001. 86 attendees joined this conference. Professor Pawlak gave a keynote talk "Applications of Rough Set Theory to Drawing Conclusions from Data". Professor Pawlak is at the 8th of the first line of the group picture of CRSSC2001 shown in Fig. 1. Figure 2 is the CRSSC2001 proceeding which is published as a special issue of the Journal of Computer Science. The Chongqing city government thought highly of the conference and supported it very much. Some high-ranking officials of the Chongqing government such as Professor Ruihua Dou, the vice Chairman of the

Fig. 1 Group picture of CRSSC2001

Fig. 2 Proceedings of CRSSC2001

Chongqing Political Consultation Committee and the former vice mayor of Chongqing, Professor Chunlin Zhai, the vice President of the Chongqing Association for Science and Technology, Professor Yuhui Qiu, the President of Southwest Normal University and the Chairman of Chongqing Computer Society, and Professor Neng Nie, the President of Chongqing University of Posts and Telecommunications, attended the conference. During this conference, I proposed a plan for setting up a Chinese rough set society. Many CRSSC2001 attendees expressed their interests of joining the society. Professor Pawlak stayed in Chongqing 4 days. He discussed with Chinese researchers about rough set related researches and introduced the development of rough set theory. He expressed his desire to support the development of rough set theory in China. Thus, we can say that Professor Pawlak witnessed and contributed to the born of Chinese rough set community.

2 Rough Set Conferences Organized in China

Since 2001, the Chinese conference on rough set and soft computing (CRSSC) is held every year. It is held together with the Chinese conference on web intelligence (CWI) and Chinese conference on granular computing (CGrC) every year since 2007. Until now, it has been held 15 times in many different cities in China. Table 1 is a list of the national rough set conferences in China. These conferences pushed the quick development of rough set research in China.

Professor Pawlak also helped Chinese researchers to join the international rough set community. With his help and support, the 9th International Conference on Rough Sets, Fuzzy Sets, Data Mining and Granular Computing (RSFDGrC2003) [5] was held in Chongqing, China. It was the 1st time to organize an international rough set conference in China. It was originally planned in May, 2003. Professor Pawlak prepared his keynote talk "Flow Graphs and Decision Algorithms" [6] and booked his ticket for attending the conference. Unfortunately, SARS happened in the whole world in 2003. RSFDGrC2003 had to be postponed to October, 2003. Professor Pawlak changed his trip plan and failed to attend the conference in October due to his health reasons at last.

In order to promote the rough set theory in the field of knowledge technology. A new international conference "Rough Sets and Knowledge Technology (RSKT)" was initialized in 2006. The 1st international conference on rough sets and knowledge technology (RSKT2006) [7] was organized in Chongqing, July, 2006. Professor Pawlak prepared his keynote talk "Conflicts and Negotations" for this conference [8]. Unfortunately, he passed away on 7 April 2006. Professors James F Peters and Andrzej Skowrn wrote a commemorative paper "Some Contributions by Zdzisław Pawlak" [3]. It should be the 1st commemorative paper for Professor Pawlak after his death. The proceeding of RSKT2006 should also be the 1st commemorative book for Professor Pawlak (Figs. 3 and 4).

Table 1 Chinese national rough set conferences

Conference name	City	Number of attendees
CRSSC2001	Chongqing	86
CRSSC2002	Suzhou	About 150
CRSSC2003	Chongqing	123
CRSSC2004	Zhoushan	Over 100
CRSSC2005	Anshan	98
CRSSC2006	Jinhua	75
CRSSC-CWI-CGrC2007	Taiyuan	356
CRSSC-CWI-CGrC2008	Xinxiang	193
CRSSC-CWI-CGrC2009	Shijiazhuang	164
CRSSC-CWI-CGrC2010	Chongqing	123
CRSSC-CWI-CGrC2011	Nanjing	218
CRSSC-CWI-CGrC2012	Hefei	About 180
CRSSC-CWI-CGrC2013	Zhangzhou	253
CRSSC-CWI-CGrC2014	Kunming	302
CRSSC-CWI-CGrC2015	Tangshan	Over 300
CRSSC-CWI-CGrC2016	Yantai	378

Note (1) CRSSC is the Chinese conference on rough set and soft computing
(2) CRSSC-CWI-CGrC the joint conference of Chinese conference on rough set and soft computing, Chinese conference on web intelligence, and Chinese conference on granular computing (CGrC)

Fig. 3 Proceeding of RSKT2006

Since 2003, many international rough set events have been organized in China. It greatly pushed the international academic exchange between Chinese rough set researchers and oversea rough set researchers. It makes the rough set research grow quickly. A list of rough set related international conferences held in China is available in Table 2.

Fig. 4 Professor Pawlak's photograph included in the proceeding of RSKT2006

Table 2 International rough set conferences held in China

Conference name	City
RSFDGrC2003	Chongqing
RSKT2006	Chongqing
IFTGrCRSP2006	Nanchang
IFKT2008	Chongqing
RSKT2008	Chengdu
RST2010	Zhoushan
RSKT2010	Beijing
JRS2012	Chengdu
RSKT2014	Shanghai
IJCRS2015	Tianjin

Note RSFDGrC is international conference on rough sets, fuzzy sets, data mining and granular computing
RSKT is international conference on rough sets and knowledge technology
IFTGrCRSP is international forum on theory of GrC from rough set perspective
IFKT is international forum on knowledge technology
RST is international workshop on theoretical rough sets
JRS is joint rough set symposium
IJCRS is international joint conference on rough sets

3 The Growing of Chinese Rough Set Community

In order to push the development of rough set research in China, Chinese Association of Artificial Intelligence (CAAI) set up a rough set and soft computing branch (CRSSC) in 2003 (http://CS.CQUPT.EDU.CN/CRSSC). Professor Xuyan Tu, the Honorary President of CAAI, wrote the poem in Fig. 5 to celebrate the establishment of CRSSC in Guangzhou, 21 November, 2003.

Professor Guoyin Wang at Chongqing University of Posts and Telecommunications had served as the Chairman of CRSSC from 2003 to 2012. Since 2012,

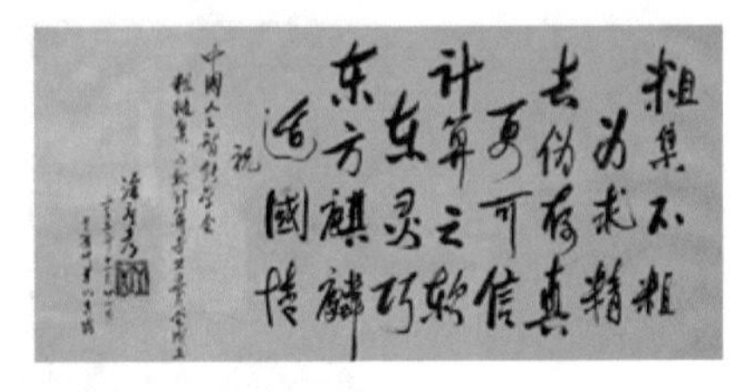

a. Original version in Chinese

Rough sets are not rough, and one moves towards precision.

One removes the "unbelievable" so that what remains is more believable.

The soft part of computing is nimble.

Rough sets imply a philosophy rooted in China.

b. Translated version in English

Fig. 5 Poem by Professor Xuyan Tu

Table 3 Rough set papers published in Chinese journals

Year	Papers	Year	Papers	Year	Papers
1991	1	2000	107	2009	1192
1992	1	2001	168	2010	1084
1993	1	2002	281	2011	853
1994	3	2003	412	2012	878
1995	1	2004	567	2013	728
1996	4	2005	756	2014	740
1997	10	2006	937	2015	738
1998	26	2007	1070		
1999	57	2008	1244		

Note It is a retrieval result from CNKI [10] on July 14, 2016

Professor Duoqian Miao at Tongji University serves as the Chairman. CRSSC sponsored all the above national and international rough set related events held in China.

The rough set research in China is growing very quickly. The information of the research on rough set theory and applications in China before 2008 is available in [9]. There are about 700–800 rough set related research papers published in Chinese academic journals in recent years. Detailed information about it is in Table 3.

Now, rough set theory has become one of the key scientific research fields in computing technology and artificial intelligence technology [11]. CRSSC is also a very active branch of CAAI, the largest national rough set society in the world. Chinese Journal of Computers is one of the key journals in the field of information technology in China. It was started in 1978. The top 15 highly cited papers published in Chinese Journal of Computers since its 1st issue is listed in Table 4. It could be found that 4 rough set papers (bold) are in the list. They are the No. 4, No. 12, No. 13, and No. 15. More and more Chinese rough set researchers are publishing a lot of high quality papers in international conferences and journals nowadays. Especially, there are a lot of young Chinese rough set researchers. They will be the future of rough set theory research.

Table 4 Top 15 most cited papers published in Chinese journal of computers

No	Author(s)	Title	Year	Cite
1	王亮, 胡卫明, 谭铁牛 Liang Wang, Weiming Hu, Tieniu Tan	人运动的视觉分析综述 A Survey of Visual Analysis of Human Motion	2002	1201
2	梁路宏, 艾海舟, 徐光祐, 张钹 Luhong Liang, Haizhou Ai, Guangyou Xu, Bo Zhang	人脸检测研究综述 A Survey of Human Face Detection	2002	1048
3	史美林, 杨光信, 向勇, 伍尚广 Meilin Shi, Guangxin Yang, Yong Xiang, Shangguang Wu	WfMS:工作流管理系统 WfMS: Workflow Management System	1999	987
4	**王国胤, 于洪, 杨大春** **Guoyin Wang, Hong Yu, Dachun Yang**	**基于条件信息熵的决策表约简** **Decision Table Reduction Based on Conditional information Entropy**	**2002**	**964**
5	丁玮, 齐东旭 Wei Ding, Dongxu Qi	数字图像变换及信息隐藏与伪装技术 Digital Image Transformation and Information Hiding and Disguising Technology	1998	623
6	徐光祐, 史元春, 谢伟凯 Guangyou Xu, Yuanchun Shi, Weikai Xie	普适计算 Pervasive/Ubiquitous Computing	2003	589
7	王珊, 王会举, 覃雄派, 周烜 Shan Wang, Huiju Wang, Xiongpai Qin, Xuan Zhou	架构大数据:挑战、现状与展望 Architecting Big Data: Challenges, Studies and Forecasts	2011	586
8	吴斌, 史忠植 Bin Wu, Zhongzhi Shi	一种基于蚁群算法的TSP问题分段求解算法 An Ant Colony Algorithm Based Partition Algorithm for TSP	2001	583
9	罗万明, 林闯, 阎保平 Wanming Luo, Chuang Lin, Baoping Yan	TCP/IP拥塞控制研究 A Survey of Congestion Control in the Internet	2001	574
10	李成法, 陈贵海, 叶懋, 吴杰 Chengfa Li, Guihai Chen, Mao Ye, Jie Wu	一种基于非均匀分簇的无线传感器网络路由协议 An Uneven Cluster-Based Routing Protocol for Wireless Sensor Networks	2007	544
11	吴健, 吴朝晖, 李莹, 邓水光 Jian Wu, Zhaohui Wu, Ying Li, Shuiguang Deng	基于本体论和词汇语义相似度的Web服务发现 Web Service Discovery Based on Ontology and Similarity of Words	2005	514
12	**王国胤, 姚一豫, 于洪** **Guoyin Wang, Yiyu Yao, Hong Yu**	**粗糙集理论与应用研究综述** **A survey on Rough Set Theory and Its Application**	**2009**	**481**

(continued)

Table 4 (continued)

No	Author(s)	Title	Year	Cite
13	王珏, 王任, 苗夺谦, 郭萌, 阮永韶, 袁小红, 赵凯 **Jue Wang, Ren Wan, Duoqian Miao, Meng Guo, Yongshao Ruan, Xiaohong Yuan, Kai Zhao**	基于**RoughSet**理论的"数据浓缩" **Data Enriching Based on Rough Set Theory**	**1998**	**481**
14	肖人彬, 王磊 Renbin Xiao, Lei Wang	人工免疫系统:原理、模型、分析及展望 Artificial Immune System: Principle, Models, Analysis and Perspectives	2002	480
15	刘少辉, 盛秋戬, 吴斌, 史忠植, 胡斐 **Shaohui Liu, Qiujian Sheng, Bin Wu, Zhongzhi Shi, Fei Hu**	**Rough**集高效算法的研究 **Research on Efficient Algorithms for Rough Set Methods**	**2003**	**472**

Note It is a search result from CNKI [10] on July 14, 2016

Nowadays, Chinese rough set researchers publish a lot of high quality research papers in international journals and conferences every year. More and more highly cited rough set papers are published by Chinese researchers.

4 Conclusions

Professor Pawlak helped the born and development of Chinese rough set community a lot. With his help, the rough set research in China is growing very quickly. Chinese rough set society becomes a key academic organization in both China and worldwide. Chinese rough set society will further push the development of rough set theory in the future together with other rough set communities.

Acknowledgements This work has been supported by the National Key Research and Development Program of China under grant 2016YFB1000905, the National Natural Science Foundation of China under Grant numbers of 61272060 and 61572091.

References

1. Pawlak, Z.: Rough sets. Int. J. Parallel Program. **11**(5), 341–356 (1982)
2. Skowron, A., Chakraborty, M.K., Grzymala-Busse, J.W., Marek, V.W., Pal, S.K., Peters, J.F., Rozenberg, G., Slezak, D., Slowinski, R., Tsumoto, S., Wakulicz-Deja, A., Wang, G., Ziarko, W., Pawlak, Z.: (1926–2006): Founder of the Polish School of Artificial Intelligence. Rough Sets Intell. Syst. (1), 1–56 (2013)

3. Peters, J.F., Skowron, A.: Some contributions by Zdzislaw Pawlak. In: Proceedings of the 1st International Conference on Rough Sets and Knowledge Technology, RSKT2006, pp. 1–11 (2006)
4. Liu, Q., Wang, G.: Proceedings of the 1st Chinese conference on rough set and soft computing, CRSSC2001. Comput. Sci. **25**(5S) (2001)
5. Wang, G., Liu, Q., Yao, Y., Skowron, A.: Proceedings of the 9th International Conference on Rough Sets, Fuzzy Sets, Data Mining and Granular Computing, RSFDGrC2003, LNCS2639. Springer (2003)
6. Pawlak, Z.: Flow graphs and decision algorithms. In: Proceedings of the 9th International Conference on Rough Sets, Fuzzy Sets, Data Mining and Granular Computing, RSFDGrC2003, pp. 1–10 (2003)
7. Wang, G., Peters, J.F., Skowron, A., Yao, Y.: Proceedings of the 1st International Conference on Rough Sets and Knowledge Technology, RSKT2006, LNCS4062. Springer (2006)
8. Pawlak, Z.: Conflicts and negotations. In: Proceedings of the 1st International Conference on Rough Sets and Knowledge Technology, RSKT2006, pp. 12–27 (2006)
9. Wang, G., Zhang, Q., Huang, H., Ye, D., Qinghua, H., Xuegang, H., Shi, Z., Li, Y., Shang, L., An, L., Sai, Y., Chen, S., Liang, J., Qin, K., Zeng, H., Xie, K., Miao, D., Min, F., Zhaocong, W., Weizhi, W., Dai, J.: Research on rough set theory and applications in China. Trans. Rough Sets VIII, LNCS **5084**, 352–395 (2008)
10. http://www.cnki.net/
11. Zhang, Q., Xie, Q., Wang, G.: A survey on rough set theory and its applications. CAAI Trans. Intell. Technol. (2016). doi:10.1016/j.trit.2016.11.001

Zdzisław Pawlak as I Saw Him and Remember Him Now

Lech Polkowski

No man is an island, Entire of itself, Every man is a piece of the continent, A part of the main (John Donne)

Abstract Zdzisław Pawlak made an impression on many people including this author due to His openness to new ideas, readiness to discuss them and the spirit of creativity He infused with. In this note, we try to sum up our experiences and also to share what we know about Him and His career on basis of what He said. We touch also some less known achievements of Him.

1 Introduction

Zdzisław was born in 1926 in the city of Łódź, in the centre of Poland. This city was founded on the marsh lands in mid-XIX century as the big centre of weaving and clothing industry, for this reason called the 'Polish Manchester'. Large fortunes were made due to the immense russian market to which most of the production went. The climate of that period is rendered in the movie by Andrzej Wajda 'The Promised Land' ('Ziemia Obiecana' in Polish) made after the novel of the same title by the Nobel laureate Władysław Reymont. Zdzisław was 13 and finished elementary school when the second World War broke out. Łódź was renamed Litzmannstadt and incorporated into Reich and Zdzisław worked in a Siemens factory. After the war He was able to pass maturity exams and He begun studies. Initially, He studied Sinology as something far from ordinary (so he said) but finally graduated from Warsaw University of Technology at the Telecommunication Department in 1951. He was lucky to work in a team building the first computing machine in Poland called GAM-1 and He had some important results like the random numbers generator (1953). It would be very difficult to relate all His achievements but it would be sufficient to mention His positional system for arithmetic with the base of −2, the Pawlak machine—a

L. Polkowski (✉)
Polish-Japanese Academy IT, Koszykowa Str. 86, 02-008 Warszawa, Poland
e-mail: polkow@pjwstk.edu.pl

G. Wang et al. (eds.), *Thriving Rough Sets*, Studies in Computational Intelligence 708, DOI 10.1007/978-3-319-54966-8_2

new model of a computing machine, the first model of DNA, and of course the idea of a rough set. It is instructive to trace these achievements and corresponding with them scientific interests. The line goes from the first computing machine GAM-1 in early 50-ties, through the work on a computing machine UMC-1 in the Warsaw University of Technology in the years 1957–1959 based on His arithmetic with the minus 2 base, which actually went to production and some dozens of it were produced and worked for about 10 years. This line of activity was crowned in 1963 by a habilitation thesis 'Organization of address-less machines'. At that time He became a professor at the Institute of Mathematics of the Polish Academy of Sciences (PAS), He became more involved in theory and His research interests shifted toward mathematical linguistics, semiotics, and scientific information. Especially the last topic proved fruitful as the work on information systems led to the idea of a rough set.

2 DNA

A striking testimony to Zdzisław's abilities and horizons is His model of DNA, regarded by Professor Solomon Marcus, an eminent specialist in mathematical linguistics, as the first in the literature model of genomic grammar. At the same time it is worthy of noticing that this model was published in a relatively little known at least off Poland series of books, 'Small Mathematical Library', published by the State Publisher of School Publications, intended as a more popular and informal in style companion to the very professional 'Mathematical Library'. The book in question was titled 'Matematyka i Gramatyka' ('Mathematics and Grammar') [3] and one chapter in it was dedicated to a model of DNA, basically as a model of genetic code which assigns to sequences of nucleic acids sequences of polypeptides. The wider reception of this model was due to the late Professor Solomon Marcus, our friend from Roumanian Academy and the University of Bucharest, who presented this model in English ('Linguistic structures and generative devices in molecular genetics') [1]. The basic facts used in the genetic language of Pawlak are: 1. DNA is a double helix built of 4 distinct amino-acids: A(denine), T(hymine), G(uanine), C(cytosine). 2. RNA is a single sequence built of 4 amino-acids: A, G, C, U(racyl). 3. Transcription from DNA to RNA follows the following productions:

$$A \rightarrow U, T \rightarrow A, G \rightarrow C, C \rightarrow G.$$

Transcription leads to RNA sequence shorter then DNA sequence. 4. Some convex subsequences of length 3 of RNA are *codons*; they code some amino-acids, hence, a sequence of codons is a code for a sequence of amino-acids—a polypeptide. 5. There are 20 amino-acids genetically valid (though some authors adopt their number as 22). In view of these facts and the one-to-one correspondence between codons and amino-acids genetically functional, Zdzisław Pawlak chose to represent active codons as equilateral triangles with sides labelled 0, 1, 2, or 3 corresponding to the sequence U, A, C, G. The rule for labelling was as follows: the left side of the triangle is labelled x, the base is labelled y, and the right side is labelled z in such a way that $x < y$ and $z \leq y$. This way of numbering produces 20 distinct codons written down in the form of a sequence xyz: 010, 011, 020, 021, 022, 10, 121, 122, 030, 031, 032, 033, 130, 13, 132, 133, 230, 231, 232, 233. We can number those codons from 1 to 20 in the order they are listed. Codons are concatenated according to the following rule in terms of their triangle representations: given already formed chain of codons X we may add to X a new codon b if there is in X a codon a whose side value is equal to the base value of b and no side of b is either a base or a side of any codon in X. For instance, if $X = 232$, then we may add 122. Codons like 020 are *terminal* because they cannot be extended; similarly any chain is terminal if it cannot be extended. The test for being terminal is clearly that each external side of such a chain is valued 0. Terminal chains code *proteins* i.e. terminal polypeptide chains. The Pawlak grammar consists of rules corresponding to triangles representing codons:

1. 1-00 2. 1-01 3. 2-00
4. 2-01 5. 2-02 6. 2-10
7. 2-11 8. 2-12 9. 3-00
10. 3-01 11. 3-02 12. 3-03
13. 3-10 14. 3-11 15. 3-12
16. 3-13 17. 3-20 18. 3-21
19. 3-22 20. 3-23

We have here some pioneering ideas like tessellations generating grammars, and graph grammars (it is easy to convert the triangle rules into graph (precisely, tree) rules). This simple genomic language projecting deep structure (codons) onto surface structure (proteins) can be regarded as an ancestor to recent results in the era when genomes are being deciphered and reveal extraordinarily complex grammars of relations between deep and surface structures [2].

3 I Meet Zdzisław

Though I knew about His existence and He was in committees for thesis defences of a few of my acquaintances including my wife Professor Maria Semeniuk-Polkowska, yet personally I did not meet Him until 1992 on my return from an American university. He took me into His group working already for about 10 years on His idea of a rough set. Prominent there were already Andrzej Skowron, Cecylia Rauszer, working in the chair of Professor Helena Rasiowa. Zdzisław proposed to investigate the problem of giving a topology to rough set spaces—He said that he tried to interest in this problem some researchers at the Mathematical Institute of the Polish Academy of Sciences but to no avail. I learned from Him that in a short time of about two weeks, Roman Słowiński was going to send to Kluwer a collective monograph on rough sets 'Handbook of Applications and Advances of Rough Sets'. I succeeded in preparing and sending to him the first note 'On convergence of rough sets' [5]. Later, in more quiet conditions, I prepared some works which were published in Bulletin of the Polish Academy of Sciences (PAS) under a common header of 'Morphology of Rough Sets'. In those papers I introduced some metrics in infinite information systems that gave topology to various spaces of rough sets. In this way, I satisfied Zdzisław's wish for a topology for rough sets.

4 Work on Mereology

Zdzisław often mentioned that when working at the Mathematical Institute of PAS, He spent time at the Library, perusing and reading works on foundations of concept and set theories. He also benefitted much from conversations and seminars with Andrzej Ehrenfeucht, the legendary logician and mathematician. When travelling once with Zdzisław to a conference in Alaska, we made a stop at Denver to meet Andrzej Ehrenfeucht at Boulder so I could see the old spirit of those discussions reenacted. Zdzisław mentioned the theory of mereology of Stanisław Le'sniewski. Mereology is a theory of parts of the whole, mentioned already by Aristotle (e.g., in his treatise 'De partibus animalium') and treated by medieval philosophers but given a formal axiomatic scheme by Leśniewski in his 'Podstawy Teoryi Zbiorów' ('Foundations of Set Theory') published in Moscow in 1916, where the author was interned during the first world war. At first glance, mereology is relevant to rough sets as set inclusion is a particular example of a part relation and basic constructs of rough set theory, i.e., approximations are defined by means of inclusion of indiscernibility classes. It was the idea of Andrzej Skowron that we consider something like a degree of containment and I found axioms for this extension called Rough Mereology. Further research led to granular computing, new classifier synthesis methods, applications to robotics and data sets. It is doubtful that all this would be done if not the creative atmosphere and free spirit which enlivened those close to Zdzisław Pawlak.

5 Boundaries

It is evident to all who study rough set idea that the most important notion and most important things that conform to that notion is the notion of a boundary and boundaries of concepts as they witness the uncertainty of the concept. The notion of a boundary has been the subject of investigation by philosophers, logicians, topologists. The latter have had an advantage of a point topology and have defined a boundary as the set of points which have the property that each neighborhood of each of them does intersect the set and its complement, so in a sense, boundary consists of points 'infinitely close' to a concept and its complement, and as a rule, boundary is disjoint to a concept and to its complement, save the case when the concept is 'closed' which means that it does contain its boundary. This is fine when we discuss imaginary boundaries in de dicto context. But the problem arises when we speak of de re boundaries existing in the real world. Typical questions are like the Leonardo question cf. Varzi [9]: 'What (...) divides the atmosphere from the water? It is necessary that there should be a common boundary which is neither air nor water but is without substance, because a body interposed between two bodies prevents their contact, and this does not happen in water with air.' We touch here the problem of impossibility of a precise delineation of the boundary. The response from mathematics could be that in such cases the boundary is a fractal dynamically changing with time. But is this fractal from water particles or from air particles? One can see here the soundness of the rough set approach: things in the world are perceived by means of their descriptions, regardless of the fact that in practical usage, the descriptions are replaced with higher level terms, e.g., 'Mount Everest' is a term describing the highest peak on earth whose description would take many attributes. And, things having the same relation to any other thing are collected in aggregates called 'indiscernibility classes' which among themselves partition the universe of things into disjoint pairwise aggregates. Any concept over this universe faces a dichotomy: either it is built of these aggregates or not. In the first case the concept is unambiguous, i.e., for each thing in the universe, every one can decide whether it falls under the concept or not. In the second case, there are aggregates which do intersect both the concept and its complement and can be ascribed to neither. Such aggregates build the boundary of the concept which is precisely defined and things in it belong to the concept and to its complement in an unambiguous way being collectively responsible for the ambiguity of the concept. We may say that indiscernibility aggregates form parts of boundaries of concepts and of their boundary-less approximations. Returning with this picture to the Leonardo question, we may say that the boundary between water and air is the foam belonging partly to water and partly to air as particles in it are closer one to another than some very small real number. One may say that this approach invented by Zdzisław Pawlak is a specimen of the pointless topology whose more general rendition is the mereotopology, i.e., topology in universa equipped with the 'part of' relation *part(.,.)*. In the generalization of Zdzisław approach, the *granular mereotopology* seems adequate. We say about it cf. Polkowski and Semeniuk-Polkowska [6].

5.1 A Granular Mereotopological Model of Boundary as a Direct Generalization of Zdzisław Pawlak's Approach

Mereology is based on the notion of a part relation, $part(x, y)$ ('x is a part to y') which satisfies over a universe U conditions: M1: For each $x \in U$ it is not true that $part(x, x)$ M2: For each triple x, y, z of things in U if $part(x, y)$ and $part(y, z)$, then $part(x, z)$. The notion of an *element* is defined as the relation $el(x, y)$ which holds true if $part(x, y)$ or $x = y$. For our purpose in this section, we modify our approach to mereology. We introduce a new version of rough mereology whose basic notion is predicate 'a part to a degree', $\mu(x, y, r)$, ('x is a part to y to a degree of r at least') on a universe U, where $r \in [0, 1]$. Conditions for μ are RM1 $\mu(x, x, 1)$; RM2 There is a partition on U such that $\mu(x, y, 1)$ and $\mu(y, x, 1)$ if and only if x and y are in the same partition class; RM3 If $\mu(x, y, 1)$ and $\mu(z, x, r)$ then $\mu(z, y, r)$; RM4 If $\mu(x, y, r)$ and $s < r$ then $\mu(x, y, s)$. The predicate $el(x, y)$ if $\mu(x, y, 1)$ defines x as an *element of* y. In the case when U is the universe of an *information system* (U, A) in the sense of Pawlak, with A the set of *attributes*, a predicate μ can be derived from Archimedean t-norms, the Łukasiewicz t-norm $t_L(x, y) = max\{0, x + y - 1\}$ and the Menger t-norm $t_M(x, y) = x \cdot y$, which admit a Hilbert-style representation $t(x, y) = g(f(x) + f(y))$, by letting $\mu^t(x, y, r)$ if and only if $g(\frac{card(Dis(x,y))}{card(A)}) \geq r$, where $Dis(x, y) = \{a \in A : a(x) \neq a(y)\}$. In particular, the *Łukasiewicz rough inclusion* $\mu^L(x, y, r)$ if $\frac{card(Ind(x,y))}{card)A)} \geq r$ satisfies RM1–RM4 with the corresponding relation induced on U partitioning the set U into indiscernibility classes, as $f(x) = 1 - x = g(x)$ for the t-norm t_L, where $Ind(x, y) = A \setminus Dis(x, y)$. The predicate μ^L satisfies the transitivity property: $\mu^L(x, y, r)$ and $\mu^L(y, z, s)$ imply $\mu^L(x, z, t_L(r, s))$. Hence, the corresponding element predicate el satisfies properties $el(x, x)$, $el(x, y)$ and $el(y, z)$ imply $el(x, z)$, $el(x, y)$ and $el(y, x)$ imply x and y are indiscernible. For a predicate μ, and $x \in U$, $r \in [0, 1]$, we define a new predicate $N(x, r)(z)$ if there exists an $s \geq r$ such that $\mu(z, x, s)$. $N(x, r)$ is the *neighborhood granular predicate about x of radius r*. Consider a predicate Ψ on U having a non-empty meaning $[\Psi]$. The *complement to* Ψ is the predicate $-\Psi$ such that $-\Psi(x)$ if not $\Psi(x)$. We define the *upper extension of* Ψ *of radius* r, denoted Ψ_r^+ by letting $\Psi_r^+(x)$ if there exists z such that $\Psi(z)$ and $N(x, r)(z)$. Similarly, we define the *lower restriction of* Ψ *of radius* r, denoted Ψ_r^- by letting $\Psi_r^-(x)$ if not $(-\Psi)_r^+(x)$. A predicate *Open* is defined on predicates on U and a predicate Φ on U is *open*, $Open(\Phi)$ in symbols if $\Phi(x)$ implies the existence of r such that $N(x, r)(z)$ implies $\Phi(z)$. We observe that $\Psi_r^+(x)$ and $\mu(x, y, 1)$ imply $\Psi_r^+(y)$, hence for *symmetric* μ (such is for instance μ^L), the predicate Ψ_r^+ is open. By duality, the complement to an open predicate is *closed*. Hence, the predicate Ψ_r^- is closed for symmetric μ. By symmetry, both predicates are open-closed for a symmetric μ. We say after Barry Smith that a granular neighborhood predicate $N(x, r)$ *straddles* a predicate Ψ if there exist y, z such that $\Psi(y)$, $(-\Psi)(z)$, $N(x, r)(y)$, and, $N(x, r)(z)$. We define the *boundary predicate* Bd on predicates on U. For a predicate Ψ, we define the boundary of Ψ, $Bd(\Psi)$ by letting $Bd(\Psi)(x)$ if each granular neighborhood predicate $N(x, r)$ straddles Ψ, equivalently, the granular neighborhood predicate $N(x, 1)$ straddles Ψ. Please observe that the boundary of Ψ is the boundary of $-\Psi$. Also, for the predicate μ^L, the boundary of Ψ is the rough set boundary, as $\mu^L(x, y, 1)$ is symmetric and partitions U into indiscernibility classes. Further results on boundaries and mereology may be found in [7, 8].

6 A Man of Many Trades

Our tale would be incomplete if we would not mention how many-talented He was. He was an accomplished tourist, in summer rowing in a kayak on rivers and lakes of Polish Pomerania and Mazury, in winters on skis in the mountains. Some 13 years ago my wife has an exhibition of her paintings in the headquarters of the Polish Tourist Organization, we also exhibited photographs submitted by Zdzisław from His trips in the 50-ties. These pictures made a sensation among people present as nobody expected that in those years such trips were possible. He with some colleagues wandered through Bieszczady montains, at that time completely desolate and wild after the second world war. He told us how once in winter in Beskidy mountains he got lost in the blizzard and only by good luck spotted lights of a mountain hostel to be saved. He was a gifted photographer: His photo 'The Polish Jungle' got a distinction at the Times of London photo conquest in 1950-ties. In later years He started painting and had exhibitions of his paintings. He painted what He liked most: water, soil, greenery, and mountains. His paintings are free of human silhouettes, animals, any form of life, He was it seems interested solely in nature's symbiosis of elements. Maybe He posed to Himself the Leonardo question about the boundary between water and air, He so often painted the two. Or, He rendered the idea of rough set in painting? With water He was in a special relation; in addition to making kayak trips and short excursions, He used to swim almost every day. In Warsaw, He used to go to the Academy of Physical Education located close to His home to the swimming pool. The same happened in hotels, every morning at six He went to a swimming pool. But He was also a carpenter, a mason as He renovated His villa in Bielany, a district of Warsaw, making a fireplace etc. pushing a wheelbarrow with lime, mortar and bricks. He told us how He went through antique shops and also read advertisements on old furniture sales to find antique furniture which He renovated. His home was equipped with those pieces of furniture. He was an indestructible voyager; in any place we were, I observed that He wanted to see everything interesting around including a perusal of a local telephone directory to find people by name of Pawlak. Usually He succeeded. He was always full of practical solutions to sudden problems. Once, when my wife had a painting exhibition at some gallery, He was also supposed to come to the opening. Unfortunately, shortly before the appointed hour, when we already were in the gallery, there came a torrential rain so we started without Him convinced that He would not come. But after some twenty minutes he appeared: He bought some newspapers and put them under the jacket so He was underneath dry. There are people who can do almost everything and do it best. He with no doubt belonged to this class. Speaking a bit on jocular side, if Arthur Conan Doyle lived in the second half of the XXth century and knew Zdzisław, He would undoubtedly model his detective on Zdzisław. After all both were masters in deduction.

7 Conclusion

He is not with us of course, but His spirit is I think with those who knew Him. By creating rough sets and making them accepted by the scientific world He gave new life to notions of old, useful but lacking a deeper semantic value, and in doing this He revealed His talent for a clear vision of ideas and ability to represent them in simple understandable to many ways. The success of His monograph on rough sets [4] is due not only to the popularity of rough sets but also to an exceptional combination of theoretical considerations with practical thinking. This seems to be characteristic of His style, avoiding abstraction and keeping in mind practice of application. This is why He was so appealing to many readers. He combined in an exceptional degree the ability to theorize with practical talents and energy to use those abilities and talents.

Acknowledgements To all who knew Zdzisław and enjoyed His goodwill: thanks for not forgetting [6].

References

1. Marcus, S.: Linguistic structures and generative devices in molecular genetics. C.L.T.A. **11**(1), 77–104 (1974)
2. Jolma, A., Yimeng, Y., Nitta, K.R., Kashyap, D., Popov,A., Taipale, M., Enge, M., Kivioja, T., Morgunova, E., Taipale, J.: DNA-dependent formation of transcription factor pairs alters their binding specificity. Nature (2015)
3. Pawlak, Z.: Matematyka i Gramatyka. PZWS Warszawa (1965)
4. Pawlak, Z.: Rough Sets. Theoretical Aspects of Reasoning about Data. Kluwer, Dordrecht (1991)
5. Polkowski, L.T.: On convergence of rough sets. In: Słowiński, R. (ed.) Intelligent Decision Support. Handbook of Applications and Advances of the Rough Sets Theory, pp. 305–311. Kluwer, Dordrecht (1992)
6. Polkowski, L., Semeniuk-Polkowska, M.: Granular Mereotopology: A First Sketch. http://ceur-ws.org/Vol-1032/paper-28.pdf
7. Polkowski, L., Semeniuk-Polkowska, M.: Boundaries, borders, fences, hedges. Fundamenta Informaticae—Dedicated to the Memory of Professor Manfred Kudlek **129**(1–2), 149–159 (2014)
8. Polkowski, L.: Approximate Reasoning by Parts. An Introduction to Rough Mereology. ISRL Series No. 20. Springer International Switzerland, Cham (2012)
9. Varzi, A.C.: Boundary. In: Stanford Encyclopedia of Philosophy. http://plato.stanford.edu/entries/boundary/

Recent Development of Rough Computing: A Scientometrics View

Jing Tao Yao and Adeniyi Onasanya

Abstract The rough set theory has been gaining popularity among researchers and scholars since its inception in 1982. We present this research in commemorating the father of rough sets, Professor Zdzisław Pawlak, and celebrating the 90th anniversary of his birth. Scientometrics is the science that quantitatively measures and analyzes sciences. We use scientometrics approach to quantitatively analyze the contents and citation trends of rough set research. The results presented in this chapter are a follow-up of Yao and Zhang's work published in 2013. We first identify prolific authors, impact authors, impact research groups, and the most impact papers based on Web of Science database. We provide comparison with previous results and analyses of the changes. We further examine features of journal articles and conference papers in terms of research topics and impacts. The third part of the chapter is to examine highly cited papers identified by Web of Science as top 1% based on the academic field and publication year. In the fourth part, we investigate the top journals of rough set publications. There are some interesting results in key indicators between 2013 and 2016 results, for instance, the number of papers published increased by 35%, the total citations increased by 83%, and the h-index values increased by over 32%, while the average citation per paper increased by about 36%. We also found that the number of publications in the recent 5 years was about one third of the total number of rough set publications. This further indicates that rough sets as a research domain is attracting more researchers and growing healthily.

1 Introduction

This research work is presented in commemorating the father of rough sets, Zdzisław Pawlak, and celebrating the 90th anniversary of his birth. In 1982, Pawlak proposed rough set theory for data analysis [49], which has been traced back to about 35 years

J.T. Yao (✉) · A. Onasanya
Department of Computer Science, University of Regina Regina,
Saskatchewan S4S 0A2, Canada
e-mail: jtyao@cs.uregina.ca

A. Onasanya
e-mail: onasanya@cs.uregina.ca

G. Wang et al. (eds.), *Thriving Rough Sets*, Studies in Computational Intelligence 708, DOI 10.1007/978-3-319-54966-8_3

since its inception. Rough set research and its applications have obtained a significant attention among researchers and practitioners. Rough set publications have featured in many journals, international workshops, seminars and conferences. Rough set theory is an extension of set theory for the study of intelligent systems that are characterized by incomplete, vague and imprecise data [53]. The study of rough sets can be classified into three groups [76], namely,

- Content based approach that focuses on the contents of rough set research,
- Method based approach that focuses on the constructive and algebraic (axiomatic) methods of rough sets, and
- Scientometrics approach that focuses on quantitatively analyzing the contents and citations of rough set publications.

This research utilizes scientometrics approach for the analysis of the current trend, development and relationship of research papers in the rough set research domain. This is intended to gain more insights in the domain of rough sets. Scientometrics is concerned with the quantitative analysis of features and characteristics of citations in academic literature and has played a major role in measuring and evaluating research performance as well as understanding the processes of citations [46, 75, 76]. We will investigate two main measures of scientometrics approach, indicators of productivity and indicators of impact. We will also analyze the current status and recent development of rough set research. In addition, we will examine the journals that publish rough set research especially highly impact research. This article can be considered as a follow-up to the previous research work by Yao and Zhang in 2013 [76].

The remaining parts of this chapter are organized as follows. Section 2 provides the methodology and database used for the analysis. Section 3 provides the search, results and analysis of the study based on productivity and impact of citations. It also provides the current status of rough sets. Section 4 presents the top 1% highly cited papers since inception and in recent 5 years. Also included is the list of all top 1% rough set papers in recent 5 years, and lastly, the comparison of highly cited papers—inception versus recent. Section 5 discusses various publication venues of rough set papers or articles based on the most cited and top 1% highly cited papers.

2 Methodology and Database Used

Scientometrics is the science that quantitatively measures and analyses sciences in academic literature. It is also viewed as a scientific measurement of the work of scientists or scholars by way of analyzing their publications and citations within them [46]. We are able to gain more understanding of a research domain by examining its productivity and number of publications through Scientometrics approach and gain more understanding of its research impact and number of citations.

Scientometrics has developed around one main idea, the citation. It has been stated that the act of citing other researcher's work provides the necessary linkages between researchers, ideas, journals, conferences institutions, and countries to constitute an empirical field that can be analyzed quantitatively. Besides, citation also provides a relationship with respect to time, between the previous publications of its references and the later appearance of its citations [46]. In recent years, scientometrics has played a major role in the measurement and evaluation of research performance as well as understanding the processes of citations [46]. It is interesting to state that through this approach, seven out of 50 most cited chemists have been awarded Noble Prize, prediction of research influences and development has been made possible, and assessments of scientific contribution has sufficed [76].

In scientometrics analysis or approach, two main bibliometric indicators are being used, i.e., indicators of productivity and indicators of impact. The indicators of productivity are expressed in terms of the number of papers produced by authors or groups for popularity of research domain while the indicators of impact are described in terms of number of citations for the influence and quality of research domain. These indicators are utilized to predict research influence and development as we examine rough set research related papers in our source of database.

The database utilized in the research work is the Web of Science or WoS for short. The URL of WoS is https://webofknowledge.com. WoS features more than 10,000 major journals since 1900 with more than 150 scientific disciplines. It is a useful research database or resource for quantitative analysis of a research area or domain.

WoS is one of the most popular databases for collecting and searching bibliographic information of research articles in high quality journals and selected international conferences. It also collects citations which reflect the relationships among research articles. WoS provides two types of searches, bibliographic search and cited reference search. The bibliographic search aims to find bibliographic information such as, document types, research areas, authors, group authors, editors, source titles, book series titles, conference titles, publication years, organizations, funding agencies, languages, countries or territories.

The cited reference search can generate citation report simply by clicking on the hyperlink Citation Report. This presents charts for published items in each year and citations in each year. It also generates summary of key parameters such as: Results found, Sum of the Times Cited; Sum of Times Cited without self-citations, Citing Articles, Citing Articles without self-citations, Average Citations per item, and h-index. The h-index, proposed by a physicist Hirsch in 2005, is an index that quantifies an individual's scientific research output with citation number of ones publications. A scientist with an h-index of h has at least h papers each of them has been cited at least h times [16]. The h-index, as a single bibliometric metric, combines both impact (number of citations) and productivity (number of papers). The use of h-index has generated interest and attention in literature and academic journals because it has been largely influential as it quantifies individual's scientific output [46].

There are two kinds of bibliometric indicators, indicators of productivity and indicators of impact on citations, that will be used in this research. The indicators of productivity are expressed in terms of the number of papers produced by authors or research units. It also includes the number of publications that is produced by journals and conference proceedings on a particular research domain. Indicators of impact are described in terms of number of citations for the influence and quality of research domain. The idea of citations is fundamental for indicators of impact.

3 Search, Results and Analysis

We define rough set papers as those containing phrases “rough sets” or “rough set” or “rough computing” or “rough computation” in Topic field in Web of Science. The Topic field is defined as the words or phrases within article titles, keywords, or abstracts in WoS. The search of rough set papers was performed during the week of July 24–30, 2016. The latest available or updated data was on July 28, 2016. We used the two bibliometric indicators or measures of productivity and impact of citation for our results. It should be noticed that not all rough set publications are included in the search results. For instance, not all papers published in Transactions on Rough Sets are recorded in the database. We also missed some rough set papers, e.g., [33–35, 39, 40], by using our search. Additionally, some search results containing rough set phrases are not considered as rough set papers in this research. For instance, a survey paper that mentioned a rough set paper is not rough set paper.

The search resulted in 9,570 rough set papers, 76,733 citations, average citations of 8.02 per paper, and h-index of 106 for the period 1982 to 2016 (July). The numbers of rough set paper by individual phases are

- Rough set: 7,389 papers
- Rough sets: 5,212 papers
- Rough computing: 18 papers
- Rough computation: 8 papers

3.1 Indicators of Productivity

We consider the following indicators for our initial search. They are, Number of Publications per Year, Prolific Authors, Top Organizations, Top Country or Territory, and Top Conferences.

The queried results of rough set publications as depicted in Table 1 and Fig. 1 show the distribution over the period of 1982 to 2016 on a yearly basis. We can deduce that 1.55% of 9,570 (i.e. 148) papers were published in the first 15 years (from 1982 to 1996). In contract, 27.01% of 9,570 (i.e. 2,585) papers were published in the following 10 years (from 1997 to 2006), which is more than 17 times of those that were published in the first 15 years. In the recent decade, between 2007 and 2016, 71.44% of 9,570 (i.e. 6,837) papers were published. This demonstrates that the productivity and popularity of rough set research have an unprecedented increase

Table 1 Number of publications per year

Year	1982	1983	1984	1985	1986	1987	1988	1989	1990	1991	1992	1993
Papers	1	0	0	4	1	1	2	4	4	11	15	10
Year	1994	1995	1996	1997	1998	1999	2000	2001	2002	2003	2004	2005
Papers	20	33	42	46	95	80	122	155	240	288	357	534
Year	2006	2007	2008	2009	2010	2011	2012	2013	2014	2015	2016	**Total**
Papers	668	766	899	929	562	663	644	721	747	618	288	**9,570**

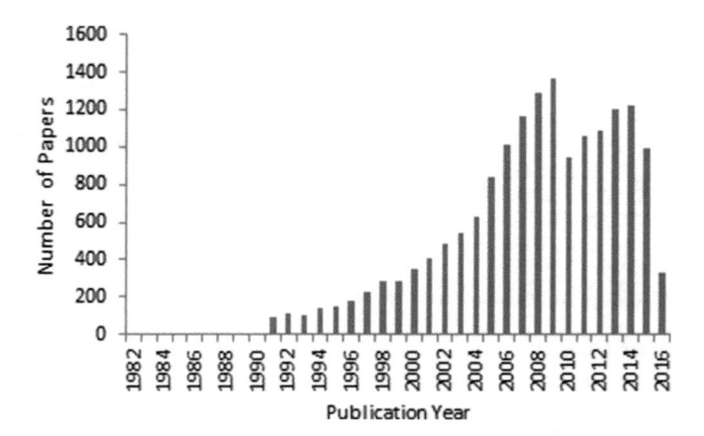

Fig. 1 Number of publications per year

in the last decade. The publication number has been doubled compared with the previous decade.

It is interesting to know that between the period of 1982–2009, rough set papers have grown exponentially. However, there was a shallow decline in 2010 but a steady increase thereafter until 2015. The superficial decrease in 2016 is due to the incomplete data in 2016 as we have about 6 months of data. The cumulative growth pattern and cumulative growth charts as presented in Figs. 2 and 3 conclude that there is consistent growth of rough set research. Overall, the number of rough set publications has grown steadily.

We next identify the most prolific authors as shown in Table 2. It is noted that the top 35 prolific authors have published at least or a minimum of 44 rough set papers. It is noticed that the top four authors, Slowinski, Yao, Skowron and Wang, each published more than 100 rough set papers, remained the same although the orders changed. Zhu was out of top 20 in 2013 research but now is the fifth. Another sign of popularity of rough set research is that there are about 1,140 authors who published at least 5 rough set papers each. The number was doubled comparing with the previous results in 2013 [76].

The top 35 organizations where the authors are affiliated are shown in Table 3.

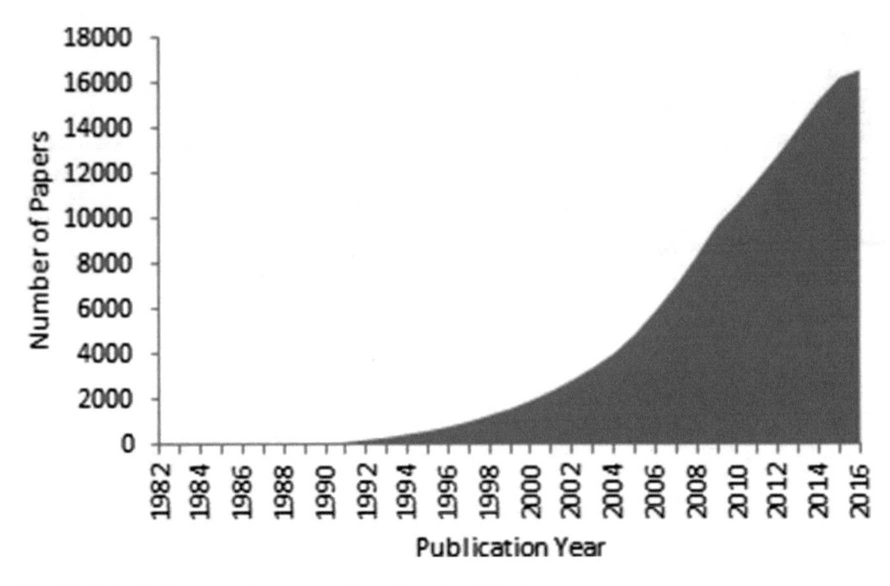

Fig. 2 Cumulative growth pattern of number of publications per year

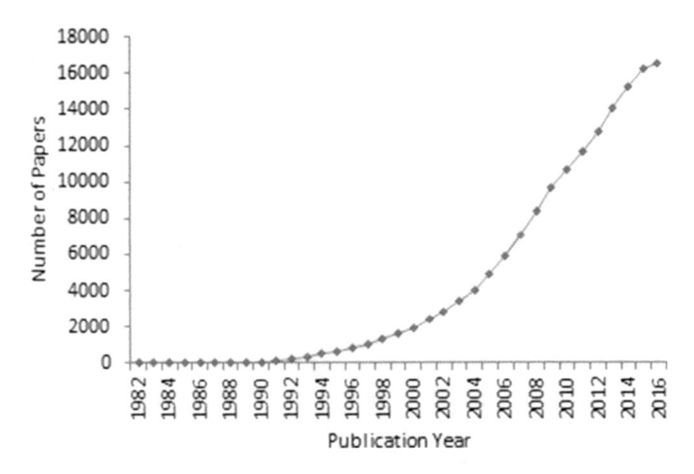

Fig. 3 Cumulative growth of number of publications per year

The top 35 countries or territories with authors publishing rough set papers are shown in Table 4. China maintains the lead with 5,127 rough set research publications which is about 55% of total publications. The results are consistent with previous research [76]. The top four countries still retain their positions and India has over taken Japan and moved to fifth. We observed that the total papers from the top 12 countries accounts for 94% total rough set publications. In fact, the most prolific authors are from these 12 countries.

Table 2 The most prolific authors (top 35)

Rank	Author	Papers	Rank	Author	Papers	Rank	Author	Papers
1	Slowinski R	111	13	Greco S	75	25	Sakai H	49
2	Yao YY	105	14	Chen DG	75	26	Liu D	48
3	Skowron A	102	15	Zhang WX	71	27	Zhang Y	48
4	Wang GY	100	16	Qian YH	69	28	Yu DR	48
5	Zhu W	95	17	Slezak D	65	29	Yang XB	48
6	Li TR	95	18	Pal SK	64	30	Tzeng GH	47
7	Wu WZ	93	19	Grzymala-Busse JW	58	31	Ramana S	47
8	Liang JY	85	20	Pedrycz W	57	32	Ziarko W	46
9	Miao DQ	81	21	Min F	57	33	Suraj Z	44
10	Hu QH	77	22	Lin TY	54	34	Shen Q	44
11	Tsumoto S	76	23	Wang J	54	35	Polkowski L	44
12	Peters JF	75	24	Jensen R	49			

We identified 5 top rough set conferences as shown in Table 5. Based on our search we found that there were 932 conference series or titles that published rough set papers. This shows that rough set is a well accepted research and applied in many domains.

3.2 *Indicators of Impact*

The second part of the results and analysis is based on the impact of rough set research. The distribution of the number of citations 1982–2016 is as depicted in Table 6, while the graphical representation is illustrated in Fig. 4. It is interesting to know that the number of citations followed the same fashion as shown in Table 1 and Fig. 1. The total numbers of citations in each year are steadily growing with a slight decrease in 2010. It is necessary to mention that the numbers shown here may include citations from non-rough set papers.

We will examine the most cited papers and the most impact or influential authors in the next part. We have identified the top 35 most cited papers as illustrated in Table 7. In order to study the impact research direction of rough sets, we have classified those rough set papers in Table 7 into three broad groups as thus:

- Theory Papers: papers about basic rough set theory;
- Hybrid papers: papers combined rough sets with other theories or methods; and
- Application papers: papers about the applications of rough sets.

Table 3 Top 35 organizations

Rank	Organizations	Papers	Rank	Organizations	Papers
1	University of Regina	223	19	Polish-Japenese Inst. of Info Tech	73
2	Southwest Jiaotong Univ.	182	20	University of Catania	72
3	Xi'an Jiaotong University	179	21	Hebei University	71
4	Chinese Academy of Sciences	174	22	Hong Kong Polytechnic Univ.	70
5	Polish Academy of Sciences	168	23	Wuhan University	68
6	Warsaw University of Tech.	161	24	University of Kansas	68
7	North China Electric Power University	157	25	North Eastern University	65
8	Harbin Institute of Technology	126	26	Huazhong Univ Sci & Tech	65
9	Tongji University	122	27	Nanjing Univ. of Sci. & Tech.	63
10	Zhejiang Ocean University	113	28	Poznan University of Techn.	61
11	Zhejiang University	105	29	Indian Institutes of Techn.	61
12	Shanxi University	103	30	Minnan Normal University	58
13	Indian Statistical Institute	99	31	University of Alberta	57
14	Chongqing Univ. Posts & Tel	95	32	Shandong University	57
15	Univ. of Elect. Sci & Tech China	94	33	Nanjing University	57
16	Shanghai Jiaotong University	87	34	Ghent University	55
17	University of Warsaw	85	35	Nanjing Univ. of Aero & Astron.	54
18	University of Manitoba	77			

Based on the above classification, the highly cited papers are grouped as follows:

- Fifteen (15) basic papers: [5, 47, 49, 50, 54–56, 58, 67, 79–82, 96, 98];
- Six (6) hybrid papers: [1, 10, 13, 48, 72, 73]; and
- Fourteen (14) application papers: [14, 20–22, 41, 43, 51–53, 68, 70, 78, 83, 95].

Table 4 Top 35 countries

Rank	Countries	Papers	Rank	Countries	Papers	Rank	Countries	Papers
1	P.R. China	5127	13	Egypt	92	25	Thailand	41
2	Poland	851	14	Turkey	86	26	Norway	39
3	USA	573	15	Wales	86	27	Cuba	39
4	Canada	529	16	Australia	84	28	Netherlands	37
5	India	505	17	South Korea	83	29	Czech Rep.	35
6	Japan	440	18	Belgium	79	30	Brazil	34
7	Taiwan	397	19	Germany	76	31	Pakistan	27
8	Italy	164	20	France	68	32	Tunisia	26
9	England	124	21	Singapore	62	33	Finland	25
10	Spain	110	22	Sweden	49	34	Mexico	21
11	Malaysia	106	23	North Ireland	47	35	Hungary	21
12	Iran	101	24	Saudi Arabia	42			

Table 5 Top 5 rough set conference series

Rank	Conference titles	Papers
1	Intl. Conf. on Rough Sets & Knowledge Technology (RSKT)	376
2	Intl. Workshop on RS Fuzzy Sets Data Mining & Granular Computing (RSFDGRC)	288
3	Intl. Conf. on Rough Sets & Current Trends in Computing (RSCTC)	238
4	IEEE Intl. Conf. on Granular Computing (GRC)	225
5	Intl. Conf. on Rough Sets and Intelligent Systems Paradigms (RSISP)	52

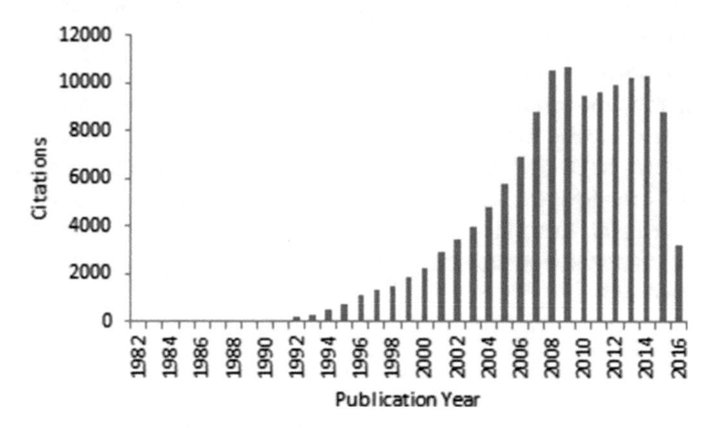

Fig. 4 Number of citations by year

Table 6 Number of citations per year

Year	1982	1983	1984	1985	1986	1987	1988	1989	1990	1991
Citations	0	0	2	7	5	6	5	11	12	22
Papers	1	0	0	4	1	1	2	4	4	11
Cites/Paper	0.00	0.00	0.00	1.75	5.00	6.00	2.50	2.75	3.00	2.00
Year	1992	1993	1994	1995	1996	1997	1998	1999	2000	2001
Citations	47	16	34	48	51	90	121	159	276	443
Papers	15	10	20	33	42	46	95	80	122	155
Cites/Paper	3.13	1.60	1.70	1.45	1.21	1.96	1.27	1.99	2.26	2.86
Year	2002	2003	2004	2005	2006	2007	2008	2009	2010	2011
Citations	661	888	1,312	1,972	2,644	3,864	5,075	5,550	4,933	6,259
Papers	240	288	357	534	668	766	899	929	562	663
Cites/Paper	2.75	3.08	3.68	3.69	3.96	5.04	5.65	5.97	8.78	9.44
Year	2012	2013	2014	2015	2016		**All**			
Citations	7,035	8,920	10,571	9,830	5,864		**76,733**			
Papers	644	721	747	618	288		**9,570**			
Cites/Paper	10.92	12.37	14.15	15.91	20.36		**8.02**			

The number of papers on basic set theory were 43%, hybrid with other theories 17%, and applications of rough sets 40%. In contract to the results in 2013, the numbers were 60, 10, and 30%. This shows a shift of research direction from basic rough set theory to applications and hybrid with other theories.

The most influential or impact authors of rough sets are listed in Table 8. We have 35 most impact authors who received citations of more than 300 while there were only 10 authors in Yao and Zhang's results. Twelve of Pawlak's papers received citations of 8,733, including his first paper that proposed the concept of rough set theory. It's obvious that Pawlak was the top highly cited researcher and his chapter has a great influence in rough set research.

In comparing the results with that of Yao and Zhang, we observe the following key changes. There are top 12 authors with more than 1,000 citations while there were 15 authors in 2013. The number of citations has been increased considerably for most authors across the board. In fact, an increase in the number of citations per author shows an increase in the level of influence of rough set research if we use the impact of citations as an instrumental for evaluating the influence of the authors and rough set research.

Table 7 Top 35 cited papers

Rank	Paper	Year	Cites	Avg/Year	Main results
1	Pawlak Z [49]	1982	5137	146.77	Theory: Rough sets seminal work
2	Ziarko W [98]	1993	880	36.67	Theory: Variable precision RS
3	Dubois D and Prade H [10]	1990	846	31.33	Hybrid: Fuzzy sets
4	Pawlak Z and Skowron A [54]	2007	767	76.70	Theory: Survey
5	Greco S et al. [14]	2001	579	36.19	App: Decision analysis
6	Kryszkiewicz M [21]	1998	541	28.47	App: Information systems
7	Pawlak Z and Skowron A [56]	2007	498	49.80	Theory: Rough sets
8	Slowinski R and Vanderpooten D [67]	2000	438	25.76	Theory: Generalized RS with similarity
9	Pawlak Z and Skowron A [55]	2007	434	43.40	Theory: Boolean reasoning
10	Yao YY [81]	1998	394	20.74	Theory: Research method of RS theory
11	Pawlak Z et al. [50]	1995	382	17.36	Theory: Survey
12	Yao YY [80]	1998	379	19.95	Theory: Relational interpretations
13	Mitra S and Hayashi Y [47]	2000	353	20.76	Theory: Survey
14	Maji PK and Roy AR [41]	2002	342	22.80	App: Decision making problem
15	Swiniarski RW and Skowron A [68]	2003	329	23.50	App: Feature selection and recognition
16	Kryszkiewicz M [22]	1999	327	18.17	App: Incomplete information systems
17	Zhu W and Wang FY [95]	2003	325	23.21	App: covering
18	Wu WZ et al. [72]	2003	310	22.14	Hybrid: Fuzzy rough sets
19	Aktas H and Cagman N [1]	2007	301	30.00	Hybrid: Soft sets and soft groups
20	Pawlak Z [52]	1998	295	15.53	App: Data analysis
21	Yao YY [79]	1996	293	13.95	Theory: Interpretation of RS
22	Pawlak Z [53]	2002	261	17.40	App: Data analysis
23	Wang XY et al. [70]	2007	253	25.30	App: Feature selection
24	Pawlak Z [51]	1997	246	12.3	App: Decision support
25	Jensen R and Shen [20]	2004	243	18.69	App: Feature selection
26	Wu WZ and Zhang WX [73]	2004	240	18.46	Hybrid: Fuzzy approximation
27	Morsi NN and Yakout MM [48]	1998	237	12.47	HybridL Fuzzy rough sets
28	Yao YY and Wong SKM [78]	1992	227	9.08	App: Decision theoretic rough sets
29	Bonikowski Z et al. [5]	1998	224	11.79	Theory: Intentions and extensions
30	Yao YY [82]	2001	221	13.81	Theory: Rough granulation
31	Polkowski L and Skowron A [58]	1996	221	10.52	Theory: Mereology
32	Zhu W [96]	2007	215	21.50	Theory: Covering
33	Mi JS et al. [43]	2004	211	16.23	App: Knowledge reduction
34	Feng F et al. [13]	2008	211	23.44	Hybrid: Soft sets
35	Yao YY and Zhao Y [83]	2008	206	22.89	App: Attribute reduction in DTRS

Table 8 35 most impact or influential authors

Rank	Authors	Cites	Papers	Rank	Authors	Cites	Papers
1	Pawlak Z	8733	12	19	Mitra S	660	3
2	Yao YY	2686	8	20	Grzymala-busse, J	517	2
3	Slowinski R	2351	8	21	Wang FY	517	2
4	Skowron A	2251	5	22	Chen DG	504	3
5	Ziarko W	1616	4	23	Vanderpooten D	438	1
6	Jensen R	1137	6	24	Xia WJ	437	2
7	Greco S	1128	4	25	Wong, SKM	419	2
8	Matarazzo B	1128	4	26	Yeung DS	371	2
9	Zhang WX	1108	5	27	Tsang ECC	371	2
10	Wu WZ	1037	5	28	Wang XZ	371	2
11	Zhu W	1030	5	29	Hayashi Y	355	1
12	Kryszkiewicz M	1022	3	30	Maji PK	344	1
13	Shen Q	883	5	31	Roy AR	344	1
14	Dubois D	847	1	32	Swiniarski RW	329	1
15	Prade H	847	1	33	Pal SK	305	2
16	Mi JS	846	4	34	Slezak D	303	2
17	Hu QH	685	4	35	Aktas H	301	1
18	Yu DR	685	4				

3.3 Current Status of Rough Sets

We examine the current development of rough sets in recent 5 years in this section. Since we only have half of 2016 data (as of July 28, 2016), we focus our analysis on papers from 2011 to 2015. Table 9 presents the top 35 cited papers in recent 5 years. Due to the fact that we have 4 papers with 39 citations, we will study 36 papers instead of 35 papers as we planed. It is observed that many of top papers are from China. This is consistent with the fact that China is rated the top country in Table 4.

With the 3,393 papers published in recent 5 years, the total citation number was 13,022. The average citations per paper was 3.80. It is also observed that 35.45% of all rough set publications are from recent 5 years. This indicates that rough sets as a research domain is attracting more researchers and growing healthily.

Following the same fashion, we classified the top 35 recent papers into 3 impact research groups,

- Twelve (12) papers are about basic rough set theory: [9, 15, 36, 62, 66, 74, 77, 84–86, 90, 97];
- Eight (8) papers are about the hybrid with other approaches or theories: [2, 11, 12, 28, 31, 60, 63, 92];
- Fifteen (15) papers are about applications of rough sets: [4, 6–8, 18, 19, 24, 25, 30, 37, 38, 42, 44, 59, 91].

Table 9 Top 36 cited papers in recent 5 years from 2011 to 2015

Rank	Authors	Year	Cites	Avg/year	Main results
1	Yao YY [84]	2011	134	22.33	Theory: Three-way decisions
2	Feng F et al. [12]	2011	127	21.17	Hybrid: Soft sets
3	Dubois D and Prade H [11]	2012	107	21.40	Hybrid: Fuzzy sets
4	Yao YY and Yao B [86]	2012	87	17.40	Theory: Covering
5	Min et al. [44]	2011	83	13.83	App: Attribute reduction
6	Liu et al. [36]	2011	73	12.17	Theory: Probabilistic criteria
7	Li H and Zhou X [24]	2011	72	12.00	App: Risk decision making
8	Herbert JP and Yao JT [15]	2011	68	11.33	Theory: Game-Theoretic RS
9	Qian Y et al. [59]	2011	67	11.17	App: Attribute reduction
10	Jia X et al. [18]	2013	64	16.00	App: Decision-theoretic RS
11	Qian Y et al. [62]	2014	63	21.00	Theory: Multigranulation decision
12	Skowron A et al. [66]	2012	58	11.60	Theory: Approximation spaces
13	Liu D et al. [37]	2011	57	9.50	App: Three-way decision
14	Zhang et al. [92]	2012	55	11.00	Hybrid: Fuzzy sets
15	Zhang et al. [90]	2012	54	10.80	Theory: Approximations
16	Lin et al. [31]	2012	52	10.40	Hybrid: Neighborhood
17	Ali MI [2]	2011	52	8.67	Hybrid: Soft sets & fuzzy soft sets
18	Chen et al. [6]	2011	49	8.17	App: Feature selection
19	Yao JT et al. [77]	2013	48	12.00	Theory: Granular computing
20	Blaszczynski J et al. [4]	2011	48	8.00	App: Sequential covering
21	Zhang J et al. [91]	2012	47	9.40	App: Data mining
22	Wu WZ and Leung Y [74]	2011	46	7.67	Theory: Granular computing
23	Li J et al. [25]	2013	42	10.50	App: Decision making
24	Dai J and Xu Q [9]	2012	42	8.40	Theory: Uncertainty measures
25	Zhu W and Wang S [97]	2011	42	7.00	Theory: Generalized RS
26	Qian Y et al. [60]	2011	42	7.00	Hybrid: Fuzzy sets
27	Chen Y et al. [8]	2011	42	7.00	App: Feature selection
28	Liu D et al. [38]	2012	41	8.20	App: Classification in DTRS
29	Chen HL et al. [7]	2011	41	6.83	App: Prediction
30	Qian J et al. [63]	2011	40	6.67	Hybrid: Attribute reduction
31	Yao YY [85]	2011	40	6.67	Theory: Semantic Issues
32	Jia X et al. [19]	2014	39	13.00	App: Optimization
32	Liang D et al. [28]	2011	39	9.75	Hybrid: Fuzzy DTRS
32	Liang J et al. [30]	2012	39	7.80	App: Feature selection
32	Medina J [42]	2012	39	7.80	App: Multi-adjoint lattices
32	Li N et al. [32]	2012	39	7.80	Hybrid: Neighbourhood and Fault Diagnosis

The numbers of papers on theory, hybrid, and applications account for 34%, 23%, and 43% respectively. There are more papers on decision-theoretic rough sets and three-way decisions which are considered as new directions. The claim on the shift in research direction towards applications and hybrid still holds.

4 Web of Science Highly Cited Papers—The Top 1% Papers

Web of Science defines highly cited papers as papers that "received enough citations to place [them] in the top 1% of the academic field based on a highly cited threshold for the field and publication year". We refer them as top 1% papers in this article. The academic field can be in the field of Computer Science, Mathematics, Economics & Business, or Engineering, for instance. The top 1% papers are identified and recognized on monthly or quarterly basis. The results in this section are based on our search conducted in November 2016, however, the data were "as of July/August 2016" according to Web of Science. On WoS Web site, the top 1% papers are presented a trophy symbol. In general, most citations to a paper come in the 2nd, 3rd, or 4th year after publication. The citations of the top 1% papers come in the year of their publication, therefore, it can be used as a sign of popularity or hotness of the topic. We may use the top 1% papers as a metric to evaluate researchers and journals.

We examined all the top cited papers presented in Table 7 and found that 8 of them were identified as the top 1% highly cited papers. They are [1, 13, 54–56, 70, 83, 96].

As the 1% papers represent some kind of research trend, it is meaningful to examine more recent publications. There are 16 top 1% papers in recent 5 year from 2011 to 2015 as shown in Table 10. In other words, 16 out of 36, or 44%, top cited papers as shown in Table 9 are highly influential papers in the field of Computer Science. These top 1% of papers received at least 39 citations.

In further analysis, we noticed that 11 out of 16 top 1% papers are focused on hybrid and application of rough set theory. This is consistent with the results of top rough set papers.

The next step is to identify all top 1% rough set papers in recent 5 years. It should be noted that new papers may have a small number of citations, however, if the citation numbers outperform citation of other papers it can still be recognized as top 1% papers. Table 11 lists all top 1% rough set papers. We have 33 top 1% papers from over 3,520 highly cited papers being ranked the top 1% status in the recent 5 years. Based on the table, most of the top 1% papers are classified in the field of Computer Science while the remaining in Engineering.

Table 10 Top 1% papers amongst top papers in recent years (2011–2015)

Rank	Authors	Year	Cites	Avg/year	Main results
1	Yao YY [84]	2011	134	22.33	Theory: Three-way decisions
2	Feng F et al. [12]	2011	127	21.17	Hybrid: Soft sets
3	Dubois D and Prade H [11]	2012	107	21.40	Hybrid: Fuzzy sets
4	Yao YY and Yao B [86]	2012	87	17.40	Theory: Covering
5	Min F et al. [44]	2011	83	13.83	App: Attribute reduction
6	Liu D et al. [36]	2011	73	12.17	Theory: Probabilistic criteria
7	Qian Y et al. [59]	2011	67	11.17	App: Attribute reduction
8	Jia X et al. [18]	2013	64	16.00	App: Decision-theoretic RS
9	Qian et al. [62]	2014	63	21.00	Theory: Multigranulation decision
10	Skowron A et al. [66]	2012	58	11.60	Theory: Approximation spaces
11	Zhang X et al. [92]	2012	55	11.00	Hybrid: Fuzzy sets
12	Lin G et al. [31]	2012	52	10.40	Hybrid: Neighborhood
13	Yao JT et al. [77]	2013	48	12.00	Theory: Granular computing
14	Li et al. [25]	2013	42	10.50	App: Decision making
15	Jia X et al. [19]	2014	39	13.00	App: Optimization
16	Liang D et al. [28]	2013	39	9.75	Hybrid: Fuzzy DTRS

5 Analysis of Journals Publishing Rough Set Papers

Aside from the analyses performed with respect to impact of individual papers and researchers, it necessary to consider the impact of journals in order to assist authors/researchers in their decisions on which journals to submit their papers. We will examine journals that publish rough set papers in this section. We use Impact Factor (IF), h-index, 1% paper, and IF Quartile rankings to evaluate these academic journals. Journal Impact Factor is an important metric and has been viewed as a pre-eminent measure or metric for choosing venues to submit ones research papers [46]. Impact Factor of an academic journal is a measure of the frequency with which the average article in a journal has been cited in a particular year or period [71]. This metric also measures the relative importance of a journal within its field. In other words, journals with higher IF are deemed to be more important than those with lower ones. Web of Science publishes two types of Impact Factor. One is the normal IF. The recent IF is in 2015, which is the number of citations in 2015 to those papers published in a journal in 2013 and 2014, divided by the number of such papers. The other is a 5-year Impact Factor, the average number of times articles from the journal published in the past five years have been cited in report year.

Table 11 All top 1% highly cited papers in recent 5 years

Rank	Authors	Title	Year	Cites
1	Yao YY [84]	Theory: Three-way decisions	2011	144
2	Feng F et al. [12]	Hybrid: Soft sets	2011	132
3	Dubois D & Prade H [11]	Hybrid: Fuzzy sets	2012	108
4	Yao YY & Yao B [86]	Theory: Covering	2012	95
5	Min F et al. [44]	App: Attribute reduction	2011	90
6	Liu D, Li TR & Ruan D [36]	Theory: Probabilistic criteria	2011	74
7	Jia XY et al. [18]	App: Decision-theoretic RS	2013	69
8	Qian YH et al. [62]	Theory: Multigranulation decision	2014	67
9	Skowron A, Stepaniuk J & Swiniarski R [66]	Theory: Approximation spaces	2012	60
10	Lin GP, Qian YH & Li JJ [31]	Hybrid: Neighborhood	2012	58
11	Yao JT, Vasilakos A & Pedrycz W [77]	Theory: Granular Computing	2013	57
12	Zhang XH, Zhou B & Li P [92]	Hybrid: Fuzzy sets	2012	57
13	Liang D et al. [28]	Hybrid: DTRS	2013	43
14	Li JH, Mei CL & Lv YJ [25]	App: Decision making	2013	43
15	Dai J & Xu Q [9]	Theory: Uncertainty measures	2012	43
16	Jia XY et al. [19]	App: Optimization	2014	42
17	Yu H, Liu Z & Wang G [87]	Hybrid: DTRS	2014	39
18	Hu BQ [17]	Theory: Three-way decisions	2014	38
19	Liang JY et al. [30]	App: Feature selection	2012	38
20	Azam N & Yao JT [3]	Hybrd: GTRS	2014	37
21	Pedrycz W [57]	App: Optimization	2014	37
22	Wang CZ et al. [69]	Theory: Covering	2014	34
23	Min F & Zhu W [45]	App: Feature selection	2012	29
24	Zhou B [94]	Hybrid: DTRS	2014	29
25	Liang DC et al. [29]	Hybrid: DTRS	2015	25
26	Zhang XH, Dai JH & Yu Y et al. [89]	Theory: Approximation spaces	2015	25
27	Li JH et al. [26]	Hybrid: Granular computing	2015	23
28	Qian J et al. [61]	App: Attribute reduction	2015	21
29	Liang D & Liu D [27]	Hybrid: Fuzzy DTRS	2014	20
30	Zhao X & Hu BQ [93]	Hybrid: Fuzzy probability measure	2015	17
31	Zhan JM, Liu Q & Davvaz B [88]	Theory: Soft hemirings	2015	15
32	Li WT & Xu WH [23]	App: DTRS	2015	13
33	Selechi S, Selamat A & Fujita H [65]	Theory: Granular computing	2015	12

Table 12 Most cited papers by journals

Rank	Journal	Cites	Papers
1	Information Sciences	5,215	14
2	International Journal of Computer and Information Sciences	5,137	1
3	European Journal of Operational Research	1,297	4
4	Journal of Computer and System Sciences	880	1
5	International Journal of General Systems	846	1
6	IEEE Transactions on Knowledge and Data Engineering	681	2
7	Pattern Recognition Letters	582	2
8	International Journal of Approximate Reasoning	514	2
9	Communications of the ACM	382	1
10	IEEE Transactions on Neural Networks	353	1
11	Computers and Mathematics with Applications	342	1
12	Cybernetics and Systems	295	1
13	Mechanical Systems and Signal Processing	256	1
14	Fuzzy Sets and Systems	237	1
15	International Journal of Man-Machine Studies	227	1
16	International Journal of Intelligent Systems	221	1

WoS categorizes journals into 4 groups according to their IF rankings. IF Quartile rankings are derived for each journal in each of its subject categories according to which quartile of the IF distribution the journal occupies for that subject category [64], where:

- Q1 denotes the top 25% of the IF distribution,
- Q2 denotes the middle-high position (between top 50% and top 25%),
- Q3 denotes the middle-low position (top 75% to top 50%),
- Q4 denotes the lowest position bottom 25% of the IF distribution.

We will examine journals based on the numbers of top rough set papers and top 1% papers as well. There are 16 journals that published rough set most cited papers based on Table 7. The results are shown in Table 12.

We further analyse journals that published top 1% rough set papers. The results are shown in Table 13.

Table 13 Top 1% highly cited papers by journals

Rank	Journals	Cites	Papers
1	Information Sciences	2420	6
2	Pattern Recognition Letters	253	1
3	Computers and Mathematics with Applications	211	1

Table 14 Most cited papers in recent 5 years by journals

Rank	Journal	Cites	Papers
1	Information Sciences	1007	14
2	Intl. Journal of Approximate Reasoning	356	7
3	Intl. Journal of Computational Intelligence Systems	138	2
4	Knowledge-Based Systems	130	3
5	Fuzzy Sets and Systems	108	1
6	Applied Soft Computing	97	2
7	Pattern Recognition	73	1
8	Fundamenta Informaticae	73	1
9	IEEE Transactions on Cybernetics	57	1
10	Computers & Mathematics with Applications	57	1
11	Expert Systems with Applications	54	1
12	IEEE Transactions on Fuzzy Systems	46	1

Similar analysis was conducted for the top 35 most cited papers and top 1% papers in recent 5 years. The results are presented in Tables 14 and 15.

It is noted that 40% (14 out of 35) top cited rough set papers and 50% (9 out of 18) of top 1% rough set papers are published in Information Sciences journal. Information Sciences constitutes the primary journal where most of the rough set related and impact papers are published.

The following is a summarization of top 10 journals we identified in Table 16 in term of IF, h-index, Quartile ranking metrics, as well as top 1% ranking metrics.

- Information Sciences
  - IF 3.364, Quartile Q1
  - Published 366 rough set papers
  - h-index 109, 22 rough set papers amongst top 109 papers
  - 166 top 1% papers, 32 of them are rough set papers

Table 15 Top 1% highly cited papers in recent 5 years by journals

Rank	Journal	Cites	Papers
1	Information Sciences	763	9
2	Intl. Journal of Approximate Reasoning	253	5
3	Fuzzy Sets and Systems	108	1
4	IEEE Transactions on Cybernetics	57	1
5	Computers & Mathematics with Applications	57	1
6	Applied Computing	43	1

Table 16 Top 10 rough set journals

Rank	h-index	IF quartile	Journals	Papers
1	109	Q1	Information Sciences	366
2	34	Q3	Fundamenta Informaticae	245
3	55	Q1	Knowledge Based Systems	171
4	32	Q1	Expert Systems with Applications	157
5	61	Q1	International Journal of Approximate Reasoning	141
6	25	Q3	Journal of Intelligent & Fuzzy Systems	86
7	68	Q1	Applied Soft Computing (ASC)	77
8	149	Q1	Fuzzy Sets and Systems	70
9	181	Q1	European Journal of Operational Research	62
10	43	Q2	Soft Computing (SC)	50

- Fundamenta Informaticae
  - IF 0.658, Quartile Q3
  - Published 245 rough set papers
  - h-index 34, 0 rough set papers amongst top 34 papers
  - 4 top 1% papers, none is rough set paper
- Knowledge Based Systems
  - IF 3.325, Quartile Q1
  - Published 171 rough set papers
  - h-index 55, 4 are rough set papers amongst to 55 papers
  - 55 top 1% papers, 11 of them are rough set papers

- Expert Systems with Applications
  - IF 2.981, Quartile Q1
  - Published 157 rough set papers
  - h-index 32, 2 rough set papers amongst top 32 papers
  - 81 top 1% papers, 2 of them are rough set papers
- International Journal of Approximate Reasoning
  - IF 2.696, Quartile Q1
  - Published 141 rough set papers
  - h-index 61, 11 rough set papers amongst top 61 papers
  - 16 top 1% papers, 11 of them are rough set papers
- Journal of Intelligent & Fuzzy Systems
  - IF 1.004, Quartile Q3
  - Published 86 rough set papers
  - h-index 25, 1 rough set paper amongst top 25 papers,
  - 15 top 1% papers, 1 of them is rough set paper
- Applied Soft Computing
  - IF 2.857, Quartile Q1
  - Published 77 rough set papers
  - h-index 68, 2 rough set papers amongst top 68 papers
  - 67 top 1% papers, 2 of them are rough set papers
- Fuzzy Sets and Systems
  - IF 2.098, Quartile Q1
  - Published 70 rough set papers
  - h-index 149, 1 rough set paper amongst top 149 papers
  - 19 top 1% papers, 1 of them is rough set paper
- European Journal of Operational Research
  - IF 2.679, Quartile Q1
  - Published 62 rough set papers
  - h-index 181, 3 rough set papers amongst top 181 papers
  - 103 top 1% papers, 3 of them are rough set papers
- Soft Computing
  - IF 1.630, Quartile Q2
  - Published 50 rough set papers
  - h-index 43, 2 rough set papers amongst top 43 papers
  - 13 top 1% papers, 2 of them are rough set papers

Once again we noticed that Information Sciences is identified as the main rough set journal due to the fact that,

- it is rated Q1 and has third highest h-index of all top 10 journals
- it has the highest top 1% highly cited papers
- 75% of top 1% highly cited papers are from IS (in Table 13)
- Most cited rough set papers are published in Information Science.

6 Conclusion

We have presented this research in commemorating the father of rough sets, Zdzisław Pawlak, and celebrating the 90th anniversary of his birth by investigating the trend and development of rough set research by using scientometrics approach. The results show that Pawlak's seminal paper has been identified the most cited paper. Pawlak has been identified as the most influential and impact author.

We have analyzed productivity and impact of rough set research domain. The top five prolific authors are: Slowinski, Yao, Skowron, Wang, and Zhu. The top five most influencial authors are: Pawlak, Yao, Slowinski, Skowron, and Ziarko. The top journal for rough set research has been identified as Information Sciences.

Comparing with the results in 2013, we found that the number of rough sets publications has increased by 35%, the total citations have increased by over 83%, and the h-index values increased by over 32%. The average citations per paper is 8.02 while the average citations per paper was 5.9 in 2013, i.e., an increase of about 36%. The results also show that about 35% of rough set publications were published in recent 5 years. The results of recent 5 years of rough set publication have demonstrated that more papers on the applications and hybrid with other theories. Decision-theoretic rough sets and three-way decisions are new research trends. The results suggest that rough set research continues growing healthy and attracting more to application oriented research. We will monitor and report the trends and development in this domain in future research.

References

1. Aktas, H., Cagman, N.: Soft sets and soft groups. Inf. Sci. **177**(13), 2726–2735 (2007)
2. Ali, M.I.: A note on soft sets, rough soft sets and fuzzy soft sets. Appl. Soft Comput. **11**(4), 3329–3332 (2011)
3. Azam, N., Yao, J.T.: Analyzing uncertainties of probabilistic rough set regions with game-theoretic rough sets. Int. J. Approx. Reason. **55**(1), 142–155 (2014)
4. Blaszczynski, J., Slowinski, R., Szelag, M.: Sequential covering rule induction algorithm for variable consistency rough set approaches. Inf. Sci. **181**(5), 987–1002 (2011)
5. Bonikowski, Z., Bryniarski, E., Wybraniec-Skardowska, U.: Extensions and intentions in the rough set theory. Inf. Sci. **107**(1–4), 149–167 (1998)
6. Chen, H.L., Yang, B., Liu, J., Liu, D.Y.: A support vector machine classifier with rough set-based feature selection for breast cancer diagnosis. Expert Syst. Appl. **38**(7), 9014–9022 (2011)

7. Chen, H.L., Yang, B., Wang, G., Liu, J., Xu, X.: Wang, SJ; Liu, DY: A novel bankruptcy prediction model based on an adaptive fuzzy k-nearest neighbor method. Knowl.-Based Syst. **24**(8), 1348–1359 (2011)
8. Chen, Y., Miao, D., Wang, R., Wu, K.: A rough set approach to feature selection based on power set tree. Knowl.-Based Syst. **24**(2), 275–281 (2011)
9. Dai, J., Xu, Q.: Approximations and uncertainty measures in incomplete Information systems. Inf. Sci. **198**, 62–80 (2012)
10. Dubois, D., Prade, H.: Rough fuzzy sets and fuzzy rough sets. Int. J. General Syst. **17**(2–3), 191–209 (1990)
11. Dubois, D., Prade, H.: Gradualness, uncertainty and bipolarity. Making sense of fuzzy sets. Fuzzy Sets Syst. **192**, 3–24 (2012)
12. Feng, F., Liu, X., Leoreanu-Fotea, V., Jun, Y.B.: Soft sets and soft rough sets. Inf. Sci. **181**(6), 1125–1137 (2011)
13. Feng, F., Jun, Y.B., Zhao, X.: Soft semirings. Comput. Math. Appl. **56**(10), 2621–2628 (2008)
14. Greco, S., Matarazzo, B., Slowinski, R.: Rough approximation of a preference relation by dominance relations. Eur. J. Oper. Res. **117**(1), 63–83 (1999)
15. Herbert, J.P., Yao, J.T.: Game-theoretic rough sets. Fundamenta Informaticae **108**(3–4), 267–286 (2011)
16. Hirsch, J.E.: An index to quantify an individual's scientific research output. Proc. Nat. Acad. Sci. **46**, 16569 (2005)
17. Hu, B.Q.: Three-way decisions space and three-way decisions. Inf. Sci. **281**, 21–52 (2014)
18. Jia, X.Y., Tang, Z., Liao, W., Shang, L.: Minimum cost attribute reduction in decision-theoretic rough set models. Inf. Sci. **219**, 151–167 (2013)
19. Jia, X.Y., Tang, Z., Liao, W., Shang, L.: On an optimization representation of decision-theoretic rough set model. Int. J. Approx. Reason. **55**(1), 156–166 (2014)
20. Jensen, R., Shen, Q.: New approaches to fuzzy-rough feature selection. IEEE Trans. Fuzzy Syst. **17**(4), 824–838 (2009)
21. Kryszkiewicz, M.: Rough set approach to incomplete information systems. Inf. Sci. **112**(1), 39–49 (1998)
22. Kryszkiewicz, M.: Rules in incomplete information systems. Inf. Sci. **113**(3), 271–292 (1999)
23. Li, W.T., Xu, W.H.: Double-quantitative decision-theoretic rough set. Inf. Sci. **316**, 54–67 (2015)
24. Li, H., Zhou, X.: Risk decision making based on decision-theoretic rough set: a three-way view decision model. Int. J. Comput. Intell. Syst. **4**(1), 1–11 (2011)
25. Li, J., Mei, C., Lv, Y.: Incomplete decision contexts: approximate concept construction, rule acquisition and knowledge reduction. Int. J. Approx. Reason. **54**(1), 149–165 (2013)
26. Li, J.H., Mei, C.L., Xu, W.H., Qian, Y.H.: Concept learning via granular computing: a cognitive viewpoint. Inf. Sci. **298**, 447–467 (2015)
27. Liang, D.C., Liu, D.: Systematic studies on three-way decisions with interval-valued decision theoretic rough sets. Inf. Sci. **276**, 186–203 (2014)
28. Liang, D., Liu, D., Pedrycz, W., Hu, P.: Triangular fuzzy decision-theoretic rough sets. Int. J. Approx. Reason. **54**(8), 1087–1106 (2013)
29. Liang, D.C., Pedrycz, W., Liu, D., Hu, P.: Three-way decisions based on decision-theoretic rough sets under linguistic assessment with the aid of group decision making. Appl. Soft Comput. **29**, 256–269 (2015)
30. Liang, J., Wang, F., Dang, C., Qian, Y.: An efficient rough feature selection algorithm with a multi-granulation view. Int. J. Approx. Reason. **53**(6), 912–926 (2012)
31. Lin, G., Qian, Y., Li, J.: NMGRS: Neighborhood-based multigranulation rough sets. Int. J. Approx. Reason. **53**(7), 1080–1093 (2012)
32. Li, N., Zhou, R., Hu, Q., Liu, X.: Mechanical fault diagnosis based on redundant second generation wavelet packet transform, neighborhood rough set and support vector machine. Mech. Syst. Signal Process. **28**, 608–621 (2012)
33. Lingras, P., Yao, Y.Y.: Time complexity of rough clustering: GAs versus K-means. Rough Sets and Current Trends in Computing Proceedings. Book Series. Lecture Notes in Artificial Intelligence **2475**, 263–270 (2002)

34. Lingras, P., Yan, R., West, C.: Comparison of conventional and rough K-means clustering. Rough Sets. Fuzzy Sets, Data Mining and Granular Computing. Book Series. Lecture Notes in Artificial Intelligence **2639**, 130–137 (2003)
35. Lingras, P., West, C.: Interval set clustering of web users with rough K-means. J. Intell. Inf. Syst. **23**(1), 5–16 (2004)
36. Liu, D., Li, T., Ruan, D.: Probabilistic model criteria with decision-theoretic rough sets. Inf. Sci. **181**(17), 3709–3722 (2011)
37. Liu, D., Yao, Y.Y., Li, T.: Three-way investment decisions with decision-theoretic rough sets. Int. J. Comput. Intell. Syst. **4**(1), 66–74 (2011)
38. Liu, D., Yao, Y.Y., Li, T.: A Multiple-category classification approach with decision-theoretic rough sets. Fundamenta Informaticae **115**(2–3), 173–188 (2012)
39. Maji, P., Pal, S.K.: Rough fuzzy c-medoids algorithm and selection of bio-basis for amino acid sequence analysis. IEEE Trans. Knowl. Data Eng. **19**(6), 859–872 (2007)
40. Maji, P., Pal, S.K.: Rough set based generalized fuzzy C-means algorithm and quantitative indices. IEEE Trans. Syst. Man Cybern. Part B-Cybern. **37**(6), 1529–1540 (2007)
41. Maji, P.K., Roy, A.R.: An application of soft sets in a decision making problem. Comput. Math. Appl. **44**(8–9), 1077–1083 (2002)
42. Medina, J.: Multi-adjoint property-oriented and object-oriented concept lattices. Inf. Sci. **190**, 95–106 (2012)
43. Mi, J.S., Wu, W.Z., Zhang, W.X.: Approaches to knowledge reduction based on variable precision rough set model. Inf. Sci. **159**(3–4), 255–272 (2004)
44. Min, F., He, H., Qian, Y., Zhu, W.: Test-cost-sensitive attribute reduction. Inf. Sci. **181**(22), 4928–4942 (2011). doi:10.1016/j.ins.2011.07.010
45. Min, F., Zhu, W.: Attribute reduction of data with error ranges and test costs. Inf. Sci. **211**, 48–67 (2012)
46. Mingers, J., Leydesdorff, L.: A Review of theory and practice in scientometrics. Eur. J. Oper. Res. **246**(1), 1–19 (2015)
47. Mitra, S., Hayashi, Y.: Neuro-fuzzy rule generation: survey in soft computing framework. IEEE Trans. Neural Netw. **11**(3), 748–768 (2000)
48. Morsi, N.N., Yakout, M.M.: Axiomatics for fuzzy rough sets. Fuzzy Sets Syst. **100**(1–3), 327–342 (1998)
49. Pawłak, Z.: Rough sets. Int. J. Parallel. Program. **11**(5), 341–356 (1982)
50. Pawłak, Z., Grzymala-Busse, J., Slowinski, R., Ziarko, W.: Rough sets. Commun. ACM **38**(11), 88–95 (1995)
51. Pawłak, Z.: Rough set approach to knowledge-based decision support. Eur. J. Oper. Res. **99**(1), 48–57 (1997)
52. Pawłak, Z.: Rough set theory and its applications to data analysis. Cybern. Syst. **29**(7), 661–688 (1998)
53. Pawłak, Z.: Rough sets and intelligent data analysis. Inf. Sci. **147**(1–4), 1–12 (2002)
54. Pawłak, Z., Showron, A.: Rudiments of rough sets. Inf. Sci. **177**(1), 3–27 (2007)
55. Pawłak, Z., Skowron, A.: Rough sets and boolean reasoning. Inf. Sci. **177**(1), 41–73 (2007)
56. Pawłak, Z., Skowron, A.: Rough sets: some extensions. Inf. Sci. **177**(1), 28–40 (2007)
57. Pedrycz, W.: Allocation of information granularity in optimization and decision-making models: towards building the foundations of granular computing. Eur. J. Oper. Res. **232**(1), 137–145 (2014)
58. Polkowski, L., Skowron, A.: Rough mereology: a new paradigm for approximate reasoning. Int. J. Approx. Reason. **15**(4), 333–365 (1996)
59. Qian, Y.H., Liang, J.Y., Pedrycz, W., Dang, C.: An efficient accelerator for attribute reduction from incomplete data in rough set framework. Pattern Recogn. **44**(8), 1658–1670 (2011)
60. Qian, Y.H., Liang, J.Y., Wu, W.Z., Dang, C.: Information granularity in fuzzy binary GrC model. IEEE Trans. Fuzzy Syst. **19**(2), 253–264 (2011)
61. Qian, J., Lv, P., Yue, X.D., Liu, C.H., Jing, Z.J.: Hierarchical attribute reduction algorithms for big data using MapReduce. Knowledge-Based Syst. **73**, 18–31 (2015)

62. Qian, Y.H., Zhang, H., Sang, Y., Liang, J.Y.: Multigranulation decision-theoretic rough sets. Int. J. Approx. Reason. **55**(1), 225–237 (2014)
63. Qian, J., Miao, D.Q., Zhang, Z.H., Li, W.: Hybrid approaches to attribute reduction based on indiscernibility and discernibility relation. Int. J. Approx. Reason. **52**(2), 212–230 (2011)
64. Research Assessment: Journal Citation Ranking and Quartile Scores (2016). https://researchassessment.fbk.eu/quartile_score. Accessed 30 July 2016
65. Salehi, S., Selamat, A., Fujita, H.: Systematic mapping study on granular computing. Knowledge-Based Syst. **80**, 78–97 (2015)
66. Skowron, A., Stepaniuk, J., Swiniarski, R.: Modeling rough granular computing based on approximation spaces. Inf. Sci. **184**(1), 20–43 (2012)
67. Slowinski, R., Vanderpooten, D.: A generalized definition of rough approximations based on similarity. IEEE Trans. Knowl. Data Eng. **12**(2), 331–336 (2000)
68. Swiniarski, R.W., Skowron, A.: Rough set methods in feature selection and recognition. Pattern Recogn. Lett. **24**(6), 833–849 (2003)
69. Wang, C.Z., He, Q., Chen, D.G., Hu, Q.H.: A novel method for attribute reduction of coveringdecision systems. Inf. Sci. **254**, 181–196 (2014)
70. Wang, X.Y., Yang, J., Teng, X.L., Xia, W.J., Jensen, R.: Feature selection based on rough sets and particle swarm optimization. Pattern Recogn. Lett. **28**(4), 459–471 (2007)
71. WoS: The Thomson Reuters Impact Factor (2014). http://wokinfo.com/essays/impact-factor/. Accessed 25 July 2016
72. Wu, W.Z., Mi, J.S., Zhang, W.X.: Generalized fuzzy rough sets. Inf. Sci. **151**, 263–282 (2003)
73. Wu, W.Z., Zhang, W.X.: Constructive and axiomatic approaches of fuzzy approximation operators. Inf. Sci. **159**(3–4), 233–254 (2004)
74. Wu, W.Z., Leung, Y.: Theory and applications of granular labelled partitions in multi-scale decision tables. Inf. Sci. **181**(18), 3878–3897 (2011)
75. Yao, J.T.: A ten-year review of granular computing. Proceedings of IEEE International Conference on Granular Computing, Silicon Valley, USA **2–5**, 734–739 (2007)
76. Yao, J.T., Zhang, Y.: A scientometrics study of rough sets in three decades. In: Proceedings of Rough Set and Knowledge Technology (RSKT 2013). Lecture Notes in Computer Science, vol. 8171, pp. 28–40 (2013)
77. Yao, J.T., Vasilakos, A.V., Pedrycz, W.: Granular computing: perspectives and challenges. IEEE Trans. Cybern. **43**(6), 1977–1989 (2013)
78. Yao, Y.Y., Wong, S.K.M.: A decision theoretic framework for approximating concepts. Int. J. Man-Mach. Stud. **37**(6), 793–809 (1992)
79. Yao, Y.Y.: Two views of the theory of rough sets in finite universes. Int. J. Approx. Reason. **15**(4), 291–317 (1996)
80. Yao, Y.Y.: Relational interpretations of neighborhood operators and rough set approximation operators. Inf. Sci. **111**(1), 239–259 (1998)
81. Yao, Y.Y.: Constructive and algebraic methods of the theory of rough sets. Inf. Sci. **109**(1), 21–47 (1998)
82. Yao, Y.Y.: Information granulation and rough set approximation. Int. J. Intell. Syst. **16**(1), 87–104 (2001)
83. Yao, Y.Y., Zhao, Y.: Attribute reduction in decision-theoretic rough set models. Inf. Sci. **178**(17), 3356–3373 (2008)
84. Yao, Y.Y.: The superiority of three-way decisions in probabilistic rough set models. Inf. Sci. **181**(6), 1080–1096 (2011)
85. Yao, Y.Y.: Two semantic issues in a probabilistic rough set model. Fundamenta Informaticae **108**(3–4), 249–265 (2011)
86. Yao, Y.Y., Yao, B.: Covering based rough set approximations. Inf. Sci. **200**, 91–107 (2012)
87. Yu, H., Liu, Z.G., Wang, G.Y.: An automatic method to determine the number of clusters using decision-theoretic rough set. Int. J. Approx. Reason. **55**(1), 101–115 (2014)
88. Zhan, J.M., Liu, Q., Davvaz, B.: A new rough set theory: rough soft hemirings. J. Intell. Fuzzy Syst. **28**(4), 1687–1697 (2015)

89. Zhang, X.H., Dai, J.H., Yu, Y.C.: On the union and intersection operations of rough sets based on various approximation spaces. Inf. Sci. **292**, 214–229 (2015)
90. Zhang, J., Li, T., Ruan, D., Liu, D.: Rough sets based matrix approaches with dynamic attribute variation in set-valued information systems. Int. J. Approx. Reason. **53**(4), 620–635 (2012)
91. Zhang, J., Li, T., Ruan, D., Gao, Z., Zhao, C.: A parallel method for computing rough set approximations. Inf. Sci. **194**, 209–223 (2012)
92. Zhang, X., Zhou, B., Li, P.: A general frame for intuitionistic fuzzy rough sets. Inf. Sci. **216**, 34–49 (2012)
93. Zhao, X.R., Hu, B.Q.: Fuzzy and interval-valued fuzzy decision-theoretic rough set approaches based on fuzzy probability measure. Inf. Sci. **298**, 534–554 (2015)
94. Zhou, B.: Multi-class decision-theoretic rough sets. Int. J. of Approx. Reason. **55**(1), 211–224 (2014)
95. Zhu, W., Wang, F.Y.: Reduction and axiomization of covering generalized rough sets. Inf. Sci. **152**(1), 217–230 (2003)
96. Zhu, W.: Topological approaches to covering rough sets. Inf. Sci. **177**(6), 1499–1508 (2007)
97. Zhu, W., Wang, S.: Matroidal approaches to generalized rough sets based on relations. Int. J. Mach. Learn. Cybern. **2**(4), 273–279 (2011)
98. Ziarko, W.: Variable precision rough set model. J. Comput. Syst. Sci. **46**(1), 39–59 (1993)

Part II
Review of Rough Set Research

Rough Sets, Rough Mereology and Uncertainty

Lech Polkowski

Abstract 35 years ago Zdzisław Pawlak published the article in which He proposed a novel idea for reasoning about uncertainty. He proposed to present knowledge as classification ability and in consequence the playground for His theory was proposed as approximation space, i.e., a set along with an equivalence relation on it, equivalence classes representing categories to which objects in the set were to be assigned. This point of view was stressed in the monograph 'Rough Sets. Theoretical Aspects of Reasoning about Data' (1992) 25 years ago. The application tint was given to rough sets by transferring center of gravity of the theory from approximation spaces to decision/information systems, i.e., data tables in the attribute—value format. In the same year the first collective monograph appeared accompanying the first workshop on rough sets: 'Intelligent Decision Support. Handbook of Applications and Advances of Rough Set Theory' edited by Roman Słowiński. The effect of those 10 years was emergence of notions like a decision rule, a reduct, a core, of algorithms for finding certain, minimal and optimal rules, for finding reducts, analyses of relations between rough sets and other paradigms describing uncertainty and emergence of hybrid approaches like rough-fuzzy sets etc. Still 10 years elapsed and a monograph on foundations of rough sets was possible (2002): 'Rough Sets. Mathematical Foundations' by this author, and, some other outlines of rough set theory appeared. Rough set research grew, extending its scope by entering realms of morphology, intelligent agents theory, linguistics, behavioral robotics, mereology, granular computing, acquiring many applications in diverse fields. In this chapter we try to sum up our personal experience and results and in a sense to unify them into a coherent conceptual scheme following the main themes of rough set theory: to understand uncertainty and to cope with it in data. In this work, we use the term 'thing' to denote a being in general, denoted with $x, y, z, ...$ and the term 'object' to denote beings in the universes of information/decision systems, denoted $u, v, w, ...$; truth values are denoted with letters $r, s, ...$

There is no uncertain language in which to discuss uncertainty (Zdzisław Pawlak)

L. Polkowski (✉)
Polish - Japanese Academy IT, Koszykowa str. 86, 02-008 Warszawa, Poland
e-mail: polkow@pjwstk.edu.pl

G. Wang et al. (eds.), *Thriving Rough Sets*, Studies in Computational Intelligence 708, DOI 10.1007/978-3-319-54966-8_4

1 The Phenomenon of Uncertainty

Uncertainty can be defined as inability to choose one optimal object in the given context from a set of more than one optional objects. A usually invoked example of uncertainty is The Uncertainty Principle of Heisenberg [22], i.e., the thesis that the precise values of position and energy of an electron cannot be known simultaneously—this is uncertainty inscribed into nature; uncertainty immanent to human thinking was revealed by Gödel, see [64]: for each formal system which contains arithmetic of natural numbers, there can be formulated statements formally correct about which one cannot decide whether they are true or false, this is uncertainty imbued into abstract thinking. There are ordinary cases for uncertainty like making decision at crossroads, investing on the stock market, forecasting a political issue, due to the lack of adequate knowledge. Due to the omnipresence of uncertainty in all venues of life, the problem of catching the essence of uncertainty and attempts at formal reasoning schemes taking uncertainty into consideration were subject of interest to many scientists. Logicians beginning from Aristotle were fully aware of difficulties with uncertain knowledge, witness the famous example in Aristotle 'De Interpretatione' [3] 'there will be sea battle tomorrow', cf., a discussion in [43], yet the logical systems up to beginning of 20th century dealt solely with definite binary–valued statements, either true or false. In 1918, Jan Łukasiewicz [32], [69], introduced the 3–valued logic with the value of 2 for statements uncertain as to their truth value, which may be labelled 'don't know' (later the value of 2 was replaced by more convenient computationally value of 1/2). The Łukasiewicz 3–valued system was defined by formulas, where $r, s \in \{0, 1, 1/2\}$,

$$r \Rightarrow s = min\{1, 1 - r + s\} \tag{1}$$

for implication and

$$\neg r = 1 - r \tag{2}$$

for negation. It was recognized immediately that Łukasiewicz formulas extend to n–valued logics as well as to logics with truth values rational in the unit interval or simply real values in the unit interval, hence, the Łukasiewicz systems allow for expressing any degree of uncertainty. Calculus of uncertain notions became possible when Lotfi Asker Zadeh [81] introduced the idea of a fuzzy set. This notion does extend the classical notion of a set characterized by the function

$$\chi(x) = 1\ x \in A\ else\ 0 \tag{3}$$

where A is a subset of the universal set U, by introducing the fuzzy membership function

$$\mu_A : U \rightarrow [0, 1] \tag{4}$$

with the interpretation that $\mu_A(x) = 1$ means that $x \in A$ is the true statement, $\mu_A(x) = 0$ means that the statement $x \in A$ is false and the value of $\mu_A(x)$ in (0, 1) indicates the degree of uncertainty whether x is in A. Applying to notions so fuzzified the logical scheme of Łukasiewicz, we can assign uncertainty degrees to complex statements about commonly used notions, e.g., the implication '*If x is high to degree of 0.8 then x will play on university basketball team with chance to degree of 0.6*' has the degree of truth of $1 - 0.8 + 0.6 = 0.8$. Topology entered the world of scientific paradigms much earlier, in the midst of XIX century, see [33] and from its beginnings it formalized uncertainty of location in the notion of the boundary of a set. Given a collection of open sets G in the universal set U and a set A a subset of the universal set U, the boundary $Bd_G A$ of A with respect to G is defined as the set of points $x \in U$ with the property $bound_A$ defined as

$$bound_A(x) \text{ if and only if } N \cap A \neq \emptyset \text{ and not } N \subseteq A \text{ for each nbhd } N \in G \textit{ of } x. \quad (5)$$

An original approach to formalization of uncertainty was proposed by Karl Menger [36–38], who considered metric spaces with the distance function known up to probability only. An upshot from his work was emergence of functions known as t–norms, each t–norm $T : [0, 1]^2 \to [0, 1]$ satisfying properties (1) $T(x, y) = T(y, x)$; (2) $T(x, T(y, z)) = T(T(x, y), z)$; (3) If $x \geq x'$ then $T(x, y) \geq T(x', y)$; (4)$T(x, 0) = 0$, $T(x, 1) = x$. Additional properties are: (5) T is continuous; (6) $T(x, x) < x$ for $x \in (0, 1)$. T–norms satisfying (5,6) satisfy the functional equation [30]:

$$T(x, y) = g(f(x) + f(y)), \quad (6)$$

where $f : [0, 1] \to [0, 1]$ is a continuous decreasing function with $f(0) = 1$, and g is the pseudo–inverse to f. Each t–norm T gives rise to the function $\Rightarrow_T : [0, 1]^2 \to [0, 1]$ called the *residuum* and defined by means of the equivalence:

$$x \Rightarrow_T y \geq r \text{ if and only if } T(x, r) \leq y. \quad (7)$$

In particular, the Łukasiewicz implication (1) is the residuum of the Łukasiewicz t–norm $L(x, y) = max\{0, x + y - 1\}$; the other t–norm $P(x, y) = x \cdot y$, the product t–norm, induces the Goguen implication:

$$x \Rightarrow_P y = \frac{y}{x} \text{ if } x > y \textit{ else } 1. \quad (8)$$

The third classical t-norm $M(x, y) = min\{x, y\}$ induces the G *o*del implication:

$$x \Rightarrow_M y = y \text{ if } x > y \textit{ else } 1. \quad (9)$$

Each of those logics combined with fuzzified notions gives a calculus of uncertain notions.

2 The Pawlak Approach to Uncertainty

Zdzisław Pawlak [44, 45] approached knowledge as ability to classify given objects into given categories; in this, he became close to definition of knowledge due to Innocenty Maria Bocheński [13]: '*knowledge is a set of relations among things, and, properties of things*'. The realization of this definition in Pawlak's approach consisted in establishing relations between things and categories of things of the form '*the thing x belongs in the category C*'. Hence, uncertain knowledge meant inability to classify certain objects into categories in a deterministic way. The simple model of this notion of knowledge was a set of objects along with a partition of this set into categories, so–called *approximation space*. Elements of the partition were decidable: for each object it was true that it belongs in the category or not and the same was true for unions of categories. Unions of categories were called exact sets. Other sets were declared rough, i.e., not decidable. It is manifest that this approach was a topological one: categories were basic open sets, their unions were open sets and due to the disjointness of categories open sets were also closed sets. Moreover, each intersection of a family of open sets was open, a peculiar property of topologies induced by partitions. Pawlak addressed also the problem of non-deterministic concepts, i.e., of uncertain knowledge: for a set X not exact, i.e., not open, he introduced approximations, the lower and the upper as, respectively, the interior and the closure in the partition topology:

$$l(X) = \bigcup\{C \in Cat : C \subseteq X\},\ u(X) = \bigcup\{C \in Cat : C \cap X \neq \emptyset\}. \tag{10}$$

A specific implementation of this idea was using the notions of an information system and a decision system. An information system (U, A, V), denoted by various authors in some distinct ways, consists of a set of objects U, a set of attributes (features) A on U and a set of attribute values V so for each pair $(a, u) \in A \times U$ (so we assume for now that those systems are complete, i.e., with no missing values) a value $a(u) \in V$ is defined; a decision system (U, A, V, d) is augmented by an additional attribute, the decision d. In case of information or decision systems, we will use the generic symbols $u, v, w, \ldots$ to denote objects in the set U. For an attribute $q \in A \cup \{d\}$, the indiscernibility relation Ind_q is a partition of the set U into categories of the form $Ind_q(u) = \{v \in U : q(u) = q(v)\}$. For a set B of attributes, the category Ind_B is defined as the intersection $\bigcap_{q \in B} Ind_q$. In this notation, relations, called decision rules, constituting knowledge represented by the decision system (U, A, V, d) are of the form:

$$B \to d = \{Ind_B(u) \to Ind_d(u) : u \in U\}. \tag{11}$$

To obtain the extended form of a relation $Ind_B(u) \to Ind_d(u)$, one lets $B = \{a_{i_1}, \ldots, a_{i_k}\}$ and represents Ind_B as the meaning of the formula

$$\bigwedge_{j=1}^{k}(a_{i_j}(u) = v_j) \to (d(u) = v), \tag{12}$$

where the meaning $[\phi]_U$ of a predicate formula ϕ is

$$[\phi]_U = \{u \in U : u \models \phi\}. \tag{13}$$

Hence, knowledge expressed by means of a set of decision rules can be presented in logical or set–theoretical formalism. A great merit of Pawlak's approach lies in reviving some of old notions of classical logic and mathematics in the new context of knowledge engineering.

3 Rough Set Spaces

Zdzisław Pawlak was a theoretician by temperament and mental constitution, as witnesssed by his many achievements. He was very interested in giving his rough sets as many formal structures as was possible, in particular he asked for topological rough set spaces. The author of this chapter was able to fulfill his desire by producing some topologies on rough sets, see [47] for summary of results, proofs and bibliography of original works. Of course, those topologies, being desirably distinct from already well known partition topologies, were to be induced from a more complex settings, viz., from infinite information systems. This approach corresponded with a well–known advice by Stan Ulam: *'if you want to discuss a finite case, go first to the infinite one'*. Let us recall the essential results, which of course have had a purely intellectual valor, as being on the side of esoteric they are not commonly known, albeit see paragraph on collage theorem. We assume given a set (a *universe*) U of *objects* along with a sequence $A = \{a_n : n = 1, 2, \dots\}$ of *attributes* where without loss of generality we may assume that $Ind_{n+1} \subseteq Ind_n$ for each n. Letting $Ind = \bigcap_n Ind_n$, we may assume that the family $\{Ind_n : n = 1, 2, \dots\}$ *separates objects*, i.e., for each pair $u \neq v$, there is a class $P \in U/Ind_n$ for some n such that $u \in P, v \notin P$, otherwise we would pass to the quotient universe U/Ind. This implies that the set U is of power of continuum. We endow U with some topologies.

3.1 Topologies Π_n, the Topology Π_0 and Exact and Rough Sets

For each n, the topology Π_n is defined as the partition topology obtained by taking as open sets unions of families of classes of the relation Ind_n. The topology Π_0 is the union of topologies Π_n for $n = 1, 2, \dots$. We apply the topology Π_0 to the task of discerning among subsets of the universe U (Cl_τ is the closure operator and Int_τ is the interior operator with respect to a topology τ):

$$\text{A set } Z \subseteq U \text{ is } \Pi_0 - \text{exact if } Cl_{\Pi_0} Z = Int_{\Pi_0} Z \text{ else } Z \text{ is } \Pi_0\text{-rough.} \tag{14}$$

3.2 The Space of Π_0-rough Sets Is Metrizable

Each Π_0-rough set can be represented as a pair (Q, T) where $Q = Cl_{\Pi_0}X, T = U \setminus Int_{\Pi_0}X$ for some $X \subseteq U$. The pair (Q, T) has to satisfy the conditions: 1. $U = Q \cup T$. 2. $Q \cap T \neq \emptyset$. 3. If $\{x\}$ is a Π_0-open singleton then $x \notin Q \cap T$. We define a metric d_n as ($[u]_n$ is the Ind_n-class of u):

$$d_n(u, v) = 1 \text{ in case } [u]_n \neq [v]_n \text{ else } d_n(u, v) = 0. \tag{15}$$

and the metric d:

$$d(u, v) = \sum_n 10^{-n} \cdot d_n(u, v). \tag{16}$$

Theorem 1 *Metric topology of d is Π_0.*

We employ the notion of the *Hausdorff metric* and apply it to pairs (Q, T) satisfying 1–3 above, i.e., representing Π_0-rough sets. For pairs $(Q_1, T_1), (Q_2, T_2)$, we let

$$D((Q_1, T_1), (Q_2, T_2)) = max\{d_H(Q_1, Q_2), d_H(T_1, T_2)\} \tag{17}$$

and

$$D^*((Q_1, T_1), (Q_2, T_2)) = max\{d_H(Q_1, Q_2), d_H(T_1, T_2), d_H(Q_1 \cap Q_2, T_1 \cap T_2)\}, \tag{18}$$

where $d_H(A, B) = max\{max_{x \in A} dist(x, B), max_{y \in B} dist(y, A)\}$ is the Hausdorff metric on closed sets and $dist(x, A) = min_{y \in A} d(x, y)$. The main result is

Theorem 2 *If each descending sequence $\{[u_n]_n : n = 1, 2, \dots\}$ of classes of relations Ind_n has a non–empty intersection, then each D^*–fundamental sequence of Π_0–rough sets converges in the metric D to a Π_0–rough set. If, in addition, each relation Ind_n has a finite number of classes, then the space of Π_0–rough sets is compact in the metric D.*

3.3 The Space of Almost Π_0-rough Sets Is Metric Complete

In notation of preceding sections, it may happen that a set X is Π_n-rough for each n but it is Π_0-exact. We call such sets *almost rough sets*. We denote those sets as Π_ω-rough. Each set X of them, is represented in the form of a sequence of pairs (Q_n, T_n) : $n = 1, 2, \dots$ such that for each n, 1. $Q_n = Cl_{\Pi_n}X, T_n = U \setminus Int_{\Pi_n}X$. 2. $Q_n \cap T_n \neq \emptyset$. 3. $Q_n \cup T_n = U$. 4. $Q_n \cap T_n$ contains no singleton $\{x\}$ with $\{x\}$ Π_n-open. To introduce a metric into the space of Π_ω-rough sets, we apply again the Hausdorff metric but in a modified way: for each n, we let $d_{H,n}$ to be the Hausdorff metric on Π_n-closed sets, and for representations (Q_n, T_n) and $(Q^*_n, T^*_n)_n$ of Π_ω-rough sets X, Y, respectively, we define the metric D' as:

$$D'(X,Y) = \sum_n 10^{-n} \cdot max\{d_{H,n}(Q_n, Q^*_n), d_{H,n}(T_n, T^*_n)\}. \tag{19}$$

It turns out that

Theorem 3 *The space of Π_ω-rough sets endowed with the metric D' is complete, i.e., each D'-fundamental sequence of Π_ω-rough sets converges to a Π_ω-rough set.*

Theoretical as are these results, yet there was an applicational tint in them.

3.4 Approximate Collage Theorem

Consider an Euclidean space E^n along with an information system $(E^n, A = \{a_k : k = 1, 2, \ldots\})$, each attribute a_k inducing the partition P_k of E^n into cubes of the form $\prod_{i=1}^n [m_i + \frac{j_i}{2^k}, m_i + \frac{j_i+1}{2^k})$, where m_i runs over integers and $j_i \in [0, 2^k - 1]$ is an integer. Hence, $P_{k+1} \subseteq P_k$, each k. We consider *fractal objects*, i.e., systems of the form $[(C_1, C_2, \ldots, C_p), f, c]$, where each C_i is a compact set and f is an affine contracting mapping on E^n with a contraction coefficient $c \in (0, 1)$. The resulting fractal is the limit of the sequence $(F_n)_n$ of compacta, where 1. $F_0 = \bigcup_{i=1}^p C_i$. 2. $F_{n+1} = f(F_n)$. In this context, fractals are classical examples of Π_0-rough sets. Assume we perceive fractals through their approximations by consecutive grids P_k, so each F_n is viewed on as its *upper approximation*, see (10), $a^+_k F_n$ for each k. As $diam(P_k) \rightarrow_{k\rightarrow\infty} 0$, it is evident that the symmetric difference $F \triangle F_n$ becomes arbitrarily close to the symmetric difference $a^+_k F \triangle a^+_k F_n$. Hence, in order to approximate F with F_n it suffices to approximate $a^+_k F$ with $a^+_k F_n$. The question poses itself: what is the least k which guarantees for a given ε, that if $a^+_k F_n = a^+_k F$ then $d_H(F, F_n) \leq \varepsilon$. We consider the metric D on fractals and their approximations. We had proposed a counterpart to Collage Theorem, by replacing fractals F_n by their grid approximations.

Theorem 4 (Approximate Collage Theorem [46]) *Assume a fractal F generated by the system $(F_0 = \bigcup_{i=1}^p C_i, f, c)$ in the space of Π_0-rough sets with the metric D. In order to satisfy the requirement $d_H(F, F_n) \leq \varepsilon$, it is sufficient to satisfy the requirement $a^+_{k_0} F_n = a^+_{k_0} F$ with $k_0 = \lceil \frac{1}{2} - log_2 \varepsilon \rceil$ and $n \geq \lceil \frac{log_2[2^{-k_0+\frac{1}{2}} \cdot K^{-1} \cdot (1-c)]}{log_2 c} \rceil$, where $K = d_H(F_0, F_1)$.*

This ends the topological chapter in rough set theory and we pass to the second large area in which the research was also in a sense provoked by Zdzisław Pawlak, i.e. to theory of rough mereology.

4 Mereology

Zdzisław Pawlak was by his education an engineer—he graduated from Department of Electrical Engineering in Warsaw Polytechnical, but by intellectual composition he was close to theory in particular to mathematics and logic. He spent his professional time in the fifties and sixties in the Mathematical Institute of the Polish Academy of Sciences (PAS) where he met first class mathematicians, in particular he often remembered seminars with Andrzej Ehrenfeucht, the world renowned logician. Much later, around 2000, when travelling with Zdzisław to a conference in Alaska, we made a stop-over at Denver and we visited Andrzej Ehrenfeucht and Jan Mycielski in Boulder where they worked. Present at the reunion, I experienced the climate of those for long passed days. So, Zdzisław was justly interested in mereology as a theory of parts; the notion of a proper subset is a particular case of the notion of a part, and, any rough set is sandwiched as a proper part between its exact approximations. Hence, basically, mereology can have the claim for being a formal base for rough set theory. In addition to this, mereology as a formal theory was first constructed by Stanislaw Leśniewski, in the years 1918–1939 professor at Warsaw University and Zdzisław also metioned that he saw in the Library of the Institute some manuscripts by him (the whole archive of Leśniewski was burned down during the Warsaw Uprising in August-September 1944, so those manuscripts could be some works submitted to Fundamenta Mathematicae). So it was our resolution to investigate the subject. Let us say a few words about it.

4.1 Classical Mereology

Mereology due to Leśniewski arose from attempts at reconciling antinomies of naïve set theory, see Leśniewski [26], [27], [29], Srzednicki et al. [73], Sobociński [71], [72]. Leśniewski [26] was the first presentation of the foundations of his theory as well as the first formally complete exposition of mereology. The primitive notion of mereology in this formalism is the notion of a *part*, mentioned already by Aristotle [4]. Given some category of things, a relation of a part is a binary relation π which is required to be

M1 *Irreflexive*: *For each thing* x, *it is not true that* $\pi(x, x)$.

M2 *Transitive*: *For each triple* x, y, z *of things, if* $\pi(x, y)$ *and* $\pi(y, z)$, *then* $\pi(x, z)$.

Remark In the original scheme of Leśniewski, the relation of parts is applied to *individual things* as defined in Ontology of Leśniewski, see Leśniewski [28], Iwanuś [24]. The relation of *part* induces the relation of an *ingredient* (the term is due to T. Kotarbiński), *ingr*, defined as

$$ingr(x, y) \Leftrightarrow \pi(x, y) \vee x = y \quad (20)$$

The relation of ingredient is a partial order on things, i.e.,

1. $ingr(x, x)$.
2. $ingr(x, y) \wedge ingr(y, x) \Rightarrow (x = y)$.
3. $ingr(x, y) \wedge ingr(y, z) \Rightarrow ingr(x, z)$.

We formulate the third axiom with a help from the notion of an ingredient.

M3 (*Inference*) For each pair of things x, y, if the property

$I(x, y)$: *For each* t, *if* $ingr(t, x)$, *then there exist* w, z *such that* $ingr(w, t)$, $ingr(w, z)$, $ingr(z, y)$ *hold*, is satisfied, then $ingr(x, y)$.

The predicate of *overlap*, *Ov* in symbols, is defined by means of

$$Ov(x, y) \Leftrightarrow \exists z.ingr(z, x) \wedge ingr(z, y). \tag{21}$$

Using the overlap predicate, one can write $I(x, y)$ down in the form

$I_{Ov}(x, y)$: *For each* t *if* $ingr(t, x)$, *then there exists* z *such that* $ingr(z, y)$ *and* $Ov(t, z)$.

The notion of a *mereological class* follows; for a non–vacuous property Φ of things, the *class of* Φ, denoted $Cls\Phi$ is defined by the conditions

C1 If $\Phi(x)$, then $ingr(x, Cls\Phi)$.

C2 If $ingr(x, Cls\Phi)$, then there exists z such that $\Phi(z)$ and $I_{Ov}(x, z)$.

In plain language, the class of Φ collects in an individual thing all things satisfying the property Φ. The existence of classes is guaranteed by an axiom.

M4 For each non–vacuous property Φ there exists a class $Cls\Phi$.

The uniqueness of the class follows by M3. M3 implies also that, for the non–vacuous property Φ, if for each thing z such that $\Phi(z)$ it holds that $ingr(z, x)$, then $ingr(Cls\Phi, x)$. The notion of an overlap allows for a succinct characterization of a class: for each non–vacuous property Φ and each thing x, it happens that $ingr(x, Cls\Phi)$ if and only if for each ingredient w of x, there exists a thing z such that $Ov(w, z)$ and $\Phi(z)$.

Remark Uniqueness of the class along with its existence is an axiom in the Leśniewski [26] scheme, from which M3 is derived. Similarly, it is an axiom in the Tarski [74–76] scheme. Please consider two examples.

1. The strict inclusion $\subset$ on sets is a part relation. The corresponding ingredient relation is the inclusion $\subseteq$. The overlap relation is the non–empty intersection. For a non–vacuous family F of sets, the class $ClsF$ is the union $\bigcup F$.
2. For reals in the interval $[0, 1]$, the strict order $<$ is a part relation and the corresponding ingredient relation is the weak order $\leq$. Any two reals overlap; for a set $F \subseteq [0, 1]$, the class of F is $supF$.

The notion of an *element* is defined as follows

$$el(x, y) \Leftrightarrow \exists \Phi.y = Cls\Phi \wedge \Phi(x). \tag{22}$$

In plain words, $el(x, y)$ means that y is a class of some property and x responds to that property. To establish some properties of the notion of an element, we begin with the property $INGR(x) = \{y : ingr(y, x)\}$, for which the identity $x = ClsINGR(x)$ holds by M3. Hence, $el(x, y)$ is equivalent to $ingr(x, y)$. Thus, each thing x is its own element. This is one of means of expressing the impossibility of the Russell paradox within the mereology, cf., Leśniewski [26], Thms. XXVI, XXVII, see also Sobociński [71]. We observe the extensionality of overlap: *For each pair x, y of things, $x = y$ if and only if for each thing z, the equivalence $Ov(z, x) \Leftrightarrow Ov(z, y)$ holds.* Indeed, assume the equivalence $Ov(z, x) \Leftrightarrow Ov(z, y)$ to hold for each z. If $ingr(t, x)$ then $Ov(t, x)$ and $Ov(t, y)$ hence by axiom M3 $ingr(t, y)$ and with $t = x$ we get $ingr(x, y)$. By symmetry, $ingr(y, x)$, hence $x = y$. The notion of a subset follows:

$$sub(x, y) \Leftrightarrow \forall z.[ingr(z, x) \Rightarrow ingr(z, y)]. \tag{23}$$

It is manifest that for each pair x, y of things, $sub(x, y)$ holds if and only if $el(x, y)$ holds if and only if $ingr(x, y)$ holds. For the property $Ind(x) \Leftrightarrow ingr(x, x)$, one calls the class *ClsInd*, *the universe*, in symbols V. It follows that 1. The universe is unique. 2. $ingr(x, V)$ holds for each thing x. 3. For each non–vacuous property Φ, it is true that $ingr(Cls\Phi, V)$. The notion of an *exterior* thing x to a thing y, $extr(x, y)$, is the following:

$$extr(x, y) \Leftrightarrow \neg Ov(x, y). \tag{24}$$

In plain words, x is exterior to y when no thing is an ingredient both to x and y. Clearly, the operator of exterior has properties 1. No thing is exterior to itself. 2. $extr(x, y)$ implies $extr(y, x)$. 3. If for a non–vacuous property Φ, a thing x is exterior to every thing z such that $\Phi(z)$ holds, then $extr(x, Cls\Phi)$. The notion of a complement to a thing, with respect to another thing, is rendered as a ternary predicate $comp(x, y, z)$, cf., Leśniewski, [26], par. 14, Def. IX, to be read: 'x is the complement to y with respect to z', and it is defined by means of the following requirements 1. $x = ClsEXTR(y, z)$. 2. $ingr(y, z)$, where $EXTR(y, z)$ is the property which holds for an thing t if and only if $ingr(t, z)$ and $extr(t, y)$ hold. This definition implies that the notion of a complement is valid only when there exists an ingredient of z exterior to y. Following are basic properties of complement 1. If $comp(x, y, z)$, then $extr(x, y)$ and $\pi(x, z)$. 2. If $comp(x, y, z)$, then $comp(y, x, z)$. We let for a thing x, $-x = ClsEXTR(x, V)$. It follows that 1. $-(-x) = x$ *for each thing* x. 2. $-V$ *does not exist*. We conclude this paragraph with two properties of classes useful in the sequel:

$$\text{If } \Phi \Rightarrow \Psi \text{ then } ingr(Cls\Phi, Cls\Psi). \tag{25}$$

and a corollary

$$\text{If } \Phi \Leftrightarrow \Psi \text{ then } Cls\Phi = Cls\Psi. \tag{26}$$

Classical mereology establishes on a given universe of objects an exact hierarchy of parts and wholes suitable for exact concepts; to account for rough concepts, we need the approximate mereology (rough mereology) in which the notion of a part would loose its objective character and would undergo a subjective evaluation. Hence, rough mereology was proposed by us. Let us mention that a parallel scheme for mereology going back to informal ideas of Alfred North Whitehead [79], [80] was established in Clarke [20] as the Calculus of Connections, based on the preducate C of *'being connected'* demanded to be reflexive and symmetric plus eventually the property of extensionality.

5 Rough Mereology

A scheme of mereology, introduced into a collection of things, sets an exact hierarchy of things of which some are (exact) parts of others; to ascertain whether a thing is an exact part of some other thing is in practical cases often difficult if possible at all, e.g., a robot sensing the environment by means of a camera or a laser range sensor, cannot exactly perceive obstacles or navigation beacons. Such evaluation can be done approximately only and one can discuss such situations up to a degree of certainty only. Thus, one departs from the exact reasoning scheme given by decomposition into parts to a scheme which approximates the exact scheme but does not observe it exactly. Such a scheme, albeit its conclusions are expressed in an approximate language, can be more reliable, as its users are aware of uncertainty of its statements and can take appropriate measures to fend off possible consequences. Imagine two robots using the language of connection mereology for describing mutual relations; when endowed with touch sensors, they can ascertain the moment when they are connected; when a robot has as a goal to enter a certain area, it can ascertain that it connected to the area or overlapped with it, or it is a part of the area, and it has no means to describe its position more precisely. Introducing some measures of overlapping, in other words, the extent to which one thing is a part to the other, would allow for a more precise description of relative position, and would add an expressional power to the language of mereology. Rough mereology answers these demands by introducing the notion of a *part to a degree* with the degree expressed as a real number in the interval $[0, 1]$. Any notion of a part by necessity relates to the general idea of *containment*, and thus the notion of a part to a degree is related to the idea of *partial containment* and it should preserve the essential intuitive postulates about the latter. The predicate of a part to a degree stems ideologically from and has as one of motivations the predicate of an element to a degree introduced by Lotfi Asker Zadeh as a basis for fuzzy set theory [81]; in this sense, rough mereology is to mereology as the fuzzy set theory is to the naive set theory. To the rough set theory, owes rough mereology the interest in concepts as things of analysis. The primitive notion of rough mereology is the notion of a *rough inclusion* which is a ternary predicate $\mu(x, y, r)$ where x, y are *things* and $r \in [0, 1]$, read *'the thing x is a part to degree at least of r of the thing y'*. Any rough inclusion is associated with a mereological scheme based on the notion of a part by postulating that $\mu(x, y, 1)$ is equivalent to $ingr(x, y)$, where the ingredient relation is defined by the adopted mereological scheme. Other postulates about rough inclusions stem from intuitions about the nature of partial containment; these intuitions can be manifold, a fortiori, postulates about rough inclusions may vary. In our scheme for rough mereology, we begin with some basic postulates which would provide a most general framework. When needed, other postulates, narrowing the variety of possible models, can be introduced.

5.1 Rough Inclusions

We have already stated that a rough inclusion is a ternary predicate $\mu(x, y, r)$. We assume that a collection of things is given, on which a part relation π is introduced with the associated ingredient relation *ingr*. We thus apply inference schemes of mereology due to Leśniewski, presented above. Predicates $\mu(x, y, r)$ were introduced in Polkowski and Skowron [62], [63]; when a predicate $\mu(x, y, r)$ is interpreted in a set of objects U, and regarded as a relation, then it does satisfy the following postulates, relative to a given part relation π on the set U and the induced by π relation *ingr* of an ingredient:

RINC1 $\mu(x, y, 1) \Leftrightarrow ingr(x, y)$. This postulate asserts that parts to degree of 1 are ingredients.

RINC2 $\mu(x, y, 1) \Rightarrow \forall z[\mu(z, x, r) \Rightarrow \mu(z, y, r)]$. This postulate does express a feature of partial containment that a 'bigger' thing contains a given thing 'more' than a 'smaller' thing. It can be called a *monotonicity condition* for rough inclusions.

RINC3 $\mu(x, y, r) \wedge s < r \Rightarrow \mu(x, y, s)$. This postulate specifies the meaning of the phrase 'a part to a degree at least of r'. From postulates RINC1–RINC3, and known properties of ingredients some consequences follow

1. $\mu(x, x, 1)$.
2. $\mu(x, y, 1) \wedge \mu(y, z, 1) \Rightarrow \mu(x, z, 1)$.
3. $\mu(x, y, 1) \wedge \mu(y, x, 1) \Leftrightarrow x = y$.
4. $x \neq y \Rightarrow \neg\mu(x, y, 1) \vee \neg\mu(y, x, 1)$.
5. $\forall z \forall r[\mu(z, x, r) \Leftrightarrow \mu(z, y, r)] \Rightarrow x = y$.

Property 5 may be regarded as an *extensionality postulate* for rough mereology. By a *model* for rough mereology, we mean a quadruple

$$M = (V_M, \pi_M, ingr_M, \mu_M)$$

where V_M is a set with a part relation $\pi_M \subseteq V_M \times V_M$, the associated ingredient relation $ingr_M \subseteq V_M \times V_M$, and a relation $\mu_M \subseteq V_M \times V_M \times [0, 1]$ which satisfies RINC1–RINC3. We now describe some models for rough mereology which at the same time give us methods by which we can define rough inclusions, see Polkowski [47, 48, 50–52], a detailed discussion may be found in Polkowski [53].

5.2 Rough Inclusions from T–norms

We resort to *continuous t–norms* which are continuous functions $T : [0,1]^2 \rightarrow [0,1]$ which are 1. symmetric. 2. associative. 3. increasing in each coordinate. 4. satisfying boundary conditions $T(x,0) = 0, T(x,1) = x$, cf., Polkowski [53], Chaps. 4 and 6, Hájek [21], Chap. 2. Classical examples of continuous t–norms are:

1. $L(x,y) = max\{0, x + y - 1\}$ (the *Łukasiewicz t–norm*).
2. $P(x,y) = x \cdot y$ (the *product t–norm*).
3. $M(x,y) = min\{x,y\}$ (the *minimum t–norm*).

The *residual implication* $\Rightarrow_T$ induced by a continuous t–norm T is defined as:

$$x \Rightarrow_T y = max\{z : T(x,z) \leq y\}. \tag{27}$$

One proves that a ternary relation μ_T defined as:

$$\mu_T(x,y,r) \Leftrightarrow x \Rightarrow_T y \geq r \tag{28}$$

is a rough inclusion; particular cases are

1. $\mu_L(x,y,r) \Leftrightarrow min\{1, 1 - x + y \geq r\}$ (the *Łukasiewicz implication*).
2. $\mu_P(x,y,r) \Leftrightarrow \frac{y}{x} \geq r$ when $x > 0$, $\mu_P(x,y,1)$ when $x = 0$ (the *Goguen implication*).
3. $\mu_M(x,y,r) \Leftrightarrow y \geq r$ when $x > 0$, $\mu_M(x,y,1)$ when $x = 0$ (the *Gödel implication*).

A particular case constitute continuous t–norms which satisfy the inequality $T(x,x) < x$ for each $x \in (0,1)$. It is well–known, see Ling [30], that each of those t–norms T admits a representation:

$$T(x,y) = g_T(f_T(x) + f_T(y)), \tag{29}$$

where the function $f_T : [0,1] \rightarrow R$ is continuous decreasing with $f_T(1) = 0$, and $g_T : R \rightarrow [0,1]$ is the *pseudo–inverse* to f_T, i.e., $g \circ f = id$. It is known, cf., e.g., Hájek [21], that up to an isomorphism there are two t–norms which satisfy conditions (5) and (6): L and P. Their representations are

$$f_L(x) = 1 - x; \; g_L(y) = 1 - y, \tag{30}$$

and,

$$f_P(x) = exp(-x); \; g_P(y) = -ln\, y. \tag{31}$$

For a t–norm T which satisfies conditions (5) and (6), we define the rough inclusion μ^T on the interval $[0,1]$ by means of

$$(ari)\; \mu^T(x,y,r) \Leftrightarrow g_T(|x-y|) \geq r, \tag{32}$$

equivalently,

$$\mu^T(x, y, r) \Leftrightarrow |x - y| \leq f_T(r). \tag{33}$$

It follows from (33), that μ^T satisfies conditions RINC1–RINC3 with *ingr* as identity =. To give a hint of proof: for RINC1: $\mu^T(x, y, 1)$ if and only if $|x - y| \leq f_T(1) = 0$, hence, if and only if $x = y$. This implies RINC2. In case $s < r$, and $|x - y| \leq f_T(r)$, one has $f_T(r) \leq f_T(s)$ and $|x - y| \leq f_T(s)$. Specific recipes are:

$$\mu^L(x, y, r) \Leftrightarrow |x - y| \leq 1 - r, \tag{34}$$

and,

$$\mu^P(x, y, r) \Leftrightarrow |x - y| \leq -ln\ r. \tag{35}$$

Both residual and induced by L and P rough inclusions satisfy the *transitivity condition*

(Trans) If $\mu(x, y, r)$ and $\mu(y, z, s)$, then $\mu(x, z, T(r, s))$. / In the way of a proof, assume, e.g., $\mu^T(x, y, r)$ and $\mu^T(y, z, s)$, i.e., $|x - y| \leq f_T(r)$ and $|y - z| \leq f_T(s)$. Hence, $|x - z| \leq |x - y| + |y - z| \leq f_T(r) + f_T(s)$, hence, $g_T(|x - z|) \geq g_T(f_T(r) + f_T(s)) = T(r, s)$, i.e., $\mu^T(x, z, T(r, s))$. Other cases go on same lines. Let us observe that rough inclusions of the form (ari) are also *symmetric*.

5.3 Rough Inclusions in Information Systems (Data Tables)

An important domain where rough inclusions will play a dominant role in our analysis of reasoning by means of parts is the realm of *information systems* of Pawlak [45], cf., Polkowski [53], Chap. 6. We will define information rough inclusions denoted with a generic symbol μ^I. We recall that an *information system* (a *data table*) is represented as a tuple (U, A, V), where U is a finite set of things and A is a finite set of *attributes*; each attribute $a : U \rightarrow V$ maps the set U into the *value set* V. For an attribute a and an object $u \in U$, $a(u)$ is the value of a on u. For objects u, v the *discernibility set* $DIS(u, v)$ is defined as:

$$DIS(u, v) = \{a \in A : a(u) \neq a(v)\}. \tag{36}$$

For an (ari) μ_T, we define a rough inclusion μ_T^I by means of

$$(airi)\ \mu_T^I(u, v, r) \Leftrightarrow g_T(\frac{|DIS(u, v)|}{|A|}) \geq r. \tag{37}$$

Then, μ_T^I is a rough inclusion with the associated ingredient relation of identity and the part relation empty. For the Łukasiewicz t–norm, the *airi* μ_L^I is given by means of the formula:

$$\mu_L^I(u, v, r) \Leftrightarrow 1 - \frac{|DIS(u, v)|}{|A|} \geq r. \tag{38}$$

We introduce the set $IND(u, v) = A \setminus DIS(u, v)$. With its help, we obtain a new form of (38):

$$\mu_L^I(u, v, r) \Leftrightarrow \frac{|IND(u, v)|}{|A|} \geq r. \tag{39}$$

The formula (39) witnesses that the reasoning based on the rough inclusion μ_L^I is the probabilistic one which goes back to Łukasiewicz [31]. Each (airi)–type rough inclusion μ_T^I satisfies the transitivity condition (Trans) and is symmetric.

5.4 Rough Inclusions on Finite Sets and Measurable Sets

Formula (39) can be abstracted to set and geometric domains. For finite sets A, B, we let:

$$\nu_L(A, B, r) \Leftrightarrow \frac{|A \cap B|}{|A|} \geq r, \tag{40}$$

where $|X|$ denotes the cardinality of X, defines a rough inclusion ν_L going back to the Łukasiewicz idea of partial truth values. The rough inclusion ν_3 is defined as

$$\begin{cases} \nu(A, B, 1) \text{ if } A \subseteq B \\ \nu(A, B, \frac{1}{2}) \text{ if } A \cap B \neq \emptyset \text{ and } (A \setminus B) \cup (B \setminus A) \neq \emptyset \\ \nu(A, B, 0) \text{ if } A \cap B = \emptyset \end{cases} \tag{41}$$

For bounded measurable sets X, Y in an Euclidean space E^n, we have:

$$\mu^G(A, B, r) \Leftrightarrow \frac{||A \cap B||}{||A||} \geq r, \tag{42}$$

where $||A||$ denotes the area (the Lebesgue measure) of the region A, defines a rough inclusion μ^G. Both μ^S, μ^G are symmetric but not transitive.

6 Mereology in Engineering: Artifacts, Design and Assembling

Zdzisław Pawlak suggested to us the direction but then we were travelling on our own finding some applications for the developed schemes. In a wider sense, this is also Zdzisław's heritage. We select here applications of mereology to artefact making, design and assembling. Mereology plays a fundamental role in problems of design and assembling as basic ingredients in those processes are parts of complex things. The process of synthesis involves sequencing of operations of fusion of parts into more complex parts until the final product—artifact. We propose a scheme for assembling and a parallel scheme for design; the difference is in the fact that design operates on *abstracta*, i.e. categories of things whereas assembling deals with *concreta*, i.e., with real things. The interplay between abstracta and concreta will be described as a result of our analysis. The term *artifact* means, etymologically, *a thing made by art*, which covers a wide specter of things from man–made things of everyday usage to abstract pieces of mathematical proofs, software modules or sonnets, or concertos. All those distinct things are unified in a scheme dependent on some common ingredients in their making, cf., e.g., a concise discussion in SEP [67]. We cannot include here a discussion of vast literature on ontological, philosophical and technological aspects of this notion, see, e.g., Baker [9], Hilpinen [23], Margolis and Laurence [34], we point only to a thorough analysis of ontological aspects of artifacts in Borgo and Vieu [16] in which authors propose also a scheme defining artifacts. It follows from discussion by many authors that important in analysis of artifacts are such aspects as: authorship, intended functionality, parthood relations. Analysis of artifacts is closely tied to design and assembly, cf., Boothroyd [14] and Boothroyd, Dewhurst and Knight [15] as well as Salustri [65] and Seibt [66]. A discussion of mereology with respect to its role in domain science and engineering and computer science can be found in Björner and Eir [12]. The present discussion comes from [61]. We attempt at a definition of an artifact as a thing obtained over a collection of things as a most complex thing in the sense of not being a part of any thing in the collection; to aspects of authorship (operator) and functionality, we add a temporal aspect, which allows for well–foundedness of the universe of parts, and seems to be a natural aspect of the assembling or design process. We regard processes leading to artifacts as *fusion processes* in which a by–product is obtained from a finite number of substrats. Though processes, e.g., of assembling a bike from its parts or a chemical reaction leading to a product obtained from a mixture of substances are very distinct to the observer, yet the formal description is identical for the two; it does require a category of *operators* P, a category of *functionalities* F, a *linear time* T with the *time origin* 0. The domain of things is a category $Things(P, F, \pi)$ of things endowed with a part relation π. The assignment operator S acts as a partial mapping on the Cartesian product $P \times F \times Things(P, F, \pi)$ with values in the category *Tree* of rooted trees. The act of assembling is expressed by means of a predicate

$$Art(p(u), < v_1(u), \cdots, v_k(u) >, u, f(u), t(u), T(u)),$$

which reads: *an operator* $p(u)$ *assembles at time* $t(u)$ *a thing* u *with functionality* $f(u)$ *according to the assembling scheme* $T(u)$ *organized by* $p(u)$ *which is a tree with the root* u*, from things* $v_1(u), \cdots, v_k(u)$ *which are leaves of* $T(u)$*. The thing* $v_i(u)$ *enters in the position* i *the assembling process for* u. The predicate *ART* is subject to the following requirements.

ART1. If $Art(p(u), <v_1(u), \cdots, v_k(u)>, u, f(u), t(u), T(u))$ and for any i in $\{1, ..., k\}$, it holds that

$$Art(p(v_i(u)), < v_{i_1}(v_i(u)), \cdots, v_{i_k}(v_i(u))) >, v_i(u), f(v_i(u)), t(v_i(u)), T(v_i(u))),$$

then $t(v_i(u)) < t(u), f(u) \subseteq f(v_i(u)), p(v_i(u)) \subseteq p(u)$, and $T(v_i(u))$ attached to $T(u)$ at the leaf $v_i(u)$ yields a tree, called an *unfolding of T(u) via the assembling tree for* $v_i(u)$.

The meaning of ART1 is that for each substrate v entering the assembly process for u, v is assembled at time earlier than time for u, functionality of u is lesser than that of v, the operator for u has a greater operating scope than that of v, and the assembly tree for u can be expanded at the leaf v by the assembly tree for

ART2. $Art(p(u), < v_1(u), \cdots, v_k(u) >, u, f(u), t(u), T(u)) \Rightarrow \pi(v_i(u), u)$ for each $v_i(u)$.

Meaning that each thing can be assembled only from its parts. We introduce an auxiliary predicate $App(v, i(v), u, t(u))$ meaning: v enters in the position i the design process for u at time $t(u)$.

ART3. $\pi(v, u) \Rightarrow \exists w_1(v, u),$

$$\cdots, w_k(v, u), t(w_2(v, u)), \cdots, t(w_k(v, u)), i(w_1(v, u)), \cdots, i(w_{k(v,u)-1}))$$

such that $v = w_1(v, u), t(w_2(v, u)) < \cdots < t(w_k(v, u), w_k(v, u)) = u,$

$$App(w_j(v, u)), i(w_j(v, u)), w_{j+1}(v, u), t(w_{j+1}(v, u))$$

for $j = 1, 2, k(v, u) - 1$.

This means that each thing which is a part of the other thing will enter the assembly tree of the thing.

ART4. Each thing used in assembling of some other thing can be used in only one such thing in only one position at only one time.

This requirement will be referred to as the *uniqueness requirement.*

Art5. Values $t(u)$ belong in the set $T = \{0, 1, 2, \dots\}$ of time moments

Corollary 1 *By ART1, ART2, ART5: The universe of assembly things is well–founded, i.e., there is no infinite sequence* $\{x_i : i = 1, 2, ...\}$ *of things with* $\pi(x_{i+1}, x_i)$ *for each i.*

From this Corollary, it follows that our notion of identity of artifacts (EA) is equivalent to extensionality notions (EP), (EC), (UC) discussed in Varzi [78].

For a tree $T(u)$, the *ART–unfolding* of $T(u)$ is the tree $T(u, 1)$ in which leaves $v_1(u), v_2(u), \dots, v_k(u)$ are expanded by attaching those trees $T(v_1(u)), \dots, T(v_k(u))$ which are distinct from their roots. For a tree $T(u)$, the *maximal ART–unfolding* $T(u, max)$ is the tree obtained from $T(u)$ by repeating the operation of ART–unfolding until no further ART–unfolding is possible.

Corollary 2 *Each leaf of the tree* $T(u, max)$ *is an atom.*

We now define an artifact: an *artifact over the category Things(P, F, π)* of assembly things is a thing u such that $\pi(u, v)$ holds for no thing v in *Things(P, F, π)*. Thus artifacts are 'final things' in a sense. We define the notion of *identity for artifacts*:

Theorem 5 (Extensionality of artifacts (EA)) *Artifacts a, b are identical if and only if trees $Tree(a, max)$, $Tree(b, max)$ are isomorphic and have identical things at corresponding under the isomorphism nodes.*

6.1 Design Artifacts

We regard the process of design as analogous to the assembly process; the only difference between the two which we introduce is that in design, the designer works with not the things but with classes of equivalent things. Thus, to begin with, we introduce an equivalence relation on things. To this end, we let:

$$u \sim v \text{ if and only if } [\pi(u, t) \text{ if and only if } \pi(v, t)] \text{ for each } thing\ t \tag{43}$$

and

$$Cat(u) = Cat(v) \text{ if and only if } u \sim v. \tag{44}$$

Things in the same category Cat are 'universally replaceable'. It is manifest that the part relation π can be factored through categories, to the relation Π of part on categories,

$$\Pi(Cat(u), Cat(v)) \text{ if and only if } \pi(u, v). \tag{45}$$

In our formalism, design will imitate assembling with things replaced with categories of things and the part relation π replaced with the factorization Π. We need only to repeat the procedure with necessary replacements. We use the designer set D, the functionality set F, and the time set T as above. The act of design is expressed by means of a predicate,

$$Des(d, < Cat_1, \cdots, Cat_k >, Cat, f(Cat), t(Cat), T(Cat))$$

which reads: a designer d designs at time t a category of things Cat with functionality $f(Cat)$ according to the design scheme $T(Cat)$ organized by d which is a tree with the root Cat, from categories $Cat_1, \ldots, Cat_k$ which are leaves of $T(Cat)$. The category Cat_i *enters in the position i* the design process for Cat. The predicate Des is subject to the following requirements.

DES1. If $Des(d, < Cat(v_1(u)), \cdots, Cat(v_k(u)) >, Cat(u), f(u), t(u), T(u))$ and for any i in $\{1, \cdots, k\}$, it holds that

$$Des(p(Cat(v_i(u))), < Cat(v_{i_1}(v_i(u))), \cdots, Cat(v_{i_k}(v_i(u)))) >,$$

$$Cat(v_i(u)), f(v_i(u)), t(v_i(u)), T(v_i(u))),$$

then $t(v_i(u)) < t(u), f(u) \subseteq f(v_i(u)), p(v_i(u)) \subseteq p(u)$, and $T(v_i(u))$ attached to $T(u)$ at the leaf $Cat(v_i(u))$ yields a tree, called the *unfolding of T(u) via the design tree for Cat*$(v_i(u))$.

DES2.

$$Des(d, < Cat(v_1(u)), \cdots, Cat(v_k(u)) >, Cat(u), f(u), t(u), T(u)) \Rightarrow$$

$$\Pi(Cat(v_i(u)), Cat(u))$$

for each $v_i(u)$.

Meaning that each thing can be designed only from its parts.

We introduce an auxiliary predicate $App(v, i(v), u, t(u))$ meaning: $Cat(v)$ enters in the position i the design process for $Cat(u)$ at time $t(u)$.

DES3. $\Pi(Cat(v), Cat(u)) \Rightarrow \exists Cat(w_1(v,u)), \ldots, Cat(w_k(v,u))$, and,

$$t(w_2(v,u)), \ldots, t(w_k(v,u)), i(w_1(v,u)), \cdots, i(w_{k(v,u)-1}))$$

such that $v = w_1(v,u)$, $t(w_2(v,u)) < \cdots < t(w_k(v,u), w_k(v,u)) = u$,

$$App(w_j(v,u)), i(w_j(v,u)), w_{j+1}(v,u), t(w_{j+1}(v,u))$$

for $j = 1, 2, \ldots, k(v,u) - 1$.

This means that for each thing which is a part of the other thing the category of the former will enter the design tree for the category of the latter.

For ART4, we may not have the counterpart in terms of DES: clearly, things of the same category may be used in many positions and at many design stages of some other category. We may only repeat our assumption about timing.

DES4. Values $t(u)$ belong in the set $T = \{0, 1, 2, \ldots\}$ of time moments.

Corollary 3 *The universe of categories is well–founded.*

We define a *design artifact* as a category $Cat(u)$ such that $\Pi(Cat(u), Cat(v))$ is true for no v. We are approaching the notion of identity for design artifacts. To begin with, for a design artifact a, denote by the symbol $art(a)$ the artifact obtained by filling in the design tree for a all positions $Cat(v)$ with things v for some choices of v. We state the identity condition for design artifacts.

Theorem 6 (Extensionality for design artifacts (ED)) *Design artifacts a, b are identical if and only if there exist artifacts $art(a), art(b)$ which are identical.*

From the principle of identity for artifacts, a corollary follows.

Corollary 4 *If design artifacts a, b are identical then a, b have isomorphic design trees and categories at corresponding nodes are identical.*

Corollary 5 *If design artifacts a, b have isomorphic design trees and categories at corresponding nodes are identical, then a, b are identical.*

Indeed, consider two design artifacts a, b which satisfy the condition in the corollary. There is at least one category $Cat(v)$ in the same position in design trees of a and b. Choose a thing x in $Cat(v)$ and let $a(x), b(x)$ be artifacts assembled according to a, b, respectively. Having a thing in common, $a(x), b(x)$ are identical hence a, b are identical.

6.2 Action of Things on Design Abstracta

The interplay between concreta and abstracta in design and assembly can be exhibited by action of things on design artifacts. We define a partial mapping ι on the product $Things(P, F, \pi) \times Design\_Artifacts$ into $Artifacts$: for a thing v and a design artifact a, we define the value $\iota(v, a)$ as NIL in case category $Cat(v)$ is not any node in the design tree for a, and, the unique artifact $a(v)$ in the contrary case. The inverse $\iota^{-1}(\iota(v, a))$ is the set $\{(u, b) : b \in Design\_Artifacts, Cat(u) \text{ a node in } b\}$; thus, abstracta are equivalent in this sense to collections of concreta.

7 Mereology in Spatial Reasoning: Mereological Theory of Shape and Orientation

Spatial orientation of a thing depends on the real world in which things are immersed, hence, to, e.g., discern among sides of a thing, one needs additional knowledge and structures. An example of this approach is found, e.g., in Aurnague, Vieu and Borillo [8], where it is proposed to exploit in determining orientation, e.g., the direction of gravity ('haut–grav', 'bas–grav') or peculiar features of things (like the neck of a bottle) suggesting direction, and usage of geometric predicates like equidistance in definitions of, e.g., orthogonal directions. It is manifest that mereology is amorphous in the sense that decomposition of a thing into parts does not depend of orientation, isometric transformations etc. Hence, to exhibit in things additional features like shape, side, one needs *augmented mereology* [54]. We mean by this adding to the Leśniewski mereology the predicate C of being connected, see Clarke [20]. Of C we require:

C1 $C(x, x)$.

C2 If $C(x, y)$, then $C(y, x)$.

C3 For each z: if $C(z, x)$ if and only if $C(z, y)$, then $(x = y)$.

C does induce predicates:

1. $P(x, y)$ if for each z: $C(z, x)$ implies $C(z, y)$.
2. $PP(x, y)$ if $P(x, y)$ and not $x = y$.
3. $Ov(x, y)$ if there is z such that $P(z, x)$ and $P(z, y)$.
4. $EC(z, x)$ if $C(z, x)$ and not $Ov(z, x)$.
5. $TP(x, y)$ if there is z such that $P(x, y)$, $EC(z, y)$ and $C(z, x)$.
6. $NTP(x, y)$ if not $TP(x, y)$.

P means 'part', PP means 'proper part', Ov means 'overlap', EC means 'externally connected', TP means 'tangential part', and, NTP means 'non-tangential part'. By means of those new predicates we may express spatial relationships.

Particular features of shape like existence of 'dents' or 'holes' in a thing resulting from removal of other things can be accounted for within mereology.

We define the predicate $hole(x, y)$ reading *a thing x constitutes a hole in a thing y* as follows:

$$hole(x, y) \Leftrightarrow \exists z.NTP(x, z) \wedge comp(y, x, z), \tag{46}$$

i.e., x is a non–tangential thing in z and y complements x in z.

The predicate $dent(x, y)$, reading *a thing x constitutes a dent in a thing y* is defined as

$$dent(x, y) \Leftrightarrow \exists z.TP(x, z) \wedge comp(y, x, z), \tag{47}$$

i.e., x is a tangential thing in z and y complements x in z. The notion of a dent may be useful in characterizing things that 'fit into a thing': the predicate $fits\_into(x, y)$ may be defined as

$$fits\_into(x, y) \Leftrightarrow \exists z.dent(z, y) \wedge ingr(x, z), \tag{48}$$

i.e., x is an ingredient of a thing which is a dent in y. A particular case of fitting is 'filling' i.e., a complete fitting of a dent. We offer a predicate *fills(x, y)*

$$fills(x, y) \Leftrightarrow \exists z.dent(z, y) \wedge z = x \cdot y, \tag{49}$$

i.e., dent–making z is the product of x and y. Following this, the notion of a *join* can be defined as

$$joins(x, y, z) \Leftrightarrow \exists w.w = x + y + z \wedge fills(x, y) \wedge fills(x, z), \tag{50}$$

i.e., x joins y and z when there is a thing $x + y + z$ and x fills both y and z. This predicate can be inductively raised to

$$join(n)(x_1, x_2, ..., x_n; y_1, y_2, ..., y_n, y_{n+1})$$

via

$$join(1)(x_1; y_1, y_2) \Leftrightarrow join(x_1, y_1, y_2)$$

and

$$join(k + 1)(x_1, x_2, ..., x_{k+1}; y_1, y_2, ..., y_{k+1}, y_{k+2}) \Leftrightarrow$$

$$join(x_{k+1}, join(k)(x_1, x_2, ..., x_k; y_1, y_2, ..., y_{k+1}), y_{k+2})$$

in which we express sequentially a possibly parallel processing. In case x joins y and z, possibility of assembling arises which may be expressed by means of modal operator $\diamond$ of 'possibility', with an extended operator Asmbl to the form $Asmbl(x, i, y, j, ...w, p, f, t)$ meaning that w can be assembled from x in position i, y in position j,... by an operator p with functionality f at time t,

$$join(x, y, z) \Rightarrow \diamond \exists w, p, f, t, i, j, k.Asmbl(x, i, y, j, z, k; w, p, f, t). \tag{51}$$

Assuming our mereology is augmented with environment endowed with directions N, S, E, W, we may represent these directions by means of mobile agents endowed with laser or infrared beams of specified width; at the moment when the beam range reaches the thing x, it marks on its boundary a region which we denote as *top* in case of N, *bottom* in case of S, *left-side* in case of W, and *right-side* in case of E. Thus we have $top(x), bottom(x), left-side(x), right-side(x)$ as areas of the boundary of x; these are not parts of x. To express relations among sides of things we need a distinct language; for the sake of this example let us adopt the language of set theory regarding sides as sets. Then we can say that the thing y

1. is *on* the thing x in case $bottom(y)$ is contained in $top(x)$.
2. is *under* the thing x when $top(y)$ is contained in $bottom(x)$.
3. *touches x on the left* when $right\text{-}side(y)$ is contained in $left\text{-}side(x)$
4. *touches x on the right* when ($left\text{-}side(y)$ is contained in $right\text{-}side(x)$).

This modus of orientation can be merged with mereological shape theory: one can say that a thing *x constitutes a dent on top/under/ on the left/on the right of the thing y* when, respectively,

1. $dent_{top}(x, y) \Leftrightarrow \exists z.TP(x, z) \wedge top(x) \subseteq top(z) \wedge comp(y, x, z)$.
2. $dent_{bottom}(x, y) \Leftrightarrow \exists z.TP(x, z) \wedge bottom(x) \subseteq bottom(z) \wedge comp(y, x, z)$.
3. $dent_{left}(x, y) \Leftrightarrow \exists z.TP(x, z) \wedge left-side(x) \subseteq left-side(z) \wedge comp(y, x, z)$.
4. $dent_{right}(x, y) \Leftrightarrow \exists z.TP(x, z) \wedge right-side(x) \subseteq right-side(z) \wedge comp(y, x, z)$.

These notions in turn allow for more precise definitions of fitting and filling; we restrict ourselves to filling as fitting is processed along same lines: we say that *a thing x fills a thing y on top/bottom/on the left-side/on the right-side*,

$$fills_{\alpha}(x, y) \Leftrightarrow \exists z.dent_{\alpha}(z, y) \wedge z = x \cdot y$$

where α is, respectively, *top, bottom, left, right*. This bears on the notion of a join which can be made more precise: we say that *a thing x (α, β)–joins things y and z*

$$joins_{.\alpha,\beta}(x, y, z) \Leftrightarrow \exists w.w = x + y + z \wedge fills_{\alpha}(x, y) \wedge fills_{\beta}(x, z)$$

where α, β=*top, bottom, left, right*. A very extensive discussion of those aspects is given in Casati and Varzi [19]. Applications are discussed in Kim et al. [25].

8 Applications of Rough Mereology: Betweenness in Spatial Problems and in Data Sets

We address geometry induced from rough mereology applying it in two areas: spatial reasoning for teams of intelligent agents and partitioning information/decision systems into specific subsystems. In either case the tool is the betweenness relation. This section introduces mereogeometry modeled on classical axiomatization of geometry by Tarski [77]. It will serve us in the sequel in building tools for defining and navigating formations of intelligent agents (robots). Elementary geometry was defined by Alfred Tarski in His Warsaw University lectures in the years 1926–27 as a part of Euclidean geometry which can be described by means of the 1st order logic. There are two main aspects in formalization of geometry: one is metric aspect dealing with the distance underlying the space of points which carries geometry and the other is affine aspect taking into account the linear structure. In Tarski axiomatization, Tarski [77], the metric aspect is expressed as a relation of *equidistance* (congruence) and the affine aspect is expressed by means of the *betweenness* relation. The only logical predicate required is the identity =. Equidistance relation denoted $Eq(x, y, u, z)$ (or, as a congruence: $xy \equiv uz$) means that the distance from x to y is equal to the distance from u to z (pairs x, y and u, z are equidistant). Betweenness relation is denoted $B(x, y, z)$, (x is between y and z). Johan Van Benthem [11] took up the subject proposing a version of betweenness predicate based on the nearness predicate and suited, hypothetically, for Euclidean spaces. We are interested in introducing into the mereological world defined by μ of a geometry in whose terms it will be possible to express spatial relations among things. We first introduce a notion of a distance κ, induced by a rough inclusion μ:

$$\kappa(x, y) = min\{max\ r, max\ s : \ \mu(x, y, r), \mu(y, x, s)\}. \tag{52}$$

Observe that the mereological distance differs essentially from the standard distance: the closer are things, the greater is the value of κ: $\kappa(x, y) = 1$ means $x = y$ whereas $\kappa(x, y) = 0$ means that x, y are either externally connected or disjoint, no matter what is the Euclidean distance between them. The notion of *betweenness in the Tarski sense* $B(z, x, y)$ in terms of κ is defined as:

$$B(z, x, y) \Leftrightarrow \text{for each thing w}, \kappa(z, w) \in [\kappa(x, w), \kappa(y, w)]. \tag{53}$$

Here, $[a, b]$ means the non–oriented interval with endpoints a, b. We use κ to define in our context the relation N of *nearness* proposed in Van Benthem [11]

$$N(x, y, z) \Leftrightarrow \kappa(x, y) \geq \kappa(z, y). \tag{54}$$

Here, $N(x, y, z)$ means that *x is closer to y than z is to y*. We introduce a *betweenness* relation in the sense of Van Benthem T_B modeled on betweenness proposed in Van Benthem [11]

$$T_B(z, x, y) \Leftrightarrow [\text{for each } t\ (z = t) \text{ or } N(z, x, t) \text{ or } N(z, y, t)]. \tag{55}$$

8.1 Betweenness in Spatial Reasoning: Autonomous Robot Navigation

Robot navigation is a main topic in behavioral robotics as an archetypical example of navigation by intelligent agents, especially in groups, see, e.g., [1, 2, 5, 10, 17, 18, 35]. The principal context bearing on our approach to robot control [39–42, 58, 59], deals with rectangles in 2D space regularly positioned, i.e., having edges parallel to coordinate axes. We model robots (which are represented in the plane as discs of the same radii in 2D space) by means of their safety regions about robots; those regions are modeled as squares circumscribed on robots. One of advantages of this representation is that safety regions can be always implemented as regularly positioned rectangles. Given two robots a, b as discs of the same radii, and their safety regions as circumscribed regularly positioned rectangles A, B, we search for a proper choice of a region X containing A, and B with the Taking the rough inclusion μ^G defined in (42), for two disjoint rectangles A, B, we define the *extent*, *ext*(A, B) of A and B as the smallest rectangle containing the union $A \cup B$. Then we have the claim.

Proposition 1 *Given two disjoint rectangles C, D, the only thing between C and D in the sense of the predicate T_B is the* extent *ext*(C, D) *of C, D.*

For a proof, as linear stretching or contracting along an axis does not change the area relations, it is sufficient to consider two unit squares A, B of which A has (0,0) as one of vertices whereas B has (a,b) with $a, b > 1$ as the lower left vertex (both squares are regularly positioned). Then the distance κ between the extent *ext*(A, B) and either of A, B is $\frac{1}{(a+)(b+1)}$. For a rectangle $R : [0, x] \times [0, y]$ with $x \in (a, a + 1), y \in (b, b + 1)$, we have that

$$\kappa(R, A) = \frac{(x - a)(y - b)}{xy} = \kappa(R, B). \tag{56}$$

For $\phi(x, y) = \frac{(x-a)(y-b)}{xy}$, we find that

$$\frac{\partial \phi}{\partial x} = \frac{a}{x^2} \cdot (1 - \frac{b}{y}) > 0, \tag{57}$$

and, similarly, $\frac{\partial \phi}{\partial y} > 0$, i.e., ϕ is increasing in x, y reaching the maximum when R becomes the extent of A, B. An analogous reasoning takes care of the case when R

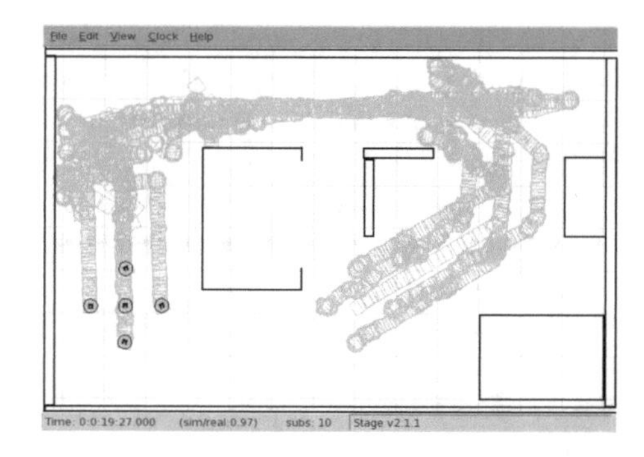

Fig. 1 Trails of robots in the restored cross formation in the free workspace after passing through the passage

has some (c,d) with $c, d > 0$ as the lower left vertex. We say henceforth that a robot C is between robots A, B if C is contained in the extent $ext(A, B)$. Further usage of the betweenness predicate is suggested by the Tarski [77] axiom of *B,Eq–upper dimension*, which implies collinearity of x, y, z. Thus,a line segment may be defined via the auxiliary notion of a pattern; we introduce this notion as a relation $Pt(u, v, z)$ which is true if and only if $T_B(z, u, v)$ or $T_B(u, z, v)$ or $T_B(v, u, z)$. We will say that a finite sequence $u_1, u_2, ..., u_n$ of things *belong in a line segment* whenever $Pt(u_i, u_{i+1}, u_{i+2})$ for $i = 1, ..., n-2$; formally, we introduce the functor *Line* of finite arity defined by means of

$$Line(u_1, u_2, ..., u_n) \textit{ if and only if } Pt(u_i, u_{i+1}, u_{i+2}) \textit{ for } i < n-1.$$

For instance, any two disjoint rectangles A, B and their extent $ext(A, B)$ form a line segment. This notion was applied to navigation of intelligent agents, e.g., mobile robots. For a team of agents F and a rough inclusion μ among them, we call a μ-formation $F(\mu)$ the team F along with the μ-betweenness relation on it. Figure 1 shows a screenshot of a cross formation of robots navigating a narrow passage in line formation and then restoring itself to the cross formation.

8.2 *Betweenness in Data Sets*

Given an information/decision system with the universe U and the attribute set A, we apply to objects in U the Łukasiewicz rough inclusion (39). Given objects u, v in U, for a choice of $\gamma \in [0, 1]$, we form objects which have $\gamma \times |A|$ attribute values in common with u and $(1-\gamma) \times |A|$ attribute values in common with v. We represent this class of things as the vector $\overline{\gamma}$ in the Euclidean plane. In this representation, u is represented as [0,1] and v is represented as [0,1], so $\overline{\gamma}$ is a convex combination of [0,1] and [0,1]. It was proved [55] that for each choice of $\gamma \in [0, 1]$, the class of things represented as $\overline{\gamma}$ is between u and v in the sense of betweenness relation T_B induced by the Łukasiewicz rough inclusion (39). This suggests a generalization.

We introduce a more general betweenness relation $GB(u, v_1, v_2, \ldots, v_n)$ (read as 'u is between $v_1, v_2, ..., v_n$'), see [55], which holds true if and only if for each object $w \in U, w \neq u$, the object u is closer than w to some v_i in the mereological sense, in formal terms

$$GB(u, v_1, v_2, ..., v_n) \text{ if and only if for each } w \neq u \text{ there is } v_i \text{ such that} \tag{58}$$

$$\kappa(u, v_i) \geq \kappa(w, v_i).$$

We consider a set $V = \{v_1, v_2, ..., v_n\}$ of objects in U. For a choice of $\gamma_1, \gamma_2, ..., \gamma_n \in [0, 1]$ with $\sum_i \gamma_i = 1$, which we summarily denote by the vector $\overline{\gamma}$, we denote as $(V, \overline{\gamma})$ the class of objects which have the fraction γ_i of attribute values in common with the object v_i. Formally,

$$u \in (V, \overline{\gamma}) \; ifInf(u) = \bigcup_{i=1}^{n} C_i \;\; C_i \subset Inf(v_i) \;\; \frac{|C_i|}{|A|} = \gamma_i \text{ for } i = 1, 2, \ldots n. \tag{59}$$

One proves

Proposition 2 *Each object u in the class $(V, \overline{\gamma})$ satisfies the relation $GB(u, v_1, v_2, ..., v_n)$. Conversely, each u satisfying the relation $GB(u, v_1, v_2, ..., v_n)$ belongs in a class $(V, \overline{\gamma})$ for some vector $\overline{\gamma}$.*

We call a maximal set of things K with the property that for each $u \in K$ there exist k and $v_1, v_2, ..., v_k \in K \setminus \{u\}$ such that $GB(u, v_1, v_2, ..., v_k)$ holds, a *kernel*. We call the set $\{v_1, v_2, ..., v_k\}$ a set of *neighbors* of u, denoted with the generic symbol $N(u)$. In order to disambiguate the notion of a neighborhood, we introduce the following structure for neighborhoods:

$$< (v_1, C_1 \subseteq Inf(v_1), q(v_1)), \ldots, (v_n, C_n \subseteq Inf(v_k), q(v_k)) >, \tag{60}$$

with neighbors $v_1, v_2, \ldots, v_n$ ordered in the descending order of the factor q, $q(v_i) = \gamma_i = \frac{|C_i|}{|A|}$.

8.3 Computing the Kernel: Dual Indiscernibility Matrix (DIM)

In order to compute the kernel of the data set [55], we introduce a matrix dual in a sense to the well–known Skowron–Rauszer Discernibility Matrix [70], whose entries consist of objects instead of attributes, which fact justifies the adjective 'dual'. The dual indiscernibility matrix for an information system $IS = (U, A, V)$, is the matrix $DIM(IS) = [c_{a,val}]_{|A|\times|V|}$, where the entry for a given $a \in A$, $val \in V$ is:

$$c_{a,val} = \{u \in U : a(u) = val\}. \tag{61}$$

ALGORITHM: COMPUTING KERNEL BY DUAL INDISCERNIBILITY MATRIX

Input: Information system $IS = (U, A, V, val)$;
set R initialized as empty, set K initialized as U;

1. Form DIM;
2. Scan DIM row by row from left to right;
3. For each cell $c_{a,val}$: if $c_{a,val} = \{u\}$ for some $u \in U$
4. then $R \leftarrow R \cup \{u\}$, $K \leftarrow K \setminus \{u\}$;
5. Return K;
6. Return R.

The set K is the kernel and the set R is said to be the *residuum* of the data set. In the kernel, each object is a convex combination of some other objects in the kernel whereas in the residuum each object is an outlier, having at least one attribute value not taken on any other object in the universe.

8.4 Kernel and Residuum in the Task of Classification

It is interesting to see how the kernel and the residuum, being on the opposite poles of betweenness, behave in tasks of classification. We recall the results from [57] in Table 1.

It follows from Table 1 that both kernel and residuum are satisfactory representatives of the whole data set as classification into decision classes is concerned as they split between themselves the data set approximately in halves with accuracy of classification not diminished.

8.5 Classifiers Based on Partial Approximation by Neighbors

The notion of a neighborhood suggests usage of neighbors in approximation of a given test object by best training neighbors [7]: for a given decision system D split into the training Trn and test Tst parts and for a given test object u^{tst}, we select the object $v_{10}^{trn} \in trn$ such that

$$v_{10}^{trn} = argmax_{v \in trn} |Inf(u^{tst}) \cap Inf(v)|, \tag{62}$$

Table 1 Classification results

Database	Set tested	Accuracy of C4.5	Accuracy of k-NN	Number of samples
Adult	Whole set	0.857 ± 0.003	0.837 ± 0.003	39074.0
	Ker_1	0.853 ± 0.004	0.835 ± 0.003	22366.0
	Res_1	0.849 ± 0.003	0.833 ± 0.003	16708.0
PID	Whole set	0.733 ± 0.027	0.723 ± 0.021	614.4
	Ker_1	0.704 ± 0.037	0.711 ± 0.032	212.9
	Res_1	0.724 ± 0.035	0.745 ± 0.030	401.5
Fertility diagnosis	Whole set	0.852 ± 0.073	0.866 ± 0.060	80.0
	Ker_1	0.846 ± 0.075	0.880 ± 0.064	71.6
	Res_1	0.852 ± 0.068	0.880 ± 0.064	8.4
German credit	Whole set	0.713 ± 0.023	0.732 ± 0.025	800.0
	Ker_1	0.671 ± 0.045	0.714 ± 0.038	98.9
	Res_1	0.712 ± 0.023	0.726 ± 0.030	701.1
Heart disease	Whole set	0.750 ± 0.054	0.825 ± 0.048	216.0
	Ker_1	0.742 ± 0.061	0.822 ± 0.051	109.2
	Res_1	0.767 ± 0.054	0.827 ± 0.041	106.8

i.e., the object v^{trn}_{10} 'covers' u^{tst} best. Then, we select the next best covering u^{tst} object v^{trn}_{11}:

$$v^{trn}_{11} = argmax_{v \in trn \setminus \{v^{trn}_{10}\}} |Inf(u^{tst}) \cap Inf(v)|. \quad (63)$$

The pair $(v^{trn}_{10}, v^{trn}_{11})$ is said to be the pair of level $L0$. We remove objects $v^{trn}_{110}, v^{trn}_{11}$ from the set trn and we repeat the process of covering by pairs from the set $trn \setminus \{(v^{trn}_{10}, v^{trn}_{11})$, obtaining the pair of level $L1$ and so on. The process is repeated up to

Table 2 Pair classifier

Database	kNN	Bayes	Pair–best	Pair-0
Adult	0.841	0.864	0.853L1	0.823
Australian	0.855	0.843	0.859L4, 5	0.859
Diabetes	0.631	0.652	0.721L0	0.710
German credit	0.730	0.704	0.722L1	0.721
Heart disease	0.837	0.829	0.822L1	0.800
Hepatitis	0.890	0.845	0.892L0	0.831
Congressional voting	0.938	0.927	0.928L0	0.928
Mushroom	1.0	0.910	1.0L0	1.0
Nursery	0.578	0.869	0.845L0	0.845
Soybean large	0.928	0.690	0.910L0	0.910

the apriori assumed level *maxLevel*. All decision values are pooled into a common set and Majority Voting is applied to select the final decision value assigned to u^{tst}. Table 2 shows results of a comparison among Pair Classifier using the best approximating pair, and k-NN and Bayes classifiers. The symbol Lx denotes the level of covering, Pair-0 is the simple pair classifier with approximations by pairs and Pair–best denotes the best result over levels studied. Those results show that Pair classifier gives satisfactory results in classification problems at least for medium size data sets mentioned above.

9 Rough Mereology in Granular Computing

Assume a rough inclusion μ on the universe U of an information system (U, A, V). For an object $u \in U$ and $r \in [0, 1]$, we call the mereological class of the property:

$$\Psi(v) \text{ if and only if } \mu(v, u, r), \tag{64}$$

the granule $g(u, r, \mu)$ of radius r about u, i.e.,

$$g(u, r, \mu) \ is \ Cls\Psi. \tag{65}$$

For symmetric transitive rough inclusions, $g(u, r, \mu)$ is simply the set $\{v \in U : \mu(v, u, r)\}$ and we will define in what follows granules as such sets.

9.1 Granular Mereotopology

Granules serve as quasi–open sets from which topologies are built [53]. The following constitute a set of basic properties of rough mereological granules:

1. If $ingr(y, x)$ then $ingr(y, g_\mu(x, r))$;
2. If $ingr(y, g_\mu(x, r))$ and $ingr(z, y)$ then $ingr(z, g_\mu(x, r))$;
3. If $\mu(y, x, r)$ then $ingr(y, g_\mu(x, r))$;
4. If $s < r$ then $ingr(g_\mu(x, r), g_\mu(x, s))$,

which follow straightforwardly from properties RINC1–RINC3 of rough inclusions and the fact that *ingr* is a partial order, in particular it is transitive, regardless of the type of the rough inclusion μ. For T–transitive rough inclusions, we can be more specific, and prove

Proposition 3 *For each T–transitive rough inclusion* μ,

1. *If* $ingr(y, g_\mu(x, r)$ *then* $ingr(g_\mu(y, s), g_\mu(x, T(r, s))$;
2. *If* $\mu(y, x, s)$ *with* $1 > s > r$, *then there exists* $\alpha < 1$ *with the property that* $ingr(g_\mu(y, \alpha), g_\mu(x, r)$.

Proof Property 1 follows by transitivity of μ with the t–norm T. Property 2 results from the fact that the inequality $T(s, \alpha) \geq r$ has a solution in α, e.g., for $T = P$, $\alpha \geq \frac{r}{s}$, and, for $T = L$, $\alpha \geq 1 - s + r$.

It is natural to regard granule system $\{g_r^{\mu_t}(x) : x \in U, r \in (0, 1)\}$ as a neighborhood system for a topology on U that may be called the *granular topology*. In order to make this idea explicit, we define classes of the form:

$$N^T(x, r) = Cls(\psi_{r,x}^{\mu_T}), \tag{66}$$

where

$$\psi_{r,x}^{\mu_T}(y) \Leftrightarrow \exists s > r.\mu_T(y, x, s). \tag{67}$$

We declare the system $\{N^T(x, r) : x \in U; r \in (0, 1)\}$ to be a neighborhood basis for a topology θ_μ. This is justified by the following

Proposition 4 *Properties of the system* $\{N^T(x, r) : x \in U; r \in (0, 1)\}$ *are as follows:*

1. $y\ ingr\ N^T(x, r) \Rightarrow \exists \delta > 0.N^T(y, \delta)\ ingr\ N^T(x, r)$;
2. $s > r \Rightarrow N^T(x, s)\ ingr\ N^T(x, r)$;
3. $z\ ingr\ N^T(x, r) \wedge z\ ingr\ N^T(y, s) \Rightarrow \exists \delta > 0\ N^T(z, \delta)\ ingr\ N^T(x, r) \wedge N^T(z, \delta)\ ingr\ N^T(y, s)$.

Proof For Property 1, $y\ ingr\ N^t(x, r)$ implies that there exists an $s > r$ such that $\mu_t(y, x, s)$. Let $\delta < 1$ be such that $t(u, s) > r$ whenever $u > \delta$; δ exists by continuity of t and the identity $t(1, s) = s$. Thus, if $z\ ingr\ N^t(y, \delta)$, then $\mu_t(z, y, \eta)$ with $\eta > \delta$ and $\mu_t(z, x, t(\eta, s))$ hence $z\ ingr\ N^t(x, r)$. Property 2 follows by RINC3 and Property 3 is a corollary to properties 1 and 2. This concludes the argument.

9.2 Granular Rough Mereological Logics

We assume that an information/decision system (U, A, V, d) is given, along with a rough inclusion ν_3 of the form (41) or the rough inclusion ν_L (40) on the subsets of the universe U; for a collection of unary predicates Pr, interpreted in the universe U (meaning that for each predicate $\phi \in Pr$ the meaning $[[\phi]]$ is a subset of U), we define the intensional logic GRM_ν, cf. [60], by assigning to each predicate ϕ in Pr its intension $I_\nu(\phi)$ defined by its extension $I_\nu^\vee(g)$ at each particular granule g as

$$I_\nu^\vee(g)(\phi) \geq r \Leftrightarrow \nu(g, [[\phi]], r). \tag{68}$$

With respect to the rough inclusion ν_L (40), the formula (68) becomes

$$I_{\nu_L}^\vee(g)(\phi) \geq r \Leftrightarrow \frac{|g \cap [[\phi]]|}{|g|} \geq r. \tag{69}$$

The counterpart for the rough inclusion ν_3 (41) comes down to:

$$I^{\vee}_{\nu_3}(g)(\phi) \geq r \Leftrightarrow \begin{cases} g \subseteq [[\phi]] \text{ and } r = 1 \\ g \cap [[\phi]] \neq \emptyset \text{ and } r \geq \frac{1}{2} \\ g \cap [[\phi]] = \emptyset \text{ and } r = 0 \end{cases} \tag{70}$$

We say that a formula ϕ interpreted in the universe U of an information system (U, A, V) is *true* at a granule g with respect to a rough inclusion ν if and only if $I^{\vee}_{\nu}(g)(\phi) = 1$. Both (40) and (41) are *regular*, i.e., their value on a pair A, B is 1 if and only if $A \subseteq B$. Hence, for every regular rough inclusion ν, a formula ϕ interpreted in the universe U, with the meaning $[[\phi]] = \{u \in U : u \models \phi\}$, is true at a granule g with respect to ν if and only if $g \subseteq [[\phi]]$. In particular, for a decision rule $r : p \Rightarrow q$, the rule r is true at a granule g with respect to a regular rough inclusion ν if and only if $g \cap [[p]] \subseteq [[q]]$. We state these facts in the following

Proposition 5 *For every regular rough inclusion ν, a formula ϕ interpreted in the universe U, with the meaning $[[\phi]]$, is true at a granule g with respect to ν if and only if $g \subseteq [[\phi]]$. In particular, for a decision rule $r : p \Rightarrow q$, the rule r is true at a granule g with respect to a regular rough inclusion ν if and only if $g \cap [[p]] \subseteq [[q]]$.*

Proof Indeed, truth of ϕ at g means that $\nu(g, [[\phi]], 1)$ which in turn, by regularity of ν is equivalent to the inclusion $g \subseteq [[\phi]]$.

We will say that a formula ϕ is a *tautology* of our intensional logic if and only if ϕ is true at every world g. The preceding proposition implies that,

Proposition 6 *For every regular rough inclusion ν, a formula ϕ is a tautology if and only if $Cls(G) \subseteq [\phi]$, where G is the property of being a granule; in the case when granules considered cover the universe U this condition simplifies to $[[\phi]] = U$. This means for a decision rule $p \Rightarrow q$ that it is a tautology if and only if $[[p]] \subseteq [[q]]$.*

Hence, the condition for truth of decision rules in the logic GRM_{ν} is the same as the truth of an implication in descriptor logic, see [43], under caveat that granules considered cover the universe U of objects.

9.3 Granular Computing in Decision Making

Rough mereological granules were proposed as objects in granular decision systems induced from given information/data systems [48], [49]. For a decision system (U, A, V, d),and a rough inclusion μ on U, for a radius of granulation r, we form the collection $\Gamma(r)$ of granules of radii of r about objects in U. From the set Γ, we select a covering Δ of the set U, so each object in U belongs in at least one granule in Δ. For each granule $g \in \Delta$, we factor through g attributes in $A \cup \{d\}$; typically we use the Majority Voting MV so we let for an attribute a, and a granule g:

$$MV(a)(g) = MV\{a(u) : u \in g\}. \tag{71}$$

The decision system $(\Delta, \{MV(a) : a \in A\}, V, MV(d))$ is the granular reflexion of the decision system (U, A, V, d). An interesting variant of granulation is the concept-dependent granulation in which granules are computed within decision classes [6], [56]. To the granular reflections one applies a classifier as to any decision system. Results are very good and show not less accuracy at much smaller usually universe and much smaller number of rules. Table 3 shows classification results for the German Credit data set from Univ. California at Irvine Data Mining Repository for four variants of Bayes Classifier, see [56], Chap. 7, for details. Table 4 shows size of granule set for four variants of Bayes classifier.

Table 3 5 x CV-5; The result of experiments for four variants of Bayes classifier; **German Credit**; Concept dependent granulation; r_{gran} = Granulation radius; nil = result for data without missing values; *Acc* = Accuracy of classification; *AccBias* = Accuracy bias; *GranSize* = The size of data set after granulation in the fixed r

r_{gran}	Acc				AccBias			
	*V*1	*V*2	*V*3	*V*4	*V*1	*V*2	*V*3	*V*4
0	0.627	0.619	0.625	0.625	0.024	0.038	0.024	0.021
0.05	0.627	0.616	0.625	0.625	0.024	0.041	0.024	0.021
0.1	0.624	0.613	0.625	0.635	0.027	0.044	0.025	0.039
0.15	0.605	0.583	0.612	0.624	0.005	0.08	0.026	0.038
0.2	0.621	0.588	0.613	0.616	0.029	0.074	0.033	0.038
0.25	0.61	0.554	0.574	0.598	0.015	0.094	0.063	0.05
0.3	0.626	0.614	0.469	0.538	0.007	0.058	0.04	0.11
0.35	0.641	0.646	0.468	0.458	0.013	0.054	0.096	0.058
0.4	0.635	0.684	0.488	0.514	0.016	0.032	0.057	0.019
0.45	0.646	0.69	0.56	0.554	0.01	0.012	0.046	0.031
0.5	0.649	0.703	0.56	0.588	0.02	0.006	0.046	0.034
0.55	0.686	0.701	0.586	0.594	0.008	0.001	0.02	0.039
0.6	0.698	0.7	0.609	0.625	0.005	0	0.016	0.013
0.65	0.706	0.7	0.636	0.667	0.022	0	0.012	0.021
0.7	0.69	0.7	0.652	0.687	0.008	0	0.019	0.018
0.75	0.677	0.7	0.666	0.7	0.007	0	0.016	0.01
0.8	0.669	0.7	0.67	0.699	0.011	0	0.005	0.012
0.85	0.679	0.7	0.67	0.703	0.005	0	0.017	0.006
0.9	0.678	0.7	0.67	0.704	0.006	0	0.014	0.01
0.95	0.679	0.7	0.671	0.705	0.005	0	0.014	0.006
1	0.677	0.7	0.671	0.704	0.005	0	0.015	0.009

Table 4 Size of the granular set for granulation radii and four variants of Bayes classifier

r_{gran}	GranSize			
	$V1$	$V2$	$V3$	$V4$
0	2	2	2	2
0.05	2	2	2	2
0.1	2	2.16	2	2.08
0.15	2.52	2.44	2.56	2.44
0.2	3.64	3.32	3.72	3.52
0.25	4.92	4.72	4.84	5.24
0.3	7.44	6.76	7.16	7.4
0.35	11.16	11.08	11.32	11.28
0.4	18.76	19.36	19.64	18.2
0.45	33.88	32.72	32.84	32.52
0.5	59.32	56.12	58.4	58.12
0.55	105.32	104.52	102.76	105.72
0.6	187.28	187.28	188.32	186.84
0.65	318.72	321.6	317.96	319.28
0.7	486.28	486	485.6	487.28
0.75	650	647.92	648.96	650.72
0.8	751.28	750.92	751.32	751.12
0.85	789.56	789.68	789.8	789.56
0.9	796.48	796.44	796.64	796.44
0.95	798.68	798.72	798.72	798.76
1	800	800	800	800

10 Conclusion

Zdzisław Pawlak left his trace on many topics, from random number generation, through architecture of computing machines, semiotics, linguistics, information systems, scientific information theory, and then reached for his greatest achievement - theory of uncertain knowledge expressed as the theory of rough sets. This author entered the inner circle of rough set devotees 10 years after rough sets were born, and he, along with others involved, rode on the crest of the vawe which carried rough set theory into the world arena of science. This growth also in management aspects was in great part thanks to Zdzisław Pawlak's energy and devotion. He was going to numerous conferences, attracted new scientists, provided funds, doing everything necessary to secure the growth of the milieu of rough sets. Due to this involvement, it was possible to organize in 1998 in Warsaw the conference RSCTC'98. From that point on, rough sets gained impetus reflected in many conferences, a number of monographs, thousands of research contributions. Rough sets constantly grow in scope and results and methods, and, this phenomenon is the Zdzisław Pawlak Heritage left us all.

Acknowledgements The author has met many people during his work in the rough set area. He wants to acknowledge the influence of Professors Helena Rasiowa and Zdzisław Pawlak to whom he owes the initiation into rough sets. He acknowledges with gratitude the help and friendship of Andrzej Skowron and remembers fine time of cooperation on mathematical morphology and mereology a well as a stimulating atmosphere of the seminar of those days. He would like to thank his colleagues from the IRS Society and his younger colleagues from the Chair of Mathematical Methods in CS at the University of Warmia and Masuria in Olsztyn, Poland: DrSc Piotr Artiemjew, Dr Przemysław Gorecki, Dr PawełDrozda, Dr Bartosz Nowak, Dr Krzysztof Sopyła for common work.

References

1. Agah, A.: Robot teams, human workgroups and animal sociobiology. A review of research on natural and artificial multi-agent autonomous systems. Adv. Robot. **10**, 523–545 (1997)
2. Agah, A., Bekey, G.A.: Tropism based cognition: a novel software architecture for agents in colonies. J. Exp. Theor. Artif. Intell. **9**(2–3), 393–404 (1997)
3. Aristotle: Categoriae et Liber de Interpretatione. In: Minio-Paluello, L. (ed.). Oxford U. Press (1949)
4. Aristotle: Metaphysics. Book Delta, 1203b. Classical Library. W. Ross transl. http://www.classicallibrary.org/aristotle/metaphysics/
5. Arkin, R.C.: Behavior-Based Robotics. MIT Press, Cambridge MA (1998)
6. Artiemjew, P.: Classifiers from granulated data sets: concept dependent and layered granulation. In: Proceedings RSKD'07. Workshop at ECML/ PKDD'07, Warsaw University Press, Warsaw, pp. 1–9 (2007)
7. Artiemjew, P., Nowak, B., Polkowski, L.: A new classifier based on the dual indiscernibility matrix. In: Proceedings ICIST 2016, Druskonninkai, Lithuania, 2016, pp. 380–391 (2016)
8. Aurnague, M., Vieu, L., Borillo, A.: Representation formelle des concepts spatiaux dans la langue. In: Denis, M. (ed.) Langage et Cognition Spatiale, pp. 69–102. Masson, Paris (1997)
9. Baker, L.R.: The ontology of artifacts. Philos. Explor. **7**(2), 99–111 (2004)
10. Balch, T., Arkin, R. C.: Behavior–based formation control for multiagent robot teams. IEEE Trans. Robot. Autom. 14(12) (1998)
11. Van Benthem, J.: The Logic of Time. Reidel, Dordrecht (1983)
12. Björner, D., Eir, A.: Compositionality: ontology and mereology of domains. In: Dams, D., Hannemann, U., Steffen, M. (eds.) Concurrency, Compositionality, and Correctness. Lecture Notes in Computer Science 5930, pp. 22–60. Springer, Berlin (2010)
13. Bocheński, I.M.: Die Zeitgenössichen Denkmethoden. A. Francke A. G, Bern (1954)
14. Boothroyd, G.: Assembly Automation and Product Design, 2nd edn. Taylor and Francis, Boca Raton, FL (2005)
15. Boothroyd, G., Dewhurst, P., Knight, W.: Product Design for Manufacture and Assembly, 2nd edn. Marcel Dekker, New York (2002)
16. Borgo, S., Vieu, L.: Artefacts in formal ontology. In: Meijers, A. (ed.) Handbook of Philosophy of Technology and Engineering Sciences, pp. 273–308. Elsevier (2009)
17. Caloud, P., Choi, W., Latombe, J.-C., Pape Le, C., Yin, M.: Indoor automation with many mobile robots. In: Proceedings IEEE/RSJ IROS, pp. 67–72 (1990)
18. Canny, J.F.: The Complexity of Robot Motion Planning. MIT Press, Cambridge, MA (1988)
19. Casati, R., Varzi, A.C.: Parts and Places. The Structures of Spatial Representations. MIT Press, Cambridge, MA (1999)

20. Clarke, B.L.: A calculus of individuals based on connection. Notre Dame J. Formal Logic **22**(2), 204–218 (1981)
21. Hájek, P.: Metamathematics of Fuzzy Logic. Kluwer, Dordrecht (1998)
22. Heisenberg, W.: Ueber den anschaulichen Inhalt der quantentheoretischen Kinematik and Mechanik. Zeitschrift fuer Physik 43, pp. 172–198 (1927). English translation: Wheeler, J.A., Zurek, W.H. (eds.) Quantum Theory and Measurement, pp. 62–84. Princeton University Press, Princeton NJ (2014)
23. Hilpinen, R.: Belief systems as artifacts. Monist **78**, 136–155 (1995)
24. Iwanuś, B.: On Leśniewski's elementary ontology. Studia Logica XXXI, pp. 73–119 (1973)
25. Kim, K.-Y., Yang, H., Kim, D.-W.: Mereotopological assembly joint information representation for collaborative product design. Robot. Comput.-Integr. Manuf. **24**(6), 744–754 (2008)
26. Leśniewski, S.: Podstawy Ogólnej Teoryi Mnogości, I (Foundations of General Set Theory, I, in Polish). Prace Polskiego Koła Naukowego w Moskwie, Sekcya Matematyczno–przyrodnicza, No. 2, Moscow (1916)
27. Leśniewski, S.: O podstawach matematyki (On foundations of mathematics, in Polish). Przegląd Filozoficzny XXX, pp. 164–206 (1927); Przegląd Filozoficzny XXXI, pp. 261–291 (1928); Przegląd Filozoficzny XXXII, pp. 60–101 (1929); Przegląd Filozoficzny XXXIII, pp. 77–105 (1930); Przegląd Filozoficzny XXXIV, pp. 142–170 (1931)
28. Leśniewski, S.: *Ü*ber die Grundlagen der Ontologie. C.R. Soc. Sci. Lett. Varsovie Cl. III, 23 Anne*é*, pp. 111–132 (1930)
29. Leśniewski, S.: On the foundations of mathematics. Topoi **2**, 7–52 (1982)
30. Ling, C.-H.: Representation of associative functions. Publ. Math. Debrecen **12**, 189–212 (1965)
31. Łukasiewicz, J.: Die Logischen Grundlagen der Warcheinlichtkeitsrechnung. Cracow (1913); cf. [English translation]: Borkowski, L. (ed.): Jan Łukasiewicz. Selected Works. North Holland, Polish Scientific Publishers. Amsterdam–Warsaw, pp. 16–63 (1970)
32. Łukasiewicz, J.: Farewell Lecture on March 17, 1918. Pro arte et studio 3, pp. 3–4 (1918). English translation: Borkowski, L. (ed.): Jan Łukasiewicz. Selected Works. North Holland, Polish Scientific Publishers. Amsterdam–Warsaw, pp. 84–86 (1970)
33. Manheim, J.H.: The Genesis of Point Set Topology. Pergamon Press, Oxford (1964)
34. Margolis, E., Laurence, S. (eds.): Creations of the Mind. Theories of Artifacts and Their Representation. Oxford University Press, Oxford and New York (2007)
35. Matarić, M.: Behavior–based control: examples from navigation, learning, and group behavior. J. Exp. Theor. Artif. Intell. 9(2, 3), 323–336 (1997)
36. Menger, K.: A logic of the doubtful on optative and imperative logic. Reports of a mathematical colloquium (Notre Dame U.), ser. 2, no. 1, pp. 53–64 (1939). Reprinted in: Karl Menger. Selected Papers in Logic and Foundations, Didactics, Economics, pp. 91–102. D. Reidel Publishing Company (1979)
37. Menger, K.: Statistical metrics. Proc. Nat. Acad. Sci. USA **28**, 535–537 (1942)
38. Menger, K.: Ensembles flous et fonctions aleatoires. C. R. Acad. Sci. Paris **232**, 2001–2003 (1951)
39. Ośmiałowski, P.: On path planning for mobile robots: introducing the mereological potential field method in the framework of mereological spatial reasoning. J. Autom. Mob. Rob. Intell. Syst. (JAMRIS) **3**(2), 24–33 (2009)
40. Ośmiałowski, P.: A case of planning and programming of a concurrent behavior: planning and navigating with formations of robots. In: Proceedings of CS & P 2009. Concurrency, Specification, Programming, Kraków, Warsaw University Press (2009)
41. Ośmiałowski, P.: Planning and Navigation for Mobile Autonomous Robots. PJIIT Publishers, Warszawa (2011)
42. Ośmiałowski, P., Polkowski, L.: Spatial reasoning based on rough mereology: path planning problem for autonomous mobile robots. In: Transactions on Rough Sets XII. Lecture Notes in Computer Science 6190, pp. 143–169. Springer, Berlin (2009)
43. Øhrstrøm, P., Hasle, P.: Future contingents. In: Zalta E.N. (ed.) The Stanford Encyclopedia of Philosophy (Winter 2015 Edition). http://plato.stanford.edu/archives/win2015/entries/future-contingents/

44. Pawłak, Z.: Rough sets. Int. J. Comput. Inf. Sci. **11**, 341–356 (1982)
45. Pawłak, Z.: Rough Sets: Theoretical Aspects of Reasoning About Data. Kluwer, Dordrecht (1991)
46. Polkowski, L.: Approximation mathematical morphology. In: Pal, S.K., Skowron, A. (eds.) Rough Fuzzy Hybridization. A New Trend in Decision Making, pp. 151–162. Springer, Singapore (1999)
47. Polkowski, L.: Rough sets. Mathematical Foundations. Physica Verlag/Springer, Heidelberg (2002)
48. Polkowski, L.: Formal granular calculi based on rough inclusions. In: Proceedings of IEEE 2005 Conference on Granular Computing GrC05, Beijing, China, pp. 57–62. IEEE Press (2005)
49. Polkowski, L.: A model of granular computing with applications. In: Proceedings of IEEE 2006 Conference on Granular Computing GrC06, Atlanta, USA, pp. 9–16. IEEE Press (2006)
50. Polkowski, L.: Granulation of knowledge in decision systems: the approach based on rough inclusions. The method and its applications. In: Proceedings RSEISP 07, Warsaw, Poland, June 2007. Lecture Notes in Artificial Intelligence 4585, pp. 271–279. Springer, Berlin (2007)
51. Polkowski, L.: A unified approach to granulation of knowledge and granular computing based on rough mereology: a survey. In: Pedrycz, W., Skowron, A., Kreinovich, V. (eds.) Handbook of Granular Computing, pp. 375–400. Wiley Ltd., Chichester, UK (2008)
52. Polkowski, L.: Granulation of knowledge: similarity based approach in information and decision systems. In: Meyers, R.A. (ed.) Springer Encyclopedia of Complexity and System Sciences, Article 00 788. Springer, Berlin (2009)
53. Polkowski, L.: Approximate reasoning by parts. An Introduction to Rough Mereology. Springer, Berlin (2011)
54. Polkowski, L.: Mereology in engineering and computer science. In: Calosi, C., Graziani, P. (eds.) Mereology and the Sciences, pp. 217–292. Springer Synthese Library, Cham, Switzerland (2015)
55. Polkowski, L.: Betweenness, Lukasiewicz rough inclusions, Euclidean representations in information systems, hyper-granules, conflict resolution. In: Proceedings CS & P 2015, Rzeszow University, Poland, 2015. http://ceur-ws.org/Vol-1492/Paper_34.pdf
56. Polkowski, L., Artiemjew, P.: Granular Computing in Decision Approximation, p. 1. Springer Verlag International, Cham, Switzerland (2015)
57. Polkowski, L., Nowak, B.: Betweenness, Lukasiewicz rough inclusions, Euclidean representations in information systems, hyper-granules. conflict resolution. Fundam. Informaticae 147(2–3), 65–80 (2016)
58. Polkowski, L., Ośmiałowski, P.: A Framework for Multi–Agent Mobile Robotics: Spatial Reasoning Based on Rough Mereology in Player/Stage system. Lecture Notes in Artificial Intelligence 5306, pp. 142–149. Springer, Berlin (2008)
59. Polkowski, L., Ośmiałowski, P.: Navigation for mobile autonomous robots and their formations: an application of spatial reasoning induced from rough mereological geometry. In: Barrera, A. (ed.) Mobile Robots Navigation, pp. 329–354. Zagreb, InTech (2010)
60. Polkowski, L., Semeniuk-Polkowska, M.: On rough set logics based on similarity relations. Fundam. Informaticae **64**, 379–390 (2005)
61. Polkowski, L., Semeniuk-Polkowska, M.: On a notion of extensionality for artifacts. Fundam. Informaticae **127**(1–4), 65–80 (2013)
62. Polkowski, L., Skowron, A.: Rough mereology. In: Proceedings of ISMIS'94. Lecture Notes in Artificial Intelligence 869, pp. 85–94. Springer, Berlin (1994)
63. Polkowski, L., Skowron, A.: Rough mereology: a new paradigm for approximate reasoning. Int. J. Approximate Reasoning **15**(4), 333–365 (1997)
64. Raatikainen, P.: Gdel's incompleteness theorems. In: Zalta E.N. (ed.) The Stanford Encyclopedia of Philosophy (Spring 2015 Edition). http://plato.stanford.edu/archives/spr2015/entries/goedel-incompleteness/
65. Salustri, F.A.: Mereotopology for product modelling. A new framework for product modelling based on logic. J. Design Res. 2 (2002)

66. Seibt, J.: Forms of emergent interaction in general process theory. Synthese **1666**, 479–512 (2009)
67. SEP (Stanford Encyclopedia of Philosophy): Artifact. http://plato.stanford.edu/entries/artifact. Accessed 01 Mar 2012
68. Simons, P.: Parts. A Sudy in Ontology. Clarendon Press, Oxford, UK (2003)
69. Simons, P.: Jan ukasiewicz. In: Zalta E.N. (ed.) The Stanford Encyclopedia of Philosophy (Summer 2014 Edition). http://plato.stanford.edu/archives/sum2014/entries/lukasiewicz/
70. Skowron, A., Rauszer, C.: The discernibility matrices and functions in decision systems. In: Słowiński, R. (ed.) Intelligent Decision Support, pp. 311–362. Handbook of Applications and Advances of the Rough Sets Theory. Kluwer, Dordrecht (1992)
71. Sobociński, B.: L'analyse de l'antinomie Russellienne par Leśniewski. Methodos I, pp. 94–107, 220–228, 308–316 (1949); Methodos II, pp. 237–257 (1950)
72. Sobociński, B.: Studies in Leśniewski's Mereology. Yearbook for 1954–55 of the Polish Society of Art and Sciences Abroad V, pp. 34–43 (1954–55)
73. Srzednicki, J., Surma, S.J., Barnett, D., Rickey, V.F. (eds.): Collected Works of Stanisław Leśniewski. Kluwer, Dordrecht (1992)
74. Tarski, A.: Les fondements de la géométrie des corps. Supplement to Annales de la Société Polonaise de Mathématique 7, pp. 29–33 (1929)
75. Tarski, A.: Zur Grundlegung der Booleschen Algebra. I. Fundam. Mathematicae **24**, 177–198 (1935)
76. Tarski, A.: Appendix E. In: Woodger J.H. (ed.) The Axiomatic Method in Biology, p. 160. Cambridge University Press, Cambridge, UK (1937)
77. Tarski, A.: What is elementary geometry? In: Henkin, L., Suppes, P., Tarski, A. (eds.) The Axiomatic Method with Special Reference to Geometry and Physics, pp. 16–29. North-Holland, Amsterdam (1959)
78. Varzi, A.C.: The extensionality of parthood and composition. Philos. Quart. **58**, 108–133 (2008)
79. Whitehead, A.N.: La théorie relationniste de l'espace. Revue de Métaphysique et de Morale 23, pp. 423–454 (1916)
80. Whitehead, A.N.: An Enquiry Concerning the Principles of Natural Knowledge. Cambridge University Press, Cambridge, UK (1919)
81. Zadeh, L.A.: Fuzzy sets. Inf. Control **8**, 338–353 (1965)

Rough Sets in Machine Learning: A Review

Rafael Bello and Rafael Falcon

Abstract This chapter emphasizes on the role played by *rough set theory* (RST) within the broad field of Machine Learning (ML). As a sound data analysis and knowledge discovery paradigm, RST has much to offer to the ML community. We surveyed the existing literature and reported on the most relevant RST theoretical developments and applications in this area. The review starts with RST in the context of *data preprocessing* (discretization, feature selection, instance selection and meta-learning) as well as the generation of both *descriptive and predictive knowledge* via decision rule induction, association rule mining and clustering. Afterward, we examined *several special ML scenarios* in which RST has been recently introduced, such as imbalanced classification, multi-label classification, dynamic/incremental learning, Big Data analysis and cost-sensitive learning.

1 Introduction

Information granulation is the process by which a collection of *information granules* are synthesized, with a granule being a collection of values (in the data space) which are drawn towards the center object(s) (in the object space) by an underlying indistinguishability, similarity or functionality mechanism. Note that the data and object spaces can actually coincide [141]. The *Granular Computing* (GrC) paradigm

R. Bello (✉)
Department of Computer Science, Universidad Central de Las Villas,
Carretera Camajuaní km 5.5, Santa Clara, Cuba
e-mail: rbellop@uclv.edu.cu

R. Falcon
Research & Engineering Division, Larus Technologies Corporation,
170 Laurier Ave West - Suite 310, Ottawa, ON, Canada
e-mail: rafael.falcon@larus.com; rfalcon@uottawa.ca

R. Falcon
School of Electrical Engineering and Computer Science, University of Ottawa,
800 King Edward Ave, Ottawa, ON, Canada

G. Wang et al. (eds.), *Thriving Rough Sets*, Studies in Computational Intelligence 708, DOI 10.1007/978-3-319-54966-8_5

[7, 183] encompasses several computational models based on fuzzy logic, Computing With Words, interval computing, rough sets, shadowed sets, near sets, etc.

The main purpose behind Granular Computing is to find a novel way to synthesize knowledge in a more human-centric fashion and from vast, unstructured, possibly high-dimensional raw data sources. Not surprisingly, Granular Computing (GrC) is closely related to Machine Learning [83, 95, 257]. The aim of a learning process is to derive a certain rule or system for either the automatic classification of the system objects or the prediction of the values of the system control variables. The key challenge with prediction lies in modeling the relationships among the system variables in such a way that it allows inferring the value of the control (target) variable.

Rough set theory (RST) [1] was developed by Zdzisław Pawlak in the early 1980s [179] as a mathematical approach to intelligent data analysis and data mining [180]. This methodology is based on the premise that lowering the degree of precision in the data makes the data pattern more visible, i.e., the rough set approach can be formally considered as a framework for pattern discovery from imperfect data [220]. Several reasons are given in [34] to employ RST in knowledge discovery, including:

- It does not require any preliminary or additional information about the data
- It provides a valuable analysis even in presence of incomplete data
- It allows the interpretation of large amounts of both quantitative and qualitative data
- It can model highly nonlinear or discontinuous functional relations to provide complex characterizations of data
- It can discover important facts hidden in the data and represent them in the form of decision rules, and
- At the same time, the decision rules derived from rough set models are based on facts, because every decision rule is supported by a set of examples.

Mert Bal [3] brought up other RST advantages, such as: (a) it performs a clear interpretation of the results and evaluation of the meaningfulness of data; (b) it can identify and characterize uncertain systems and (c) the patterns discovered using rough sets are concise, strong and sturdy.

Among the main components of the knowledge discovery process we can mention:

- PREPROCESSING
  - Discretization
  - Training set edition (instance selection)
  - Feature selection
  - Characterization of the learning problem (data complexity, metalearning)
- KNOWLEDGE DISCOVERY
  - Symbolic inductive learning methods
  - Symbolic implicit learning methods (a.k.a. lazy learning)

- KNOWLEDGE EVALUATION
  - Evaluation of the discovered knowledge

All of the above stages have witnessed the involvement of rough sets in their algorithmic developments. Some of the RST applications are as follows:

- Analysis of the attributes to consider
  - Feature selection
  - Inter-attribute dependency characterization
  - Feature reduction
  - Feature weighting
  - Feature discretization
  - Feature removal
- Formulation of the discovered knowledge
  - Discovery of decision rules
  - Quantification of the uncertainty in the decision rules.

RST's main components are an *information system* and an *indiscernibility relation*. An information system is formally defined as follows. Let $A = \{A_1, A_2, \ldots, A_n\}$ be a set of attributes characterizing each example (object, entity, situation, state, etc.) in non-empty set U called the universe of discourse. The pair (U, A) is called an *information system*. If there exists an attribute $d \notin A$, called the *decision attribute*, that represents the decision associated with each example in U, then a *decision system* $(U,\ A \cup \{d\})$ is obtained.

The fact that RST relies on the existence of an information system allows establishing a close relationship with data-driven knowledge discovery processes given that these information or decision systems can be employed as training sets for unsupervised or supervised learning models, respectively.

A binary *indiscernibility relation* I_B is associated with each subset of attributes $B \subseteq A$. This relation contains the pairs of objects that are inseparable from each other given the information expressed in the attributes in B, as shown in Eq. (1).

$$I_B = \{(x, y) \in U \times U\ :\ f(x, A_i) = f(y, A_i)\ \ \forall A_i \in B\}. \quad (1)$$

where $f(x, A_i)$ returns the value of the i-th attribute in object $x \in U$.

The indiscernibility relation induces a granulation of the information system. The classical RST leaned on a particular type of indiscernibility relations called equivalence relations (i.e., those that are simmetric, reflexive and transitive). An equivalence relation induces a granulation of the universe in the form of a partition. This type of relation works well when there are only nominal attributes and no missing values in the information system.

Information systems having incomplete, continuous, mixed or heterogeneous data are in need of a more flexible type of indiscernibility relation. Subsequent RST formulations relaxed the stringent requirement of having an equivalence relation by considering either a tolerance or a similarity relation [61, 68, 181, 207, 212, 231, 283, 284, 305, 306]; these relations will induce a *covering* of the system. Another relaxation avenue is based on the probabilistic approach [65, 182, 210, 259, 264, 267, 307]. A third alternative is the hybridization with fuzzy set theory [54, 55, 172, 258, 280]. These different approaches have contributed to positioning RST as an important component within Soft Computing [12].

All of the aforementioned RST formulations retain some basic definitions, such as the lower and upper approximations; however, they defined it in multiple ways. The canonical RST definition for the *lower approximation* of a concept X is given as $B_*(X) = \{x \in U : B(x) \subseteq X\}$ whereas its *upper approximation* is calculated as $B^*(X) = \{x \in U : B(x) \cap X \neq \emptyset\}$. From these approximations we can compute the *positive region* $POS(X) = B_*(X)$, the *boundary region* $BND(X) = B^*(X) - B_*(X)$ and the *negative region* $NEG(X) = U - B^*(X)$. These concepts serve as building blocks for developing many problem-solving approaches, including data-driven learning.

RST and Machine Learning are also related in that both take care of removing irrelevant/redundant attributes. This process is termed *feature selection* and RST approaches it from the standpoint of calculating the system reducts. Given an information system $S = (U, A)$, where U is the universe and A is the set of attributes, a *reduct* is a minimum set of attributes $B \subseteq A$ such that $I_A = I_B$.

This chapter emphasizes on the role played by RST within the broad field of Machine Learning (ML). As a sound data analysis and knowledge discovery paradigm, RST has much to offer to the ML community. We surveyed the existing literature and reported on the most relevant RST theoretical developments and applications in this area. The review starts with RST in the context of *data preprocessing* (discretization, feature selection, instance selection and meta-learning) as well as the generation of both *descriptive and predictive knowledge* via decision rule induction, association rule mining and clustering. Afterward, we examined *several special ML scenarios* in which RST has been recently introduced, such as imbalanced classification, multi-label classification, dynamic/incremental learning, Big Data analysis and cost-sensitive learning.

The rest of the chapter is structured as follows. Section 2 reviews ML methods and processes from an RST standpoint, with emphasis on data preprocessing and knowledge discovery. Section 3 unveils special ML scenarios that are being gradually permeated by RST-based approaches, including imbalanced classification, multi-label classification, dynamic/incremental learning, Big Data analysis and cost-sensitive learning. Section 5 concludes the chapter.

2 Machine Learning Methods and RST

This section briefly goes over reported studies showcasing RST as a tool in data preprocessing and descriptive/predictive knowledge discovery.

2.1 Preprocessing

2.1.1 Discretization

As mentioned in [195], *discretization* is the process of converting a numerical attribute into a nominal one by applying a set of cuts to the domain of the numerical attribute and treating each interval as a discrete value of the (now nominal) attribute. Discretization is a mandatory step when processing information systems with the canonical RST formulation, as there is no provisioning for handling numerical attributes there. Some RST extensions avoid this issue by, for example, using similarity classes instead of equivalence classes and building a similarity relation that encompasses both nominal and numerical attributes.

It is very important that any discretization method chosen in the context of RST-based data analysis preserves the underlying discernibility among the objects. The level of granularity at which the cuts are performed in the discretization step will have a direct impact on any ensuing prediction, i.e., generic (wider) intervals (cuts) will likely avoid overfitting when predicting the class for an unseen object.

Dougherty et al. [53] categorize discretization methods along three axes:

- *global versus local*: indicates whether an approach simultaneously converts all numerical attributes (global) or is restricted to a single numerical attribute (local). For instance, the authors in [174] suggest both local and global handling of numerical attributes in large data bases.
- *supervised versus unsupervised*: indicates whether an approach considers values of other attributes in the discretization process or not. A simple example of an unsupervised approach is an "equal width" interval method that works by dividing the range of continuous attributes into k equal intervals, where k is given. A supervised discretization method, for example, will consider the correlation between the numerical attribute and the label (class) attribute when choosing the location of the cuts.
- *static versus dynamic*: indicates whether an approach requires a parameter to determine the number of cut values or not. Dynamic approaches automatically generate this number along the discretization process whereas static methods require an a priori specification of this parameter.

Lenarcik and Piasta [128] introduced an RST-based discretization method that leans on the concepts of a random information system and of an expected value of classification quality. The method of finding suboptimal discretizations based on these concepts is presented and is illustrated with data from concretes' frost resistance investigations.

Nguyen [173] considers the problem of searching for a minimal set of cuts that preserves the discernibility between objects with respect to any subset of s attributes, where s is a user-defined parameter. It was shown that this problem is NP-hard and its heuristic solution is more complicated than that for the problem of searching for an optimal, consistent set of cuts. The author proposed a scheme based on Boolean reasoning to solve this problem.

Bazan [5] put forth a method to search for an irreducible sets of cuts of an information system. The method is based on the notion of *dynamic reduct*. These reducts are calculated for the information system and the one with the best stability coefficient is chosen. Next, as an irreducible set of cuts, the author selected cuts belonging to the chosen dynamic reduct.

Bazan et al. [6] proposed a discretization technique named *maximal discernibility* (MD), which is based on rough sets and Boolean reasoning. MD is a greedy heuristic that searches for cuts along the domains of all numerical attributes that discern the largest number of object pairs in the dataset. These object pairs are removed from the information system before the next cut is sought. The set of cuts obtained that way is optimal in terms of object indiscernibility; however this procedure is not feasible since computing one cut requires $O(|A| \cdot |U|^3)$. Locally optimal cuts [6] are computed in $O(|A| \cdot |U|)$ steps using only $O(|A| \cdot |U|)$ space.

Dai and Li [46] improved Nguyen's discretization techniques by reducing the time and space complexity required to arrive at the set of candidate cuts. They proved that all *bound cuts* can discern the same object pairs as the entire set of initial cuts. A strategy to select candidate cuts was proposed based on that proof. They obtained identical results to Nguyen's with a lower computational overhead.

Chen et al. [26] employ a genetic algorithm (GA) to derive the minimal cut set in a numerical attribute. Each gene in a binary chromosome represents a particular cut value. Enabling this gene means the corresponding cut value has been selected as a member of the minimal cut set. Some optimization strategies such as elitist selection and father-offspring combined selection helped the GA converge faster. The experimental evidence showed that the GA-based scheme is more efficient than Nguyen's basic heuristic based on rough sets and Boolean reasoning.

Xie et al. [249] defined an information entropy value for every candidate cut point in their RST-based discretization algorithm. The final cut points are selected based on this metric and some RST properties. The authors report that their approach outperforms other discretization techniques and scales well with the number of cut points.

Su and Hsu [219] extended the modified Chi2 discretizer by learning the predefined misclassification rate (input parameter) from data. The authors additionally considered the effect of variance in the two adjacent intervals. In the modified Chi2, the inconsistency check in the original Chi2 is replaced with the "quality of approximation" measure from RST. The result is a more robust, parameterless discretization method.

Singh and Minz [205] designed a hybrid clustering-RST-based discretizer. The values of each numerical attribute are grouped using density-based clustering algorithms. This produces a set of (possibly overlapping) intervals that naturally reflect the data distribution. Then, the rough membership function in RST is employed to refine these intervals in a way that maximizes class separability. The proposed scheme yielded promising results when compared to seven other discretizers.

Jun and Zhou [116] enhanced existing RST-based discretizers by (i) computing the candidate cuts with an awareness of the decision class information; in this way, the scales of candidate cuts can be remarkably reduced, thus considerably saving time and space and (ii) introducing a notion of cut selection probability that is defined to measure cut significance in a more reasonable manner. Theoretical analyses and simulation experiments show that the proposed approaches can solve the problem of data discretization more efficiently and effectively.

2.1.2 Feature Selection

The purpose behind feature selection is to discard irrelevant features that are generally detrimental to the classifier's performance, generate noise, increase the amount of information to be stored and the computational cost of the classification process [222, 302]. Feature selection is a computationally expensive problem that requires searching for a subset of the n original features in a space of $2^n - 1$ candidate subsets according to a predefined evaluation criterion. The main components of a feature selection algorithm are: (1) an *evaluation function* (EF), used to calculate the fitness of a feature subset and (2) a *generation procedure* that is responsible for generating different subsets of candidate features.

Different feature selection schemes that integrate RST into the feature subset evaluation function have been developed. The *quality of the classification* γ is the most frequently used RST metric to judge the suitability of a candidate feature subset, as shown in [9–11, 64] etc. Other indicators are *conditional independence* [208] and *approximate entropy* [209].

The concept of *reduct* is the basis for these results. Essentially, a reduct is a minimal subset of features that generates the same granulation of the universe as that induced by all features. Among these works we can list [37, 38, 85, 89, 111, 136, 168, 196, 221, 223, 239, 247, 248, 255, 270, 302]. One of the pioneer methods is the *QuickReduct* algorithm, which is typical of those algorithms that resort to a greedy search strategy to find a relative reduct [136, 202, 247]. Generally speaking, feature selection algorithms are based on heuristic search [97, 164, 302]. Other RST-based methods for reduct calculation are [98, 209].

More advanced methods employ metaheuristic algorithms (such as Genetic Algorithms, Ant Colony Optimization or Particle Swarm Optimization) as the underlying feature subset generation engine [8–11, 15, 64, 102, 119, 241, 242, 245, 246, 268, 274, 297]. Feature selection methods based on the hybridization between fuzzy and rough sets have been proposed in [13, 28, 42–44, 51, 75, 87, 90, 92, 101, 103–105, 125, 193, 197, 203, 225, 299]. Some studies aim at calculating all possible reducts of a decision system [27, 28, 206, 225, 299].

Feature selection is arguably the Machine Learning (ML) area that has witnessed the most influx of rough-set-based methods. Other RST contributions to ML are concerned with providing metrics to calculate the inter-attribute dependence and the importance (weight) of any attribute [120, 222].

2.1.3 Instance Selection

Another important data preprocessing task is the editing of the training sets, also referred to as *instance selection*. The aim is to reduce the number of examples in order to bring down the size of the training set while maintaining the system efficiency. By doing that, a new training set is obtained that will bring forth a higher efficiency usually also produces a reduction of the data.

Some training set edition approaches using rough sets have been published in [16, 19]. The simplest idea is to remove all examples in the training set that are not contained in the lower approximation of any of the decision classes. A more thorough investigation also considers those examples that lie in the boundary region of any of the decision classes. Fuzzy rough sets have been also applied to the instance selection problem in [99, 232, 233].

2.1.4 Meta-Learning

An important area within knowledge discovery is that of *meta-learning*, whose objective is to learn about the underlying learning processes in order to make them more efficient or effective [234]. These methods may consider measures related to the complexity of the data [79]. The study in [18] explores the use of RST-based metrics to estimate the quality of a data set. The relationship between the "quality of approximation" measure and the performance of some classifiers is investigated in [17]. This measure describe the inexactness of the rough-set-based classification and denotes the percentage of examples that were correctly classified employing the attributes included in the indiscernibility relationship [224]. The authors in [251] analyze the inclusion degree as a perspective on measures for rough set data analysis (RSDA). Other RSDA measures are the "accuracy of the approximation" and the rough membership function [120]; for example, in [108, 109], the rough membership function and other RST-based measures are employed to detect outliers (i.e., examples that behave in an unexpected way or have abnormal properties).

2.2 Descriptive and Predictive Knowledge Discovery

2.2.1 Decision Rule Induction

The knowledge uncovered by the different data analysis techniques can be either *descriptive* or *predictive*. The former characterizes the general properties of the data in the data set (e.g., association rules) while the latter allows performing inferences from the available data (e.g., decision rules). A decision rule summarizes the relationship between the properties (features) and describes a causal relationship among them. For example, IF Headache = Yes AND Weakness = YES THEN Influenza =

YES. The most common **rule induction** task is to generate a rule base R that is both consistent and complete.

According to [161], RST-based rule induction methods provide the following benefits:

- Better explanation capabilities
- Generate a simple and useful set of rules.
- Work with sparse training sets.
- Work even when the underlying data distribution significantly deviates from the normal distribution.
- Work with incomplete, inaccurate, and heterogeneous data.
- Usually faster execution time to generate the rule base compared to other methods.
- No assumptions made on the size or distribution of the training data.

Among the most popular RST-based rule induction methods we can cite LERS [67, 215], which includes the LEM1 (Learn from examples model v1) and LEM2 methods (Learn from examples model v2); the goal is to extract a minimum set of rules to cover the examples by exploring the attribute-value pairs search space of while taking into account possible data inconsistency issues. MODLEM [214, 215] is based on sequentially building coverings of the training data and generating minimal decision rule sets for each decision class. Each of these sets aims at covering all positive examples that belong to a concept and none from any other concept. The EXPLORE algorithm [216] extracts from data all the decision rules satisfying certain requirements. It can be adapted to handle inconsistent examples. The LEM2, EXPLORE and MODLEM algorithms rule induction algorithms are implemented in the ROSE2 software [3]. Filiberto et al. proposed the IRBASIR method [62], which generates decision rules using an RST extension rooted on similarity relations; another technique is put forth in [121] to discover rules using similarity relations for incomplete data sets. This learning problem in presence of missing data is also addressed in [80].

Other **RST-based rule induction algorithms** available in the literature using rough sets are [3, 14, 63, 110, 118, 129, 154, 179, 228, 229]. The use of hybrid models based on rough sets and fuzzy sets for rule induction and other knowledge discovery methods is illustrated in [2, 24, 41, 100, 123, 159, 201, 298, 300], which includes working with the so called "fuzzy decision information systems" [2].

One of the most popular rule induction methods based on rough sets is the so-called *three-way decisions* model [81, 260–263]. This methodology is strongly related to *decision making*. Essentially, for each decision alternative, this method defines three rules based on the RST's positive, negative and boundary regions. They respectively indicate acceptance, rejection or abstention (non-commitment, denotes weak or insufficient evidence).

This type of rules, derived from the basic RST concepts, is a suitable knowledge representation vehicle in a plethora of application domains. Hence, it has been integrated into common machine learning tasks to facilitate the knowledge engineering process required for a successful modeling of the domain under consideration. The three-way decisions model has been adopted in feature selection [106, 107, 133, 163, 265, 293], classification [273, 281, 282, 293], clustering [276, 277] and face recognition [132, 289].

2.2.2 Association Rule Mining

The discovery of **association rules** is one of the classical data mining tasks. Its goal is to uncover relationships among attributes that frequently appear together; i.e., the presence of one implies the presence of the other. One of the typical examples is the purchase of beer and diapers during the weekends. Association rules are representative of descriptive knowledge. A particular case are the so called "class association rules", which are used to build classifiers. Several methods have been developed for discovering association rules using rough sets, including [49, 70, 94, 111, 127, 134, 211, 266].

2.2.3 Clustering

The **clustering** problem is another learning task that has been approached from a rough set perspective. Clustering is a landmark unsupervised learning problem whose main objective is to group similar objects in the same cluster and separate objects that are different from each other by assigning them to different clusters [96, 167]. The objects are grouped in such a way that those in the same group exhibit a high degree of association among them whereas those in different groups show a low degree of association. Clustering algorithms map the original N-dimensional feature space to a 1-dimensional space describing the cluster each object belongs to. This is why clustering is considered both an important dimensionality reduction technique and also one of the most prevalent Granular Computing [183] manifestations.

One of the most popular and efficient clustering algorithms for conventional applications is *K-means clustering* [71]. In the K-means approach, randomly selected objects serve as initial cluster centroids. The objects are then assigned to different clusters based on their distance to the centroids. In particular, an object gets assigned to the cluster with the nearest centroid. The newly modified clusters then employ this information to determine new centroids. The process continues iteratively until the cluster centroids are stabilized. K-means is a very simple clustering algorithm, easy to understand and implement. The underlying alternate optimization approach iteratively converges but might get trapped into a local minimum of the objective function. K-means' best performance is attained in those applications where clusters are well separated and a crisp (bivalent) object-to-cluster decision is required. Its disadvantages include the sensitivity to outliers and the initial cluster centroids as well as the a priori specification of the desired number of clusters k.

Pawan Lingras [142, 145] found that the K-means algorithm often yields clustering results with unclear, vague boundaries. He pointed out that the "hard partitioning" performed by K-means does not meet the needs of grouping vague data. Lingras then proposed to combine K-means with RST and in the so-called "Rough K-means" approach. In this technique, each cluster is modeled as a rough set and each object belongs either to the lower approximation of a cluster or to the upper approximation of multiple clusters. Instead of building each cluster, its lower and upper approximations are defined based on the available data. The basic properties of the Rough K-means method are: (i) an object can be a member of at most a lower approximation; (ii) an object that is a member of the lower approximation of a cluster is also a member of its upper approximation and (iii) an object that does not belong to the lower approximation of any cluster is a member of at least the upper approximation of two clusters. Other pioneering works on rough clustering methods are put forth in [78, 192, 235, 236].

Rough K-means has been the subject of several subsequent studies aimed at improving its clustering capabilities. Georg Peters [187] concludes that rough clustering offers the possibility of reducing the number of incorrectly clustered objects, which is relevant to many real-world applications where minimizing the number of wrongly grouped objects is more important than maximizing the number of correctly grouped objects. Hence in these scenarios, Rough K-means arises as a powerful and stronger alternative to K-means. The same author proposes some improvements to the method regarding the calculation of the centroids, thus aiming to make the method more stable and robust to outliers [184, 185]. The authors in [291] proposed a Rough K-means improvement based on a variable weighted distance measure. Another enhancement brought forward in [186] suggested that well-defined objects must have a greater impact on the cluster centroid calculation rather than having this impact be governed by the number of cluster boundaries an object belongs to, as proposed in the original method. An extension to Rough K-means based on the decision-theoretic rough sets model was developed in [130]. An evolutionary approach for rough partitive clustering was designed in [168, 189] while [45, 190] elaborate on dynamic rough clustering approaches.

Other works that tackle the clustering problem using rough sets are [35, 72, 76, 77, 122, 124, 135, 143, 144, 162, 177, 178, 213, 271, 272, 275, 292]. These methods handle more specific scenarios (such as sequential, imbalanced, categorical and ordinal data), as well as applications of this clustering approach to different domains. The rough-fuzzy K-means method is put forward in [88, 170] whereas the fuzzy-rough K-means is unveiled in [169, 188]. Both approaches amalgamate the main features of Rough K-means and Fuzzy C-means by using the fuzzy membership of the objects to the rough clusters. Other variants of fuzzy and rough set hybridization for the clustering problem are presented in [56, 126, 160, 171].

3 Special Learning Cases Based on RST

This section elaborates on more recent ML scenarios tackled by RST-based approaches. In particular, we review the cases of imbalanced classification, multi-label classification, dynamic/incremental learning and Big Data analysis.

3.1 Imbalanced Classification

The traditional knowledge discovery methods presented in the previous section have to be adapted if we are dealing with an *imbalanced dataset* [21]. A dataset is balanced if it has an approximately equal percentage of positive and negative examples (i.e., those belonging to the concept to be classified and those belonging to other concepts, respectively). However, there are many application domains where we find an imbalanced dataset; for instance, in healthcare scenarios there are usually a plethora of patients that do not have a particularly rare disease. When learning a normalcy model for a certain environment, the number of labeled anomalous events is often scarce as most of the data corresponds to normal behaviour. The problem with imbalanced classes is that the classification algorithms have a tendency towards favoring the majority class. This occurs because the classifier attempts to reduce the overall error, hence the classification error does not take into account the underlying data distribution [23].

Several solutions have been researched to deal with this kind of situations. Two of the most popular avenues are either resampling the training data (i.e., oversampling the minority class or undersampling the majority class) or modifying the learning method [153]. One of the classical methods for learning with imbalanced data is SMOTE (*synthetic minority oversampling technique*) [22]. Different learning methods for imbalanced classification have been developed from an RST-based standpoint. For instance, Hu et al. [91] proposed models based on probabilistic rough sets where each example has an associated probability $p(x)$ instead of the default $1/n$. Ma et al. [158] introduced weights in the variable-precision rough set model (VPRS) to denote the importance of each example. Liu et al. [153] bring about some weights in the RST formulation to balance the class distribution and develop a method based on weighted rough sets to solve the imbalanced class learning problem. Ramentol et al. [194] proposed a method that integrates SMOTE with RST.

Stefanowski et al. [217] introduced filtering techniques to process inconsistent examples of the majority class (i.e., those lying in the boundary region), thereby adapting the MODLEM rule extraction method for coping with imbalanced learning problems. Other RST-based rule induction methods in the context of imbalanced data are also presented in [152, 243]. The authors in [218] proposed the EXPLORE method that generates rules for the minority class with a minimum coverage equal to a user-specified threshold.

3.2 Multi-label Classification

Normally, in a typical classification problem, a class (label) c_i from a set $C = \{c_1, \ldots, c_k\}$ is assigned to each example. However, in **multi-label classification**, a subset $S \subseteq C$ is assigned to each example, which means that an example could belong to multiple classes. Some applications of this type of learning emerge from text classification and functional genomics, namely, assigning functions to genes [226]. This gives rise to the so-called *multi-label learning* problem. The two avenues envisioned for solving this new class of learning problems have considered either converting the multi-label scenario to a single-label (classical) scenario or adapting the learning methods. Examples of the latter trend are the schemes proposed in [47, 198, 227, 290]. Similar approaches have been proposed for multi-label learning using rough sets. A first alternative is to transform the multi-label problem into a traditional single-label case and use classical RST-based learning methods to derive the rules (or any other knowledge); the other option is to adapt the RST-based learning methods, as shown in [240, 278, 279, 288].

In the first case, a decision system can be generated where some instances could belong to multiple classes. Multi-label classification can be regarded as an inconsistent decision problem, in which two objects having the same predictive attribute values do not share the same decision class. This leads to the modification of the definition of the lower/upper approximations through a probabilistic approach that facilitates modeling the uncertainty generated by the inconsistent system. This idea gives rise to the so-called *multi-label rough set model*, which incorporates a probabilistic approach such as the decision-theoretic rough set model. Some RST-based feature selection methods in multi-label learning scenarios have been enunciated [131], where the reduct concept was reformulated for the multi-label case.

3.3 Dynamic/Incremental Learning

Data are continuously being updated in nowadays' information systems. New data are added and obsolete data are purged over time. Traditional batch-learning methods lean on the principle of running these algorithms on all data when the information is updated, which obviously affects the system efficiency while ignoring any previous learning. Instead, learning should occur as new information arrives. Managing this learning while adapting the previous knowledge learned is the essence behind *incremental learning*. This term refers to an efficient strategy for the analysis of data in dynamic environments that allows acquiring additional knowledge from an uninterrupted information flow. The advantage of incremental learning is not to have to analyze the data from scratch but to utilize the learning process' previous outcomes as much as possible [57, 73, 112, 176, 200]. The continuous and massive acquisition of data becomes a challenge for the discovery of knowledge; especially in the context of Big Data, it becomes very necessary to develop capacities to assimilate the continuous data streams [29].

As an information-based methodology, RST is not exempt from being scrutinized in the context of dynamic data. The fundamental RST concepts and the knowledge discovery methods ensuing from them are geared towards the analysis of static data; hence, they need to be thoroughly revised in light of the requirements posed by data stream mining systems [151]. The purpose of the incremental learning strategy in rough sets is the development of incremental algorithms to quickly update the concept approximations, the reduct calculation or the discovered decision rules [40, 284]. The direct precursor of these studies can be found in [175]. According to [149], in recent years RST-based incremental learning approaches have become "hot topics" in knowledge extraction from dynamic data given their proven data analysis efficiency.

The study of RST in the context of learning with dynamic data can be approached from two different angles: what kind of information is considered to be dynamic and what type of learning task must be carried out. In the first case, the RST-based incremental updating approach could be further subdivided into three alternatives: (i) object variation (insertion or deletion of objects in the universe), (ii) attribute variation (insertion/removal of attributes) and (iii) attribute value variation (insertion/deletion of attribute values). In the second case, we can mention (i) incremental learning of the concept approximations [33, 139]; (ii) incremental learning of attribute reduction [52, 140, 237, 238, 250] and (iii) incremental learning of decision rules [59, 66, 148, 301].

Object variations include so-called object immigration and emigration [148]. Variations of the attributes include feature insertion or deletion [138, 287]. Variations in attribute values are primarily manifested via the refinement or scaling of the attribute values [32, 146]. Other works that propose modifications to RST-based methods for the case of dynamic data are [147, 149, 157].

The following studies deal with dynamic object variation:

- The update of the lower and upper approximations of the target concept is analyzed in [33, 137, 156].
- The update in the reduction of attributes is studied in [82, 250].
- The update of the decision rule induction mechanism is discussed in [4, 40, 59, 93, 148, 199, 230, 244, 269, 301].

If the variation occurs in the set of attributes, its effects have been studied with respect to these aspects:

- The update of the lower and upper approximations of the target concept is analyzed in [20, 36, 138, 139, 150, 287].
- The update of the decision rule induction mechanism is discussed in [39].

The effect of the variations in the attribute values (namely, via refinement or extension of the attribute domains) with respect to the update of the lower and upper approximations of the target concept is analyzed in [30–32, 50, 237, 308].

The calculation of reducts for dynamic data has also been investigated. The effect when the set of attributes varies is studied in [39]. The case of varying the attribute values is explored in [50, 69] whereas the case of dynamic object update is dissected in [199, 244]. Other studies on how dynamic data affect the calculation of reducts appear in [140, 204, 237, 238].

3.4 Rough Sets and Big Data

On the other hand, the accelerated pace of technology has led to an exponential growth in the generation and collection of digital information. This growth is not only limited to the amount of data available but to the plethora of diverse sources that emit these data streams. It becomes paramount then to efficiently analyze and extract knowledge from many dissimilar information sources within a certain application domain. This has led to the emergence of the *Big Data* era [25], which has a direct impact on the development of RST and its applications. Granular Computing, our starting point in this chapter, has a strong relation to Big Data [25], as its inherent ability to process information at multiple levels of abstraction and interpret information from different perspectives greatly facilitates the efficient management of large data volumes.

Simply put, Big Data can be envisioned as a large and complex data collection. These data are very difficult to analyze through traditional data management and processing tools. Big Data scenarios require new architectures, techniques, algorithms and processes to manage and extract value and knowledge hidden in the data streams. Big Data is often characterized by the 5 V's vector: Volume, Velocity, Variety, Veracity and Value. Big Data includes both structured and unstructured data, including images, videos, textual reports, etc. Big Data frameworks such as MapReduce and Spark have been recently developed and constitute indispensable tools for the accurate and seamless knowledge extraction from an array of disparate data sources. For more information on the Big Data paradigm, the reader is referred to the following articles: [25, 48, 60, 117].

As a data analysis and information extraction methodology, RST needs to adapt and evolve in order to cope with this new phenomenon. A major motivation to do so lies in the fact that the sizes of nowadays' decision systems are already extremely large. This poses a significant challenge to the efficient calculation of the underlying RST concepts and the knowledge discovery methods that emanate from them. Recall that the computational complexity of computing the target concept's approximations is $\mathbf{O}(lm^2)$, the computational cost of finding a reduct is bounded by $\mathbf{O}(l^2m^2)$ and the time complexity to find all reducts is $\mathbf{O}(2^lJ)$, where l is the number of attributes characterizing the objects, m is the number of objects in the universe and J is the computational cost required to calculate a reduct.

Some researchers have proposed RST-based solutions to the Big Data challenge [191, 286]. These methods are concerned with the design of parallel algorithms to compute equivalence classes, decision classes, associations between equivalence classes and decision classes, approximations, and so on. They are based on partitioning the universe, concurrently processing those information subsystems and then integrating the results. In other words, given the decision system $S = (U, C \cup D)$, generate the subsystems $\{S_1, S_2, \dots, S_m\}$, where $S_i = (U_i, C \cup D)$ and $U = \bigcup U_i$, then process each subsystem S_i, $i \in \{1, 2, \dots, m\}$, $U_i/B, B \subseteq C$. Afterwards, the results are amalgamated. This MapReduce-compliant workflow is supported by several theorems stating that (a) equivalence classes can be independently computed for each subsystem and (b) the equivalence classes from different subsystems can be merged if they are based on the same underlying attribute set. These results enable the parallel computation of the equivalence classes of the decision system S. Zhang et al. [286] developed the PACRSEC algorithm to that end.

Analogously, RST-based knowledge discovery methods, including reduct calculation and decision rule induction, have been investigated in in the context of Big Data [58, 256, 285].

3.5 Cost-Sensitive Learning

Cost is an important property inherent to real-world data. Cost sensitivity is an important problem which has been addressed from different angles. *Cost-sensitive learning* [252, 294, 303, 304] emerged when an awareness of the learning context was brought into Machine Learning. This is one of the most difficult ML problems and was listed as one of the top ten challenges in the Data Mining/ML domain [296].

Two types of learning costs have been addressed through RST: *misclassification cost* and *test cost* [253]. Test cost has been studied by Min et al. [163, 165, 166, 295] using the classical rough set approach, i.e., using a single granulation; a test-cost-sensitive multigranulation rough set model is presented in [253]. Multigranulation rough set is an extension of the classical RST that leans on multiple granular structures.

A recent cost-sensitive rough set approach was put forward in [115]. The crux of this method is that the information granules are sensitive to test costs while approximations are sensitive to decision costs, respectively; in this way, the construction of the rough set model takes into account both the test cost and the decision cost simultaneously. This new model is called *cost-sensitive rough set* and is based on decision-theoretic rough sets. In [132], the authors combine sequential three-way decisions and cost-sensitive learning to solve the face recognition problem; this is particularly interesting since in real-world face recognition scenarios, different kinds of misclassifications will lead to different costs [155, 294].

Other studies focused on the cost-sensitive learning problem from an RST perspective are presented in [84, 113, 253, 254]; these works have considered both the test cost and the decision cost. Attribute reduction based on test-cost-sensitivity has been quite well investigated [74, 86, 106, 114, 115, 133, 163, 164, 166, 296].

4 Reference Categorization

Table 1 lists the different RST studies according to the ML tasks they perform.

Table 1 Rough sets in machine learning

Task	Subtask	Subproblem/Approach	References
Data preprocessing	Discretization	Rough sets and Boolean reasoning	[5, 6, 46, 173, 174]
		Other approaches	[26, 116, 128, 205, 219, 249]
	Feature selection	Reduct calculation	[37, 38, 85, 89, 98, 106, 107, 111, 133, 136, 163, 168, 196, 202, 208, 209, 221, 223, 239, 247, 248, 255, 265, 270, 293, 302]
		Heuristic search	[97, 164, 302]
		Metaheuristic search	[8–11, 15, 64, 102, 119, 241, 242, 245, 246, 268, 274, 297]
		Fuzzy and rough set hybridization	[13, 28, 42–44, 51, 75, 87, 90, 92, 101, 103–105, 125, 193, 197, 203, 225, 299]
	Instance selection	Rough sets	[16, 19]
		Fuzzy and rough set hybridization	[99, 232, 233]
	Meta-learning	Quality of training set	[17, 18, 251]
		Outlier detection	[108, 109]
	Miscellaneous	Inter-attribute dependence/attribute importance	[120, 222]
Knowledge discovery	Decision rule induction	Rough sets	[3, 14, 62, 63, 67, 110, 118, 129, 154, 179, 214–216, 228, 229]
		Fuzzy and rough set hybridization	[2, 24, 41, 100, 123, 159, 201, 298, 300]
		Three-way decisions	[81, 260–263, 273, 281, 282, 293]
	Association rule mining	Rough sets	[49, 70, 94, 111, 127, 134, 211, 266]
	Clustering	Rough clustering	[78, 130, 142, 145, 184–187, 192, 235, 236, 291]
		Evolutionary/dynamic rough clustering	[45, 168, 189, 190]
		Other rough clustering scenarios	[35, 56, 72, 76, 77, 122, 124, 135, 143, 144, 162, 177, 178, 213, 271, 272, 275, 292]
		Fuzzy and rough set hybridization	[56, 88, 126, 160, 169–171, 188]
		Three-way decisions	[276, 277]

(continued)

Table 1 (continued)

Task	Subtask	Subproblem/Approach	References
Special learning cases	Imbalanced classification	Rough set theory adaptation	[91, 152, 153, 158, 194, 217, 218, 243]
	Multi-label classification	Rough set theory adaptation	[131, 240, 278, 279, 288]
	Dynamic/incremental learning	Dynamic objects	[4, 33, 40, 58, 82, 93, 137, 146, 148, 156, 199, 230, 244, 250, 269, 301]
		Dynamic attributes	[138, 287]
		Dynamic attribute values	[30–32, 32, 50, 146, 237, 308]
		Incremental learning of concept approximations	[20, 33, 137–139, 150, 156, 287]
		Incremental learning of attribute reduction	[52, 82, 140, 237, 238, 250, 250]
		Incremental learning of decision rules	[4, 39, 40, 59, 66, 93, 146, 199, 230, 244, 269, 301, 301]
		Dynamic reducts	[39, 50, 69, 140, 199, 204, 237, 238, 244]
		Miscellaneous	[40, 147, 149, 151, 157, 175, 284]
	Big Data analysis	Rough set theory adaptation	[58, 191, 256, 285, 286]
	Cost-sensitive learning	Test-cost-aware	[74, 86, 106, 114, 132, 133, 163–166, 166, 253, 295, 296]
		Test-and-decision-cost aware	[84, 113, 115, 254]

5 Conclusions

We have reported on hundreds of successful attempts to tackle different ML problems using RST. These approaches touch all components of the knowledge discovery process, ranging from data preprocessing to descriptive and predictive knowledge induction. Aside from the well-known RST strengths in identifying inconsistent information systems, calculating reducts to reduce the dimensionality of the feature space or generating an interpretable rule base, we have walked the reader through more recent examples that show the redefinition of some of the RST's building blocks to make it a suitable approach for handling special ML scenarios characterized by an imbalance in the available class data, the requirement to classify a pattern into one or more predefined labels, the dynamic processing of data streams, the need to manage large volumes of static data or the management of misclassification/test costs. All of these efforts bear witness to the resiliency and adaptability of the rough set approach, thus making it an appealing choice for solving non-conventional ML problems.

References

1. Abraham, A., Falcon, R., Bello, R.: Rough Set Theory: A True Landmark in Data Analysis. Springer, Berlin, Germany (2009)
2. Bai, H., Ge, Y., Wang, J., Li, D., Liao, Y., Zheng, X.: A method for extracting rules from spatial data based on rough fuzzy sets. Knowl. Based Syst. **57**, 28–40 (2014)
3. Bal, M.: Rough sets theory as symbolic data mining method: an application on complete decision table. Inf. Sci. Lett. **2**(1), 111–116 (2013)
4. Bang, W.C., Bien, Z.: New incremental learning algorithm in the framework of rough set theory. Int. J. Fuzzy Syst. **1**, 25–36 (1999)
5. Bazan, J.G.: A comparison of dynamic and non-dynamic rough set methods for extracting laws from decision tables. Rough Sets Knowl Discovery **1**, 321–365 (1998)
6. Bazan, J.G., Nguyen, H.S., Nguyen, S.H., Synak, P., Wróblewski, J.: Rough set algorithms in classification problem. In: Rough Set Methods and Applications, pp. 49–88. Springer (2000)
7. Bello, R., Falcon, R., Pedrycz, W., Kacprzyk, J.: Granular Computing: At the Junction of Rough Sets and Fuzzy Sets. Springer, Berlin, Germany (2008)
8. Bello, R., Gómez, Y., Caballero, Y., Nowe, A., Falcon, R.: Rough sets and evolutionary computation to solve the feature selection problem. In: Abraham, A., Falcon, R., Bello, R. (eds.) Rough Set Theory: A True Landmark in Data Analysis. Studies in Computational Intelligence, vol. 174, pp. 235–260. Springer, Berlin (2009)
9. Bello, R., Nowe, A., Gómez, Y., Caballero, Y.: Using ACO and rough set theory to feature selection. WSEAS Trans. Inf. Sci. Appl. **2**(5), 512–517 (2005)
10. Bello, R., Puris, A., Falcon, R., Gómez, Y.: Feature selection through dynamic mesh optimization. In: Ruiz-Shulcloper, J., Kropatsch, W. (eds.) Progress in Pattern Recognition, Image Analysis and Applications. Lecture Notes in Computer Science, vol. 5197, pp. 348–355. Springer, Berlin (2008)
11. Bello, R., Puris, A., Nowe, A., Martínez, Y., García, M.M.: Two step ant colony system to solve the feature selection problem. In: Iberoamerican Congress on Pattern Recognition, pp. 588–596. Springer (2006)
12. Bello, R., Verdegay, J.L.: Rough sets in the soft computing environment. Inf. Sci. **212**, 1–14 (2012)

13. Bhatt, R.B., Gopal, M.: On fuzzy-rough sets approach to feature selection. Pattern Recogn. Lett. **26**(7), 965–975 (2005)
14. Błaszczyński, J., Słowiński, R., Szelkag, M.: Sequential covering rule induction algorithm for variable consistency rough set approaches. Inf. Sci. **181**(5), 987–1002 (2011)
15. Caballero, Y., Bello, R., Alvarez, D., Garcia, M.M.: Two new feature selection algorithms with rough sets theory. In: IFIP International Conference on Artificial Intelligence in Theory and Practice, pp. 209–216. Springer (2006)
16. Caballero, Y., Bello, R., Alvarez, D., Gareia, M.M., Pizano, Y.: Improving the k-nn method: rough set in edit training set. In: Professional Practice in Artificial Intelligence, pp. 21–30. Springer (2006)
17. Caballero, Y., Bello, R., Arco, L., García, M., Ramentol, E.: Knowledge discovery using rough set theory. In: Advances in Machine Learning I, pp. 367–383. Springer (2010)
18. Caballero, Y., Bello, R., Arco, L., Márquez, Y., León, P., García, M.M., Casas, G.: Rough set theory measures for quality assessment of a training set. In: Granular Computing: At the Junction of Rough Sets and Fuzzy Sets, pp. 199–210. Springer (2008)
19. Caballero, Y., Joseph, S., Lezcano, Y., Bello, R., Garcia, M.M., Pizano, Y.: Using rough sets to edit training set in k-nn method. In: ISDA, pp. 456–463 (2005)
20. Chan, C.C.: A rough set approach to attribute generalization in data mining. Inf. Sci. **107**(1), 169–176 (1998)
21. Chawla, N.V.: Data mining for imbalanced datasets: an overview. In: Data Mining and Knowledge Discovery Handbook, pp. 853–867. Springer (2005)
22. Chawla, N.V., Bowyer, K.W., Hall, L.O., Kegelmeyer, W.P.: Smote: synthetic minority over-sampling technique. J. Artif. Intell. Res. **16**, 321–357 (2002)
23. Chawla, N.V., Cieslak, D.A., Hall, L.O., Joshi, A.: Automatically countering imbalance and its empirical relationship to cost. Data Min. Knowl. Discovery **17**(2), 225–252 (2008)
24. Chen, C., Mac Parthaláin, N., Li, Y., Price, C., Quek, C., Shen, Q.: Rough-fuzzy rule interpolation. Inf. Sci. **351**, 1–17 (2016)
25. Chen, C.P., Zhang, C.Y.: Data-intensive applications, challenges, techniques and technologies: a survey on big data. Inf. Sci. **275**, 314–347 (2014)
26. Chen, C.Y., Li, Z.G., Qiao, S.Y., Wen, S.P.: Study on discretization in rough set based on genetic algorithm. In: 2003 International Conference on Machine Learning and Cybernetics, vol. 3, pp. 1430–1434. IEEE (2003)
27. Chen, D., Hu, Q., Yang, Y.: Parameterized attribute reduction with gaussian kernel based fuzzy rough sets. Inf. Sci. **181**(23), 5169–5179 (2011)
28. Chen, D., Zhang, L., Zhao, S., Hu, Q., Zhu, P.: A novel algorithm for finding reducts with fuzzy rough sets. IEEE Trans. Fuzzy Syst. **20**(2), 385–389 (2012)
29. Chen, H., Chiang, R.H., Storey, V.C.: Business intelligence and analytics: from big data to big impact. MIS Q. **36**(4), 1165–1188 (2012)
30. Chen, H., Li, T., Qiao, S., Ruan, D.: A rough set based dynamic maintenance approach for approximations in coarsening and refining attribute values. Int. J. Intell. Syst. **25**(10), 1005–1026 (2010)
31. Chen, H., Li, T., Ruan, D.: Dynamic maintenance of approximations under a rough-set based variable precision limited tolerance relation. J. Multiple-Valued Log. Soft Comput. **18** (2012)
32. Chen, H., Li, T., Ruan, D.: Maintenance of approximations in incomplete ordered decision systems while attribute values coarsening or refining. Knowl. Based Syst. **31**, 140–161 (2012)
33. Chen, H., Li, T., Ruan, D., Lin, J., Hu, C.: A rough-set-based incremental approach for updating approximations under dynamic maintenance environments. IEEE Trans. Knowl. Data Eng. **25**(2), 274–284 (2013)
34. Chen, Y.S., Cheng, C.H.: A delphi-based rough sets fusion model for extracting payment rules of vehicle license tax in the government sector. Expert Syst. Appl. **37**(3), 2161–2174 (2010)
35. Cheng, X., Wu, R.: Clustering path profiles on a website using rough k-means method. J. Comput. Inf. Syst. **8**(14), 6009–6016 (2012)
36. Cheng, Y.: The incremental method for fast computing the rough fuzzy approximations. Data Knowl. Eng. **70**(1), 84–100 (2011)

37. Choubey, S.K., Deogun, J.S., Raghavan, V.V., Sever, H.: A comparison of feature selection algorithms in the context of rough classifiers. In: Proceedings of the Fifth IEEE International Conference on Fuzzy Systems, 1996, vol. 2, pp. 1122–1128. IEEE (1996)
38. Chouchoulas, A., Shen, Q.: A rough set-based approach to text classification. In: International Workshop on Rough Sets, Fuzzy Sets, Data Mining, and Granular-Soft Computing, pp. 118–127. Springer (1999)
39. Ciucci, D.: Attribute dynamics in rough sets. In: International Symposium on Methodologies for Intelligent Systems, pp. 43–51. Springer (2011)
40. Ciucci, D.: Temporal dynamics in information tables. Fundamenta Informaticae **115**(1), 57–74 (2012)
41. Coello, L., Fernandez, Y., Filiberto, Y., Bello, R.: Improving the multilayer perceptron learning by using a method to calculate the initial weights with the similarity quality measure based on fuzzy sets and particle swarms. Computación y Sistemas **19**(2), 309–320 (2015)
42. Cornelis, C., Jensen, R.: A noise-tolerant approach to fuzzy-rough feature selection. In: IEEE International Conference on Fuzzy Systems, 2008. FUZZ-IEEE 2008. (IEEE World Congress on Computational Intelligence), pp. 1598–1605. IEEE (2008)
43. Cornelis, C., Jensen, R., Hurtado, G., Śle, D., et al.: Attribute selection with fuzzy decision reducts. Inf. Sci. **180**(2), 209–224 (2010)
44. Cornelis, C., Verbiest, N., Jensen, R.: Ordered weighted average based fuzzy rough sets. In: International Conference on Rough Sets and Knowledge Technology, pp. 78–85. Springer (2010)
45. Crespo, F., Peters, G., Weber, R.: Rough clustering approaches for dynamic environments. In: Rough Sets: Selected Methods and Applications in Management and Engineering, pp. 39–50. Springer (2012)
46. Dai, J.H., Li, Y.X.: Study on discretization based on rough set theory. In: 2002 International Conference on Machine Learning and Cybernetics, 2002. Proceedings, vol. 3, pp. 1371–1373. IEEE (2002)
47. De Comité, F., Gilleron, R., Tommasi, M.: Learning multi-label alternating decision trees from texts and data. In: International Workshop on Machine Learning and Data Mining in Pattern Recognition, pp. 35–49. Springer (2003)
48. Dean, J., Ghemawat, S.: Mapreduce: simplified data processing on large clusters. Commun. ACM **51**(1), 107–113 (2008)
49. Delic, D., Lenz, H.J., Neiling, M.: Improving the quality of association rule mining by means of rough sets. In: Soft Methods in Probability, Statistics and Data Analysis, pp. 281–288. Springer (2002)
50. Deng, D., Huang, H.: Dynamic reduction based on rough sets in incomplete decision systems. In: International Conference on Rough Sets and Knowledge Technology, pp. 76–83. Springer (2007)
51. Derrac, J., Cornelis, C., García, S., Herrera, F.: Enhancing evolutionary instance selection algorithms by means of fuzzy rough set based feature selection. Inf. Sci. **186**(1), 73–92 (2012)
52. Dey, P., Dey, S., Datta, S., Sil, J.: Dynamic discreduction using rough sets. Appl. Soft Comput. **11**(5), 3887–3897 (2011)
53. Dougherty, J., Kohavi, R., Sahami, M., et al.: Supervised and unsupervised discretization of continuous features. Machine Learning: Proceedings of the Twelfth International Conference **12**, 194–202 (1995)
54. Dubois, D., Prade, H.: Twofold fuzzy sets and rough sets some issues in knowledge representation. Fuzzy Sets Syst. **23**(1), 3–18 (1987)
55. Dubois, D., Prade, H.: Rough fuzzy sets and fuzzy rough sets*. Int. J. Gen. Syst. **17**(2–3), 191–209 (1990)
56. Falcon, R., Jeon, G., Bello, R., Jeong, J.: Rough clustering with partial supervision. In: Rough Set Theory: A True Landmark in Data Analysis, pp. 137–161. Springer (2009)
57. Falcon, R., Nayak, A., Abielmona, R.: An Online shadowed clustering algorithm applied to risk visualization in territorial security. In: IEEE Symposium on Computational Intelligence for Security and Defense Applications (CISDA), pp. 1–8. Ottawa, Canada (2012)

58. Fan, Y.N., Chern, C.C.: An agent model for incremental rough set-based rule induction: a Big Data analysis in sales promotion. In: 2013 46th Hawaii International Conference on System Sciences (HICSS), pp. 985–994. IEEE (2013)
59. Fan, Y.N., Tseng, T.L.B., Chern, C.C., Huang, C.C.: Rule induction based on an incremental rough set. Expert Syst. Appl. **36**(9), 11439–11450 (2009)
60. Fernández, A., del Río, S., López, V., Bawakid, A., del Jesus, M.J., Benítez, J.M., Herrera, F.: Big data with cloud computing: an insight on the computing environment, mapreduce, and programming frameworks. Wiley Interdisc. Rev. Data Min. Knowl. Discovery **4**(5), 380–409 (2014)
61. Filiberto, Y., Caballero, Y., Larrua, R., Bello, R.: A method to build similarity relations into extended rough set theory. In: 2010 10th International Conference on Intelligent Systems Design and Applications, pp. 1314–1319. IEEE (2010)
62. Filiberto Cabrera, Y., Caballero Mota, Y., Bello Pérez, R., Frías, M.: Algoritmo para el aprendizaje de reglas de clasificación basado en la teoría de los conjuntos aproximados extendida. Dyna; vol. 78, núm. 169 (2011); 62-70 DYNA; vol. 78, núm. 169 (2011); 62-70 2346-2183 0012-7353 (2011)
63. Gogoi, P., Bhattacharyya, D.K., Kalita, J.K.: A rough set-based effective rule generation method for classification with an application in intrusion detection. Int. J. Secur. Netw. **8**(2), 61–71 (2013)
64. Gómez, Y., Bello, R., Puris, A., Garcia, M.M., Nowe, A.: Two step swarm intelligence to solve the feature selection problem. J. UCS **14**(15), 2582–2596 (2008)
65. Greco, S., Matarazzo, B., Słowiński, R.: Parameterized rough set model using rough membership and bayesian confirmation measures. Int. J. Approximate Reasoning **49**(2), 285–300 (2008)
66. Greco, S., Słowiński, R., Stefanowski, J., Żurawski, M.: Incremental versus non-incremental rule induction for multicriteria classification. In: Transactions on Rough Sets II, pp. 33–53. Springer (2004)
67. Grzymala-Busse, J.W.: LERS—a system for learning from examples based on rough sets. In: Intelligent decision support, pp. 3–18. Springer (1992)
68. Grzymała-Busse, J.W.: Characteristic relations for incomplete data: A generalization of the indiscernibility relation. In: International Conference on Rough Sets and Current Trends in Computing, pp. 244–253. Springer (2004)
69. Grzymala-Busse, J.W., Grzymala-Busse, W.J.: Inducing better rule sets by adding missing attribute values. In: International Conference on Rough Sets and Current Trends in Computing, pp. 160–169. Springer (2008)
70. Guan, J., Bell, D.A., Liu, D.: The rough set approach to association rule mining. In: Third IEEE International Conference on Data Mining, 2003. ICDM 2003, pp. 529–532. IEEE (2003)
71. Hartigan, J.A., Wong, M.A.: Algorithm as 136: a k-means clustering algorithm. J. Roy. Stat. Soc. Ser. C (Appl. Stat.) **28**(1), 100–108 (1979)
72. Hassanein, W., Elmelegy, A.A.: An algorithm for selecting clustering attribute using significance of attributes. Int. J. Database Theory Appl. **6**(5), 53–66 (2013)
73. He, H., Chen, S., Li, K., Xu, X.: Incremental learning from stream data. IEEE Trans. Neural Netw. **22**(12), 1901–1914 (2011)
74. He, H., Min, F., Zhu, W.: Attribute reduction in test-cost-sensitive decision systems with common-test-costs. In: Proceedings of the 3rd International Conference on Machine Learning and Computing, vol. 1, pp. 432–436 (2011)
75. He, Q., Wu, C., Chen, D., Zhao, S.: Fuzzy rough set based attribute reduction for information systems with fuzzy decisions. Knowl. Based Syst. **24**(5), 689–696 (2011)
76. Herawan, T.: Rough set approach for categorical data clustering. Ph.D. thesis, Universiti Tun Hussein Onn Malaysia (2010)
77. Herawan, T., Deris, M.M., Abawajy, J.H.: A rough set approach for selecting clustering attribute. Knowl. Based Syst. **23**(3), 220–231 (2010)

78. Hirano, S., Tsumoto, S.: Rough clustering and its application to medicine. J. Inf. Sci. **124**, 125–137 (2000)
79. Ho, T.K., Basu, M.: Complexity measures of supervised classification problems. IEEE Trans. Pattern Anal. Mach. Intell. **24**(3), 289–300 (2002)
80. Hong, T.P., Tseng, L.H., Wang, S.L.: Learning rules from incomplete training examples by rough sets. Expert Syst. Appl. **22**(4), 285–293 (2002)
81. Hu, B.Q.: Three-way decisions space and three-way decisions. Inf. Sci. **281**, 21–52 (2014)
82. Hu, F., Wang, G., Huang, H., Wu, Y.: Incremental attribute reduction based on elementary sets. In: International Workshop on Rough Sets, Fuzzy Sets, Data Mining, and Granular-Soft Computing, pp. 185–193. Springer (2005)
83. Hu, H., Shi, Z.: Machine learning as granular computing. In: IEEE International Conference on Granular Computing, 2009, GRC'09, pp. 229–234. IEEE (2009)
84. Hu, Q., Che, X., Zhang, L., Zhang, D., Guo, M., Yu, D.: Rank entropy-based decision trees for monotonic classification. IEEE Trans. Knowl. Data Eng. **24**(11), 2052–2064 (2012)
85. Hu, Q., Liu, J., Yu, D.: Mixed feature selection based on granulation and approximation. Knowl. Based Syst. **21**(4), 294–304 (2008)
86. Hu, Q., Pan, W., Zhang, L., Zhang, D., Song, Y., Guo, M., Yu, D.: Feature selection for monotonic classification. IEEE Trans. Fuzzy Syst. **20**(1), 69–81 (2012)
87. Hu, Q., Xie, Z., Yu, D.: Hybrid attribute reduction based on a novel fuzzy-rough model and information granulation. Pattern Recogn. **40**(12), 3509–3521 (2007)
88. Hu, Q., Yu, D.: An improved clustering algorithm for information granulation. In: International Conference on Fuzzy Systems and Knowledge Discovery, pp. 494–504. Springer (2005)
89. Hu, Q., Yu, D., Liu, J., Wu, C.: Neighborhood rough set based heterogeneous feature subset selection. Inf. Sci. **178**(18), 3577–3594 (2008)
90. Hu, Q., Yu, D., Xie, Z.: Information-preserving hybrid data reduction based on fuzzy-rough techniques. Pattern Recogn. Lett. **27**(5), 414–423 (2006)
91. Hu, Q., Yu, D., Xie, Z., Liu, J.: Fuzzy probabilistic approximation spaces and their information measures. IEEE Trans. Fuzzy Syst. **14**(2), 191–201 (2006)
92. Hu, Q., Zhang, L., An, S., Zhang, D., Yu, D.: On robust fuzzy rough set models. IEEE Trans. Fuzzy Syst. **20**(4), 636–651 (2012)
93. Huang, C.C., Tseng, T.L.B., Fan, Y.N., Hsu, C.H.: Alternative rule induction methods based on incremental object using rough set theory. Appl. Soft Comput. **13**(1), 372–389 (2013)
94. Huang, Z., Hu, Y.Q.: Applying AI technology and rough set theory to mine association rules for supporting knowledge management. In: 2003 International Conference on Machine Learning and Cybernetics, vol. 3, pp. 1820–1825. IEEE (2003)
95. Hüllermeier, E.: Granular computing in machine learning and data mining. In: Handbook of Granular Computing, pp. 889–906 (2008)
96. Jain, A.K., Murty, M.N., Flynn, P.J.: Data clustering: a review. ACM Comput. Surv. (CSUR) **31**(3), 264–323 (1999)
97. Janusz, A., Slezak, D.: Rough set methods for attribute clustering and selection. Appl. Artif. Intell. **28**(3), 220–242 (2014)
98. Janusz, A., Stawicki, S.: Applications of approximate reducts to the feature selection problem. In: International Conference on Rough Sets and Knowledge Technology, pp. 45–50. Springer (2011)
99. Jensen, R., Cornelis, C.: Fuzzy-rough instance selection. In: 2010 IEEE International Conference on Fuzzy Systems (FUZZ), pp. 1–7. IEEE (2010)
100. Jensen, R., Cornelis, C., Shen, Q.: Hybrid fuzzy-rough rule induction and feature selection. In: IEEE International Conference on Fuzzy Systems, 2009. FUZZ-IEEE 2009, pp. 1151–1156. IEEE (2009)
101. Jensen, R., Shen, Q.: Fuzzy-rough sets for descriptive dimensionality reduction. In: Proceedings of the 2002 IEEE International Conference on Fuzzy Systems, 2002. FUZZ-IEEE'02, vol. 1, pp. 29–34. IEEE (2002)
102. Jensen, R., Shen, Q.: Finding rough set reducts with ant colony optimization. In: Proceedings of the 2003 UK Workshop on Computational Intelligence, vol. 1, pp. 15–22 (2003)

103. Jensen, R., Shen, Q.: Fuzzy-rough attribute reduction with application to web categorization. Fuzzy Sets Syst. **141**(3), 469–485 (2004)
104. Jensen, R., Shen, Q.: Semantics-preserving dimensionality reduction: rough and fuzzy-rough-based approaches. IEEE Trans. Knowl. Data Eng. **16**(12), 1457–1471 (2004)
105. Jensen, R., Shen, Q.: New approaches to fuzzy-rough feature selection. IEEE Trans. Fuzzy Syst. **17**(4), 824–838 (2009)
106. Jia, X., Liao, W., Tang, Z., Shang, L.: Minimum cost attribute reduction in decision-theoretic rough set models. Inf. Sci. **219**, 151–167 (2013)
107. Jia, X., Shang, L., Zhou, B., Yao, Y.: Generalized attribute reduct in rough set theory. Knowl. Based Syst. **91**, 204–218 (2016)
108. Jiang, F., Sui, Y., Cao, C.: Outlier detection based on rough membership function. In: International Conference on Rough Sets and Current Trends in Computing, pp. 388–397. Springer (2006)
109. Jiang, F., Sui, Y., Cao, C.: Some issues about outlier detection in rough set theory. Expert Syst. Appl. **36**(3), 4680–4687 (2009)
110. Jiang, Y.C., Liu, Y.Z., Liu, X., Zhang, J.K.: Constructing associative classifier using rough sets and evidence theory. In: International Workshop on Rough Sets, Fuzzy Sets, Data Mining, and Granular-Soft Computing, pp. 263–271. Springer (2007)
111. Jiao, X., Lian-cheng, X., Lin, Q.: Association rules mining algorithm based on rough set. In: International Symposium on Information Technology in Medicine and Education, Print ISBN, pp. 978–1 (2012)
112. Joshi, P., Kulkarni, P.: Incremental learning: areas and methods—a survey. Int. J. Data Min. Knowl. Manage. Process **2**(5), 43 (2012)
113. Ju, H., Yang, X., Song, X., Qi, Y.: Dynamic updating multigranulation fuzzy rough set: approximations and reducts. Int. J. Mach. Learn. Cybern. **5**(6), 981–990 (2014)
114. Ju, H., Yang, X., Yang, P., Li, H., Zhou, X.: A moderate attribute reduction approach in decision-theoretic rough set. In: Rough Sets, Fuzzy Sets, Data Mining, and Granular Computing, pp. 376–388. Springer (2015)
115. Ju, H., Yang, X., Yu, H., Li, T., Yu, D.J., Yang, J.: Cost-sensitive rough set approach. Inf. Sci. **355**, 282–298 (2016)
116. Jun, Z., Zhou, Y.H.: New heuristic method for data discretization based on rough set theory. J. China Univ. Posts Telecommun. **16**(6), 113–120 (2009)
117. Kambatla, K., Kollias, G., Kumar, V., Grama, A.: Trends in big data analytics. J. Parallel Distrib. Comput. **74**(7), 2561–2573 (2014)
118. Kaneiwa, K.: A rough set approach to mining connections from information systems. In: Proceedings of the 2010 ACM Symposium on Applied Computing, pp. 990–996. ACM (2010)
119. Ke, L., Feng, Z., Ren, Z.: An efficient ant colony optimization approach to attribute reduction in rough set theory. Pattern Recogn. Lett. **29**(9), 1351–1357 (2008)
120. Komorowski, J., Pawlal, Z., Polkowski, L., Skowron, A.: A rough set perspective on data and knowledge. In: The Handbook of Data Mining and Knowledge Discovery. Oxford University Press, Oxford (1999)
121. Kryszkiewicz, M.: Rough set approach to incomplete information systems. Inf. Sci. **112**(1), 39–49 (1998)
122. Kumar, P., Krishna, P.R., Bapi, R.S., De, S.K.: Rough clustering of sequential data. Data Knowl. Eng. **63**(2), 183–199 (2007)
123. Kumar, P., Vadakkepat, P., Poh, L.A.: Fuzzy-rough discriminative feature selection and classification algorithm, with application to microarray and image datasets. Appl. Soft Comput. **11**(4), 3429–3440 (2011)
124. Kumar, P., Wasan, S.K.: Comparative study of k-means, pam and rough k-means algorithms using cancer datasets. In: Proceedings of CSIT: 2009 International Symposium on Computing, Communication, and Control (ISCCC 2009), vol. 1, pp. 136–140 (2011)
125. Kuncheva, L.I.: Fuzzy rough sets: application to feature selection. Fuzzy Sets Syst. **51**(2), 147–153 (1992)

126. Lai, J.Z., Juan, E.Y., Lai, F.J.: Rough clustering using generalized fuzzy clustering algorithm. Pattern Recogn. **46**(9), 2538–2547 (2013)
127. Lee, S.C., Huang, M.J.: Applying ai technology and rough set theory for mining association rules to support crime management and fire-fighting resources allocation. J. Inf. Technol. Soc. **2**(65), 65–78 (2002)
128. Lenarcik, A., Piasta, Z.: Discretization of condition attributes space. In: Intelligent Decision Support, pp. 373–389. Springer (1992)
129. Leung, Y., Fischer, M.M., Wu, W.Z., Mi, J.S.: A rough set approach for the discovery of classification rules in interval-valued information systems. Int. J. Approximate Reasoning **47**(2), 233–246 (2008)
130. Li, F., Ye, M., Chen, X.: An extension to rough c-means clustering based on decision-theoretic rough sets model. Int. J. Approximate Reasoning **55**(1), 116–129 (2014)
131. Li, H., Li, D., Zhai, Y., Wang, S., Zhang, J.: A variable precision attribute reduction approach in multilabel decision tables. Sci. World J. **2014** (2014)
132. Li, H., Zhang, L., Huang, B., Zhou, X.: Sequential three-way decision and granulation for cost-sensitive face recognition. Knowl. Based Syst. **91**, 241–251 (2016)
133. Li, H., Zhou, X., Zhao, J., Liu, D.: Non-monotonic attribute reduction in decision-theoretic rough sets. Fundamenta Informaticae **126**(4), 415–432 (2013)
134. Li, J., Cercone, N.: A rough set based model to rank the importance of association rules. In: International Workshop on Rough Sets, Fuzzy Sets, Data Mining, and Granular-Soft Computing, pp. 109–118. Springer (2005)
135. Li, M., Deng, S., Wang, L., Feng, S., Fan, J.: Hierarchical clustering algorithm for categorical data using a probabilistic rough set model. Knowl. Based Syst. **65**, 60–71 (2014)
136. Li, M., Shang, C., Feng, S., Fan, J.: Quick attribute reduction in inconsistent decision tables. Inf. Sci. **254**, 155–180 (2014)
137. Li, S., Li, T., Liu, D.: Dynamic maintenance of approximations in dominance-based rough set approach under the variation of the object set. Int. J. Intell. Syst. **28**(8), 729–751 (2013)
138. Li, S., Li, T., Liu, D.: Incremental updating approximations in dominance-based rough sets approach under the variation of the attribute set. Knowl. Based Syst. **40**, 17–26 (2013)
139. Li, T., Ruan, D., Geert, W., Song, J., Xu, Y.: A rough sets based characteristic relation approach for dynamic attribute generalization in data mining. Knowl. Based Syst. **20**(5), 485–494 (2007)
140. Liang, J., Wang, F., Dang, C., Qian, Y.: A group incremental approach to feature selection applying rough set technique. IEEE Trans. Knowl. Data Eng. **26**(2), 294–308 (2014)
141. Lin, T.Y., Yao, Y.Y., Zadeh, L.A.: Data mining, rough sets and granular computing. Physica **95** (2013)
142. Lingras, P.: Unsupervised rough set classification using gas. J. Intell. Inf. Syst. **16**(3), 215–228 (2001)
143. Lingras, P., Chen, M., Miao, D.: Rough cluster quality index based on decision theory. IEEE Trans. Knowl. Data Eng. **21**(7), 1014–1026 (2009)
144. Lingras, P., Chen, M., Miao, D.: Qualitative and quantitative combinations of crisp and rough clustering schemes using dominance relations. Int. J. Approximate Reasoning **55**(1), 238–258 (2014)
145. Lingras, P., West, C.: Interval set clustering of web users with rough k-means. J. Intell. Inf. Syst. **23**(1), 5–16 (2004)
146. Liu, D., Li, T., Liu, G., Hu, P.: An approach for inducing interesting incremental knowledge based on the change of attribute values. In: IEEE International Conference on Granular Computing, 2009, GRC'09, pp. 415–418. IEEE (2009)
147. Liu, D., Li, T., Ruan, D., Zhang, J.: Incremental learning optimization on knowledge discovery in dynamic business intelligent systems. J. Glob. Optim. **51**(2), 325–344 (2011)
148. Liu, D., Li, T., Ruan, D., Zou, W.: An incremental approach for inducing knowledge from dynamic information systems. Fundamenta Informaticae **94**(2), 245–260 (2009)
149. Liu, D., Li, T., Zhang, J.: A rough set-based incremental approach for learning knowledge in dynamic incomplete information systems. Int. J. Approximate Reasoning **55**(8), 1764–1786 (2014)

150. Liu, D., Li, T., Zhang, J.: Incremental updating approximations in probabilistic rough sets under the variation of attributes. Knowl. Based Syst. **73**, 81–96 (2015)
151. Liu, D., Liang, D.: Incremental learning researches on rough set theory: status and future. Int. J. Rough Sets Data Anal. (IJRSDA) **1**(1), 99–112 (2014)
152. Liu, J., Hu, Q., Yu, D.: A comparative study on rough set based class imbalance learning. Knowl. Based Syst. **21**(8), 753–763 (2008)
153. Liu, J., Hu, Q., Yu, D.: A weighted rough set based method developed for class imbalance learning. Inf. Sci. **178**(4), 1235–1256 (2008)
154. Liu, Y., Xu, C., Zhang, Q., Pan, Y.: Rough rule extracting from various conditions: Incremental and approximate approaches for inconsistent data. Fundamenta Informaticae **84**(3, 4), 403–427 (2008)
155. Lu, J., Tan, Y.P.: Cost-sensitive subspace analysis and extensions for face recognition. IEEE Trans. Inf. Forensics Secur. **8**(3), 510–519 (2013)
156. Luo, C., Li, T., Chen, H., Liu, D.: Incremental approaches for updating approximations in set-valued ordered information systems. Knowl. Based Syst. **50**, 218–233 (2013)
157. Luo, C., Li, T., Yi, Z., Fujita, H.: Matrix approach to decision-theoretic rough sets for evolving data. Knowl. Based Syst. **99**, 123–134 (2016)
158. Ma, T., Tang, M.: Weighted rough set model. In: Sixth International Conference on Intelligent Systems Design and Applications, vol. 1, pp. 481–485. IEEE (2006)
159. Maji, P., Garai, P.: Fuzzy-rough simultaneous attribute selection and feature extraction algorithm. IEEE Trans. Cybern. **43**(4), 1166–1177 (2013)
160. Maji, P., Pal, S.K.: RFCM: a hybrid clustering algorithm using rough and fuzzy sets. Fundamenta Informaticae **80**(4), 475–496 (2007)
161. Mak, B., Munakata, T.: Rule extraction from expert heuristics: a comparative study of rough sets with neural networks and ID3. Eur. J. Oper. Res. **136**(1), 212–229 (2002)
162. Miao, D., Chen, M., Wei, Z., Duan, Q.: A reasonable rough approximation for clustering web users. In: International Workshop on Web Intelligence Meets Brain Informatics, pp. 428–442. Springer (2006)
163. Min, F., He, H., Qian, Y., Zhu, W.: Test-cost-sensitive attribute reduction. Inf. Sci. **181**(22), 4928–4942 (2011)
164. Min, F., Hu, Q., Zhu, W.: Feature selection with test cost constraint. Int. J. Approximate Reasoning **55**(1), 167–179 (2014)
165. Min, F., Liu, Q.: A hierarchical model for test-cost-sensitive decision systems. Inf. Sci. **179**(14), 2442–2452 (2009)
166. Min, F., Zhu, W.: Attribute reduction of data with error ranges and test costs. Inf. Sci. **211**, 48–67 (2012)
167. Mirkin, B.: Mathematical classification and clustering: from how to what and why. In: Classification, Data Analysis, and Data Highways, pp. 172–181. Springer (1998)
168. Mitra, S.: An evolutionary rough partitive clustering. Pattern Recogn. Lett. **25**(12), 1439–1449 (2004)
169. Mitra, S., Banka, H.: Application of rough sets in pattern recognition. In: Transactions on Rough Sets VII, pp. 151–169. Springer (2007)
170. Mitra, S., Banka, H., Pedrycz, W.: Rough-fuzzy collaborative clustering. IEEE Trans. Syst. Man, Cybern. Part B (Cybern.) **36**(4), 795–805 (2006)
171. Mitra, S., Barman, B.: Rough-fuzzy clustering: an application to medical imagery. In: International Conference on Rough Sets and Knowledge Technology, pp. 300–307. Springer (2008)
172. Nanda, S., Majumdar, S.: Fuzzy rough sets. Fuzzy Sets Syst. **45**(2), 157–160 (1992)
173. Nguyen, H.S.: Discretization problem for rough sets methods. In: International Conference on Rough Sets and Current Trends in Computing, pp. 545–552. Springer (1998)
174. Nguyen, H.S.: On efficient handling of continuous attributes in large data bases. Fundamenta Informaticae **48**(1), 61–81 (2001)
175. Orlowska, E.: Dynamic information systems. Institute of Computer Science, Polish Academy of Sciences (1981)

176. Ozawa, S., Pang, S., Kasabov, N.: Incremental learning of chunk data for online pattern classification systems. IEEE Trans. Neural Netw. **19**(6), 1061–1074 (2008)
177. Park, I.K., Choi, G.S.: Rough set approach for clustering categorical data using information-theoretic dependency measure. Inf. Syst. **48**, 289–295 (2015)
178. Parmar, D., Wu, T., Blackhurst, J.: MMR: an algorithm for clustering categorical data using rough set theory. Data Knowl Eng. **63**(3), 879–893 (2007)
179. Pawlak, Z.: Rough sets. Int. J. Comput. Inf. Sci. **11**(5), 341–356 (1982)
180. Pawlak, Z.: Rough sets and intelligent data analysis. Inf. Sci. **147**(1), 1–12 (2002)
181. Pawlak, Z., Skowron, A.: Rough sets: some extensions. Inf. Sci. **177**(1), 28–40 (2007)
182. Pawlak, Z., Wong, S.K.M., Ziarko, W.: Rough sets: probabilistic versus deterministic approach. Int. J. Man-Mach. Stud. **29**(1), 81–95 (1988)
183. Pedrycz, W.: Granular Computing: An Emerging Paradigm, vol. 70. Springer Science & Business Media (2001)
184. Peters, G.: Outliers in rough k-means clustering. In: International Conference on Pattern Recognition and Machine Intelligence, pp. 702–707. Springer (2005)
185. Peters, G.: Some refinements of rough k-means clustering. Pattern Recogn. **39**(8), 1481–1491 (2006)
186. Peters, G.: Rough clustering utilizing the principle of indifference. Inf. Sci. **277**, 358–374 (2014)
187. Peters, G.: Is there any need for rough clustering? Pattern Recogn. Lett. **53**, 31–37 (2015)
188. Peters, G., Crespo, F., Lingras, P., Weber, R.: Soft clustering-fuzzy and rough approaches and their extensions and derivatives. Int. J. Approximate Reasoning **54**(2), 307–322 (2013)
189. Peters, G., Lampart, M., Weber, R.: Evolutionary rough k-medoid clustering. In: Transactions on Rough Sets VIII, pp. 289–306. Springer (2008)
190. Peters, G., Weber, R., Nowatzke, R.: Dynamic rough clustering and its applications. Appl. Soft Comput. **12**(10), 3193–3207 (2012)
191. Pradeepa, A., Selvadoss ThanamaniLee, A.: Hadoop file system and fundamental concept of mapreduce interior and closure rough set approximations. Int. J. Adv. Res. Comput. Commun. Eng. **2** (2013)
192. do Prado, H.A., Engel, P.M., Chaib Filho, H.: Rough clustering: an alternative to find meaningful clusters by using the reducts from a dataset. In: International Conference on Rough Sets and Current Trends in Computing, pp. 234–238. Springer (2002)
193. Qian, Y., Wang, Q., Cheng, H., Liang, J., Dang, C.: Fuzzy-rough feature selection accelerator. Fuzzy Sets Syst. **258**, 61–78 (2015)
194. Ramentol, E., Caballero, Y., Bello, R., Herrera, F.: Smote-rsb*: a hybrid preprocessing approach based on oversampling and undersampling for high imbalanced data-sets using smote and rough sets theory. Knowl. Inf. Syst. **33**(2), 245–265 (2012)
195. Riza, L.S., Janusz, A., Bergmeir, C., Cornelis, C., Herrera, F., Śle, D., Benítez, J.M., et al.: Implementing algorithms of rough set theory and fuzzy rough set theory in the R package "roughsets". Inf. Sci. **287**, 68–89 (2014)
196. Salamó, M., López-Sánchez, M.: Rough set based approaches to feature selection for case-based reasoning classifiers. Pattern Recogn. Lett. **32**(2), 280–292 (2011)
197. Salido, J.F., Murakami, S.: Rough set analysis of a general type of fuzzy data using transitive aggregations of fuzzy similarity relations. Fuzzy Sets Syst. **139**(3), 635–660 (2003)
198. Schapire, R.E., Singer, Y.: Boostexter: a boosting-based system for text categorization. Mach. Learn. **39**(2–3), 135–168 (2000)
199. Shan, N., Ziarko, W.: Data-based acquisition and incremental modification of classification rules. Comput. Intell. **11**(2), 357–370 (1995)
200. Shen, F., Yu, H., Kamiya, Y., Hasegawa, O.: An online incremental semi-supervised learning method. JACIII **14**(6), 593–605 (2010)
201. Shen, Q., Chouchoulas, A.: Combining rough sets and data-driven fuzzy learning for generation of classification rules. Pattern Recogn. **32**(12), 2073–2076 (1999)
202. Shen, Q., Chouchoulas, A.: A modular approach to generating fuzzy rules with reduced attributes for the monitoring of complex systems. Eng. Appl. Artif. Intell. **13**(3), 263–278 (2000)

203. Shen, Q., Jensen, R.: Selecting informative features with fuzzy-rough sets and its application for complex systems monitoring. Pattern Recogn. **37**(7), 1351–1363 (2004)
204. Shu, W., Shen, H.: Incremental feature selection based on rough set in dynamic incomplete data. Pattern Recogn. **47**(12), 3890–3906 (2014)
205. Singh, G.K., Minz, S.: Discretization using clustering and rough set theory. In: International Conference on Computing: Theory and Applications, 2007. ICCTA'07, pp. 330–336. IEEE (2007)
206. Skowron, A., Rauszer, C.: The discernibility matrices and functions in information systems. In: Intelligent Decision Support, pp. 331–362. Springer (1992)
207. Skowron, A., Stepaniuk, J.: Tolerance approximation spaces. Fundamenta Informaticae **27**(2, 3), 245–253 (1996)
208. Slezak, D.: Approximate bayesian networks. In: Technologies for Constructing Intelligent Systems 2, pp. 313–325. Springer (2002)
209. Ślezak, D.: Approximate entropy reducts. Fundamenta Informaticae **53**(3–4), 365–390 (2002)
210. Slezak, D., Ziarko, W., et al.: The investigation of the bayesian rough set model. Int. J. Approximate Reasoning **40**(1), 81–91 (2005)
211. Slimani, T.: Class association rules mining based rough set method. arXiv preprint arXiv:1509.05437 (2015)
212. Slowinski, R., Vanderpooten, D., et al.: A generalized definition of rough approximations based on similarity. IEEE Trans. Knowl. Data Eng. **12**(2), 331–336 (2000)
213. Soni, R., Nanda, R.: Neighborhood clustering of web users with rough k-means. In: Proceedings of 6th WSEAS International Conference on Circuits, Systems, Electronics, Control & Signal Processing, pp. 570–574 (2007)
214. Stefanowski, J.: The rough set based rule induction technique for classification problems. In: In Proceedings of 6th European Conference on Intelligent Techniques and Soft Computing EUFIT, vol. 98 (1998)
215. Stefanowski, J.: On combined classifiers, rule induction and rough sets. In: Transactions on Rough Sets VI, pp. 329–350. Springer (2007)
216. Stefanowski, J., Vanderpooten, D.: Induction of decision rules in classification and discovery-oriented perspectives. Int. J. Intell. Syst. **16**(1), 13–27 (2001)
217. Stefanowski, J., Wilk, S.: Rough sets for handling imbalanced data: combining filtering and rule-based classifiers. Fundamenta Informaticae **72**(1–3), 379–391 (2006)
218. Stefanowski, J., Wilk, S.: Extending rule-based classifiers to improve recognition of imbalanced classes. In: Advances in Data Management, pp. 131–154. Springer (2009)
219. Su, C.T., Hsu, J.H.: An extended Chi2 algorithm for discretization of real value attributes. IEEE Trans. Knowl. Data Eng. **17**(3), 437–441 (2005)
220. Su, C.T., Hsu, J.H.: Precision parameter in the variable precision rough sets model: an application. Omega **34**(2), 149–157 (2006)
221. Susmaga, R.: Reducts and constructs in classic and dominance-based rough sets approach. Inf. Sci. **271**, 45–64 (2014)
222. Świniarski, R.W.: Rough sets methods in feature reduction and classification. Int. J. Appl. Math. Comput. Sci. **11**(3), 565–582 (2001)
223. Swiniarski, R.W., Skowron, A.: Rough set methods in feature selection and recognition. Pattern Recogn. Lett. **24**(6), 833–849 (2003)
224. Tay, F.E., Shen, L.: Economic and financial prediction using rough sets model. Eur. J. Oper. Res. **141**(3), 641–659 (2002)
225. Tsang, E.C., Chen, D., Yeung, D.S., Wang, X.Z., Lee, J.W.: Attributes reduction using fuzzy rough sets. IEEE Trans. Fuzzy Syst. **16**(5), 1130–1141 (2008)
226. Tsoumakas, G., Katakis, I.: Multi-label classification: an overview. Aristotle University of Thessaloniki, Greece, Deparment of Informatics (2006)
227. Tsoumakas, G., Vlahavas, I.: Random k-labelsets: an ensemble method for multilabel classification. In: European Conference on Machine Learning, pp. 406–417. Springer (2007)
228. Tsumoto, S.: Automated extraction of medical expert system rules from clinical databases based on rough set theory. Inf. Sci. **112**(1), 67–84 (1998)

229. Tsumoto, S.: Automated extraction of hierarchical decision rules from clinical databases using rough set model. Expert Syst. Appl. **24**(2), 189–197 (2003)
230. Tsumoto, S.: Incremental rule induction based on rough set theory. In: International Symposium on Methodologies for Intelligent Systems, pp. 70–79. Springer (2011)
231. Vanderpooten, D.: Similarity relation as a basis for rough approximations. Adv. Mach. Intell. Soft Comput. **4**, 17–33 (1997)
232. Verbiest, N.: Fuzzy rough and evolutionary approaches to instance selection. Ph.D. thesis, Ghent University (2014)
233. Verbiest, N., Cornelis, C., Herrera, F.: FRPS: a fuzzy rough prototype selection method. Pattern Recogn. **46**(10), 2770–2782 (2013)
234. Vilalta, R., Drissi, Y.: A perspective view and survey of meta-learning. Artif. Intell. Rev. **18**(2), 77–95 (2002)
235. Voges, K., Pope, N., Brown, M.: A rough cluster analysis of shopping orientation data. In: Proceedings Australian and New Zealand Marketing Academy Conference, Adelaide, pp. 1625–1631 (2003)
236. Voges, K.E., Pope, N., Brown, M.R.: Cluster analysis of marketing data examining on-line shopping orientation: a comparison of k-means and rough clustering approaches. In: Heuristics and Optimization for Knowledge Discovery, pp. 207–224 (2002)
237. Wang, F., Liang, J., Dang, C.: Attribute reduction for dynamic data sets. Applied Soft Computing **13**(1), 676–689 (2013)
238. Wang, F., Liang, J., Qian, Y.: Attribute reduction: a dimension incremental strategy. Knowl. Based Syst. **39**, 95–108 (2013)
239. Wang, G., Yu, H., Li, T., et al.: Decision region distribution preservation reduction in decision-theoretic rough set model. Inf. Sci. **278**, 614–640 (2014)
240. Wang, X., An, S., Shi, H., Hu, Q.: Fuzzy rough decision trees for multi-label classification. In: Rough Sets, Fuzzy Sets, Data Mining, and Granular Computing, pp. 207–217. Springer (2015)
241. Wang, X., Yang, J., Peng, N., Teng, X.: Finding minimal rough set reducts with particle swarm optimization. In: International Workshop on Rough Sets, Fuzzy Sets, Data Mining, and Granular-Soft Computing, pp. 451–460. Springer (2005)
242. Wang, X., Yang, J., Teng, X., Xia, W., Jensen, R.: Feature selection based on rough sets and particle swarm optimization. Pattern Recogn. Lett. **28**(4), 459–471 (2007)
243. Wei, M.H., Cheng, C.H., Huang, C.S., Chiang, P.C.: Discovering medical quality of total hip arthroplasty by rough set classifier with imbalanced class. Qual. Quant. **47**(3), 1761–1779 (2013)
244. Wojna, A.: Constraint based incremental learning of classification rules. In: International Conference on Rough Sets and Current Trends in Computing, pp. 428–435. Springer (2000)
245. Wróblewski, J.: Finding minimal reducts using genetic algorithms. In: Proceedings of the Second Annual Join Conference on Information Science, pp. 186–189 (1995)
246. Wróblewski, J.: Theoretical foundations of order-based genetic algorithms. Fundamenta Informaticae **28**(3, 4), 423–430 (1996)
247. Wróblewski, J.: Ensembles of classifiers based on approximate reducts. Fundamenta Informaticae **47**(3–4), 351–360 (2001)
248. Wu, Q., Bell, D.: Multi-knowledge extraction and application. In: International Workshop on Rough Sets, Fuzzy Sets, Data Mining, and Granular-Soft Computing, pp. 274–278. Springer (2003)
249. Xie, H., Cheng, H.Z., Niu, D.X.: Discretization of continuous attributes in rough set theory based on information entropy. Chin. J. Comput. Chin. Ed. **28**(9), 1570 (2005)
250. Xu, Y., Wang, L., Zhang, R.: A dynamic attribute reduction algorithm based on 0–1 integer programming. Knowl. Based Syst. **24**(8), 1341–1347 (2011)
251. Xu, Z., Liang, J., Dang, C., Chin, K.: Inclusion degree: a perspective on measures for rough set data analysis. Inf. Sci. **141**(3), 227–236 (2002)
252. Yang, Q., Ling, C., Chai, X., Pan, R.: Test-cost sensitive classification on data with missing values. IEEE Trans. Knowl. Data Eng. **18**(5), 626–638 (2006)

253. Yang, X., Qi, Y., Song, X., Yang, J.: Test cost sensitive multigranulation rough set: model and minimal cost selection. Inf. Sci. **250**, 184–199 (2013)
254. Yang, X., Qi, Y., Yu, H., Song, X., Yang, J.: Updating multigranulation rough approximations with increasing of granular structures. Knowl. Based Syst. **64**, 59–69 (2014)
255. Yang, Y., Chen, D., Dong, Z.: Novel algorithms of attribute reduction with variable precision rough set model. Neurocomputing **139**, 336–344 (2014)
256. Yang, Y., Chen, Z., Liang, Z., Wang, G.: Attribute reduction for massive data based on rough set theory and mapreduce. In: International Conference on Rough Sets and Knowledge Technology, pp. 672–678. Springer (2010)
257. Yao, J., Yao, Y.: A granular computing approach to machine learning. FSKD **2**, 732–736 (2002)
258. Yao, Y.: Combination of rough and fuzzy sets based on α-level sets. In: Rough sets and Data Mining, pp. 301–321. Springer (1997)
259. Yao, Y.: Decision-theoretic rough set models. In: International Conference on Rough Sets and Knowledge Technology, pp. 1–12. Springer (2007)
260. Yao, Y.: Three-way decision: an interpretation of rules in rough set theory. In: International Conference on Rough Sets and Knowledge Technology, pp. 642–649. Springer (2009)
261. Yao, Y.: Three-way decisions with probabilistic rough sets. Inf. Sci. **180**(3), 341–353 (2010)
262. Yao, Y.: The superiority of three-way decisions in probabilistic rough set models. Inf. Sci. **181**(6), 1080–1096 (2011)
263. Yao, Y.: An outline of a theory of three-way decisions. In: International Conference on Rough Sets and Current Trends in Computing, pp. 1–17. Springer (2012)
264. Yao, Y., Greco, S., Słowiński, R.: Probabilistic rough sets. In: Springer Handbook of Computational Intelligence, pp. 387–411. Springer (2015)
265. Yao, Y., Zhao, Y.: Attribute reduction in decision-theoretic rough set models. Inf. Sci. **178**(17), 3356–3373 (2008)
266. Yao, Y., Zhao, Y., Maguire, R.B.: Explanation oriented association mining using rough set theory. In: International Workshop on Rough Sets, Fuzzy Sets, Data Mining, and Granular-Soft Computing, pp. 165–172. Springer (2003)
267. Yao, Y., Zhou, B.: Two bayesian approaches to rough sets. Eur. J. Oper. Res. **251**(3), 904–917 (2016)
268. Ye, D., Chen, Z., Ma, S.: A novel and better fitness evaluation for rough set based minimum attribute reduction problem. Inf. Sci. **222**, 413–423 (2013)
269. Yong, L., Congfu, X., Yunhe, P.: An incremental rule extracting algorithm based on pawlak reduction. In: 2004 IEEE International Conference on Systems, Man and Cybernetics, vol. 6, pp. 5964–5968. IEEE (2004)
270. Yong, L., Wenliang, H., Yunliang, J., Zhiyong, Z.: Quick attribute reduct algorithm for neighborhood rough set model. Inf. Sci. **271**, 65–81 (2014)
271. Yu, H., Chu, S., Yang, D.: Autonomous knowledge-oriented clustering using decision-theoretic rough set theory. Fundamenta Informaticae **115**(2–3), 141–156 (2012)
272. Yu, H., Liu, Z., Wang, G.: An automatic method to determine the number of clusters using decision-theoretic rough set. Int. J. Approximate Reasoning **55**(1), 101–115 (2014)
273. Yu, H., Su, T., Zeng, X.: A three-way decisions clustering algorithm for incomplete data. In: International Conference on Rough Sets and Knowledge Technology, pp. 765–776. Springer (2014)
274. Yu, H., Wang, G., Lan, F.: Solving the attribute reduction problem with ant colony optimization. In: Transactions on Rough Sets XIII, pp. 240–259. Springer (2011)
275. Yu, H., Wang, Y.: Three-way decisions method for overlapping clustering. In: International Conference on Rough Sets and Current Trends in Computing, pp. 277–286. Springer (2012)
276. Yu, H., Wang, Y., Jiao, P.: A three-way decisions approach to density-based overlapping clustering. In: Transactions on Rough Sets XVIII, pp. 92–109. Springer (2014)
277. Yu, H., Zhang, C., Hu, F.: An incremental clustering approach based on three-way decisions. In: International Conference on Rough Sets and Current Trends in Computing, pp. 152–159. Springer (2014)

278. Yu, Y., Miao, D., Zhang, Z., Wang, L.: Multi-label classification using rough sets. In: International Workshop on Rough Sets, Fuzzy Sets, Data Mining, and Granular-Soft Computing, pp. 119–126. Springer (2013)
279. Yu, Y., Pedrycz, W., Miao, D.: Multi-label classification by exploiting label correlations. Expert Syst. Appl. **41**(6), 2989–3004 (2014)
280. Zhai, J., Zhang, S., Zhang, Y.: An extension of rough fuzzy set. J. Intell. Fuzzy Syst. (Preprint), 1–10 (2016)
281. Zhai, J., Zhang, Y., Zhu, H.: Three-way decisions model based on tolerance rough fuzzy set. Int. J. Mach. Learn. Cybern. 1–9 (2016)
282. Zhang, H.R., Min, F.: Three-way recommender systems based on random forests. Knowl. Based Syst. **91**, 275–286 (2016)
283. Zhang, J., Li, T., Chen, H.: Composite rough sets. In: International Conference on Artificial Intelligence and Computational Intelligence, pp. 150–159. Springer (2012)
284. Zhang, J., Li, T., Chen, H.: Composite rough sets for dynamic data mining. Inf. Sci. **257**, 81–100 (2014)
285. Zhang, J., Li, T., Pan, Y.: Parallel rough set based knowledge acquisition using mapreduce from big data. In: Proceedings of the 1st International Workshop on Big Data, Streams and Heterogeneous Source Mining: Algorithms, Systems, Programming Models and Applications, pp. 20–27. ACM (2012)
286. Zhang, J., Li, T., Ruan, D., Gao, Z., Zhao, C.: A parallel method for computing rough set approximations. Inf. Sci. **194**, 209–223 (2012)
287. Zhang, J., Li, T., Ruan, D., Liu, D.: Rough sets based matrix approaches with dynamic attribute variation in set-valued information systems. Int. J. Approximate Reasoning **53**(4), 620–635 (2012)
288. Zhang, L., Hu, Q., Duan, J., Wang, X.: Multi-label feature selection with fuzzy rough sets. In: International Conference on Rough Sets and Knowledge Technology, pp. 121–128. Springer (2014)
289. Zhang, L., Li, H., Zhou, X., Huang, B., Shang, L.: Cost-sensitive sequential three-way decision for face recognition. In: International Conference on Rough Sets and Intelligent Systems Paradigms, pp. 375–383. Springer (2014)
290. Zhang, M.L., Zhou, Z.H.: Ml-knn: A lazy learning approach to multi-label learning. Pattern Recogn. **40**(7), 2038–2048 (2007)
291. Zhang, T., Chen, L., Ma, F.: An improved algorithm of rough k-means clustering based on variable weighted distance measure. Int. J. Database Theory Appl. **7**(6), 163–174 (2014)
292. Zhang, T., Chen, L., Ma, F.: A modified rough c-means clustering algorithm based on hybrid imbalanced measure of distance and density. Int. J. Approximate Reasoning **55**(8), 1805–1818 (2014)
293. Zhang, X., Miao, D.: Three-way weighted entropies and three-way attribute reduction. In: International Conference on Rough Sets and Knowledge Technology, pp. 707–719. Springer (2014)
294. Zhang, Y., Zhou, Z.H.: Cost-sensitive face recognition. IEEE Trans. Pattern Anal. Mach. Intell. **32**(10), 1758–1769 (2010)
295. Zhao, H., Min, F., Zhu, W.: Test-cost-sensitive attribute reduction based on neighborhood rough set. In: 2011 IEEE International Conference on Granular Computing (GrC), pp. 802–806. IEEE (2011)
296. Zhao, H., Wang, P., Hu, Q.: Cost-sensitive feature selection based on adaptive neighborhood granularity with multi-level confidence. Inf. Sci. **366**, 134–149 (2016)
297. Zhao, M., Luo, K., Liao, X.X.: Rough set attribute reduction algorithm based on immune genetic algorithm. Jisuanji Gongcheng yu Yingyong (Comput. Eng. Appl.) **42**(23), 171–173 (2007)
298. Zhao, S., Chen, H., Li, C., Du, X., Sun, H.: A novel approach to building a robust fuzzy rough classifier. IEEE Trans. Fuzzy Syst. **23**(4), 769–786 (2015)
299. Zhao, S., Tsang, E.C., Chen, D.: The model of fuzzy variable precision rough sets. IEEE Trans. Fuzzy Syst. **17**(2), 451–467 (2009)

300. Zhao, S., Tsang, E.C., Chen, D., Wang, X.: Building a rule-based classifier–a fuzzy-rough set approach. IEEE Trans. Knowl. Data Eng. **22**(5), 624–638 (2010)
301. Zheng, Z., Wang, G., Wu, Y.: A rough set and rule tree based incremental knowledge acquisition algorithm. In: International Workshop on Rough Sets, Fuzzy Sets, Data Mining, and Granular-Soft Computing, pp. 122–129. Springer (2003)
302. Zhong, N., Dong, J., Ohsuga, S.: Using rough sets with heuristics for feature selection. J. Intell. Inf. Syst. **16**(3), 199–214 (2001)
303. Zhou, Z.H.: Cost-sensitive learning. In: International Conference on Modeling Decisions for Artificial Intelligence, pp. 17–18. Springer (2011)
304. Zhou, Z.H., Liu, X.Y.: Training cost-sensitive neural networks with methods addressing the class imbalance problem. IEEE Trans. Knowl. Data Eng. **18**(1), 63–77 (2006)
305. Zhu, W.: Generalized rough sets based on relations. Inf. Sci. **177**(22), 4997–5011 (2007)
306. Zhu, W.: Topological approaches to covering rough sets. Inf. Sci. **177**(6), 1499–1508 (2007)
307. Ziarko, W.: Variable precision rough set model. J. Comput. Syst. Sci. **46**(1), 39–59 (1993)
308. Zou, W., Li, T., Chen, H., Ji, X.: Approaches for incrementally updating approximations based on set-valued information systems while attribute values' coarsening and refining. In: 2009 IEEE International Conference on Granular Computing (2009)

Application of Tolerance Rough Sets in Structured and Unstructured Text Categorization: A Survey

Sheela Ramanna, James Francis Peters and Cenker Sengoz

Abstract Text categorization or labelling methods assign unseen documents or unknown linguistic entities to pre-defined categories or labels. This is an essential preprocessing step in web mining. Text categorization is popularly referred to as document classification/clustering. In this chapter, we present a survey of literature, where tolerance rough set model (TRSM) is used as a text categorization and learning model. The approach taken is to consider tolerance relations instead of equivalence relations where the binary relation is both symmetric and reflexive but not transitive. A very brief overview of the history of tolerance rough sets from an axiomatic point of view is also presented. Various approaches to text categorization of both structured information such as documents as well as unstructured information such as nouns and relations based on TRSM are presented. This survey is meant to demonstrate the versatility of the tolerance form of rough sets and its successful application in text categorization and labelling.

1 Introduction

Rough Set theory was introduced by Zdzisław Pawlak during the early 1980s [1] (elaborated in [2–6]) as a mathematical framework for reasoning about ill-defined objects. It is of fundamental importance to artificial intelligence (AI) and cognitive sciences, especially in the areas of machine learning, knowledge acquisition, decision analysis, knowledge discovery from databases, expert systems, decision support

S. Ramanna (✉)
University of Winnipeg, Applied Computer Science Winnipeg, Winnipeg, MB, Canada
e-mail: s.ramanna@uwinnipeg.ca

J.F. Peters
Computational Intelligence Laboratory Winnipeg, Department of Electrical and Computer Engineering, University of Manitoba, Winnipeg, MB, Canada
e-mail: james.peters3@umanitoba.ca

C. Sengoz
Mexia Interactive, Winnipeg, MB R3B 2E9, Canada
e-mail: cenker.sengoz@gmail.com

G. Wang et al. (eds.), *Thriving Rough Sets*, Studies in Computational Intelligence 708, DOI 10.1007/978-3-319-54966-8_6

systems, inductive reasoning, and pattern recognition [5]. Considerable work has been done on combining rough sets and tolerance relations [7] to obtain a realistic model (see for ex: [8–11]) leading to the tolerance rough sets model (TRSM). Tolerance relations provide the most general tool for studying indiscernibility phenomena [10]. The idea of tolerance first appeared in Poincaré's work in 1905 [12] and can be traced back to his work on the similarity of sensations [13]. Subsequently, tolerance relations were considered by Zeeman [14] in his study of the topology of the brain where he was concerned with different cells of the brain which, when they are very *near*, are perceived as identical or indistinguishable. Tolerance spaces as a framework for studying the concept of resemblance was presented in [15] and in [13].

Text categorization with classical rough sets where the concept of indiscernibility (similarity) formed by an equivalence relation is plausible but too restrictive. The approach taken is to consider tolerance relations instead of equivalence relations where the binary relation is both symmetric and reflexive but not transitive. We use the broader term of text categorization in this work to admit both structured (documents) and unstructured (nouns and relations) information from web sources. Applications of TRSM can be found in (i) document clustering [16–19], (ii) document retrieval [20], an, (iii) unstructured text labelling [21].

The challenges with using TRSM for information representation and its subsequent use in information retrieval are multi-fold. Firstly, a thesaurus (which is a collection of documents) must be *constructed* using some tolerance value ε, since this forms the very basis for queries used to retrieve related documents. So the challenge is to determine the optimal value of ε. With unstructured text, a thesaurus (for nouns and relations) must be created. Specifically, in a semi-supervised setting, seed values for nouns (relations) are promoted based on some tolerance value ε. The second challenge is the *optimization* of a thesaurus with co-occurrence information between elements of the thesaurus. In this context, it means *constructing* some form of tolerance matrix. This phase involves the use of approximation (upper) operator which forms a basis of a weighting scheme (see Sect. 4). The third challenge is to employ appropriate retrieval methods such as clustering or semi-supervised algorithms. In this survey, we restrict our discussion to text categorization rather than information retrieval in general. The contribution of the chapter is to present both the foundations and applications of tolerance rough sets in text categorization in an attempt to interest the readers to the tolerance form of rough sets as a useful tool in mining both structured text such as documents as well as unstructured text such as nouns and relations.

The chapter is organized as follows: In Sect. 2, we first start with an informal presentation of classical and tolerance forms of rough sets followed by the formal definitions of rough and tolerance rough sets. In Sect. 4, TRSM as a document representation model is discussed. Section 5 presents various approaches to document clustering using TRSM. Section 6 presents a detailed discussion of TRSM as an unstructured information representation model.

2 Rough Sets

In classical rough sets theory, a universe of objects is *partitioned* into indiscernible classes (i.e. granules) by means of an indiscernibility relation. Indiscernible classes form basic granules of knowledge about the universe. Given a concept that is determined to be vague (not precise), this theory makes it possible to express the vague concept by a pair of precise concepts called the lower and the upper approximation. A vague concept is defined as a class (or decision) that cannot be properly classified. The difference between the upper and the lower approximation constitutes the boundary region of the vague concept. Hence, rough set theory expresses vagueness not by means of membership, but by employing a boundary region [5]. Figure 1 shows the regions that emerge with set approximation. The regions are depicted as squares only for the sake of illustration, but they can be of arbitrary shape. We should note that each granule can contain an arbitrary number of objects or may be empty.

However, there are some cases where the disjoint granules are not desired. Particularly, when it comes to natural language processing and information retrieval, overlapping classes would better fit to describe this universe and the desired outcome is shown in Fig. 2 first presented by Ho and Nguyen [18]. Consider the universe U of words {*account, agency, antecedent, backbone, backing, bottom, basis, cause, center, derivation, motive, root*} excerpted from Roget's thesaurus. Assume we would like to define an indiscernibility (equivalence) relation R over those words based on their semantic affinity. Each of those words seem to share a meaning with one or more of the concepts *Root, Cause* and *Basis* and their meanings are not transitive.

Since classical rough set theory relies on an equivalence relation $R \subseteq U \times U$ to approximate a target concept, R has the following 3 properties:

- Reflexivity: $(x, x) \in R, \forall x \in U$.
- Symmetry: $(x, y) \in R \Rightarrow (y, x) \in R, \forall x, y \in U$.
- Transitivity: $(x, y) \in R \wedge (y, z) \in R \Rightarrow (x, z) \in R, \forall x, y, z \in U$.

In practice, R partitions the universe into disjoint (non-overlapping) equivalence classes which are regarded as information granules. Particularly, when it comes to text categorization or document clustering, a non-transitive binary relation that is reflexive and symmetric is necessary.

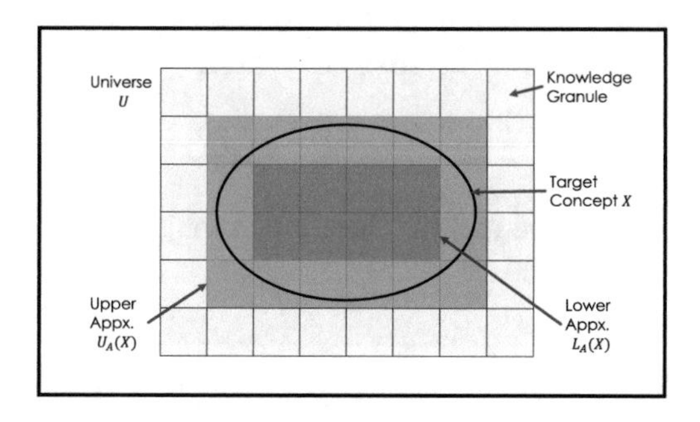

Fig. 1 Rough sets and set approximation

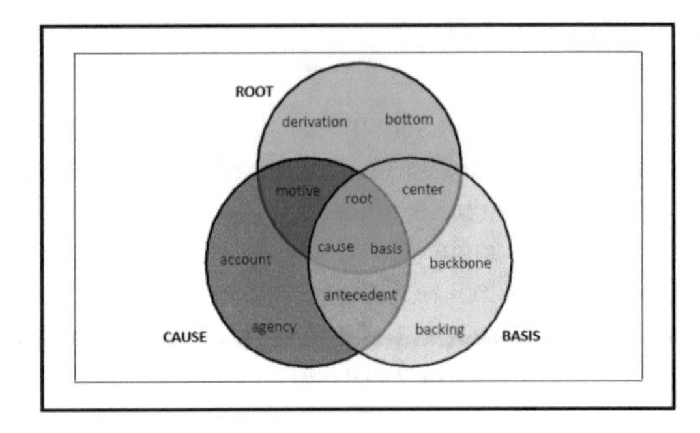

Fig. 2 Overlapping classes of words [18]

2.1 Formal Framework for Rough Sets

Let U be a finite, non-empty universe of objects and let $R \subseteq U \times U$ denote a binary relation on the universe U. R is called an *indiscernibility relation* and for rough sets, it has to be an *equivalence relation*. The pair $(U, R) = \mathcal{A}$ constitutes an *approximation space* $\mathcal{A}$. Assume we have $X \subseteq U$ as a target concept in this universe. Then the task is to create an approximated representation for X in U with the help of R. Let $[x]_R$ denote the indiscernibility class of x i.e. $y \in [x]_R \iff (x, y) \in R$. Then, every equivalence class forms a *granule* or *partition* which, as the name implies, contains objects that are indiscernible for this approximation space $\mathcal{A}$. Therefore, every single item in a granule is considered identical and inseparable. These granules are approximated by the following means:

- **Lower approximation**. Intuitively, these are the objects which *certainly* belong to X with respect to $\mathcal{A}$.

$$\mathcal{L}_{\mathcal{A}}(X) = \{x \in U : [x]_R \subseteq X\}.$$

- **Upper approximation**. Intuitively, these are the objects which *may* belong to X with respect to $\mathcal{A}$.

$$\mathcal{U}_{\mathcal{A}}(X) = \{x \in U : [x]_R \cap X \neq \emptyset\}.$$

These two approximations will also form the following two regions:

- **Boundary region**. These are the objects occurring in the upper approximation but not in lower approximation of X.

$$\mathcal{B}_{\mathcal{A}}(X) = \mathcal{U}_{\mathcal{A}}(X) - \mathcal{L}_{\mathcal{A}}(X).$$

- **Negative region**. These are the objects that certainly don't belong to X.

$$U - \mathcal{U}_{\mathcal{A}}(X).$$

with this framework, we end up with two different types of sets: a set X is called a *crisp set* if and only if $\mathcal{B}_{\mathcal{A}}(X) = \emptyset$. Otherwise, it is called a *rough set*. The pair $(\mathcal{U}_{\mathcal{A}}(X), \mathcal{L}_{\mathcal{A}}(X))$ forms the *rough approximation* for X.

3 Tolerance Rough Sets

What follows is a very brief overview of the history of tolerance rough sets from an axiomatic point of view leading upto the TRSM model that is currently used in text categorization.

3.1 History of Tolerance Rough Sets: An Axiomatic View

Nieminen [8] introduce what are known as *tolerance black boxes* where a tolerance box is defined as $B = (I, O, f)$ with I as a finite non-empty set of inputs and O as a finite non-empty set of outputs and $f : \mathbb{P}(I) \rightarrow O$. A black box B is a *tolerance black box* if there is a similarity relation $\sim$ on O such that $o_1 \sim o_2 \iff o_1$ and o_2 are approximately the same [8, §3, p. 295]. From a practical point of view, most real control systems are tolerance blackboxes. The authors give conditions for tolerance admissibility and goodness for tolerance black boxes similar to rough black boxes given in [9].

Polkowski et al. [10, §3, pp. 56–57] introduce a tolerance information and decision system as well as a theoretical adaptive decision algorithm to discover rules. In general, a tolerance information system denoted by $\mathcal{A} = (U, A, \tau)$ where τ is a tolerance relation on information function $Inf_B(x) = \{(a, a(x) : a \in B\}$ with $x \in U$ and $B \subseteq A$. In particular, a tolerance information system can be *realized* as a pair $(\mathcal{A}, D)$ where D is a discernibility relation which also satisfies the conditions for a tolerance relation (see [13] for the details). Formally, $D = (D_B)$ with $B \subseteq A$ and $D_B \subseteq INF(B) \times INF(B)$. The decision algorithm is termed *adaptive*, since a choice of tolerance τ (measure of degree of similarity) as well as information function Inf_B has a bearing on the performance of the algorithm, which can be tuned.

Marcus [11, §4, p. 475] introduce a tolerance form of rough sets where the emphasis is on attributes with unsharp borders (including real-valued attributes) which admits different degrees of indiscernibility. A tolerance rough set is defined as an ordered pair $\left(T^*(A), T_*(A)\right)$, where (X, τ) is a tolerance space and X is endowed with a tolerance relation τ. Let $A \subset X$. For any $x \in A$, $\tau(x)$ is its tolerance class. Here, tolerance rough sets are presented in the language of Čech topology. This characterization leads to an hierarchical interval of approximations and permits a learning process. Readers may refer to this chapter for an extensive bibliography dealing with early work on topology and the two forms of rough sets *equivalence and tolerance* [11, §12, pp. 483–486].

Skowron and Stepaniuk [22, §2, p. 3] generalize the notion of approximation space to tolerance approximation spaces and introduce a tolerance relation defined by an uncertainty function for attribute reduction with Boolean reasoning. Specifically, the authors generalize the rough membership function by introducing three functions: (i) vague inclusion function (μ), (ii) uncertainty function (I), (iii) and a structurality function (P). Formally a tolerance approximation space is defined as $\mathcal{A} = (U, I, \mu, P)$ and the attribute reduction process leads to tolerance reducts and relative tolerance reducts useful in decision-making.

3.2 Formal Framework for Tolerance Rough Sets

In order to define overlapping classes or granules, we need *tolerance relations*. Tolerance relations are reflexive and symmetric but they are not necessarily transitive so the classes induced by such relations can overlap. Let U be a finite, non-empty universe of objects, $\mathbb{P}(U)$ be the power set of U, and let $\mathcal{T} : U \rightarrow \mathbb{P}(U)$ be a binary relation such that $x\mathcal{T}y \iff y \in \mathcal{T}(x)$ holds for any $x, y \in U$. $\mathcal{T}$ implicitly defines a tolerance relation and $\mathcal{T}(x)$ defines a tolerance class of x. Then a tolerance membership function (also known as vague inclusion function) $\nu : \mathbb{P}(U) \times \mathbb{P}(U) \rightarrow [0, 1]$ is defined as $\nu(X, Y) = \frac{|X \cap Y|}{|X|}$. The lower and upper approximations of set X can be defined as

$$\mathcal{L}_{\mathcal{A}}(X) = \{x \in U : \frac{|\mathcal{T}(x) \cap X|}{|\mathcal{T}(x)|} = 1\},$$

$$\mathcal{U}_{\mathcal{A}}(X) = \{x \in U : \frac{|\mathcal{T}(x) \cap X|}{|\mathcal{T}(x)|} > 0\},$$

where $\mathcal{A}$ defines a simplified form of tolerance approximation space $\mathcal{A} = (U, \mathcal{T}, \nu)$ [22]. More recently, tolerance rough sets have been applied for pattern classification in classical machine learning data sets [23].

4 Structured Information Representation with TRSM

Structured information such as *documents* were first modeled using the tolerance form of rough sets by Kawasaki in 2000. In [16, 18, 19], the authors used TRSM for text clustering and document clustering/classification and to model relations between terms and documents. Briefly, TRSM introduces a vectorial representation of documents where each vector dimension corresponds to a term weight that is to be enhanced by means of rough sets and tolerance approximation, by relating terms across documents. This is useful particularly when each document is characterized by only a small number of terms along with many zero-valued entries in a high dimensional term vector space. So TRSM promises a richer representation for documents to be clustered.

Definition 1 Let $D = \{d_1, d_2, \dots, d_M\}$ be a set of documents and let $T = \{t_1, t_2, \dots, t_N\}$ be a universe of index terms that occur in those documents. Let each document d_j be represented as a weight vector of its terms $\hat{d}_j = \langle (t_1, w_{1j}), (t_2, w_{2j}), \dots, (t_N, w_{Nj}) \rangle$ where $w_{ij} \in [0, 1]$ shows the significance of term i in document j.

Let a query Q be a cluster representative in the form of $\hat{Q} = \langle (q_1, w_{1q}), (q_2, w_{2q}), \dots, (q_s, w_{sq}) \rangle$ where $q_i \in T$ and $w_{iq} \in [0, 1]$,

The task in hand is to find ordered documents $d_j \in D$ that are relevant to Q [18]. The tolerance approximation space $\mathcal{A} = (U, I, \nu, P)$ for documents is reconstituted as follows:

- The *universe* is the set of index terms: $U = T = \{t_1, t_2, \dots, t_N\}$.
- The *uncertainty function* $I \subseteq T \times T$ aims to capture the affinity amongst the terms and defines the tolerance class for each index term. It is based on a tolerance relation that binds two terms if they co-occur frequently across documents. So the function becomes

$$I_\theta(t_i) = \{t_j | f_D(t_i, t_j) \geq \theta\} \cup \{t_i\}.$$

  It is parametrized over a threshold value θ where $f_D(t_i, t_j)$ denotes the number of terms in which t_i and t_j co-occur. Note that $t_j \in I_\theta(t_i) \iff t_i I_\theta t_j$ and that I_θ is reflexive ($t_i \in I_\theta(t_i)$) and symmetric ($t_j \in I_\theta(t_i) \iff t_i \in I_\theta(t_j)$) for all $t_i, t_j \in T$, satisfying the tolerance relation requirements.
- The *vague inclusion function* is $\nu(X, Y) = \frac{|X \cap Y|}{|X|}$. It is monotonous w.r.t the second argument, as required. It can now be regarded as the membership function μ for term $t_i \in T$ to target concept $X \subseteq T$

$$\mu(t_i, X) = \nu(I_\theta(t_i), X) = \frac{|I_\theta(t_i) \cap X|}{|I_\theta(t_i)|},$$

  provided that T is a closed set and Q consists exclusively of terms from T, the structurality function is simply $P = 1$ for TRSM.

  The lower and upper approximations of X are defined as follows:

$$\mathcal{L}_\mathcal{A}(X) = \{t_i \in T : \frac{|I_\theta(t_i) \cap X|}{|I_\theta(t_i)|} = 1\},$$

$$\mathcal{U}_\mathcal{A}(X) = \{t_i \in T : \frac{|I_\theta(t_i) \cap X|}{|I_\theta(t_i)|} > 0\}.$$

Weight Adjustment via TF-IDF scheme:
Tolerance approximation is used to enhance the document representation by adjusting the term weights. In the absence of such enhancement, term weights are assigned

by using the term frequency-inverse document frequency (tf-idf) scheme

$$w_{ij}^{\text{tfidf}} = \begin{cases} (1 + log(f_{d_j}(t_i))) \times log\frac{M}{f_D(t_i)}, & \text{if } t_i \in d_j, \\ 0, & \text{if } t_i \notin (d_j), \end{cases}$$

where $f_{d_j}(t_i)$ denotes the number of times t_i occurs in d_j (term frequency) and $f_D(t_i)$ denotes the number of documents in D that accommodates t_i (document frequency) [18]. In such a model, a term t_i acquires a nonzero weight for $\hat{d}_j$ if and only if, it directly occurs in the document d_j.

Weight Adjustment via Tolerance Rough Sets:
TRSM uses the following weighting scheme which also takes boundary terms into account and assigns non-zero weights. Note that the upper approximation of a document $\mathcal{U}_A(d_j)$ covers the *tolerant* terms for all of its own terms as well, creating the enriched representation.

$$w_{ij}^{\text{trs}} = \begin{cases} (1 + log(f_{d_j}(t_i))) \times log\frac{M}{f_D(t_i)}, & \text{if } t_i \in d_j, \\ min_{t_h \in d_j} w_{hj} \times \frac{log(M/f_D(t_i))}{1+log(M/f_D(t_i))}, & \text{if } t_i \in \mathcal{U}_A(d_j) \backslash d_j, \\ 0, & \text{if } t_i \notin \mathcal{U}_A(d_j). \end{cases}$$

Weight Adjustment via Tolerance Rough Sets with linear neurons:
In this work [24, §3, p. 389], weights w_{ij}^{ln} for $t_j \in \mathcal{U}_A(d_i)$ are determined using a neural network (training a set of linear neurons) with the assumption that $\beta_{kj} > 0$.

$$w_{ij}^{\text{ln}} = \sum_{k=1}^{N} \delta(i,j,k)\beta_{kj}w_{ik},$$

where

$$\delta(i,j,k) = \begin{cases} 1, \text{ for } t_k \in d_i \wedge t_j \in I_\theta(t_k), \\ 0, \text{ otherwise,} \end{cases}$$

which is rewritten as

$$w_{ij}^{\text{ln}} = \sum_{k=1}^{N} \delta(i,j,k)e^{\alpha_{kj}}w_{ik}.$$

The problem is framed as one of determining weights α_{kj} resulting in a TRSM-WL model.
Similarity Measure
Once the weights are adjusted within the framework of tolerance rough sets, one can measure the similarity between a query vector Q where $\hat{Q} = \langle(q_1, w_{1q}), (q_2, w_{2q}), \ldots,$ $(q_s, w_{sq})\rangle$ and a document vector d_j by using the following formula

$$\text{Similarity}(Q, d_j) = \frac{2 \times \sum_{k=1}^{N}(w_{kq} \times w_{kj})}{\sum_{k=1}^{N} w_{kq}^2 + \sum_{k=1}^{N} w_{kj}^2},$$

and ultimately, cluster documents that are similar. A query vector may represent an actual query in the context of information retrieval or a class of documents in the context of document classification.

5 Structured Information Categorization with TRSM

TRSM has been gaining popularity in one form of text categorization method which is document clustering. In this section, we discuss the popular TRSM-based work for document clustering.

5.1 Hierarchical Document Clustering

The earliest known work on the application of TRSM as a document representation model was proposed by Kawasaki et al. They introduced a TRSM-based hierarchical document clustering that is an extension of the hierarchical agglomerative (bottom-up) clustering algorithm [16]. In this model, every document is represented as a weight vector of its terms and the upper approximation calculated by using a tolerance relation over the terms, as described by the TRSM framework by w_{ij}^{trs}. As before, it aims to minimize the number of zero-valued coefficients in document vectors as well as to increase the degree of similarity between documents with few common terms. Once the representation is established, the clustering algorithm takes place. It first assigns each document to a different cluster and defines cluster representatives as supersets of popular terms of the constituting documents'. Subsequently, it finds the most similar pair of clusters (by using a similarity method such as Dice, Jaccard or Cosine) and merges them in an iterative fashion, until all the clusters are merged into an ultimate single cluster. The advantage of using a hierarchy is that it allows the use of document cluster representatives to calculate the similarity between clusters instead of averaging similarities of all document pairs included in clusters, which aids the execution time [16]. The results of validation and evaluation of this method suggest that this clustering algorithm can be well adapted to text mining.

5.2 Non-hierarchical Document Clustering

Soon after, Ho et al. introduced a non-hierarchical document clustering method using TRSM [18]. The authors pointed out that hierarchical methods become unsuitable for large document corpora, due to exploding time and space requirements of the underlying algorithms. This model also uses the TRSM framework described in Sect. 4 and forms a pre-specified number of possibly overlapping document clusters. First, the TRSM-based document representation is established (documents are approximated using the upper approximation operator, term weights are adjusted using w_{ij}^{trs}). Then, the cluster representatives R_k are formed by randomly assigning a document to each cluster. Similar to the hierarchical approach, this is done by using the popular terms of the constituting documents. Next, the similarity between each cluster representative and the upper approximation of each document is calculated using Similarity (Q, d_j) given in Sect. 4. If the similarity is above a given threshold, the document is assigned to that cluster, and the cluster representative is recalculated. The authors use a normalized Dice coefficient which is applied to the upper approximation of cluster representative [18, §3.22, p. 206]. This process continues until there is no more change in the clusters. The algorithm has been evaluated and validated by experiments on test collections.

5.3 Lexicon-Based Document Clustering

More recently, a new method for document clustering, named a lexicon-based document representation (LBDR) was introduced by Virginia et al. [20]. This model uses TRSM in the form of a lexicon with the intention of creating an enhanced and compact document representation. First of all, LBDR creates a term weight vector for each document and then enhances the representation by means of TRSM, just like the hierarchical [16] and non-hierarchical [18] methods. Next, the terms are mapped to a lexicon and the ones which do not occur in the lexicon (i.e. irrelevant, non-informative terms) are filtered out reducing the number of dimensions in the vectors, creating a compact but yet enhanced representation. The intuition behind this approach can be demonstrated via Fig. 3 [20, §1, p. 29]. In Fig. 3a, we can see how document d_1 and the lexicon overlap. The intersection is compact, but limited. In Fig. 3b, the dashed line shows the upper approximated TRSM representation of d_1. LBDR combines the two and creates the dense and enhanced representation of d_1 in lower dimensional space, as shown in the dark shaded area in Fig. 3c. Eventually,

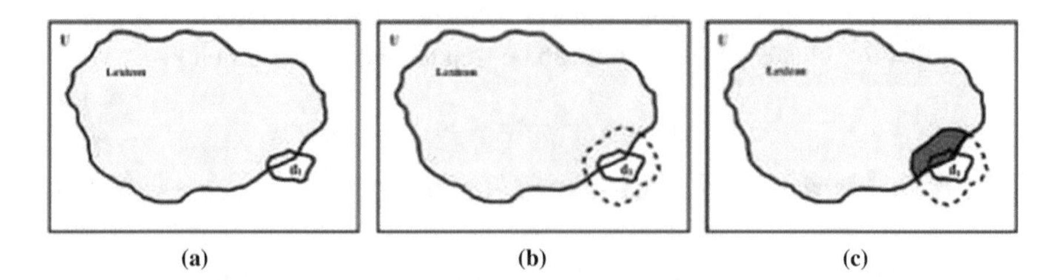

Fig. 3 Lexicon-based document representation [20]

the authors conclude that the effectiveness of lexicon-representation is comparable with TRSM-representation while the efficiency of lexicon-representation should be better than the existing TRSM-representation.

5.4 K-Means Based Tolerance Rough Set Clustering

Ngo and Nguyen [19] focused on a more specific type of document clustering. They proposed a web search results clustering method which is based on tolerance rough sets model. Their goals were the same as in [18] and [16], creating an enriched representation for the web documents in order to reveal the subtle inter-document similarities and to boost the clustering quality. They proposed a Tolerance Rough set Clustering (TRC) algorithm [19, §5.3, p. 42], which is based on k-means clustering. First, each document is pre-processed to create an index term-based vectorial representation. After that, those vectors are combined and a term-document matrix is formed. Then, they enhance the term weights of the documents by using TRSM and upper approximation. Ultimately, TRC clusters the search results and labels them on a given query. Their experiments have shown that tolerance rough sets and upper approximation it offers can indeed improve the representations, with positive effects on the clustering quality.

5.5 Two-Class Document Classification with Ensemble Learning

Shi et al. [25] proposed a tolerance-based semi-supervised two-class ensemble classifier for documents with only positive and unlabeled examples i.e. in the absence of labeled negative examples. TRSM model (discussed in Sect. 4) is used as the formal model. The term weighting is done with a popular TF * IDF (term frequency times inverse document frequency) weighting scheme w_{ij}^{tfidf} to assign weight values for a document vector. The methodology for generating a reliable negative set of examples can be found in [25, §4.2, p. 6303]. There are four key steps: (i) selecting a positive feature set (from positive and unlabeled examples) based on a frequency threshold, (ii) generating tolerance classes for terms from the *unlabeled set*, (iii) expanding the *positive feature set* with the aid of tolerance classes and, (iv) generating a reliable negative set by filtering out possible positive documents from the unlabeled set whose upper approximation does not have any positive feature in *positive feature set*. Support Vector Machines, Rocchio and Naive Bayes algorithms are used as base classifiers to construct an ensemble classifier, which runs iteratively and exploits margins between positive and negative data to progressively improve the approximation of negative data. Experimental results indicate that the proposed method achieves significant performance improvement.

5.6 Indonesian Text Retrieval with TRSM

Virginia and Nguyen propose a framework for efficient retrieval of text specific to the Indonesian language based on TRSM [26]. They propose an alternative to the classical document representation by mapping index terms to terms in the lexicon (termed LEX-representation) thus resulting in a compact representation and yielding better retrieval results. This chapter by far, gives the most in depth discussion of the challenges of employing TRSM for information retrieval. Three challenges are addressed: creation of the thesaurus, optimization the thesaurus and finally the retrieval process. A classical vector model is used where document and query are represented as vectors in a high-dimensional space and each vector corresponds to a term in the vocabulary of the collection. The framework includes a high-dimensional vectorial space with standard linear algebra operations on vectors. The association degree of documents with regard to the query is quantified by the cosine of the angle between these two vectors.

In the discussion thus far, documents were the target entities and the index terms were the features describing the documents. In Sect. 6, the focus is on text categorization where the text consists of *categorical noun phrases* and *relation instances*. Instead of documents, the target entities are the noun phrases and instead of index terms, the features are the contextual patterns. In other words, this model grew out of the observation that there is a natural affinity between the document clustering problem and the context-based noun phrase clustering problem.

6 Unstructured Information Representation with TRSM

In this section, we discuss the problem of representation and categorization of unstructured information typically gleaned from the web. The representation includes definition of approximation spaces to provide the framework for the categorical and relational information extraction, respectively. The categorization involves employing semi-supervised learning algorithms.

6.1 Unstructured Web Information

Figure 4 illustrates a typical unstructured information categorization problem where the information needs to be categorized into noun and relational phrases. This categorization is typically accomplished using a co-occurrence matrix.

The structured information resulting from this process is represented as:

- **categorical** noun phrase instances
  Sport(Ice Hockey), Country(Canada)

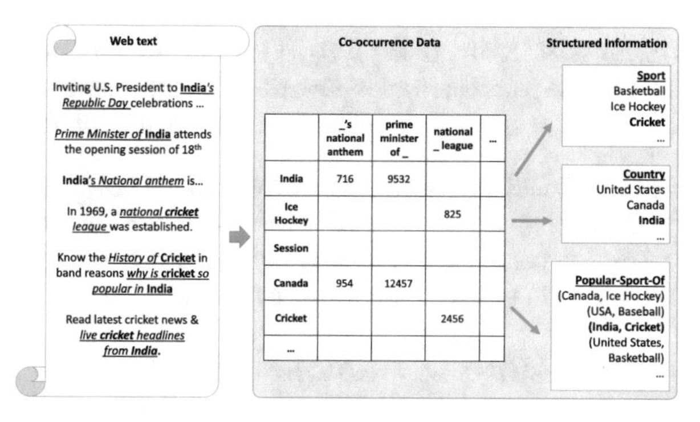

Fig. 4 Web information labelling

- **relational** noun phrase pairs
  Popular-Sport-Of(Canada, Ice Hockey)

In other words, the noun phrase Ice Hockey belongs to the *category* of Sport. On the other hand, a relational phrase "Canada, Ice-hockey" belongs to the *category* of Popular-Sport-Of. A relational phrase is composed two noun phrases. In keeping with the terminology used in current literature, we will refer to these as categorical instances and relational instances respectively [27]. An illustration of the contextual patterns and co-occurrence statistics are given as follows:

- **contextual extraction patterns**

  e.g. "_ league", "_ and other sports", "_ is popular in _"

- **co-occurrence statistics**

  e.g. f*("Ice Hockey", "_ league")* $= n$
  e.g. f*("Ice Hockey", "Canada", "_ is popular in _")* $= n$

6.2 *TRSM for Noun Phrases*

A tolerance form of rough sets model that labels categorical noun phrase instances from a given corpus representing unstructured web pages was proposed in [28].

- $\mathcal{N} = \{n_1, n_2, \ldots, n_M\}$ is the universe of noun phrases. This set will accommodate every single noun phrase to be parsed from the source web documents.
- $\mathcal{C} = \{c_1, c_2, \ldots, c_P\}$ is the universe of categorical (unary) contextual patterns. These contexts are to yield the individual noun phrases to be extracted as category instances.
- $\mathcal{R} = \{r_1, r_2, \ldots, r_Q\}$ is the universe of relational (binary) contextual patterns. These contexts are to yield the noun phrase pairs to be extracted for relations.

- $\mathcal{T} = \{t_{ij} = (n_i, n_j) \in \mathcal{N}^2 : \exists r_k \in \mathcal{R} \mid f_{\mathcal{T}}(t_{ij}, r_k) > 0\}$ is the universe of co-occurring noun phrase pairs (i.e. tuples) described via the relational co-occurrence function $f_{\mathcal{T}}(t_{ij}, r_k) = \{\kappa \in \mathbb{N} : t_{ij}$ occurs κ times within the context $r_k\}$

Definition 2 A *categorical noun-context tolerance model* [28] is described by the tolerance approximation space $\mathcal{A} = (C, \mathcal{N}, I, \omega, \nu)$ where $\mathcal{N}$ and C are as defined previously. $I = I_\theta$ is the parametrized uncertainty function describing the tolerance classes for the contexts, in terms of contextual overlaps:

$$I_\theta(c_i) = \{c_j : \omega(N(c_i), N(c_j)) \geq \theta\}.$$

Here, θ is the tolerance threshold and ω is the overlap index which is the Sorensen-Dice index [29]:

$$\omega(A, B) = \frac{2|A \cap B|}{|A| + |B|}.$$

The degree of inclusion is measured by $\nu : \mathbb{P}(C) \times \mathbb{P}(C) \to [0, 1]$ and is defined as $\nu(X, Y) = \frac{|X \cap Y|}{|X|}$. Within the framework of $\mathcal{A}$, a context-described noun phrase can now be approximated using the lower approximation:

$$\mathcal{L}_{\mathcal{A}}(n_i) = \{c_j \in C : \nu(I_\theta(c_j), C(n_i)) = 1\},$$

giving us its *closely* related contexts; or else it can be approximated with the upper approximation to its *somewhat* related contexts:

$$\mathcal{U}_{\mathcal{A}}(n_i) = \{c_j \in C : \nu(I_\theta(c_j), C(n_i)) > 0\}.$$

6.3 TRSM for Relational Phrases

A tolerance form of rough sets model that labels relational phrase instances from a given corpus representing unstructured web pages was proposed in [30]. The following cross-mapping functions to represent every noun phrase (and noun phrase pair) by means of their contexts, and vice versa [28, 30] is defined as:

- $C : \mathcal{N} \to \mathbb{P}(C)$ maps each noun phrase to its set of co-occurring categorical contexts: $C(n_i) = \{c_j : f_{\mathcal{N}}(n_i, c_j) > 0\}$ where $f_{\mathcal{N}}(n_i, c_j) = \{\kappa \in \mathbb{N} : n_i$ occurs κ times within context $c_j\}$
- $N : C \to \mathbb{P}(\mathcal{N})$ maps each categorical context to its set of co-occurring noun phrases: $N(c_j) = \{n_i : f_{\mathcal{N}}(n_i, c_j) > 0\}$
- $R : \mathcal{T} \to \mathbb{P}(\mathcal{R})$ maps each noun phrase pair to its set of co-occurring relational contexts: $R(t_{ij}) = \{r_k : f_{\mathcal{T}}(t_{ij}, r_k) > 0\}$
- $T : \mathcal{R} \to \mathbb{P}(\mathcal{T})$ maps each relational context to its set of co-occurring noun phrase pairs: $T(r_k) = \{t_{ij} : f_{\mathcal{T}}(t_{ij}, r_k) > 0\}$

Definition 3 A *relational noun-context tolerance model* [21] is the analogous model to extract related pairs. It is described by the approximation space $\mathcal{A} = (\mathcal{R}, \mathcal{T}, I, \omega, \nu)$ where $\mathcal{T}$, $\mathcal{R}$, ω and ν are defined as previously. I_θ is again the uncertainty function with the tolerance threshold θ:

$$I_\theta(r_i) = \{r_j : \omega(T(r_i), T(r_j)) \geq \theta\}.$$

Within the framework of $\mathcal{A}$, a context-described noun phrase pair can now be lower approximated to its closely related contexts:

$$\mathcal{L}_\mathcal{A}(t_{ij}) = \{r_k \in \mathcal{R} : \nu(I_\theta(r_k), R(t_{ij})) = 1\},$$

or else it can be upper approximated to its somewhat related contexts:

$$\mathcal{U}_\mathcal{A}(t_{ij}) = \{r_k \in \mathcal{R} : \nu(I_\theta(r_k), R(t_{ij})) > 0\}.$$

7 Semi-supervised Text Categorization Algorithms

In this section, we give a brief overview of the semi-supervised text categorization (TPL) algorithms. The categorical extractor and relational extractor algorithms are based on the two TRSM models discussed in Sects. 6.2 and 6.3 for noun and relation phrase labelling. The algorithm(s) were experimentally compared with Coupled Bayesian Sets (CBS) [31] and Coupled Pattern Learner (CPL) algorithms [27] respectively. TPL (tolerant pattern learner) does not use a vector-space model since it describes noun phrases as sets of co-occurring contexts, instead of vectors. In accordance, every trusted instance n_i of a given category *cat* is associated with the following three descriptor sets: $C(n_i)$, $\mathcal{U}_\mathcal{A}(n_i)$ and $\mathcal{L}_\mathcal{A}(n_i)$. These sets are employed to calculate a *micro-score* for the candidate noun phrase n_j, against the trusted instance n_i of the category *cat*:

$$micro(n_i, n_j) = \omega(C(n_i), C(n_j))\alpha + \omega(\mathcal{U}_\mathcal{A}(n_i), C(n_j))\beta + \omega(\mathcal{L}_\mathcal{A}(n_i), C(n_j))\gamma.$$

An overlap index function ω given in Definition 1 is used for this calculation. α, β and γ are the contributing factors of the scoring components and they may be adjusted for the particular application domain.

The intuition behind this approach is illustrated in Fig. 5. A trusted instance n_i has the universe of contexts partitioned by its descriptors $\mathcal{L}_\mathcal{A}(n_i)$, $C(n_i)$ and $\mathcal{U}_\mathcal{A}(n_i)$ into four zones of recognition. For a candidate n_j, each zone will represent a different degree of similarity. When calculating the micro-score, the candidate's contexts falling in zone 1 (lower approximation) will be covered by all three descriptors and will thus make a high contribution to its score. Contexts in zone 2 will be covered by $C(n_i)$ and $\mathcal{U}_\mathcal{A}(n_i)$ so they will make medium contribution. Zone 3 contexts will only be covered by $\mathcal{U}_\mathcal{A}(n_i)$ and they will make low contribution. Contexts in zone 4 will

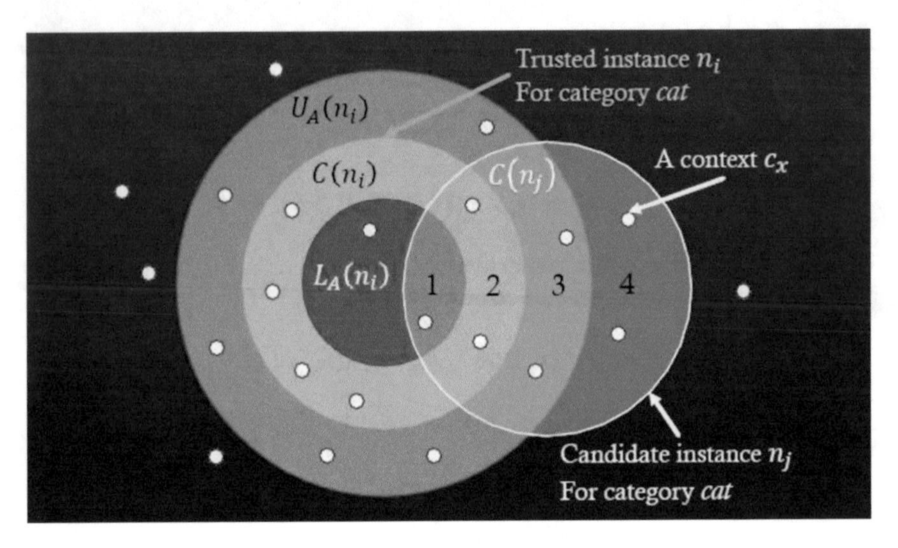

Fig. 5 Four zones of recognition for contexts emerging from approximations of n_i

not to contribute at all since they suggest no resemblance between n_i and n_j. An analogous scoring mechanism is also employed for learning relations. These descriptors are used to calculate a micro-score for a candidate pair t_{kl}, by the trusted pair t_{ij}:

$$micro(t_{ij}, t_{kl}) = \omega(C(t_{ij}), C(t_{kl}))\alpha + \omega(\mathcal{U}_A(t_{ij}), C(t_{kl}))\beta + \omega(\mathcal{L}_A(t_{ij}), C(t_{kl}))\gamma.$$

Algorithm 1 outlines the semi-supervised mechanism for learning categories. The input for the categorical extractor is an ontology which is formed by a set of categories (e.g. *City*) and a handful of seed noun phrases (e.g. *Winnipeg, New Delhi, Ankara*). Furthermore, it expects a large co-occurrence matrix representing the noun phrases and the contextual patterns cropped from the world wide web. The output consists of trusted instances assigned to their respective categories within the ontology. TPL employs a score-based ranking and the scoring mechanism uses tolerance approximation spaces. For a given category *cat*, a macro-score (i.e. an accumulated micro-score of proxies) for the candidate n_j is maintained:

$$macro_{cat}(n_j) = \sum_{\forall n_i \in Trusted_{cat}}^{n} micro(n_i, n_j).$$

After calculating the score for every candidate of *cat*, the candidates are ranked by their macro-scores (normalized by the number of trusted instances of *cat*). Eventually, the top new candidates are promoted as trusted instances. Overall, TPL managed to achieve a comparable performance with CBS, by means of the precision metric as shown in Table 1. A detailed discussion of the experiment can be found in [28].

Algorithm 1: Tolerant Pattern Learner for Categories

Input : An ontology O defining categories and a small set of seed examples; a large corpus U
Output: Trusted instances for each category
1 **for** $r = 1 \rightarrow \infty$ **do**
2 **for** each category *cat* **do**
3   **for** each new trusted noun phrase n_i of *cat* **do**
4   Calculate the approximations $U_A(n_i)$ and $L_A(n_i)$;
5   **for** each candidate noun phrase n_j **do**
6   Calculate $micro(n_i, n_j)$;
7   **for** each candidate noun phrase n_j **do**
8    $macro_{cat}(n_j) = \sum_{\forall n_i \in cat} micro(n_i, n_j)$;
9   Rank instances by $macro_{cat}/|cat|$;
10   Promote top instances as trusted;

Table 1 Precision@30 of TPL and CBS per category. CBS results are as seen in [31]

Categories	Iteration 5		Iteration 10	
	TPL (%)	CBS (%)	TPL (%)	CBS (%)
Company	100	100	100	100
Disease	100	100	100	100
KitchenItem	100	94	100	94
Person	100	100	100	100
PhysicsTerm	93	100	90	100
Plant	100	100	97	100
Profession	100	100	100	87
Sociopolitics	100	48	100	34
Sport	97	97	100	100
Website	90	94	90	90
Vegetable	93	83	63	48
Average	**97.5**	**92**	**94.5**	**87**

Similarly, the input for the relational extractor (algorithm given in [21]) an ontology formed by a set of relations (e.g. *City-Country*) as well as a few seed noun phrase pairs per relation (e.g. *(Winnipeg, Canada), (New Delhi, India), (Ankara, Turkey)*). It also expects a large co-occurrence matrix representing the noun phrase pairs, and the contextual patterns. The output are trusted relation instances, in forms of ordered noun phrase pairs, assigned to their respective relations.

As shown in Table 2, for most relations, TPL maintained high quality extractions and high precision values throughout the iterations, steering clear from the concept drift problem. Overall, TPL is able to demonstrate comparable performance with

Table 2 Precision@30 results of TPL and CPL as seen in [27] (%)

Evaluation	Ranking-based			Promotion-based			
Iterations	TPL			TPL			CPL
	1	5	10	1	5	10	10
Relations							
Athlete-Team	100	90	87	100	96	87	100
CEO-Company	100	100	100	100	100	100	100
City-Country	100	100	100	100	100	100	93
City-State	100	100	100	100	100	100	100
Coach-Team	93	93	93	100	100	93	100
Company-City	83	90	93	40	84	97	50
Stadium-City	97	93	80	80	92	70	100
State-Capital	100	97	73	100	100	63	60
State-Country	100	100	100	100	100	100	97
Team-versus-Team	93	83	80	100	84	80	100
Average	96.6	94.6	90.6	92.0	95.6	89.0	90.0

CBS and CPL in terms of precision [21]. Experimental details can be found in [30] and can be downloaded from.[1]

8 Concluding Remarks

In this chapter, we present a survey of the literature where the tolerance rough set model serves as a text categorization and learning model. Particularly when it comes to natural language processing and information retrieval tasks such as text categorization or document clustering, a non-transitive binary relation that is reflexive and symmetric is necessary. A brief overview of the history of tolerance rough sets that led to the model that is widely used in document classification is presented. The four representative papers are also an ideal source of broader related works dealing with the theoretical aspects of tolerance form of rough sets. Document clustering appears to be a more popular form of text categorization with the tolerance form of rough sets. However, the more recent work on categorizing unstructured text in a semi-supervised learning environment with tolerance form of rough sets points to another fruitful area of application of TRSM. Future work includes categorizing unstructured text on a large dataset as well as comparison with rough-fuzzy models. This survey is meant to demonstrate the versatility of the tolerance form of rough sets and its successful application in structured and unstructured text categorization.

[1] http://winnspace.uwinnipeg.ca/handle/10680/821.

Acknowledgements This research has been supported by the Natural Sciences and Engineering Research Council of Canada (NSERC) Discovery grants.

References

1. Pawłak, Z.: Rough sets. Int. J. Comput. Inf. Sci. **11**(5), 341–356 (1982)
2. Pawłak, Z.: Rough Sets: Theoretical Aspects of Reasoning About Data. Kluwer Academic Publishers, Norwell, MA, USA (1992)
3. Pawłak, Z., Skowron, A.: Rough sets and boolean reasoning. Inf. Sci. **177**(1), 41–73 (2007)
4. Pawłak, Z., Skowron, A.: Rough sets: some extensions. Inf. Sci. **177**(1), 28–40 (2007)
5. Pawłak, Z., Skowron, A.: Rudiments of rough sets. Inf. Sci. **177**(1), 3–27 (2007)
6. Peters, J., Skowron, A.: Zdzisław Pawlak life and work: 1926–2006. Inf. Sci. **177**(1), 1–2 (2007)
7. Schroeder, M., Wright, M.: Tolerance and weak tolerance relations. J. Comb. Math. Comb. Comput. **11**, 123–160 (1992)
8. Nieminen, J.: Rough tolerance equality and tolerance black boxes. Fundam. Inf. **11**, 289–296 (1988)
9. Novotný, M., Pawłak, Z.: Black box analysis and rough top equality. Bull. Pol. Acad. Sci. Tech. Sci. **33**, 105–113 (1985)
10. Polkowski, L., Skowron, A., Zytkow, J.: Tolerance based rough sets. In: Lin, T.Y., Wildberger, M. (eds.) Soft Computing: Rough Sets, Fuzzy Logic, Neural Networks, Uncertainty Management, Knowledge Discovery, pp. 55–58. Simulation Councils Inc., San Diego (1994)
11. Marcus, S.: Tolerance rough sets, cech topologies, learning processes. Bull. Pol. Acad. Sci. Tech. Sci. **42**(3), 471–487 (1994)
12. Poincaré, H.: Science and Hypothesis. The Mead Project. Brock University, L. G. Ward's translation (1905)
13. Peters, J., Wasilewski, P.: Tolerance spaces: origins, theoretical aspects and applications. Inf. Sci. **195**(1–2), 211–225 (2012)
14. Zeeman, E.: The topology of the brain and visual perception. In: Fort Jr., M.K., (ed.) Topology of 3-Manifolds and Related Topics. University of Georgia Institute Conference Proceedings, pp. 240–256. Prentice-Hall, Inc. (1962)
15. Sossinsky, A.B.: Tolerance space theory and some applications. Acta Appl. Math. Int Surv. J. Appl. Math. Math. Appl. **5**(2), 137–167 (1986)
16. Kawasaki, S., Nguyen, N.B., Ho, T.B.: Hierarchical document clustering based on tolerance rough set model. In: Proceedings of the 4th European Conference on Principles of Data Mining and Knowledge Discovery, pp. 458–463 (2000)
17. Thanh, N.C., Yamada, K., Unehara, M.: A similarity rough set model for document representation and document clustering. J. Adv. Comput. Intell. Intell. Inf. **15**(2), 125–133 (2011)
18. Ho, T.B., Nguyen, N.B.: Nonhierarchical document clustering based on a tolerance rough set model. Int. J. Intell. Syst. **17**, 199–212 (2002)
19. Ngo, C.L.: A tolerance rough set approach to clustering web search results. Master's thesis, Warsaw University (2003)
20. Virginia, G., Nguyen, H.S.: Lexicon-based document representation. Fundam. Inf. **124**(1–2), 27–46 (2013)
21. Sengoz, C., Ramanna, S.: Learning relational facts from the web: a tolerance rough set approach. Pattern Recogn. Lett. **67**(P2), 130–137 (2015)
22. Skowron, A., Stepaniuk, J.: Tolerance approximation spaces. Fundam. Inf. **27**(2,3), 245–253 (1996)
23. Hu, Y.C.: Flow-based tolerance rough sets for pattern classification. Appl. Soft Comput. **177**(27), 322–331 (2015)
24. Swieboda, W., Meina, M., Nguyen, H.: Weight learning for document tolerance rough set model. In: RSKT 2013, LNAI 8171, pp. 386–396. Springer (2013)

25. Shi, L., Ma, X., Xi, L., Duan, Q., Zhao, J.: Rough set and ensemble learning based semi-supervised algorithm for text classification. Expert Syst. Appl. **38**(5), 6300–6306 (2011)
26. Virginia, G., Nguyen, H.S.: A semantic text retrieval for indonesian using tolerance rough sets models. Transactions on Rough Sets LNCS 8988(XIX), pp. 138–224 (2015)
27. Carlson, A., Betteridge, J., Wang, R.C., Hruschka Jr., E.R., Mitchell, T.M.: Coupled semi-supervised learning for information extraction. In: Proceedings of the Third ACM International Conference on Web Search and Data Mining, pp. 101–110 (2010)
28. Sengoz, C., Ramanna, S.: A semi-supervised learning algorithm for web information extraction with tolerance rough sets. In: AMT 2014, LNCS 8610, pp. 1–10 (2014)
29. Sørensen, T.: A method of establishing groups of equal amplitude in plant sociology based on similarity of species content and its application to analyses of the vegetation on Danish commons. Biologiske skrifter. I kommission hos E. Munksgaard (1948)
30. Sengoz, C.: A granular-based approach for semi-supervised web information labeling. Master's thesis, University of Winnipeg, supervisor: S. Ramanna (2014)
31. Verma, S., Hruschka Jr., E.R.: Coupled bayesian sets algorithm for semi-supervised learning and information extraction. In: ECML PKDD Part II LNCS **7524**, 307–322 (2012)

Medical Diagnosis: Rough Set View

Shusaku Tsumoto

Abstract This chapter dicusses formalization of medical diagnosis from the viewpoint of rule reasoning based on rough sets. Medical diagnosis consists of the following three procedures. First, screening process selects the diagnostic candidates, where rules from upper approximations are used. Then, from the selected candidates, differential diagnosis is evoked, in which rules from lower approximations are used. Finally, consistency of the diagnosis will be checked with all the inputs: inconsistent symptoms suggest the existence of complications of other diseases. The final process can be viewed as complex relations between rules. The proposed framework successfully formalizes the representation of three types of reasoning styles.

1 Introduction

Classical medical diagnosis of a disease assumes that a disease is defined as a set of symptoms, in which the basic idea is *symptomatology*. Symptomatology had been a major diagnostic rules before laboratory and radiological examinations. Although the power of symptomatology for differential diagnosis is now lower, it is true that change of symptoms are very important to evaluate the status of chronic status. Even when laboratory examinations cannot detect the change of patient status, the set of symptoms may give important information to doctors.

Symptomatological diagnostic reasoning is conducted as follows. First, doctors make physical examinations to a patient and collect the observed symptoms. If symptoms are observed enough, a set of symptoms give some confidence to diagnosis of a corresponding disease. Thus, correspondence between a set of manifestations and a disease will be useful for differential diagnosis. Moreover, similarity of diseases will be infered by sets of symptoms.

The author has been discussed modeling of symptomatological diagnostic reasoning by using the core ideas of rough sets since [16]: selection of candidates (screen-

S. Tsumoto (✉)
Faculty of Medicine, Department of Medical Informatics, Shimane University,
Matsue, Japan
e-mail: tsumoto@med.shimane-u.ac.jp

G. Wang et al. (eds.), *Thriving Rough Sets*, Studies in Computational
Intelligence 708, DOI 10.1007/978-3-319-54966-8_7

ing) and differential diagnosis are closely related with diagnostic rules obtained by upper and lower approximations of a given concept. Thus, this chapter dicusses formalization of medical diagnostic rules which is closely related with rough set rule model. The important point is that medical diagnostic reasoning is characterized by focusing mechanism, composed of screening and differential diagnosis, which corresponds to upper approximation and lower approximation of a target concept. Furthremore, this chapter focuses on detection of complications, which can be viewed as relations between rules of different diseases.

The chapter is organized as follows. Section 2 shows charateristics of medical diagnostic process. Section 3 introduces rough sets and basic definition of probabilistic rules. Section 4 gives two style of formalization of medical diagnostic rules. The first one is a deterministic model, which correspond to Pawlak's rough set model. And the other one gives an extention of the above ideas in probabilistic domain, which can be viewed as application of variable precision rough set model [18]. Section 5 proposes a new rule induction model, which includes formalization of rules for detection of complications. Section 6 shows how to induce the above formalized rules from data. Section 7 discussed what has not been achieved yet. Finally, Sect. 8 concludes this chapter.

2 Background: Medical Diagnostic Process

This section focuses on medical diagnostic process as rule-based reasoning. The fundamental discussion of medical diagnostic reasoning related with rough sets is given in [11].

2.1 RHINOS

RHINOS is an expert system which diagnoses clinical cases on headache or facial pain from manifestations. In this system, a diagnostic model proposed by Matsumura [1] is applied to the domain, which consists of the following three kinds of reasoning processes: exclusive reasoning, inclusive reasoning, and reasoning about complications.

First, exclusive reasoning excludes a disease from candidates when a patient does not have a symptom which is necessary to diagnose that disease. Secondly, inclusive reasoning suspects a disease in the output of the exclusive process when a patient has symptoms specific to a disease. Finally, reasoning about complications suspects complications of other diseases when some symptoms which cannot be explained by the diagnostic conclusion are obtained.

Each reasoning is rule-based and all the rules needed for diagnostic processes are acquired from medical experts in the following way.

2.1.1 Exclusive Rules

These rule correspond to exclusive reasoning. In other words, the premise of this rule is equivalent to the necessity condition of a diagnostic conclusion. From the discussion with medical experts, the following six basic attributes are selected which are minimally indispensable for defining the necessity condition: *1. Age, 2. Pain location, 3. Nature of the pain, 4. Severity of the pain, 5. History since onset, 6. Existence of jolt headache.* For example, the exclusive rule of common migraine is defined as:

```
In order to suspect common migraine,
the following symptoms are required:
pain location: not eyes,
nature :throbbing or persistent
or radiating,
history: paroxysmal or sudden and
jolt headache: positive.
```

One of the reasons why the six attributes are selected is to solve an interface problem of expert systems: if all attributes are considered, all the symptoms should be input, including symptoms which are not needed for diagnosis. To make exclusive reasoning compact, we chose the minimal requirements only. It is notable that this kind of selection can be viewed as the ordering of given attributes, which is expected to be induced from databases. This issue is discussed later in Sect. 6.

2.1.2 Inclusive Rules

The premises of inclusive rules are composed of a set of manifestations specific to a disease to be included. If a patient satisfies one set, this disease should be suspected with some probability. This rule is derived by asking the medical experts about the following items for each disease: *1. a set of manifestations by which we strongly suspect a disease. 2. the probability that a patient has the disease with this set of manifestations: SI (Satisfactory Index) 3. the ratio of the patients who satisfy the set to all the patients of this disease: CI (Covering Index) 4. If the total sum of the derived CI (tCI) is equal to 1.0 then end. Otherwise, goto 5. 5. For the patients with this disease who do not satisfy all the collected set of manifestations, goto 1.* Therefore a positive rule is described by a set of manifestations, its satisfactory index (SI), which corresponds to *accuracy measure*, and its covering index (CI), which corresponds to *total positive rate*. Note that SI and CI are given empirically by medical experts.

For example, one of three positive rules for common migraine is given as follows.

```
If history: paroxysmal,
jolt headache: yes,
nature: throbbing or persistent,
```

```
prodrome: no, intermittent symptom: no,
persistent time: more than 6 hours,
and location: not eye,
then common migraine is suspected with
accuracy 0.9 (SI=0.9) and
this rule covers
60 percent of the total cases (CI=0.6).
```

2.1.3 Disease Image: Complications Detection

This rule is used to detect complications of multiple diseases, acquired by all the possible manifestations of the disease. By the use of this rule, the manifestations which cannot be explained by the conclusions will be checked, which suggest complications of other diseases. For example, the disease image of common migraine is:

```
The following symptoms can be
explained by common migraine:
pain location: any or depressing:
not or jolt headache: yes or ...
```

Therefore, when a patient who suffers from common migraine is depressing, it is suspected that he or she may also have other disease.

2.2 Focusing Mechanism

The most important process in medical differential diagnosis shown above is called a focusing mechanism [7, 17]. Even in differential diagnosis of headache, medical experts should check possibilities of more than 100 candidates, though frequent diseases are 5 or 6. These candidates will be checked by past and present history, physical examinations, and laboratory examinations. In diagnostic procedures, a candidate is excluded one by one if symptoms necessary for diagnosis are not observed.

Focusing mechanism consists of the following two styles: exclusive reasoning and inclusive reasoning. Relations of this diagnostic model with another diagnostic model are discussed in [5, 11], which is summarized in Fig. 1: First, exclusive reasoning excludes a disease from candidates when a patient does not have symptoms that is necessary to diagnose that disease. Second, inclusive reasoning suspects a disease in the output of the exclusive process when a patient has symptoms specific to a disease. Based on the discussion with medical experts, these reasoning processes are modeled as two kinds of rules, negative rules (or exclusive rules) and positive

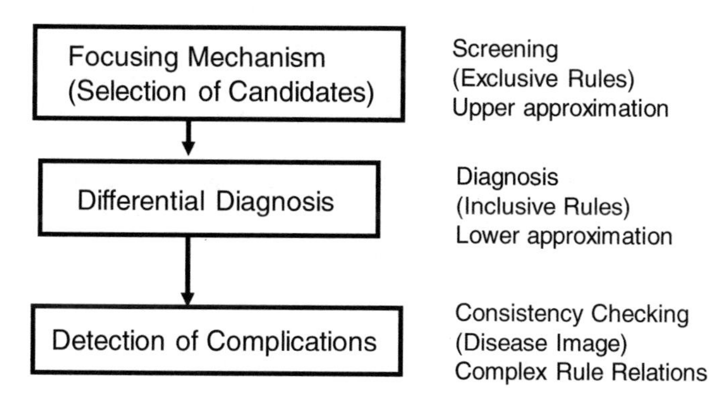

Fig. 1 Focusing mechanism

rules; the former corresponds to exclusive reasoning, the latter to inclusive reasoning [1].[1]

2.3 *Medical Diagnosis = Set Classification?*

Most of the conventional rule learning scheme assumes that medical diagnosis is based on conventional set classification scheme. That is, it is assumed that once the final conclusion is given, its classification is mutual exclusive. However, it is not a correct assumption if the etiologies of complicated diseases is different. For example, in the case of differential diagnosis of headache, differentiation of muscle tention headache and vascular headache is very important since the corresponding therapy is completely different. However, complication of both type of headache is possible, since one is due to muscle pain and the other is due to the problems with arteries and the etiologies are completely different.

Then, is it "fuzzy classification"? The author says that it may be under some specific condition. If the diagnostic time is fixed, we can think its fuzzy classification, because the degree can be easily quantified: which type of headache is dominant or not. Both diseases should be treated and the order of the treatment may depend on the applied situation. In some contexts, preference should be considered. For example, the status of one disease is in emergency, then this disease should be treated immediately. In other contexts, the disease which is easy to treat may be selected at first. Thus, preference depends on a given clinical context, which may not be included in datasets.

Thus, we should think in the following way. Here we assume that there are many binary decision attributes. That is, each diagnostic candidate corresponds to one

[1] Implementation of detection of complications is not discussed here because it is derived after main two process, exclusive and inclusive reasoning. The way to deal with detection of complications is discussed in Sect. 5.

decision attribute, and we will think about decision rules for each. For each decision attribute, set-classification based scheme can be applied. We do no think about the preference of decisions or the degree of decisions because choice of decisions may not be needed in general.

For this task, we may have to convert ordinary datasets, but we do not get into the details of data preprocessing. Here we only focus on representations of rules for such multiple decision attributes and their rule induction algorithms.

3 Basics of Rule Definitions

3.1 Rough Sets

In the following sections, we use the following notation introduced by Grzymala-Busse and Skowron [4], based on rough set theory [2]. Let U denote a nonempty finite set called the universe and A denote a nonempty, finite set of attributes, i.e., $a : U \rightarrow V_a$ for $a \in A$, where V_a is called the domain of a, respectively. Then a decision table is defined as an information system, $A = (U, A \cup \{d\})$. The atomic formulas over $B \subseteq A \cup \{d\}$ and V are expressions of the form $[a = v]$, called descriptors over B, where $a \in B$ and $v \in V_a$. The set $F(B, V)$ of formulas over B is the least set containing all atomic formulas over B and closed with respect to disjunction, conjunction, and negation.

For each $f \in F(B, V)$, f_A denotes the meaning of f in A, i.e., the set of all objects in U with property f, defined inductively as follows:

1. If f is of the form $[a = v]$, then $f_A = \{s \in U | a(s) = v\}$.
2. $(f \wedge g)_A = f_A \cap g_A$; $(f \vee g)_A = f_A \vee g_A$; $(\neg f)_A = U - f_a$.

3.2 Classification Accuracy and Coverage

3.2.1 Definition of Accuracy and Coverage

By use of the preceding framework, classification accuracy and coverage, or true positive rate are defined as follows.

Definition 1 Let R and D denote a formula in $F(B, V)$ and a set of objects that belong to a decision d. Classification accuracy and coverage(true positive rate) for $R \rightarrow d$ is defined as:

$$\alpha_R(D) = \frac{|R_A \cap D|}{|R_A|} (= P(D|R)), \tag{1}$$

$$\kappa_R(D) = \frac{|R_A \cap D|}{|D|} (= P(R|D)), \tag{2}$$

where $|S|$, $\alpha_R(D)$, $\kappa_R(D)$, and $P(S)$ denote the cardinality of a set S, a classification accuracy of R as to classification of D, and coverage (a true positive rate of R to D), and probability of S, respectively.

It is notable that $\alpha_R(D)$ measures the degree of the sufficiency of a proposition, $R \rightarrow D$, and that $\kappa_R(D)$ measures the degree of its necessity. For example, if $\alpha_R(D)$ is equal to 1.0, then $R \rightarrow D$ is true. On the other hand, if $\kappa_R(D)$ is equal to 1.0, then $D \rightarrow R$ is true. Thus, if both measures are 1.0, then $R \leftrightarrow D$.

3.3 *Probabilistic Rules*

By use of accuracy and coverage, a probabilistic rule is defined as:

$$R \overset{\alpha,\kappa}{\rightarrow} d \quad s.t.\, R = \wedge_j [a_j = v_k], \alpha_R(D)\, \delta_\alpha \text{ and } \kappa_R(D)\, \delta_\kappa, \tag{3}$$

where D denotes a set of samples that belong to a class d. If the thresholds for accuracy and coverage are set to high values, the meaning of the conditional part of probabilistic rules corresponds to the highly overlapped region. This rule is a kind of probabilistic proposition with two statistical measures, which is an extension of Ziarko's variable precision model (VPRS) [18].[2]

It is also notable that both a positive rule and a negative rule are defined as special cases of this rule, as shown in the next sections.

4 Formalization of Medical Diagnostic Rules

4.1 *Deterministic Model*

4.1.1 Positive Rules

A positive rule is defined as a rule supported by only positive examples. Thus, the accuracy of its conditional part to a disease is equal to 1.0. Each disease may have many positive rules. If we focus on the supporting set of a rule, it corresponds to a subset of the lower approximation of a target concept, which is introduced in rough sets [2]. Thus, a positive rule is defined as:

$$R \rightarrow d \quad s.t. \qquad R = \wedge_j [a_j = v_k], \quad \alpha_R(D) = 1.0 \tag{4}$$

where D denotes a set of samples that belong to a class d.

[2]This probabilistic rule is also a kind of *rough modus ponens* [3].

This positive rule is often called a deterministic rule. However, we use the term, positive (deterministic) rules, because a deterministic rule supported only by negative examples, called a negative rule, is introduced below.

4.1.2 Negative Rules

The important point is that a negative rule can be represented as the contrapositive of an exclusive rule [17]. An exclusive rule is defined as a rule whose supporting set covers all the positive examples. That is, the coverage of the rule to a disease is equal to 1.0. That is, an exclusive rule represents the necessity condition of a decision. The supporting set of an exclusive rule corresponds to the upper approximation of a target concept, which is introduced in rough sets [2]. Thus, an exclusive rule is defined as:

$$R \rightarrow d \quad s.t. \qquad R = \vee_j [a_j = v_k], \quad \kappa_R(D) = 1.0, \tag{5}$$

where D denotes a set of samples that belong to a class d.

Next, let us consider the corresponding negative rules in the following way. An exclusive rule should be described as:

$$d \rightarrow \vee_j [a_j = v_k],$$

because the condition of an exclusive rule corresponds to the necessity condition of conclusion d. Since a negative rule is equivalent to the contrapositive of an exclusive rule, it is obtained as:

$$\wedge_j \neg [a_j = v_k] \rightarrow \neg d,$$

which means that if a case does not satisfy any attribute value pairs in the condition of a negative rule, then we can exclude a decision d from candidates.

Thus, a negative rule is represented as:

$$\wedge_j \neg [a_j = v_k] \rightarrow \neg d \quad s.t. \quad \forall [a_j = v_k] \kappa_{[a_j = v_k]}(D) = 1.0, \tag{6}$$

where D denotes a set of samples that belong to a class d.

Negative rules should also be included in a category of deterministic rules, because their coverage, a measure of negative concepts, is equal to 1.0. It is also notable that the set supporting a negative rule corresponds to a subset of negative region, which is introduced in rough sets [2].

In summary, positive and negative rules correspond to positive and negative regions defined in rough sets. Figure 2 shows the Venn diagram of those rules.

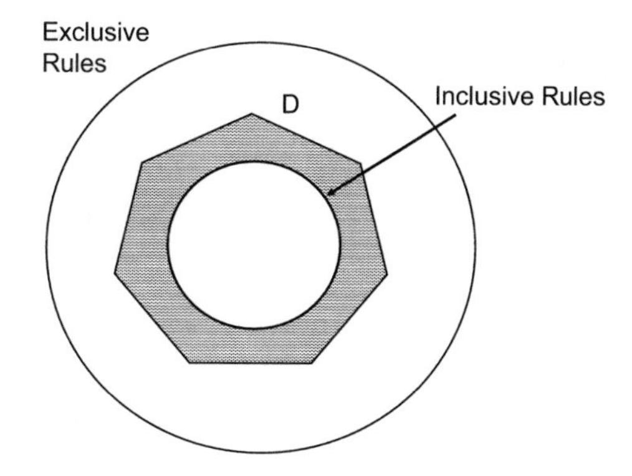

Fig. 2 Venn diagram of exclusive and positive rules

4.2 Probabilistic Model

Although the above deterministic model exactly corresponds to original Pawlak rough set model, rules for differential diagnosis is strict for clinical setting, because clinical diagnosis may include elements of uncertainty.[3] Tsumoto [5] relaxes the condition of positive rules and defines an inclusive rules, which models the inclusive rules of RHINOS model. The definition is almost the same as probabilistic rules defined in Sect. 3, except for the constraints for accuracy: the threshold for accuracy is sufficiently high. Thus, the definitions of rules are summarized as follows.

4.2.1 Exclusive Rules

$$\begin{aligned} R \to d \qquad & s.t.\ R = \vee_j [a_j = v_k], \\ & (s.t. \quad \kappa_{[}a_j = v_k](D) > \delta_\kappa) \\ & \kappa_R(D) = 1.0. \end{aligned} \tag{7}$$

4.2.2 Inclusive Rules

$$\begin{aligned} R \overset{\alpha,\kappa}{\to} d \qquad & s.t.\ R = \wedge_j [a_j = v_k], \\ & \alpha_R(D) > \delta_\alpha \text{ and } \kappa_R(D) > \delta_\kappa. \end{aligned} \tag{8}$$

[3]However, deterministic rule induction model is still powerful in knowledge discovery context as shown in [8].

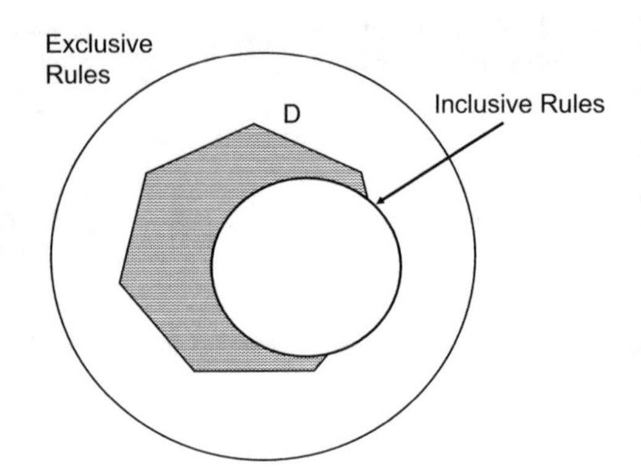

Fig. 3 Venn diagram of exclusive and inclusive rules

In summary, positive and negative rules correspond to positive and negative regions defined in variable rough set model [18]. Figure 3 shows the Venn diagram of those rules.

Tsumoto introduces an algorithm for induction of exclusive and inclusive rules as PRIMEROSE-REX and conducted experimental validation and compared induced results with rules manually acquired from medical experts [5]. The results show that the rules do not include components of hierarchical diagnostic reasoning. Medical experts classify a set of diseases into groups of similar diseases and their diagnostic reasoning is multi-staged: first, different groups of diseases are checked, then final differential diagnosis is performed with the selected group of diseases. In order to extend the method into induction of hierarchical diagnostic rules, one of the authors proposes several approach to mining taxonomy from a dataset in [6, 9, 10].

5 New Rule Induction Model

The former rule induction models do not include reasoning about detection of complications, which is introduced as *disease image* as shown in Sect. 1. The core idea is that medical experts detect the symptoms which cannot be frequently occurred in the final diagnostic candidates. For example, let us assume that a patient suffering from muscle contraction headache, who usually complains of persistent pain, also complains of paroxysmal pain, say he/she feels a strong pain every 1 month. The situation is unusual and since paroxysmal pain is frequently observed by migraine, medical experts suspect that he/she suffers from muscle contraction headache and common migraine. Thus, a set of symptoms which are not useful for diagnosis of a disease may be important if they belong to the set of symptoms frequently manifested in other diseases. In other means, such set of symptoms will be elements of detection of complications. Based on these observations, complications detection rules can be defined as follows.

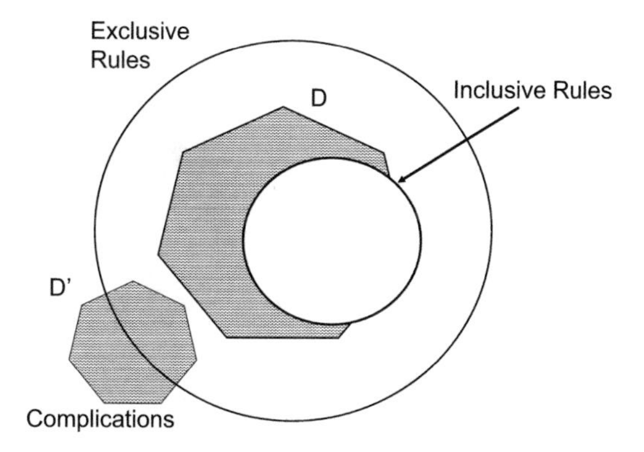

Fig. 4 Venn diagram of exclusive, inclusive and complications detection rules. Type 1

5.1 Complications Detection Rules

Complications detection rule of a diseases are defined as a set of rules each of which is included into inclusive rules of other diseases.[4]

$$\{R \to d \quad s.t. \quad R = [a_i = v_j], 0 \le \alpha_R(D) \le \delta_\alpha, 0 \le \kappa_R(D) \le \delta_\kappa$$
$$\exists D', \alpha_R(D') > \delta_\alpha, \kappa_R(D') > \delta_\kappa\} \tag{9}$$

Figures 4 and 5 depict the relations between exclusive, inclusive and complications detection rules. The first type shows when a symptom for complicated disease can be observed in the main diagnosis. On the other hand, in the second type, a symptom will not be observed in the main diagnosis. Compared with the first case, the second one may be more difficult, because complicated diseases may be filtered out from diagnostic candidates.

The relations between three types of rules can be visualized in a two dimensional plane, called (α, κ)*-plane*, as shown in Fig. 6. The vertical and horizontal axis denotes the values of accuracy and coverage, respectively. Then, each rule can be plotted in the plane with its accuracy and coverage values. The region for inclusive rules is shown in upper right, whereas the region for candidates of detection of complications is in lower left. When a rule of that region belongs to an inclusive rule of other disease, it is included into complications detection rule of the target diseases.

Figure 7 shows the relations of rules for complications of disease D and $D2$. Two (α, κ)*-plane* should be considered in this case, but the regions for both sides are complimentary.

[4]The first term $R = [a_i = v_j]$ may not be needed theoretically. However, since deriving conjunction in an exhaustive way is sometimes computationally expensive, here this constraint is imposed for computational efficiency.

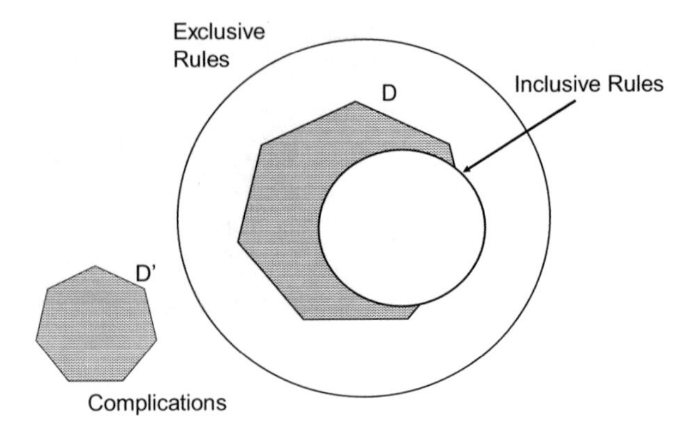

Fig. 5 Venn diagram of exclusive, inclusive and complications detection rules. Type 2

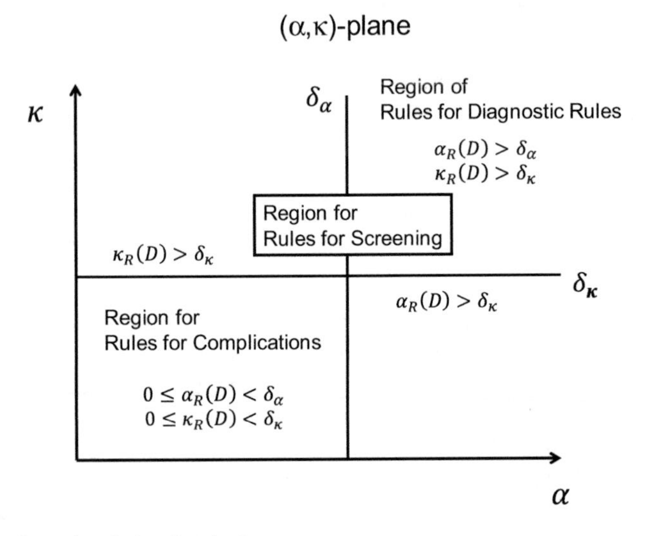

Fig. 6 Two dimensional plot: (α, κ)-*plane*

6 Rule Induction Algorithm

From Eqs. 8 to 9, rule induction algorithms can be described as search algorithms using the inequalities of accuracy and coverage as follows.

Algorithm 1 classifies a formula R in terms of a given decision D in which a parameter *Level* denotes the number of attribute-value pairs in R. First, calculate accuracy $\alpha_R(D)$ and coverage $\kappa_R(D)$ from Eqs. (1) and (2). Then, if both of the values are larger than given thresholds, the formula R will be included into the list of the candidates for rules, denoted by $List_{out}(Level)$.

Algorithm 2 is a main routine of induction of inclusive rules. First, a set of elementary formula, that is, a formula which has only a single attribute-value pair is

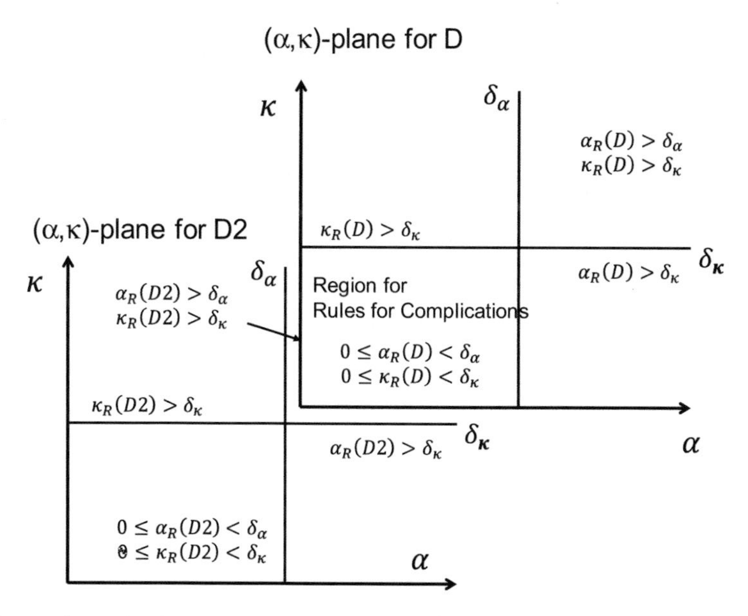

Fig. 7 Two dimensional plot: (α, κ)-*plane* for disease image

given. For each formula in a given *Level*, select one formula from a elementary formula set, say $[a = v]$ and make conjunction of R and $[a = v]$. For example, if *Level* $= 2$, R is of form $[a_1 = v_1] \wedge [a_2 = v_2]$, and if a selected elementary formula is $[a_3 = v_3]$, then a new conjunctive formula will be $[a_1 = v_1] \wedge [a_2 = v_2] \wedge [a_3 = v_3]$. Then, accuracy and coverage of the formula will be checked whether it satisfies the condition for inclusive rules. After all the conjunctive procedures are finished, a member of $List_{rule}$(*Level*) is used for a conditional part of a rule. Algorithm 3 shows how to induce exclusive rules. A set of elementary formula and a decision D is given, for each member of a set, accuracy and coverage will be calculated. If the value of a coverage is larger than a threshold, it is included into an output list and calculate the total coverage. If the total is equal to 1.0, then the list of R will be outputed.

Algorithm 4 shows how to induce disease image. First, select one elementary formula and calculate accuracy and coverage. Then, if both of the values satisfies the inequalities, the algorithm checks whether this formula may have high values of accuracy and coverage for another disease. If so, the formula is included into a list of disease image. Since there are described as a heuristic search algorithm with the constraints of accuracy and coverage, it is easy to extend them into incremental rule induction as shown in [14]. Algorithms 5 and 6 shows how to induce probabilistic rules incrementally. Algorithm 5 can be viewed as an extension of Algorithm 1. Here, a formula R will be classified into three parts: first, regular candidates for probabilistic rules, an element of $List_{rule}$(*Level*). Secondly, a member of subrule layer out, $List_{sub\_out}$ which may be deleted from the candidate if a unsupportive case is appended. Finally, a member of subrule layer in $List_{sub\_in}$, whose element may be included into the candidate if a supportive case is appended.

Algorithm 6 is a main part of incremental rule induction, an extension of Algorithm 2. Here, based on the classification obtained by Algorithm 5, a set of formula is classified into rule layers and two subrule layers.

Algorithm 1 Checking inequalities for Probabilistic Rules

---

procedure CLASSIFICATION OF FORMULA(R:formula, D:decision,Level)
  $Level \leftarrow$ Number of attribute-value pairs in R
  Calculate $\alpha_R(D)$ and $\kappa_R(D)$
  if $\alpha_R(D) > \delta_\alpha$ $\kappa_R(D) > \delta_\kappa$ **then**
    $List_{out}(Level) \leftarrow List_{out}(Level) + \{R\}$
  end if
end procedure

Algorithm 2 Induction of Probabilistic Rules

---

procedure RULE INDUCTION($List_{rule}(0)$: A Set of Elementary Formula, D:decision)
  for $Level = 1$ to Number of Attributes **do**
    for all $R \in List_{rule}(Level - 1)$ **do**
                                        $\triangleright$ $List_{rule}(0) = 1$: $[x]_1 = U$
      for all $[a = v] \in List_{rule}(1)$ **do**
        $R_n \leftarrow R \wedge [a = v]$
        Execute **Procedure**
          Classification_of_Formula(R_n,D,Level)
      end for
    end for
    for all $R \in List_{rule}(Level)$ **do**
      Register $R \rightarrow D$ as a Rule
    end for
  end for
end procedure

---

7 Discussion: What Has Not Been Achieved?

In [11], one of the authors discusses the characteristics of differential diagnosis of headache as follows: (a) Hierarchical classification is used. (b) A set of symptoms is used to describe each disease. (c) Description is based on specificity weighted over sensitivity, which shows that reasoning about frequency is implicitly included. (d) For coverage, exceptions are described. (e) Diagnostic criteria gives temporal information about episodes of headache. In the previous studies, automated extraction of knowledge with respect to (a), (b), (c) has been solved.

Algorithm 3 Induction of Exclusive Rule

procedure EXCLUSIVE RULE INDUCTION($List_{rule}(exclusive, 0)$, A Set of Elementary Formula, D:decision)
  $List_{out}(exclusive, 0) \leftarrow \{\}$
  for all R in A Set of Elementary Formula **do**
    Calculate $\alpha_R(D)$ and $\kappa_R(D)$
    if $\kappa_R(D) > \delta_\kappa$ **then**
      $List_{out}(exclusive, 0) \leftarrow List_{out}(exclusive, 0) + \{R\}$
      Calculate $\kappa_{List_{out}(exclusive,0)}(D)$
      if $\kappa_{List_{out}(exclusive,0)}(D) = 1.0$ **then**
        quit
      end if
    end if
  end for
  Output $List_{out}(exclusive, 0)$
end procedure

Algorithm 4 Induction of Disease Image

procedure DISEASE IMAGE INDUCTION($List_{rule}(0)$: A Set of Elementary Formula, D:decision, $List(D)$: List of Diseases)
  $List_{out}(image, 0) \leftarrow \{\}$
  for all R in A Set of Elementary Formula **do**
    Calculate $\alpha_R(D)$ and $\kappa_R(D)$
    if $0 \leq \alpha_R(D) \leq \delta_\alpha$ $0 \leq \kappa_R(D) \leq \delta_\kappa$ **then**
      for all $D' \in List(D)$ **do**
        if $\alpha_R(D') > \delta_\alpha$ $\kappa_R(D') > \delta_\kappa$ **then**
          $List_{out}(image, 0)$
            $\leftarrow List_{out}(image, 0) + \{(R, D')\}$
        end if
      end for
    end if
  end for
  Output $List_{out,image}(0)$
end procedure

However, (d) and (e) still remains. Dealing with exceptions is related with complications detection, so partially (d) is solved. However, in some cases, exceptions are used for case-based reasoning by medical experts. Thus, combination of rule-based and case-based reasoning should be introduced.

Acquisition of temporal knowledge is important because medical experts use temporal reasoning in a flexible way. When one of the author interviewed the domain expert for RHINOS, he found that temporal reasoning is very important for complicated cases. For example, one patient suffers from both common migraine and tension headache. According to the diagnostic rules, RHINOS diagnoses the case as migraine. However, the main complaint came from tension headache. Since the onset of tension headache is persistent but the severity is mild, the patient focuses on the symptoms of migraine. If the system can focus on the differences in temporal natures of headaches, then it can detect the complications of migraine and tension

Algorithm 5 Construction of Rule Layer

```
procedure CLASSIFICATION OF FORMULA(R:formula, D:decision,Level)
    Level ← Number of attribute-value pairs in R
    Calculate α_R(D) and κ_R(D)
    if α_R(D) > δ_α κ_R(D) > δ_κ then
        List_rule(Level) ← List_rule(Level) + {R}
        if δ_α < α_R(D)(t) < δ_α(n_R+1)/n_R and
              δ_κ < κ_R(D)(t) < δ_κ(n_D+1)/n_D then
            List_sub_out(Level) ← List_sub_out(Level) + {R}
        end if
    else if (δ_α(n_R+1)−1)/n_R < α_R(D)(t) ≤ δ_α and
              (δ_κ(n_D+1)−1)/n_D < κ_R(D)(t) ≤ δ_κ then
        List_sub_in(Level) ← List_sub_in(Level) + {R}
    end if
    List_out(Level) ← List_out(Level) + {R}
end procedure
```

Algorithm 6 Incremental Rule Induction

```
procedure INCREMENTAL RULE INDUCTION(Table, D:decision)
    List_rule(0) ← a Set of Elementary Formula of Table
    Execute Procedure Rule Induction(List_rule(0), D)
    List_rule ← ∪_{i=1} List_rule(i)
    List_sub_in ← ∪_{i=1} List_sub_in(i)
    List_sub_out ← ∪_{i=1} List_sub_out(i)
    List_out ← ∪_{i=1} List_out(i)
    repeat
        Read a New Case x
        for all R ∈ List_rule do
            Execute Procedure Classification of Formula(R:formula,
              D:decision,Level)
        end for
        for all R ∈ New_List_rule do
            Register R → D as a Rule
        end for
        for all R ∈ New_List_sub_in do
            Delete R → D from a set of Rule
            Register R → D as a SubRule (in)
        end for
        for all R ∈ New_List_out do
            Delete R → D from a set of Rule
            Register R → D as a set of SubRule (out)
        end for
    until Abort
end procedure
```

headache. Thus, temporal reasoning is a key to diagnose completed cases especially when all the symptoms may give a contradict interpretation.

Research on temporal data mining is ongoing, and now the authors show that temporal data mining is very important for risk management in several fields [12, 13, 15]. It will be our future work to develop methodologies for combination of rule-based and case-base reasoning and temporal rule mining in clinical data.

8 Conclusion

Formalization of medical diagnostic reasoning based on symptomatology is discussed. Reasoning consists of three processes, exclusive reasoning, inclusive reasoning and complications detection, the former two of which belongs to a focusing mechanism. In exclusive reasoning, a disease is ruled out from diagnostic candidates when a patient does not have symptoms necessary for diagnosis. The process corresponds to screening. Second, in inclusive reasoning, a disease out of selected candidates is suspected when a patient has symptoms specific to a disease, which corresponds to differential diagnosis. Finally, if symptoms which are rarely observed in the final candidate, complication of other diseases will be suspected.

Previous studies are surveyed: one of the author concentrate on the focusing mechanism. First, in a deterministic version, two steps are modeled as two kinds of rules obtained from representations of upper and lower approximation of a given disease. Then, he extends it into probabilistic rule induction, which can be viewed as an application of VPRS.

Then, the authors formalize complications detection rules in this chapter. The core idea is that the rules are not simply formalized by the relations between a set of symptoms and a disease, but by those between a symptoms, a target disease and other diseases. The next step will be to introduce an efficient algorithm to generate complication detection rules from data.

Acknowledgements The author would like to thank past Professor Pawlak for all the comments on my research and his encouragement. Without his influence, the author would neither have received Ph.D. on computer science, nor become a professor of medical informatics. The author also would like to thank Professor Jerzy Grzymala-Busse, Andrezj Skowron, Roman Slowinski, Yiyu Yao, Guoyin Wang, Wojciech Ziarko for their insightful comments. This research is supported by Grant-in-Aid for Scientific Research (B) 15H2750 from Japan Society for the Promotion of Science (JSPS).

References

1. Matsumura, Y., Matsunaga, T., Maeda, Y., Tsumoto, S., Matsumura, H., Kimura, M.: Consultation system for diagnosis of headache and facial pain: "rhinos". In: Wada E. (ed.) LP. Lecture Notes in Computer Science, vol. 221, pp. 287–298. Springer (1985)

2. Pawłak, Z.: Rough Sets. Kluwer Academic Publishers, Dordrecht (1991)
3. Pawłak, Z.: Rough modus ponens. In: Proceedings International Conference on Information Processing and Management of Uncertainty in Knowledge-Based Systems 98, Paris (1998)
4. Skowron, A., Grzymala-Busse, J.: From rough set theory to evidence theory. In: Yager, R., Fedrizzi, M., Kacprzyk, J. (eds.) Advances in the Dempster-Shafer Theory of Evidence, pp. 193–236. John Wiley & Sons, New York (1994)
5. Tsumoto, S.: Automated induction of medical expert system rules from clinical databases based on rough set theory. Inf. Sci. **112**, 67–84 (1998)
6. Tsumoto, S.: Extraction of experts' decision rules from clinical databases using rough set model. Intell. Data Anal. **2**(3) (1998)
7. Tsumoto, S.: Modelling medical diagnostic rules based on rough sets. In: Rough Sets and Current Trends in Computing, pp. 475–482 (1998)
8. Tsumoto, S.: Automated discovery of positive and negative knowledge in clinical databases based on rough set model. IEEE Eng. Med. Biol. Mag. **19**, 56–62 (2000)
9. Tsumoto, S.: Extraction of hierarchical decision rules from clinical databases using rough sets. Inf. Sci. (2003)
10. Tsumoto, S.: Extraction of structure of medical diagnosis from clinical data. Fundam. Inform. **59**(2–3), 271–285 (2004)
11. Tsumoto, S.: Rough sets and medical differential diagnosis. In: Skowron, A., Suraj, Z. (eds.) Rough Sets and Intelligent Systems (1), Intelligent Systems Reference Library, vol. 42, pp. 605–621. Springer (2013)
12. Tsumoto, S., Hirano, S.: Risk mining in medicine: application of data mining to medical risk management. Fundam. Inform. **98**(1), 107–121 (2010)
13. Tsumoto, S., Hirano, S.: Detection of risk factors using trajectory mining. J. Intell. Inf. Syst. **36**(3), 403–425 (2011)
14. Tsumoto, S., Hirano, S.: Incremental induction of medical diagnostic rules based on incremental sampling scheme and subrule layers. Fundam. Inform. **127**(1–4), 209–223 (2013)
15. Tsumoto, S., Hong, T.P.: Special issue on data mining for decision making and risk management. J. Intell. Inf. Syst. **36**(3), 249–251 (2011)
16. Tsumoto, S., Tanaka, H.: Induction of probabilistic rules based on rough set theory. In: Jantke, K.P., Kobayashi, S., Tomita, E., Yokomori, T. (eds.) Proceedings of the Algorithmic Learning Theory, 4th International Workshop, ALT '93, Tokyo, Japan, 8–10 Nov 1993. Lecture Notes in Computer Science, vol. 744, pp. 410–423. Springer (1993). doi:10.1007/3-540-57370-4_64
17. Tsumoto, S., Tanaka, H.: Automated discovery of medical expert system rules from clinical databases based on rough sets. In: Proceedings of the Second International Conference on Knowledge Discovery and Data Mining 96, pp. 63–69. AAAI Press, Palo Alto (1996)
18. Ziarko, W.: Variable precision rough set model. J. Comput. Syst. Sci. **46**, 39–59 (1993)

Rough Set Analysis of Imprecise Classes

Masahiro Inuiguchi

Abstract Lower approximations of single decision classes have been mainly treated in the classical rough set approaches. Attribute reduction and rule induction have been developed based on the lower approximations of single classes. In this chapter, we propose to use the lower approximations of unions of k decision classes instead of the lower approximations of single classes. We first show various kinds of attribute reduction are obtained by the proposed approach. Then we consider set functions associated with attribute reduction and demonstrate that the attribute importance degrees defined from set functions are very different depending on k. Third, we consider rule induction based on the lower approximations of unions of k decision classes and show that the classifiers with rules for unions of k decision classes can perform better than the classifiers with rules for single decision classes. Finally, utilization of rules for unions of k decision classes in privacy protection is proposed. Throughout this chapter, we demonstrate that the consideration of lower approximations of unions of k classes enriches the applicability of rough set approaches.

1 Introduction

Rough set theory [25, 26] provides useful tools for reasoning from data. Attribute reduction and rule induction are well developed techniques based on rough set theory. They are applied to various fields including data analysis, signal processing, knowledge discovery, machine learning, artificial intelligence, medical informatics, decision analysis, granular computing, Kansei engineering, and so forth [3, 20, 28].

In rough set approach, the lower approximation (a set of objects whose classification is consistent in all given data) and upper approximation (a set of possible members in view of given data) are calculated for a set of objects. Lower and upper approximations of the classical rough sets [25, 26] are defined based on an equivalence relation called an 'indiscernibility relation'. They are extended in many ways

M. Inuiguchi (✉)
Graduate School of Engineering Science, Osaka University, Toyonaka, Osaka 560-8531, Japan
e-mail: inuiguti@sys.es.osaka-u.ac.jp

G. Wang et al. (eds.), *Thriving Rough Sets*, Studies in Computational Intelligence 708, DOI 10.1007/978-3-319-54966-8_8

depending on the necessity in applications. For example, the equivalence relation is replaced with a similarity relation as a natural generalization [33], with a dominance relation in decision making problems [6, 7], with a tolerance relation for treating decision tables with missing data [22], with a fuzzy relation for a generalized setting that a degree of similarity is available [2, 11]. Moreover, using a precision degree and a consistency degree, the rough sets are generalized to variable precision rough sets [37] and variable consistency rough sets [1] are proposed. Under those generalized rough sets, various techniques for attribute reduction and rule induction are developing.

However, the usage of rough sets has not yet investigated considerably. The lower approximation of each decision class has been majorly used for obtaining attribute reduction and rule induction, so far, although some studies using upper approximation of each decision class (see [19, 30]). Other sets of objects have not yet used actively. In the dominance based rough set approach [6, 7], upward and downward unions are used because they match well to the dominance relation (in other words, a single decision class does not work well for obtaining lower and upper approximation under dominance relations). Inuiguchi et al. [18] proposed to use upward and downward unions when the decision attribute is ordinal and showed its advantage in the classification accuracy of the obtained classifier.

Recently, in the classical rough set setting, the authors [10, 12, 14–17] proposed to use the lower approximations of unions of k decision classes instead of lower approximations of single decision classes and demonstrated the interesting and useful results. This approach can be seen as a rough set approach to imprecise modeling because it provides the analysis based on the preservation of imprecise classification, i.e., correct classification up to k possible decision classes. After a brief introduction of the classical rough set approaches, we describe the following recent results obtained by the replacement of the lower approximation of each decision class with that of each union of k decision classes:

(1) In the first part of Sect. 3, the attribute reduction based on lower approximations of unions of k decision classes provides an intermediate between two extreme attribute reductions using lower and upper approximations of single decision classes. These two extremes are obtained by special parameter settings of k.
(2) In the last part of Sect. 3, it shows that the evaluation of attribute importance changes drastically by the selection of parameter k. It implies that the attribute importance cannot be evaluated univocally.
(3) In the major part of Sect. 4, the classifier with rules induced for unions of k decision classes achieves a better performance than the classifier with rules induced for single decision classes.
(4) In the last part of Sect. 4, we describe the possible utilization of rules for k decision classes in the protection of data privacy.

Before the main part of this chapter, we briefly introduce the classical rough set approaches, and after describing (1)–(4) shown above, we conclude this chapter with giving some remarks for future investigation. In this chapter, as we consider unions of k decision classes ($k \geq 2$), we assume that a decision table with multiple decision classes (more than two decision classes) is given.

2 Rough Sets in Decision Tables

The classical rough sets are defined under an equivalence relation which is often called an indiscernibility relation. In this chapter, we restrict ourselves to discussions of the classical rough sets under decision tables. A decision table is characterized by four-tuple $\mathscr{I} = \langle U, C \cup \{d\}, V, \rho \rangle$, where U is a finite set of objects, C is a finite set of condition attributes, d is a decision attribute, $V = \bigcup_{a \in C \cup \{d\}} V_a$ and V_a is a domain of the attribute a, and $\rho : U \times C \cup \{d\} \to V$ is an information function such that $\rho(x, a) \in V_a$ for every $a \in C \cup \{d\}$, $x \in U$. A condition attribute value vector $\boldsymbol{\rho}(u, A) = (\rho(u, a_1), \rho(u, a_2), \ldots, \rho(u, a_l))$ of an object $u \in U$ is called a profile of u in A, where $A = \{a_1, a_2, \ldots, a_l\} \subseteq C$. The profile of u in C is simply called the profile of u. Multiple objects can have a common profile. Let Ψ the set of all profiles appearing in the decision table. Let $fr : \Psi \times V_d \to \mathbf{N} \cup \{0\}$ be a function showing the frequency of objects having $v \in V_d$ in the set of objects having a profile $\boldsymbol{\rho} \in \Psi$, where $\mathbf{N}$ be a set of natural numbers. A set of frequency vector $(fr(\boldsymbol{\rho}, v_1), fr(\boldsymbol{\rho}, v_2), \ldots, fr(\boldsymbol{\rho}, v_p))$, $\boldsymbol{\rho} \in \Psi$ is denoted by Fr. When the distinction between objects having a same profile is not significant, a decision table $\mathscr{I} = \langle U, C \cup \{d\}, V, \rho \rangle$ can be rewritten by a table $\hat{\mathscr{I}} = \langle \Psi, C \cup d, V, fr \rangle$. An example of the decision table and its representation by profiles are shown in Table 1.

Given a set of attributes $A \subseteq C \cup \{d\}$, we define an equivalence relation I_A referred to as an indiscernibility relation by $I_A = \{(x, y) \in U \times U \mid \rho(x, a) = \rho(y, a), \forall a \in A\}$. From I_A, we have an equivalence class, $[x]_A = \{y \in U \mid (y, x) \in I_A\}$. When $A = \{d\}$, we define

Table 1 Decision table and its representation by profiles

(a) Decision table

Object	a_1	a_2	a_3	d
u_1	modern	modern	round	class 1
u_2	modern	modern	round	class 1
u_3	modern	modern	round	class 2
u_4	modern	classic	round	class 3
u_5	modern	classic	round	class 1
u_6	modern	modern	cubed	class 2
u_7	modern	modern	cubed	class 2
u_8	modern	modern	cubed	class 2
u_9	classic	classic	round	class 3
u_{10}	classic	classic	round	class 3

⇒

(b) Representation by profiles

Profile	a_1	a_2	a_3	fr-vector
$\boldsymbol{\rho}_1$	modern	modern	round	$(2, 1, 0)$
$\boldsymbol{\rho}_2$	modern	classic	round	$(1, 0, 1)$
$\boldsymbol{\rho}_3$	modern	modern	cubed	$(0, 3, 0)$
$\boldsymbol{\rho}_4$	classic	classic	round	$(0, 0, 2)$

$$\mathscr{D} = \{D_j, j = 1, 2, \ldots, p\} = \{[x]_{\{d\}} \mid x \in U\},\ D_i \neq D_j\ (i \neq j). \tag{1}$$

D_j is called a 'decision class'. There exists a unique $v_j \in V_d$ such that $\rho(x, d) = v_j$ for each $x \in D_j$, i.e., $D_j = \{x \in U \mid \rho(x, d) = v_j\}$. Moreover, since $D_i \cap D_j = \emptyset\ (i \neq j)$ and $\bigcup \mathscr{D} = U$ hold, $\mathscr{D}$ forms a partition.

For a set of condition attributes $A \subseteq C$, the lower and upper approximations of an object set $X \subseteq U$ are defined as follows:

$$A_*(X) = \{x \mid [x]_A \subseteq X\},\ \ A^*(X) = \{x \mid [x]_A \cap X \neq \emptyset\}. \tag{2}$$

A pair $(A_*(X), A^*(X))$ is called a rough set of X. The boundary region of X is defined by

$$BN_A(X) = A^*(X) - A_*(X). \tag{3}$$

Since $[x]_A$ can be seen as a set of objects indiscernible from $x \in U$ in view of condition attributes in A, $A_*(X)$ is interpreted as a collection of objects whose membership to X is noncontradictive in view of condition attributes in A. $BN_A(X)$ is interpreted as a collection of objects whose membership to X is doubtful in view of condition attributes in A. $A^*(X)$ is interpreted as a collection of possible members. For $x \in U$, the generalized decision attribute value $\partial_A(x)$ of x with respect to a condition attribute set $A \subseteq C$ is defined as follows (see [27, 30, 31]):

$$\partial_A(x) = \{\rho(y, d) \mid y \in [x]_A\}. \tag{4}$$

Let $X, Y \subseteq U$. The following fundamental properties are satisfied with rough sets:

$$A_*(X) \subseteq X \subseteq A^*(X), \tag{5}$$

$$A \subseteq B \Rightarrow A_*(X) \subseteq B_*(X),\ A^*(X) \supseteq B^*(X), \tag{6}$$

$$A_*(X \cap Y) = A_*(X) \cap A_*(Y),\ \ A^*(X \cup Y) = A^*(X) \cup A^*(Y), \tag{7}$$

$$A_*(X \cup Y) \supseteq A_*(X) \cup A_*(Y),\ \ A^*(X \cap Y) \subseteq A^*(X) \cap A^*(Y), \tag{8}$$

$$BN_A(X) = A^*(X) \cap A^*(U - X), \tag{9}$$

$$A_*(X) = X - BN_A(X), \tag{10}$$

$$A^*(X) = X \cup BN_A(X) = U - A_*(U - X), \tag{11}$$

$$A_*(X) = A^*(X) - A^*(U - X) = U - A^*(U - X). \tag{12}$$

Let X_i, $i = 1, 2, \ldots, q$ forms a partition, i.e., $\bigcup_{i=1,2,\ldots,q} X_i = U$, $X_i \cap X_j = \emptyset$ for $i, j \in \{1, 2, \ldots, q\}$ such that $i \neq j$. The following properties show the interpretation among lower and upper approximations and boundary regions:

$$A^*(X_j) = X_j \cup BN_A(X_j),\ j = 1, 2, \dots, q, \tag{13}$$

$$BN_A(X_j) = A^*(X_j) \cap \bigcup_{i \neq j} A^*(X_i),\ j = 1, 2, \dots, q, \tag{14}$$

$$A_*(X_j) = A^*(X_j) - \bigcup_{i \neq j} A^*(X_i),\ j = 1, 2, \dots, q, \tag{15}$$

$$A_*(X_j) = X_j - BN_A(X_j),\ j = 1, 2, \dots, q. \tag{16}$$

Equations (13) and (16) show that upper and lower approximations of X_j can be obtained from the boundary region of X_j. Equations (14) and (15) show that the boundary region and lower approximation of X_j can be obtained from upper approximations of X_i, $i = 1, 2, \dots, q$.

3 Attribute Reduction and Importance

3.1 *Attribute Reduction*

3.1.1 Conventional Attribute Reduction

A given decision table can include superfluous condition attribute to the decision attribute. It is significant to find the necessary condition attributes for the determination of decision attribute values. The selection of necessary condition attributes is called 'feature selection' while the elimination of unnecessary condition attributes is called 'attribute reduction'. By utilizing rough sets, we can find sets of minimally necessary condition attributes to classify objects without the deterioration of classification accuracy. A set of minimally necessary attributes is called a 'reduct'. Finding reducts is one of the major topics in rough set approaches. Finding all reducts reveals indispensable condition attributes.

In the classical rough set analysis of decision tables, reducts preserving lower approximations of decision classes D_j, $j = 1, 2, \dots, p$ are frequently used. The attribute reduction called a reduct is defined as follows (see [26, 36]).

Definition 1 A set of condition attributes, $A \subseteq C$ is called a reduct if and only if it satisfies

(L1) $A_*(D_j) = C_*(D_j), j = 1, 2, \dots, p$, and
(L2) $\forall a \in A, (A - \{a\})_*(D_j) \neq C_*(D_j), j = 1, 2, \dots, p$.

Since we discuss several kinds of reducts, we call this reduct, a 'reduct preserving lower approximations' or an 'L-reduct' for short. Let $\mathscr{R}^{\mathrm{L}}$ be a set of L-reducts. Then $\bigcap \mathscr{R}^{\mathrm{L}}$ is called the 'core preserving lower approximation' or the 'L-core'. Attributes in the L-core are important because we cannot preserve all lower approximations of decision classes without any of them. Set $A \subseteq C$ satisfying (L1) is called a 'superreduct preserving lower approximation' or an 'L-superreduct'.

We can consider reducts preserving upper approximations or equivalently, preserving boundary regions [19, 30].

Definition 2 A set of condition attributes, $A \subseteq C$ is called a 'reduct preserving upper approximations' or a 'U-reduct' for short if and only if it satisfies

(U1) $A^*(D_j) = C^*(D_j), j = 1, 2, \ldots, p$, and
(U2) $\forall a \in A, (A - \{a\})^*(D_j) \neq C^*(D_j), j = 1, 2, \ldots, p$.

On the other hand, a set of condition attributes, $A \subseteq C$ is called a 'reduct preserving boundary regions' or a 'B-reduct' for short if and only if it satisfies

(B1) $BN_A(D_j) = BN_C(D_j), j = 1, 2, \ldots, p$, and
(B2) $\forall a \in A, BN_{(A-\{a\})}(D_j) \neq BN_C(D_j), j = 1, 2, \ldots, p$.

For those reducts, we have

(R1) A U-reduct is also a B-reduct and vice versa,
(R2) There exists an L-reduct A for a U-reduct B such that $B \supseteq A$, and
(R3) There exists an L-reduct A for a B-reduct B such that $B \supseteq A$.

Those relations can be proved easily from (13) to (16). Since B-reduct is equivalent to U-reduct, we describe only U-reduct in what follows. Let $\mathscr{R}^{\mathrm{U}}$ be a set of U-reducts. Then $\bigcap \mathscr{R}^{\mathrm{U}}$ is called the 'core preserving upper approximation' or the 'U-core'. Attributes in the U-core are important because we cannot preserve all upper approximations of decision classes without any of them.

To obtain a part or all of reducts, many approaches have been proposed in the literature [26, 32]. Among them, we mention an approach based on a discernibility matrix [27, 32]. In this approach, we construct a Boolean function which characterizes the preservation of the lower approximations to obtain L-reducts. Each L-reduct is obtained as a prime implicant of the Boolean function. For the detailed discussion of the discernibility matrix for L-reducts, see references [27, 32].

Remark 1 Reducts preserving generalized decision attribute values, $\partial_C(u), \forall u \in U$ is also proposed and called '∂-reduct' (see [30]). a set of condition attributes, $A \subseteq C$ is a ∂-reduct if and only if it satisfies

(∂1) $\partial_A(u) = \partial_C(u), \forall u \in U$, and
(∂2) $\forall a \in A, \partial_{(A-\{a\})}(u) \neq \partial_C(u), \forall u \in U$.

∂-reduct is also equivalent to U-reduct as well as to B-reduct. This are understood from the following equations.

$$A^*(D_j) = \{u \mid v_j \in \partial_A(u)\} \quad \text{and} \quad \partial_A(u) = \{v_j \mid u \in A^*(D_j)\}, \tag{17}$$

where we remind you of the definition of D_j, i.e., $D_j = \{u \in U \mid \rho(u, d) = v_j\}, j = 1, 2, \ldots, p$.

3.1.2 Refinement of Attribute Reduction

In the previous subsubsection, we find that L-reducts are smaller than U-reduct. It is interesting to investigate the existence of intermediate reducts between L- and U-reducts. For such intermediate reducts, we assume that all decision classes are treated equally. This has been studied by Inuiguchi [14]. We describe the result.

Consider a cover $\mathscr{F}_k = \{D_{i_1} \cup D_{i_2} \cup \cdots \cup D_{i_k} \mid 1 \leq i_1 < i_2 < \cdots < i_k \leq p\}$ for $k \in \{1, 2, \ldots, p-1\}$. We define attribute reduction based on $\mathscr{F}_k$ as follows.

Definition 3 A condition attribute set A is called an $\mathscr{F}_k$-reduct if and only if

(F1(k)) $A_*(F) = C_*(F)$ for all $F \in \mathscr{F}_k$, and

(F2(k)) $\forall a \in A, (A - \{a\})_*(F) \neq C_*(F)$ for all $F \in \mathscr{F}_k$.

$\mathscr{F}_k$-reducts have the following properties:

(i) From (11) and (12), we know that an $\mathscr{F}_k$-reduct A is a minimal set such that $A^*(F) = C^*(F)$ for all $F \in \mathscr{F}_{p-k}$.
(ii) Because for any $F \in \mathscr{F}_l$, there exists $F_1, F_2 \in \mathscr{F}_k$ such that $F = F_1 \cap F_2$ and $l < k$, from (7), an $\mathscr{F}_k$-reduct A satisfies (F1(l)) for all $l \leq k$, i.e., $A_*(F) = C_*(F)$, for all $F \in \mathscr{F}_l$ for all $l \leq k$.
(iii) From (i) and (ii), an $\mathscr{F}_k$-reduct A satisfies $A^*(F) = C^*(F)$ for all $F \in \mathscr{F}_{p-l}$ and for all $l \leq k$.
(iv) In particular, $\mathscr{F}_1$-reducts are equivalent to L-reducts and $\mathscr{F}_{p-1}$-reducts are equivalent to U-reducts.

From this observation the strong-weak relations among $\mathscr{F}_k$-reducts for $1 \leq k \leq p-1$ can be depicted as in Fig. 1. The reducts located on the upper side of Fig. 1 are strong, i.e., the condition to be the upper reduct is stronger than the lower. On the contrary, the reducts located on the lower side of Fig. 1 are weak, i.e., the condition to be the lower reduct is weaker than the upper. Therefore, for any reduct A located on the upper side, there exists a reduct B located on the lower side such that $B \subseteq A$.

Let $\mathscr{R}(k)$ be a set of $\mathscr{F}_k$-reducts. Then $\bigcap \mathscr{R}(k)$ is called the '$\mathscr{F}_k$-core'. Attributes in the $\mathscr{F}_k$-core are important because we cannot preserve $C_*(F)$ for all $F \in \mathscr{F}_k$ without any of them.

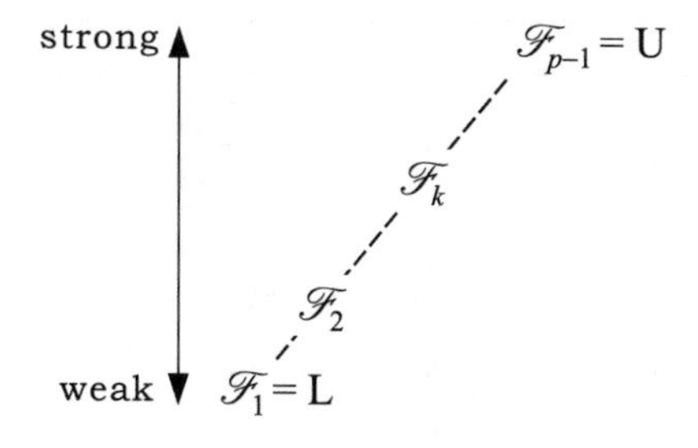

Fig. 1 The strong-weak relation among reducts

As all L-reducts can be calculated using a discernibility matrix [27, 32], all $\mathscr{F}_k$-reducts for $1 \le k \le p-1$ can be calculated by a discernibility matrix. The (i,j)-component $\mathscr{D}_{ij}^k$ of the discernibility matrix $\mathscr{D}^k$ for calculating $\mathscr{F}_k$-reducts is obtained as the following set of attributes:

$$\mathscr{D}_{ij}^k = \begin{cases} \{a \in C \mid \rho(x_i, a) \neq \rho(x_j, a)\}, \\ \qquad\qquad \text{if } \partial_C(x_i) \neq \partial_C(x_j) \text{ and } \min(|\partial_C(x_i)|, |\partial_C(x_j)|) \le k, \\ C \text{ (don't care), otherwise.} \end{cases} \tag{18}$$

While the discernibility matrix shown in [14] is asymmetric, $\mathscr{D}^k$ defined by (18) is symmetric. Therefore, we need to know only the upper triangular portion of $\mathscr{D}^k$, i.e., $\mathscr{D}_{ij}^k$ such that $i < j$.

Then all $\mathscr{F}_k$-reducts are obtained as prime implicants of a Boolean function,

$$f^k = \bigwedge_{i,j:\ x_i,x_j \in U, i<j} \bigvee \mathscr{D}_{ij}^k, \tag{19}$$

where we regard $a \in \mathscr{D}_{ij}^k$ as a statement that 'the reduct includes a'. The computational complexity is NP-hard as in the classical decision matrix method [32].

Note that $\mathscr{D}_{ij}^k$ can be obtained from $\mathscr{D}_{ij}^l$ with $l > k$ by replacing (i,j)-component such that $|\partial_C(x_i)| > k$ and $|\partial_C(x_j)| > k$ with C (don't care). Then, once $\mathscr{D}_{ij}^{p-1}$ is obtained, the other decision matrices can be obtained easily.

Example 1 Consider the decision table given in Table 2 with $C = \{a_1, a_2, a_3, a_4\}$. The decision table is represented by profiles and the corresponding generalized decision attribute values are also given. Namely, *fr*-vector $(1,0,1,0)$ at the row of w_1 implies that there are two objects of profile w_1 and one of them takes decision attribute value v_1 and the other takes decision attribute value v_3. Therefore, the generalized decision attribute value becomes $\{v_1, v_3\}$. Similarly, *fr*-vector $(2,0,0,0)$ at the

Table 2 A decision table

Profile	a_1	a_2	a_3	a_4	*fr*-vector	∂_C
w_1	1	1	1	1	$(1,0,1,0)$	$\{v_1, v_3\}$
w_2	1	1	2	2	$(1,0,0,1)$	$\{v_1, v_4\}$
w_3	2	2	3	1	$(2,0,0,0)$	$\{v_1\}$
w_4	3	3	4	2	$(0,1,0,0)$	$\{v_2\}$
w_5	4	1	2	2	$(0,1,1,0)$	$\{v_2, v_3\}$
w_6	2	5	3	2	$(1,0,0,0)$	$\{v_1\}$
w_7	2	4	4	2	$(0,0,2,0)$	$\{v_3\}$
w_8	4	1	5	5	$(0,1,1,1)$	$\{v_2, v_3, v_4\}$
w_9	4	1	5	4	$(1,0,1,1)$	$\{v_1, v_3, v_4\}$

Table 3 The upper triangular portion of discernibility matrix $\mathscr{D}^3$

	w_2	w_3	w_4	w_5	w_6	w_7	w_8	w_9
w_1	$\{a_3,a_4\}$ $(k \geq 2)$	$\{a_1,a_2,a_3\}$ $(k \geq 1)$	C $(k \geq 1)$	$\{a_1,a_3,a_4\}$ $(k \geq 2)$	C $(k \geq 1)$	C $(k \geq 1)$	$\{a_1,a_3,a_4\}$ $(k \geq 2)$	$\{a_1,a_3,a_4\}$ $(k \geq 2)$
w_2	—	C $(k \geq 1)$	$\{a_1,a_2,a_3\}$ $(k \geq 1)$	$\{a_1\}$ $(k \geq 2)$	$\{a_1,a_2,a_3\}$ $(k \geq 1)$	$\{a_1,a_2,a_3\}$ $(k \geq 1)$	C $(k \geq 2)$	$\{a_1,a_3,a_4\}$ $(k \geq 2)$
w_3	—	—	C $(k \geq 1)$	C $(k \geq 1)$	C $(k \geq 1)$	$\{a_2,a_3,a_4\}$ $(k \geq 1)$	C $(k \geq 1)$	C $(k \geq 1)$
w_4	—	—	—	$\{a_1,a_2,a_3\}$ $(k \geq 1)$	$\{a_1,a_2,a_3\}$ $(k \geq 1)$	$\{a_1,a_2\}$ $(k \geq 1)$	C $(k \geq 1)$	C $(k \geq 1)$
w_5	—	—	—	—	$\{a_1,a_2,a_3\}$ $(k \geq 1)$	$\{a_1,a_2,a_3\}$ $(k \geq 1)$	$\{a_3,a_4\}$ $(k \geq 2)$	$\{a_3,a_4\}$ $(k \geq 2)$
w_6	—	—	—	—	—	$\{a_2,a_3\}$ $(k \geq 1)$	C $(k \geq 1)$	C $(k \geq 1)$
w_7	—	—	—	—	—	—	C $(k \geq 1)$	C $(k \geq 1)$
w_8	—	—	—	—	—	—	—	$\{a_4\}$ $(k \geq 3)$

row of w_3 implies that there are two objects of profile w_1 and both of them take decision attribute value v_1. Accordingly, the generalized decision attribute value becomes $\{v_1\}$.

For this decision table, we calculate all $\mathscr{F}_k$ reducts for $k = 1, 2, 3$. We may apply the discernibility matrix defined by (18) with substitution of w_i for u_i. The upper triangular portion of $\mathscr{D}^k$, $k = 1, 2, 3$ is shown in Table 3. Each entry of Table 3 is composed of a set of condition attributes and the condition of k. It implies that $\mathscr{D}^k_{ij}(i < j)$ takes the set of condition attributes if k satisfies the condition and C otherwise. From (19), applying absorption laws, we obtain

$$f^1 = (a_1 \vee a_2) \wedge (a_2 \vee a_3) = a_2 \vee (a_1 \wedge a_3), \tag{20}$$

$$f^2 = a_1 \wedge (a_2 \vee a_3) \wedge (a_3 \vee a_4) = (a_1 \wedge a_3) \vee (a_1 \wedge a_2 \wedge a_4), \tag{21}$$

$$f^3 = a_1 \wedge (a_2 \vee a_3) \wedge a_4 = (a_1 \wedge a_2 \wedge a_4) \vee (a_1 \wedge a_3 \wedge a_4). \tag{22}$$

Therefore, we find $\mathscr{F}_k$-reducts and $\mathscr{F}_k$-core for $k = 1, 2, 3$ as follows:

$$\begin{array}{llll} \mathscr{F}_1\text{-reducts}: & \{a_2\}, \{a_1, a_3\}; & \mathscr{F}_1\text{-core}: & \emptyset, \\ \mathscr{F}_2\text{-reducts}: & \{a_1, a_3\}, \{a_1, a_2, a_4\}; & \mathscr{F}_2\text{-core}: & \{a_1\}, \\ \mathscr{F}_3\text{-reducts}: & \{a_1, a_2, a_4\}, \{a_1, a_3, a_4\}; & \mathscr{F}_2\text{-core}: & \{a_1, a_4\}. \end{array}$$

As shown above, $\mathscr{F}_k$-reducts as well as $\mathscr{F}_k$-cores are different by $k \in \{1, 2, 3\}$.

Remark 2 Given a family $\mathscr{G}$ on U, we can define $\mathscr{G}$-reducts as follows: A condition attribute set A is called an $\mathscr{G}$-reduct if and only if

(G1) $A_*(F) = C_*(F)$ for all $F \in \mathscr{G}$, and
(G2) $\forall a \in A, (A - \{a\})_*(F) \neq C_*(F)$ for all $F \in \mathscr{G}$.

Let $\mathscr{R}^{\mathscr{G}}$ be a set of U-reducts. Then $\bigcap \mathscr{R}^{\mathscr{G}}$ is called the '$\mathscr{G}$-core'. $A \subseteq C$ satisfying (G1) is called a '$\mathscr{G}$-superreduct'.

3.2 Set Functions Associated with Reducts

Consider the following set function μ^{Q}:

$$\mu^{\mathrm{Q}}(A) = \frac{\sum_{i=1}^{p} |A_*(D_i)|}{|U|}, \tag{23}$$

where μ^{Q} is called a 'quality of approximation' of partition $\mathscr{D}$ and evaluates to what extent the set of condition attributes clearly classifies the objects into decision classes (see [5, 26]). L-reducts can be defined by μ^{Q} as follows: $A \subseteq C$ is an L-reduct if and only if A satisfies

(L1) $\mu^{\mathrm{Q}}(A) = \mu^{\mathrm{Q}}(C)$, and
(L2) $\forall a \in A,\ \mu^{\mathrm{Q}}(A - \{a\}) \neq \mu^{\mathrm{Q}}(C)$.

Because L-reducts can be defined by using μ_Q, we call it an "associated set function" with L-reducts. For a kind of reduct, the associated set function is not unique. For example, $\mu_1^{\mathrm{IG}}(A) = \log p - \left(\sum_{i:|\partial_C(u_i)|=1} \log|\partial_A(u_i)|\right) / |\{i : |\partial_C(u_i)| = 1\}|$ is also the associated set function with L-reducts.

Similarly, associated with U-reducts, the following set functions μ^{sp}, μ^{∂} are conceivable (see [14, 19]):

$$\mu^{\mathrm{sp}}(A) = \frac{\sum_{i=1}^{p} |U - A^*(D_i)|}{(p-1)|U|} = \frac{\sum_{u_i \in U} \left(p - |\partial_A(u_i)|\right)}{(p-1)|U|}, \tag{24}$$

$$\mu^{\partial}(A) = \frac{\sum_{i=1}^{|U|} \left[\partial_C(u_i) = \partial_A(u_i)\right]}{|U|}, \tag{25}$$

where [*statement*] takes 1 if *statement* is true and 0 otherwise. μ^{sp} shows the degree of specificity and evaluates to what extent the set of condition attributes decreases the possible classes of objects. On the other hand, μ^{∂} shows the ratio of objects u whose generalized decision attribute value $\partial_C(u)$ with respect to C is preserved in that with respect to the reduced set $A \subseteq C$. Those are not all of associated set functions with

U-reducts. U-reducts can be defined by μ^{sp} and μ^{∂} as follows: For $\mu = \mu^{\text{sp}}, \mu^{\partial}, A \subseteq C$ is a U-reduct if and only if A satisfies

(U1) $\mu(A) = \mu(C)$, and

(U2) $\forall a \in A$, $\mu(A - \{a\}) \neq \mu(C)$.

Associated set functions with $\mathscr{F}_k$-reducts can be also considered. The following set functions are given by Inuiguchi [12, 14]:

$$\mu_k^{\text{re}}(A) = \begin{cases} 1, \text{ if } A_*(F) = C_*(F),\ \forall F \in \mathscr{F}_k, \\ 0, \text{ otherwise}, \end{cases} \tag{26}$$

$$\mu_k^{\text{sp}}(A) = \frac{\sum_{F \in \mathscr{F}_k} |A_*(F)|}{\binom{p-1}{k-1}|U|} = \frac{\sum_{u_i \in U_k} \binom{p - |\partial_A(u_i)|}{k - |\partial_A(u_i)|}}{\binom{p-1}{k-1}|U|}, \tag{27}$$

$$\mu_k^{\partial}(A) = \frac{\sum_{u_i \in U_k} \left[\partial_C(u_i) = \partial_A(u_i)\right]}{|U|}, \tag{28}$$

where we define

$$U_k = \{u_i \in U : |\partial_C(u_i)| \leq k\}. \tag{29}$$

μ_k^{re} is a characteristic function of $\mathscr{F}_k$-superreduct. μ_k^{sp} and μ_k^{∂} are the generalizations of μ^{sp} and μ^{∂}. Indeed, we have $\mu_p^{\text{sp}}(A) = \mu^{\text{sp}}(A)$ and $\mu_p^{\partial}(A) = \mu^{\partial}(A)$ for all $A \subseteq C$, but we assume $k \leq p - 1$. Even when $k = p - 1$, we have $\mu_{p-1}^{\text{sp}}(A) = \mu^{\text{sp}}(A)$ and $\mu_{p-1}^{\partial}(A) = \mu^{\partial}(A) - \left|\{x \in U \mid |\partial_C(x)| = p\}\right| / |U|$ for all $A \subseteq C$, where we note that $\left|\{x \in U \mid |\partial_C(x)| = p\}\right| / |U|$ is independent of $A \subseteq C$. Moreover, we have $\mu_1^{\text{sp}}(A) = \mu_1^{\partial}(A) = \mu^{Q}(A)$ for all $A \subseteq C$. For any $A \subseteq C$, $\mu_k^{\text{re}}(A)$, $\mu_k^{\text{sp}}(A)$ and $\mu_k^{\partial}(A)$ takes a value between $[0, 1]$ for $k = 1, 2, \ldots, p - 1$. Especially when $\partial_A(u) = p$ for all $u \in U$, we have $\mu_k^{\text{re}}(A) = \mu_k^{\text{sp}}(A) = \mu_k^{\partial}(A) = 0$, $k = 1, 2, \ldots, p - 1$, i.e., set functions μ_k^{re}, μ_k^{sp} and μ_k^{∂} are 'grounded' [4].

Ślęzak [31] defined the following set functions to define approximate reducts:

$$g_{\partial}(A) = \frac{1}{|U|} \sum_{u_i \in U} \frac{1}{|\partial_A(u_i)|}, \tag{30}$$

$$e_{\partial}(A) = \frac{1}{|U|} \sum_{u_i \in U} \frac{1}{2^{|\partial_A(u_i)|-1}}, \tag{31}$$

$$h_{\partial}(A) = \frac{1}{|U|} \sum_{u_i \in U} \log |\partial_A(u_i)|, \tag{32}$$

where the base of a logarithm log is 2. Functions g_∂ and h_∂ are related gini index and non-specificity measure (Hartley measure), respectively (see [21, 31]). $e_\partial(A)$ satisfies the following equation:

$$e_\partial(A) = 1 - \frac{1}{2^p} \sum_{k=1}^{p-1} \sum_{F \in \mathscr{F}_k} \left(\frac{|A^*(F)|}{|U|} - \frac{|A_*(F)|}{|U|} \right). \tag{33}$$

We note that $2^p = \sum_{k=1}^{p-1} |\mathscr{F}_k| + 2$, i.e., the number of subsets of $\{D_j, j = 1, 2, \dots, p\}$.

By modifying those functions to be monotonously increasing and zero-normalized and introducing the size parameter k of generalized decision class, we consider the following set functions:

$$\mu_k^g(A) = \frac{1}{|U|} \sum_{u_i \in U_k} \left(\frac{1}{|\partial_A(u_i)|} - \frac{1}{p} \right), \tag{34}$$

$$\mu_k^e(A) = \frac{1}{|U|} \sum_{u_i \in U_k} \left(\frac{1}{2^{|\partial_A(u_i)|-1}} - \frac{1}{2^{p-1}} \right), \tag{35}$$

$$\mu_k^h(A) = \frac{1}{|U|} \sum_{u_i \in U_k} (\log p - \log |\partial_A(u_i)|). \tag{36}$$

We note that $\mu_1^g(A) = (p-1)\mu^Q(A)/p$, $\mu_1^e(A) = (2^{p-1}-1)\mu^Q(A)/2^{p-1}$ and $\mu_1^h(A) = (\log p)\mu^Q(A)$. Moreover, we have $\mu_{p-1}^g(A) = g_\partial(A) - |U|/p$, $\mu_{p-1}^e(A) = e_\partial(A) - |U|/2^{p-1}$ and $\mu_{p-1}^h(A) = \log p - h_\partial(A)$.

Those set functions, μ_k^α, $\alpha \in \{\text{re}, \text{sp}, \partial, g, s, h\}$ are associated set functions with $\mathscr{F}_k$-reduct, i.e., $A \subseteq U$ is a $\mathscr{F}_k$-reduct if and only if

(F1(k)) $\mu_k^\alpha(A) = \mu_k^\alpha(C)$, and
(F2(k)) $\forall a \in A$, $\mu_k^\alpha(A - \{a\}) \neq \mu_k^\alpha(C)$.

We note that set functions, μ_k^α, $\alpha \in \{\text{re}, \text{sp}, \partial, g, s, h\}$ are monotonously increasing with respect to set-inclusion, i.e., if $A_1 \subseteq A_2 \subseteq C$, we have $\mu_k^\alpha(A_1) \leq \mu_k^\alpha(A_2)$. Such a zero-normalized set function can be seen as a fuzzy measure [24].

Remark 3 Similar to (33), we have

$$\mu_k^e(A) = \frac{|U_k|}{|U|} \left(1 - \frac{1}{2^p} \sum_{j=1}^{k} \sum_{F \in \mathscr{F}_k} \left(\frac{|A^*(F)|}{|U|} - \frac{|A_*(F)|}{|U|} \right) - \frac{1}{2^{p-1}} \right). \tag{37}$$

3.3 Attribute Importance Based on the Associated Set Functions

Regarding a fuzzy measure $\mu : 2^C \to \mathbf{R}$ as a characteristic function of cooperative game theory, the attribute importance and interactions are evaluated by the Shapley value [29] and Harsanyi dividend (called also, Möbius transform) defined respectively as (see [5, 23])

$$I_\mu^S(A) = \sum_{K \subseteq C-A} \frac{(|C|-|K|-|A|)!|K|!}{(|C|-|A|+1)!} \sum_{L \subseteq A} (-1)^{|A|-|L|} \mu(K \cup L), \tag{38}$$

$$m_\mu(A) = \sum_{B \subseteq A} (-1)^{|A-B|} \mu(B). \tag{39}$$

Shapley index $I_\mu^S(\{a_i\})$ called Shapley value shows the average contribution of a_i in μ, i.e., the average importance degree of a_i in the sense of μ. Shapley index $I_\mu^S(A)$ shows the interaction among condition attributes in A. Let $A = \{a_i, a_j\}$, $i \neq j$. Fact $I_\mu^S(\{a_i, a_j\}) < 0$ implies that condition attributes a_i and a_j are compensative in μ. Fact $I_\mu^S(\{a_i, a_j\}) = 0$ implies that condition attributes a_i and a_j are additive in μ. Fact $I_\mu^S(\{a_i, a_j\}) > 0$ implies that condition attributes a_i and a_j are synergic in μ. Harsanyi dividends (Möbius transform) $m_\mu(A)$ shows the additional contribution of a coalition A in itself, i.e., the change of importance degree by a coalition A in itself. Especially, $m_\mu(\{a_i\})$ shows the individual contribution of condition attribute a_i in itself, i.e., the individual importance degree of a_i. Therefore, we have

$$\mu(A) = \sum_{B \subseteq A} m_\mu(B), \tag{40}$$

where we define $m_\mu(\emptyset) = 0$. For Shapley value $I_\mu(\{a_i\})$, we have

$$I_\mu(\{a_i\}) = \sum_{A : A \ni a_i} \frac{1}{|A|} m_\mu(A). \tag{41}$$

Applying this idea to an associated set function, μ_k^α, $\alpha \in \{\text{re}, \text{sp}, \partial, g, s, h\}$, we can analyze the degrees of attribute importance and interactions in the sense of μ_k^α. In next example, we calculate the degrees of attribute importance and interactions in decision table given by Table 2.

Example 2 Consider a decision table shown in Table 2. As shown in Example 1, we obtained $\{a_2\}$ and $\{a_1, a_3\}$ as $\mathscr{F}_1$-reducts, $\{a_1, a_3\}$ and $\{a_1, a_2, a_4\}$ as $\mathscr{F}_2$-reducts, and $\{a_1, a_2, a_4\}$ and $\{a_1, a_3, a_4\}$ as $\mathscr{F}_3$-reducts. We have $\{a_1\}$ as $\mathscr{F}_2$-core and $\{a_1, a_4\}$ as $\mathscr{F}_3$-core. We have no $\mathscr{F}_1$-core, i.e., $\mathscr{F}_1$-core is the empty set. Set functions μ_k^{re}, μ_k^{sp} and μ_k^∂ as well as their Shapley interaction indices and Harsanyi dividends are shown in Table 4. Set functions μ_k^g, μ_k^e and μ_k^h as well as their Shapley interaction indices and Harsanyi dividends are shown in Table 5.

Table 4 Shapley interaction indices and Harsanyi dividends for μ_k^{re}, μ_k^{sp} and μ_k^{∂}

A	μ_1^{re}	$I^S_{\mu_1^{\mathrm{re}}}$	$m_{\mu_1^{\mathrm{re}}}$	μ_2^{re}	$I^S_{\mu_2^{\mathrm{re}}}$	$m_{\mu_2^{\mathrm{re}}}$	μ_3^{re}	$I^S_{\mu_3^{\mathrm{re}}}$	$m_{\mu_3^{\mathrm{re}}}$
a_1	0	0.1667	0	0	0.5833	0	0	0.4167	0
a_2	1	0.6667	1	0	0.0833	0	0	0.0833	0
a_3	0	0.1667	0	0	0.25	0	0	0.0833	0
a_4	0	0	0	0	0.0833	0	0	0.4167	0
a_1a_2	1	−0.5	0	0	0.1667	0	0	0.1667	0
a_1a_3	1	0.5	1	1	0.6667	1	0	0.1667	0
a_1a_4	0	0	0	0	0.1667	0	0	0.6667	0
a_2a_3	1	−0.5	0	0	−0.3333	0	0	−0.3333	0
a_2a_4	1	0	0	0	0.1667	0	0	0.1667	0
a_3a_4	0	0	0	0	−0.3333	0	0	0.1667	0
$a_1a_2a_3$	1	−1	−1	1	−0.5	0	0	−0.5	0
$a_1a_2a_4$	1	0	0	1	0.5	1	1	0.5	1
$a_1a_3a_4$	1	0	0	1	−0.5	0	1	0.5	1
$a_2a_3a_4$	1	0	0	0	−0.5	0	0	−0.5	0
C	1	0	0	1	−1	−1	1	−1	−1
A	μ_1^{sp}	$I^S_{\mu_1^{\mathrm{sp}}}$	$m_{\mu_1^{\mathrm{sp}}}$	μ_2^{sp}	$I^S_{\mu_2^{\mathrm{sp}}}$	$m_{\mu_2^{\mathrm{sp}}}$	μ_3^{sp}	$I^S_{\mu_3^{\mathrm{sp}}}$	$m_{\mu_3^{\mathrm{sp}}}$
a_1	0.0556	0.0556	0.0556	0.1481	0.1204	0.1481	0.3148	0.2099	0.3148
a_2	0.3333	0.1759	0.3333	0.3333	0.1512	0.3333	0.3333	0.1265	0.3333
a_3	0.1667	0.0926	0.1667	0.2593	0.1265	0.2593	0.3519	0.1481	0.3519
a_4	0	0.0093	0	0.0741	0.0463	0.0741	0.2593	0.1821	0.2593
a_1a_2	0.3333	−0.1296	−0.0556	0.3333	−0.1358	−0.1481	0.4074	−0.1420	−0.2407
a_1a_3	0.3333	0.0370	0.1111	0.4444	0.0123	0.0370	0.5556	−0.0494	−0.1111
a_1a_4	0.1667	0.0370	0.1111	0.3333	0.0494	0.1111	0.6111	0.0247	0.0370
a_2a_3	0.3333	−0.1852	−0.1667	0.3704	−0.1728	−0.2222	0.4074	−0.1605	−0.2778

(continued)

Table 4 (continued)

A	μ_1^{re}	$I^S_{\mu_1^{re}}$	$m_{\mu_1^{re}}$	μ_2^{re}	$I^S_{\mu_2^{re}}$	$m_{\mu_2^{re}}$	μ_3^{re}	$I^S_{\mu_3^{re}}$	$m_{\mu_3^{re}}$
a_2a_4	0.3333	−0.0185	0	0.3704	−0.0247	−0.0370	0.5185	−0.0309	−0.0741
a_3a_4	0.1667	−0.0185	0	0.2593	−0.0988	−0.0741	0.4630	−0.1420	−0.1481
$a_1a_2a_3$	0.3333	−0.0556	−0.1111	0.4444	0.0556	0.0370	0.5556	0.1667	0.1852
$a_1a_2a_4$	0.3333	−0.0556	−0.1111	0.4444	−0.0185	−0.0370	0.6667	0.0185	0.0370
$a_1a_3a_4$	0.3333	−0.0556	−0.1111	0.4444	−0.0926	−0.1111	0.6667	−0.0556	−0.0370
$a_2a_3a_4$	0.3333	0.0556	0	0.3704	0.0556	0.0370	0.5185	0.0556	0.0741
C	0.3333	0.1111	0.1111	0.4444	0.0370	0.0370	0.6667	−0.0370	−0.0370
A	μ_1^{∂}	$I^S_{\mu_1^{\partial}}$	$m_{\mu_1^{\partial}}$	μ_2^{∂}	$I^S_{\mu_2^{\partial}}$	$m_{\mu_2^{\partial}}$	μ_3^{∂}	$I^S_{\mu_3^{\partial}}$	$m_{\mu_3^{\partial}}$
a_1	0.0556	0.0556	0.0556	0.0556	0.2037	0.0556	0.0556	0.2037	0.0556
a_2	0.3333	0.1759	0.3333	0.3333	0.1759	0.3333	0.3333	0.1759	0.3333
a_3	0.1667	0.0926	0.1667	0.2778	0.1852	0.2778	0.2778	0.1852	0.2778
a_4	0	0.0093	0	0.1111	0.1019	0.1111	0.4444	0.4352	0.4444
a_1a_2	0.3333	−0.1296	−0.0556	0.3333	−0.1296	−0.0556	0.3333	−0.1296	−0.0556
a_1a_3	0.3333	0.0370	0.1111	0.6667	0.1481	0.3333	0.6667	0.1481	0.3333
a_1a_4	0.1667	0.0370	0.1111	0.5000	0.1481	0.3333	0.8333	0.1481	0.3333
a_2a_3	0.3333	−0.1852	−0.1667	0.4444	−0.1852	−0.1667	0.4444	−0.1852	−0.1667
a_2a_4	0.3333	−0.0185	0	0.4444	−0.0185	0	0.7778	−0.0185	0
a_3a_4	0.1667	−0.0185	0	0.2778	−0.2407	−0.1111	0.6111	−0.2407	−0.1111
$a_1a_2a_3$	0.3333	−0.0556	−0.1111	0.6667	−0.0556	−0.1111	0.6667	−0.0556	−0.1111
$a_1a_2a_4$	0.3333	−0.0556	−0.1111	0.6667	−0.0556	−0.1111	1	−0.0556	−0.1111
$a_1a_3a_4$	0.3333	−0.0556	−0.1111	0.6667	−0.2778	−0.3333	1	−0.2778	−0.3333
$a_2a_3a_4$	0.3333	0.0556	0	0.4444	0.0556	0	0.7778	0.0556	0
C	0.3333	0.1111	0.1111	0.6667	0.1111	0.1111	1	0.1111	0.1111

Table 5 Shapley interaction indices and Harsanyi dividends for μ_k^g, μ_k^e and μ_k^h

A	μ_1^g	$I^S_{\mu_1^g}$	$m_{\mu_1^g}$	μ_2^g	$I^S_{\mu_2^g}$	$m_{\mu_2^g}$	μ_3^g	$I^S_{\mu_3^g}$	$m_{\mu_3^g}$
a_1	0.0417	0.0417	0.0417	0.1111	0.0903	0.1111	0.1296	0.0965	0.1296
a_2	0.25	0.1319	0.25	0.25	0.1134	0.25	0.25	0.1134	0.25
a_3	0.125	0.0694	0.125	0.1944	0.0949	0.1944	0.1944	0.0918	0.1944
a_4	0	0.0069	0	0.0556	0.0347	0.0556	0.0833	0.0594	0.0833
a_1a_2	0.25	−0.0972	−0.0417	0.25	−0.1019	−0.1111	0.2685	−0.1019	−0.1111
a_1a_3	0.25	0.0278	0.0833	0.3333	0.0093	0.0278	0.3333	0	0.0093
a_1a_4	0.125	0.0278	0.0833	0.25	0.0370	0.0833	0.2778	0.0278	0.0648
a_2a_3	0.25	−0.1389	−0.125	0.2778	−0.1296	−0.1667	0.2778	−0.1296	−0.1667
a_2a_4	0.25	−0.0139	0	0.2778	−0.0185	−0.0278	0.3056	−0.0185	−0.0278
a_3a_4	0.125	−0.0139	0	0.1944	−0.0741	−0.0556	0.2222	−0.0648	−0.0556
$a_1a_2a_3$	0.25	−0.0417	−0.0833	0.3333	0.0417	0.0278	0.3333	0.0417	0.0278
$a_1a_2a_4$	0.2500	−0.0417	−0.0833	0.3333	−0.0139	−0.0278	0.3611	−0.0139	−0.0278
$a_1a_3a_4$	0.25	−0.0417	−0.0833	0.3333	−0.0694	−0.0833	0.3611	−0.0509	−0.0648
$a_2a_3a_4$	0.25	0.0417	0	0.2778	0.0417	0.0278	0.3056	0.0417	0.0278
C	0.25	0.0833	0.0833	0.3333	0.0278	0.0278	0.3611	0.0278	0.0278
A	μ_1^e	$I^S_{\mu_1^e}$	$m_{\mu_1^e}$	μ_2^e	$I^S_{\mu_2^e}$	$m_{\mu_2^e}$	μ_3^e	$I^S_{\mu_3^e}$	$m_{\mu_3^e}$
a_1	0.0486	0.0486	0.0486	0.1528	0.1215	0.1528	0.1806	0.1308	0.1806
a_2	0.2917	0.1539	0.2917	0.2917	0.1262	0.2917	0.2917	0.1262	0.2917
a_3	0.1458	0.0810	0.1458	0.25	0.1192	0.25	0.25	0.1146	0.25
a_4	0	0.0081	0	0.0833	0.0498	0.0833	0.125	0.0868	0.125
a_1a_2	0.2917	−0.1134	−0.0486	0.2917	−0.1204	−0.1528	0.3194	−0.1204	−0.1528
a_1a_3	0.2917	0.0324	0.0972	0.4167	0.0046	0.0139	0.4167	−0.0093	−0.0139
a_1a_4	0.1458	0.0324	0.0972	0.3333	0.0463	0.0972	0.375	0.0324	0.0694
a_2a_3	0.2917	−0.1620	−0.1458	0.3333	−0.1481	−0.2083	0.3333	−0.1481	−0.2083

(continued)

Table 5 (continued)

A	μ_1^g	$I^S_{\mu_1^g}$	$m_{\mu_1^g}$	μ_2^g	$I^S_{\mu_2^g}$	$m_{\mu_2^g}$	μ_3^g	$I^S_{\mu_3^g}$	$m_{\mu_3^g}$
a_2a_4	0.2917	−0.0162	0	0.3333	−0.0231	−0.0417	0.375	−0.0231	−0.0417
a_3a_4	0.1458	−0.0162	0	0.25	−0.1065	−0.0833	0.2917	−0.0926	−0.0833
$a_1a_2a_3$	0.2917	−0.0486	−0.0972	0.4167	0.0764	0.0694	0.4167	0.0764	0.0694
$a_1a_2a_4$	0.2917	−0.0486	−0.0972	0.4167	−0.0069	−0.0139	0.4583	−0.0069	−0.0139
$a_1a_3a_4$	0.2917	−0.0486	−0.0972	0.4167	−0.0903	−0.0972	0.4583	−0.0625	−0.0694
$a_2a_3a_4$	0.2917	0.0486	0	0.3333	0.0486	0.0417	0.375	0.0486	0.0417
C	0.2917	0.0972	0.0972	0.4167	0.0139	0.0139	0.4583	0.0139	0.0139
A	μ_1^h	$I^S_{\mu_1^h}$	$m_{\mu_1^h}$	μ_2^h	$I^S_{\mu_2^h}$	$m_{\mu_2^h}$	μ_3^h	$I^S_{\mu_3^h}$	$m_{\mu_3^h}$
a_1	0.1111	0.1111	0.1111	0.3889	0.3056	0.3889	0.4811	0.3363	0.4811
a_2	0.6667	0.3519	0.6667	0.6667	0.2778	0.6667	0.6667	0.2778	0.6667
a_3	0.3333	0.1852	0.3333	0.6111	0.2870	0.6111	0.6111	0.2717	0.6111
a_4	0	0.0185	0	0.2222	0.1296	0.2222	0.3606	0.2526	0.3606
a_1a_2	0.6667	−0.2593	−0.1111	0.6667	−0.2778	−0.3889	0.7589	−0.2778	−0.3889
a_1a_3	0.6667	0.0741	0.2222	1	0	0	1	−0.0461	−0.0922
a_1a_4	0.3333	0.0741	0.2222	0.8333	0.1111	0.2222	0.9717	0.0650	0.1300
a_2a_3	0.6667	−0.3704	−0.3333	0.7778	−0.3333	−0.5	0.7778	−0.3333	−0.5
a_2a_4	0.6667	−0.0370	0	0.7778	−0.0556	−0.1111	0.9161	−0.0556	−0.1111
a_3a_4	0.3333	−0.0370	0	0.6111	−0.2778	−0.2222	0.7495	−0.2317	−0.2222
$a_1a_2a_3$	0.6667	−0.1111	−0.2222	1	0.2222	0.2222	1	0.2222	0.2222
$a_1a_2a_4$	0.6667	−0.1111	−0.2222	1	0	0	1.1383	0	0.0000
$a_1a_3a_4$	0.6667	−0.1111	−0.2222	1	−0.2222	−0.2222	1.1383	−0.1300	−0.1300
$a_2a_3a_4$	0.6667	0.1111	0.0000	0.7778	0.1111	0.1111	0.9161	0.1111	0.1111
C	0.6667	0.2222	0.2222	1	0	0	1.1383	0	0

As shown in Tables 4 and 5, the attribute importance is very different by k and by set function. We observe signs of $I^S_{\mu_k^\alpha}(A)$ and $m_{\mu_k^\alpha}(A)$ are similar. From Shapley indices and Harsanyi dividends of $\{a_i, a_j\}$, $i \neq j$ in μ_k^α, $\alpha \in \{\mathrm{cl}, \partial, g, e, h\}$, we found that a_2 is compensative with other condition attributes. Therefore, Shapley value (the average importance degree) $I^S_{\mu_k^\alpha}(\{a_2\})$ is smaller than Harsanyi dividend (individual importance degree) $m_\mu(\{a_2\})$. On the other hand, a_1 and a_4 are synergic. When $k = 1$, except μ_1^{re}, the behaviors of set function values $\mu_1^\alpha(A)$, Shapley interaction indices and Harsanyi dividends with respect to $A \subseteq C$ are similar because those set functions are proportional to μ^Q.

In what follows, we compare mainly Shapley indices $I_{\mu_k^\alpha}(\{a_i\})$, $i = 1, 2, 3, 4$ because it is most understandable. From $\mathscr{F}_1$-reducts, we feel that a_2 is more important than a_1 because a_2 itself forms an $\mathscr{F}_1$-reduct although a_1 does not. However, considering $\mathscr{F}_2$- and $\mathscr{F}_3$-reducts, we found a_1 is more important because a_1 is in $\mathscr{F}_2$- and $\mathscr{F}_3$-cores although a_2 is not. These facts are well captured by Shapley indices of a_i, $i = 1, 2, 3, 4$ with respect to μ_k^{re}, $k = 1, 2, 3$ as shown in Table 4. When $k = 1$, in Shapley index $I_{\mu_1^{\mathrm{re}}}(\{a_i\})$, a_2 is four times more important than a_1 and a_3, and a_4 is not important at all. When $k = 2$, in Shapley index $I_{\mu_1^{\mathrm{re}}}(\{a_i\})$, the average importance degree of a_1 is significantly increased because it composes $\mathscr{F}_2$-core by itself. The average importance degree of a_2 decreases by 0.5833 while those of a_3 and a_4 increase by 0.0833. As the result, a_1 becomes the most important attribute, a_3 becomes the second and a_2 as well as a_4 are the least important attributes. The roles of a_2 and a_4 are same in the sense of μ_2^{re}. When $k = 3$, in Shapley index $I_{\mu_3^{\mathrm{re}}}(\{a_i\})$, a_4 as well as a_1 become the most important because $\{a_1, a_4\}$ is $\mathscr{F}_3$-core. a_2 and a_3 take the smallest average importance degree because a_1 and a_4 compose an $\mathscr{F}_3$-reduct with one of a_2 and a_3. The roles of a_1 and a_4 are same and the roles of a_2 and a_3 are also same. However, a_1 and a_4 are complementary while a_2 and a_3 are substitute. The attribute importance degree is changed significantly by the required precision level k of classification.

From the viewpoint of the specificity μ_k^{sp} of the class estimation, the evaluation of the attribute importance is a little different because it considers the precision of class estimation for all objects in U. Namely, the value of $\mu_k^{\mathrm{sp}}(A)$ is evaluated by the sum of the scores of objects whose classes are estimated to be one of $l(\leq k)$ possible classes by A, i.e., objects $x \in U$ such that $|\partial_A(x)| \leq k$. The smaller $|\partial_A(x)|$, the larger the score of $x \in U$. When $k = 1$, the value of $\mu_1^{\mathrm{sp}}(A)$ depends on the number of objects such that their classes are uniquely determined by A. Because a_3 itself determines uniquely the classes of 16.67% of objects in U which is three times more than a_1, as shown in μ_1^{sp} values in Table 4. On the other hand, together with a_4, a_1 improves μ_1^{sp} value as much as a_3. As the result, $I_{\mu_1^{\mathrm{sp}}}(\{a_3\})$ is only 1.7 times bigger than $I_{\mu_1^{\mathrm{sp}}}(\{a_1\})$. This fact is very different from the evaluation by μ_1^{re}. When $k = 2$, the value of $\mu_2^{\mathrm{sp}}(A)$ becomes larger as $|\partial_A(x)| \leq 2$ becomes smaller. Because of this fact, a_2 still takes the highest average importance degree among a_i, $i = 1, 2, 3, 4$. This is different from the evaluation by μ_2^{re}. When $k = 3$, as the value of $\mu_3^{\mathrm{sp}}(A)$ is influenced by scores of objects x such that $|\partial_A(x)| = 3$, the Shapley index $I_{\mu_3^{\mathrm{sp}}}(\{a_2\})$ of a_2 becomes the lowest. The Shapley index decreases in the order of a_1, a_4, a_3, a_2. Similarly to the evaluation by μ_3^{re}, a_1 and a_4 are more important than a_2 and a_3.

From the viewpoint of generalized decision class preservation μ_k^{∂}, the evaluation of the attribute importance is further different. When $k = 1$, the evaluation of attribute importance is the same as that by the specificity. When $k = 2$, a_1 takes the highest Shapley index. However, the difference of Shapley indices between a_1 and a_2 are not very big. This is very different from the evaluation by μ_2^{re}. This is also different from the evaluation by μ_2^{sp} which implies that a_2 takes the highest Shapley index although the differences of Shapley indices from a_1 and a_3 are small. When $k = 3$, a_4 takes the highest Shapley index and the differences from the Shapley indices of other attributes are significantly big. Those results are obtained from the fact that the evaluation of μ_k^{∂} does not depend on the size of the preserved generalized decision class $|\partial_C(u)|$ while others except μ_k^{re} does.

From the viewpoint of a set function μ_k^g related to gini index, the evaluation of the attribute importance is different again. When $k = 1$, the evaluation of attribute importance is the same as that by the specificity and the generalized decision class preservation but the values are 0.75 times of values in μ_1^{sp}. In this set function, a_2 takes the highest Shapley index $I_{\mu_3^g}(\{a_i\})$ among a_i, $i = 1, 2, 3, 4$ for all $k \in \{1, 2, 3\}$. When $k \geq 2$, the Shapley value (average importance degree) of a_4 is much smaller than those of other condition attributes a_i, $i = 1, 2, 3$. Attribute a_4 contributes to the preservation of some generalized decision classes whose sizes are 2 and 3. The preservation of big generalized decision classes does not influence very much to μ_k^g because it is discounted by the inverse number of the size of generalized decision class.

The behaviors of the Shapley indices with respect to μ_k^e is similar to those with respect to μ_k^g although a_2 does not take the highest Shapley index when $k = 3$. The behavior of the Shapley indices with respect to μ_k^h is similar to that with respect to μ_k^{sp}. This can be understood from the fact that both set functions evaluate the specificity of decision attribute value. However, because the contribution of a_4 in the preservations of some generalized decision classes is also discounted by logarithmic function, the Shapley indices of a_4 are evaluated relatively small values in μ_k^h.

The results in Example 2 imply the following significant fact: to evaluate the attribute importance we should determine

(i) the required precision of classification, and
(ii) in what sense we evaluate the attribute importance.

(i) means the selection of parameter k and (ii) means the selection of set function, i.e., reducibility, specificity, generalized decision class preservation, gini-index, etc.

4 Rule Induction

4.1 The Conventional Approach

The other major topic in rough set approaches is the minimal rule induction, i.e., inducing rules inferring the membership to D_j with minimal conditions which can differ members of $C_*(D_j)$ from non-members, are investigated well. In this chapter, we use minimal rule induction algorithms proposed in the field of rough sets, i.e., LEM2 and MLEM2 algorithms [8, 9]. By those algorithms, we obtain minimal set of rules with minimal conditions which can explain all objects in lower approximations of X under a given decision table. LEM2 algorithm [8] and MLEM2 algorithm [9] are different in their forms of condition parts of rules: by LEM2 algorithm, we obtain rules of the form of "if $f(u, a_1) = v_1, f(u, a_2) = v_2, \ldots$ and $f(u, a_s) = v_s$ then $u \in X$", while by MLEM2 algorithm, we obtain rules of the form of "if $v_1^{\mathrm{L}} \leq f(u, a_1) \leq v_1^{\mathrm{R}}$, $v_2^{\mathrm{L}} \leq f(u, a_2) \leq v_2^{\mathrm{R}}, \ldots$ and $v_s^{\mathrm{L}} \leq f(u, a_s) \leq v_s^{\mathrm{R}}$ then $u \in X$". Namely, MLEM2 algorithm is a generalized version of LEM2 algorithm to cope with numerical/ordinal condition attributes. For each decision class D_i we induce rules inferring the membership of D_i. Using all those rules, we build a classifier system as proposed in LERS [8, 9]. Namely, The classification of a new object u is made by the following two steps:

⟨1⟩ When the condition attribute values of u match to all conditions of at least one of the induced rules, for each D_i, we calculate

$$S(D_i) = \sum_{\text{matching rules } r \text{ for } D_i} \mathit{Stren}(r) \times \mathit{Spec}(r), \tag{42}$$

where r is called a *matching rule* if the condition part of r is satisfied with u. $\mathit{Stren}(r)$ is the total number of objects in given decision table correctly classified by rule r. $\mathit{Spec}(r)$ is the total number of condition attributes in the condition part of rule r. For convenience, when rules for D_i are not matched by the object, we define $S(D_i) = 0$. Then u is classified into D_i with highest $S(D_i)$. If D_j such that $S(D_j) > 0$ exists, class D_i with the largest $S(D_i)$ is selected. However, if class D_i with the largest $S(D_i)$ is not unique, class D_i with smallest index i is selected from them and terminate the procedure.

⟨2⟩ When the condition attribute values of u do not match totally to the condition part of any rule composing the classifier system, for each D_i, we calculate

$$M(D_i) = \sum_{\text{partially matching rules } r \text{ for } D_i} \mathit{Mat\_f}(r) \times \mathit{Stren}(r) \times \mathit{Spec}(r), \tag{43}$$

where r is called a *partially matching rule* if a part of the premise of r is satisfied with u. $\mathit{Mat\_f}(r)$ is the ratio of the number of matched conditions of rule r to the total number of conditions of rule r. Then class D_i with the largest $M(D_i)$

is selected. If a tie occurs, class D_i with smallest index i is selected from tied classes.

Here we note that D_i is regarded not only as a subset of U but also as a conclusion indicating a member of decision class D_i.

4.2 Classification with Imprecise Rules

As described above, in the conventional rough set approaches, rules inferring the memberships to single decision classes have been induced and used to build the classifier system. However, rules inferring the memberships to unions of multiple decision classes can be also induced based on the rough set model. We call rules inferring the memberships to single decision classes 'precise rules' and rules inferring the memberships to unions of multiple decision classes 'imprecise rules'.

Each of imprecise rules cannot give a conclusion univocally while each of precise rules can give a conclusion univocally. However, if we have imprecise rules for many kinds of unions, we can give a univocal conclusion. For example, if we have a rule for $D_1 \cup D_2$ and a rule for $D_1 \cup D_3$ and if a new object satisfy the conditions of both rules, we know that the object is in $D_1 \cup D_2$ and at the same time in $D_1 \cup D_3$, and thus it is in D_1. Therefore, many kinds of imprecise rules can work well for classification of objects.

From this point of view, Inuiguchi and Hamakawa [10, 15, 16] investigated the induction of imprecise rules and classifier based on imprecise rules. They induced rules for every $F \in \mathscr{F}_k$ under fixed $k \in \{1, 2, \ldots, p-1\}$ and build a classifier using all induced imprecise rules. Here, we note that $F \in \mathscr{F}_k$ is regarded not only as a subset of U but also as a conclusion indicating a member of one of k decision classes $D_{i_1}, \ldots, D_{i_k}$. We can induce imprecise rules for $F \in \mathscr{F}_k$, $k \in \{1, 2, \ldots, p-1\}$ in the same way as the induction method for rules about D_i (see [10, 15, 16]). Namely, LEM2-based algorithms can be applied to the induction of imprecise rules. Moreover, in the same way, we can build a classifier by induced imprecise rules.

Two classifiers under imprecise rules for $F \in \mathscr{F}_k$ have been investigated. In the first classifier Cla_1, a new object u is classified by the following procedure:

⟨1⟩ When u matches to at least one of the conditions of the rule, we calculate

$$\hat{S}(D_i) = \sum_{\substack{\text{matching rule } r \\ \text{for } F \supseteq D_i}} Stren(r) \times Spec(r), \tag{44}$$

where r is called a *matching rule* if the condition part of r is satisfied with u. The strength $Stren(r)$ is the total number of objects in the given dataset correctly classified by rule r. The specificity $Spec(r)$ is the total number of condition attributes in the condition part of rule r. F is a variable set such that $F \in \mathscr{F}_k$. For convenience, when there is no matching rules about $F \supseteq D_i$, we define $\hat{S}(D_i) =$

0. If there exists D_j such that $\hat{S}(D_j) > 0$, the class D_i with the largest $\hat{S}(D_i)$ is selected. If a tie occurs, class D_i with smallest index i is selected from tied classes.

⟨2⟩ When u does not match totally to any rule, for each D_i, we calculate

$$\hat{M}(D_i) = \sum_{\substack{\text{partially matching} \\ \text{rules } r \text{ for } F \supseteq D_i}} Mat\_f(r) \times Stren(r) \times Spec(r), \tag{45}$$

where r is called a *partially matching rule* if a part of the premise of r is satisfied. The matching factor $Mat\_f(r)$ is the ratio of the number of matched conditions of rule r to the total number of conditions of rule r. Then the class D_i with the largest $\hat{M}(D_i)$ is selected. If a tie occurs, class D_i with smallest index i is selected from tied classes.

In the other classifier Cla_2, a new object u is classified by the following procedure:

⟨1⟩ For all $F \in \mathscr{F}_k$, we calculate $S(F)$ of (42). Let $W = \bigcup\{F \in \mathscr{F}_k \mid S(F) = 0\}$. If $W = U$, go to ⟨3⟩.

⟨2⟩ We calculate $\hat{S}(D_i)$ for D_i such that $D_i \cap W = \emptyset$. The class D_i with the largest $\hat{S}(D_i)$ is selected. If a tie occurs, class D_i with smallest index i is selected from tied classes.

⟨3⟩ For each D_i, we calculate $\hat{M}(D_i)$. Then the class D_i with the largest $\hat{M}(D_i)$ is selected. If a tie occurs, class D_i with smallest index i is selected from tied classes.

In Cla_2, $D_i \subseteq W$ cannot be a candidate of the decision class for u if $W \neq U$ (W includes all decision classes). We note that these classification methods Cla_1 and Cla_2 are reduced to the conventional one when $k = 1$, because $F \in \mathscr{F}_1$ becomes a decision class D_i.

4.3 Numerical Experiment

We examined the classification accuracy of classifiers Cl_1 and Cl_2 with imprecise rules by using eight datasets shown in Table 6. Those datasets are obtained from UCI machine learning repository [35], and consistent, i.e., $U = U_1 = \{u \in U \mid |\partial(u)| \leq 1\}$. For the evaluation, we apply a 10-fold cross validation method. Namely we divide the dataset into 10 subsets and 9 subsets are used for training dataset and the remaining subset is used for checking dataset. Changing the combination of 9 subsets, we obtain 10 different evaluations. We calculate the averages and the standard deviations in number of obtained rules and classification accuracy. We execute this procedure 10 times with different divisions.

The results are shown in Table 7. In column '# rules', the numbers of rules are shown. In columns 'Accuracy (Cla_1)' and 'Accuracy (Cla_2)', the classification accuracy scores (%) of classifiers Cla_1 and Cla_2 are shown, respectively. Each entry in

Table 6 Eight datasets

Dataset	$\lvert U \rvert$	$\lvert C \rvert$	$p = \lvert V_d \rvert$	Attribute type
car	1,728	6	4	Ordinal
dermatology	358	34	6	Numerical
ecoli	336	7	8	Numerical
glass	214	9	6	Numerical
hayes-roth	159	4	3	Nominal
iris	150	4	3	Numerical
wine	178	13	3	Numerical
zoo	101	16	7	Nominal

those columns shows the average *ave* and the standard deviation *dev* in the form of $ave \pm dev$. Asterisk $*$ and two asterisks $**$ imply the significant differences from the corresponding classification accuracy scores of classifiers with precise rules ($k = 1$) in the paired t-test with significance levels $\alpha = 0.05$ and $\alpha = 0.01$, respectively. Scores with superscript asterisks are significantly bigger than those of classifiers with precise rules ($k = 1$) while scores with subscripts asterisks are significantly smaller than those of classifiers with precise rules ($k = 1$). Underline $\underline{\ \ }$ and double underline $\underline{\underline{\ \ }}$ imply the significant differences between classification accuracy scores of classifiers Cla_1 and Cla_2 with same imprecise rules in the paired t-test with significance levels $\alpha = 0.05$ and $\alpha = 0.01$, The better scores are underlined.

As shown in Table 7, when $p = |V_d|$ is larger than 2, the classification accuracy is improved by using imprecise rules except $k = p - 1$. In attribute reduction, $\mathscr{F}_{p-1}$-reducts preserve the classification ability more than other reducts ($\mathscr{F}_k$-reducts, $k < p - 1$). From this fact, we may expect the classifier with imprecise rules for $F \in \mathscr{F}_{p-1}$ works well although given datasets are consistent. Namely the obtained results for $k = p - 1$ are counter-intuitive.

An important issue is the selection of k. The best performed k can depend on the given dataset. However we cannot know the best performed k in advance, yet. As far as the results in Table 7 show, the classifiers with imprecise rules for $F \in \mathscr{F}_k$ with $k \approx p/2$ perform well. As k approaches to $p/2$, the number of possible combinations of k decision classes, i.e., $|\mathscr{F}_k|$, increases. Accordingly, the number of induced rules attains the maximum around $k = p/2$. Since we used consistent datasets in the numerical experiment, having many rules may be advantageous for the classifier in making robust estimation.

There is no big difference in the classification ability between classifiers Cla_1 and Cla_2, as far as in Table 7, Cla_2 is a bit better.

To sum up, classifiers with imprecise rules work well although the number of rules is increased significantly. This can be understood, even from imprecise rules, we can obtain a correct conclusion if you have many of them. The big volume of

Table 7 Classification accuracy (%)

Dataset	k	# rules	Accuracy (Cla_1)	Accuracy (Cla_2)
car	1	57.22 ± 1.74	98.67 ± 0.97	98.67 ± 0.97
	2	128.02 ± 3.16	98.96** ± 0.75	99.16** ± 0.76
	3	69.55 ± 1.37	99.68** ± 0.49	99.57** ± 0.54
dermatology	1	12.09 ± 1.27	92.32 ± 4.42	92.32 ± 4.42
	2	61.32 ± 4.07	94.58** ± 3.59	95.72** ± 3.15
	3	103.58 ± 6.11	96.03** ± 3.26	96.40** ± 3.14
	4	77.28 ± 4.45	95.58** ± 3.69	95.78** ± 3.67
	5	23.84 ± 1.81	91.87 ± 4.75	88.83$_{**}$ ± 5.73
ecoli	1	35.89 ± 2.03	77.52 ± 6.21	77.52 ± 6.21
	2	220.67 ± 8.93	83.20** ± 5.66	83.42** ± 5.56
	3	565.67 ± 21.48	84.66** ± 5.64	84.54** ± 5.75
	4	781.36 ± 28.42	84.87** ± 5.71	84.84** ± 5.65
	5	617.06 ± 23.06	83.74** ± 6.26	83.53** ± 6.38
	6	269.27 ± 10.5	82.56** ± 6.26	82.76** ± 6.27
	7	54.09 ± 2.86	78.38 ± 6.70	77.17 ± 6.71
glass	1	25.38 ± 1.5	68.34 ± 10.18	68.34 ± 10.18
	2	111.40 ± 4.33	72.57** ± 8.81	73.59** ± 8.77
	3	178.35 ± 5.41	73.44** ± 9.19	74.28** ± 9.93
	4	130.14 ± 4.96	71.16* ± 9.91	72.71** ± 9.45
	5	39.59 ± 2.18	65.04$_{**}$ ± 9.96	63.55$_{**}$ ± 10.79
hayes-roth	1	23.17 ± 1.41	81.38 ± 7.95	81.38 ± 7.95
	2	39.25 ± 2.2	72.94** ± 10.42	70.81** ± 10.5
iris	1	7.40 ± 0.72	92.87 ± 5.52	92.87 ± 5.52
	2	8.52 ± 0.78	92.93 ± 5.32	94.60** ± 4.96
wine	1	4.65 ± 0.5	93.25 ± 5.87	93.25 ± 5.87
	2	7.31 ± 0.59	88.83$_{**}$ ± 7.15	89.15$_{**}$ ± 6.75
zoo	1	9.67 ± 0.55	95.84 ± 6.63	95.84 ± 6.63
	2	48.5 ± 2.1	95.55 ± 7.15	95.74 ± 6.33
	3	105.37 ± 4.25	96.74* ± 5.45	96.74* ± 5.45
	4	113.78 ± 3.74	96.84* ± 5.22	96.84* ± 5.22
	5	66.76 ± 2.69	97.24** ± 5.07	97.44** ± 4.97
	6	17.72 ± 0.66	96.05 ± 6.51	96.05 ± 6.51

imprecise rules can affect the interpretability of results as well as the computational time. The reduction of number of rules is investigated in [10, 16].

4.4 Rule Anonymization

When a classifier with rules is used in public, it may be required to unfold the underlying rules. The publication of rules makes the classification fair and impartial. However, by the publication of rule can invade the privacy of individuals if some of rules are supported only a few objects. Namely, from such a rule, some sensitive personal data can be revealed. From this point of view, the concept of K-anonymity [34][1] has been investigated. In the case of a rule, the K-anonymity implies that the rule is supported by at least K objects. However, if we restrict rule adoption to K-anonymous rules, we will not obtain a sufficient number of rules to perform good classification. Inuiguchi et al. [17] proposed to use K-anonymous imprecise rules.

The procedure for K-anonymous rule induction proposed by Inuiguchi et al. [17] is as follows:

⟨1⟩ Let $\mathscr{R}$ be the set of K-anonymous rules and initialize $\mathscr{R} = \emptyset$. Let $l = 1$.
⟨2⟩ Induce a set $\mathscr{S}_1$ of precise rules by MLEM2 algorithm.
⟨3⟩ Select rules $r \in \mathscr{S}_l$ satisfying $\mathrm{Supp}(r) \geq K$ and put them in $\mathscr{R}$.
⟨4⟩ If $\mathscr{S}_l - \mathscr{R} \neq \emptyset$ and $l < n$, update $l = l + 1$. Otherwise, terminate this procedure.
⟨5⟩ Define object set B by objects match $r \in \mathscr{S}_l$ such that $\mathrm{Supp}(r) < K$.
⟨6⟩ Induce a set $\mathscr{S}_l$ of imprecise rules for each possible union $F \in \mathscr{F}_l$ by MLEM2 algorithm inputting $B \cap F \neq \emptyset$ as a set of objects uncovered by presently induced rules. Return to ⟨3⟩. We note that we skip F such that $B \cap F = \emptyset$ in this procedure.

In order to examine the performances of this K-anonymous rule induction procedure, we apply it to eight datasets in Table 6 and compare classifiers with K-anonymous rules and the classifier with MLEM2 rules. For classifiers with K-anonymous rules, we use classifier Cla_2. By 10 times run of 10-fold cross validation method described in the previous subsection, we obtain the results as shown in Table 8. In Table 8, the average av and the standard deviation sd is shown in the style of $av \pm sd$ in each cell of table. The underlined numbers are average scores of classification accuracy obtained by the classifiers composed of K-anonymous imprecise rules better than those obtained by the classifiers composed of MLEM2 precise rules. Asterisk $*$ shown in the columns of classification accuracy of K-anonymous imprecise rules stands for the value is significantly different from the case of MLEM2 precise rules by the paired t-test with significance level $\alpha = 0.05$.

As shown in Table 8, the numbers of K-anonymous rules are smaller than those of MLEM2 precise rules for $K = 5$, 10 and 15 in datasets 'hayes-roth', 'iris' and

[1] We use capital letter K because we already used lower case letter k to show the number of decision classes to be combined by union.

Table 8 MLEM2 precise rules versus K-anonymous imprecise rules (from [17])

Data-set	MLEM2 precise rules		5-anonymous imprecise rules	
	Number of rules	Accuracy	Number of rules	Accuracy
car	57.22 ± 1.74	98.67 ± 0.97	58.94 ± 2.64	98.76 ± 0.88
dermatology	12.09 ± 1.27	92.32 ± 4.42	13.78 ± 5.63	92.38 ± 4.28
ecoli	35.89 ± 2.03	77.52 ± 6.21	1343.84 ± 123.08	82.88* ±5.64
glass	25.38 ± 1.5	68.34 ± 10.18	84.61 ± 28.87	66.17 ± 10.32
hayes-roth	23.17 ± 1.41	81.38 ± 7.95	18.01 ± 1.40	74.56* ± 9.24
iris	7.40 ± 0.72	92.87 ± 5.52	6.53 ± 0.90	93.07 ± 5.65
wine	4.65 ± 0.50	93.25 ± 5.87	4.46 ± 0.50	93.25 ± 5.87
zoo	9.67 ± 0.55	95.84 ± 6.63	138.63 ± 29.08	94.55 ± 8.00
Data-set	10-anonymous imprecise rules		15-anonymous imprecise rules	
	Number of rules	Accuracy	Number of rules	Accuracy
car	66.48 ± 4.19	98.94* ± 0.85	66.26 ± 8.75	98.21* ± 1.34
dermatology	30.69 ± 18.06	91.73 ± 4.80	41.2 ± 31.14	92.77 ± 4.74
ecoli	1351.04 ± 58.85	83.68* ± 6.31	1094.66 ± 47.43	82.48* ± 6.52
glass	254.90 ± 17.07	72.52* ± 8.76	200.32 ± 5.95	70.88 ± 9.2
hayes-roth	14.25 ± 1.44	59.44* ± 16.64	1.90 ± 1.46	39.63* ± 12.38
iris	6.87 ± 1.05	93.20 ± 5.33	5.11 ± 1.03	92.47 ± 6.16
wine	4.09 ± 0.47	92.92 ± 6.06	4.05 ± 0.46	92.92 ± 6.06
zoo	196.08 ± 14.05	93.98 ± 7.06	198.22 ± 9.66	95.17 ± 6.55

'wine'. This implies that in those datasets, with high probability, no K-anonymous rules are induced for some of objects in U. Indeed we observed some objects uncovered by induced rules in those datasets. Especially for dataset 'hayes-roth', no rule is induced in some sets of training data when $K = 15$. We note that any new object is classified into a default class D_1 (the decision class of the first decision attribute value) when no rule is induced. It is very hard to protect the privacy in dataset 'hayes-roth'. We also observe that the three datasets 'hayes-roth', 'iris' and 'wine' have only three classes. This observation can be understood by the following reason. If we have only a few classes, we obtain a limited number of unions of classes and we cannot make the unions large enough to have many K-anonymous rules. Although the classification accuracy scores of K-anonymous imprecise rules are comparable to those of MLEM2 precise rules in datasets 'iris' and 'wine', we find that the proposed approach is not always very efficient in datasets with a few classes. For such datasets we need a lot of samples to induce a sufficient number of K-anonymous rules.

On the contrary, the proposed approach works well in dataset 'ecoli' having eight classes. We observe that very many K-anonymous imprecise rules are induced in this dataset. In datasets 'car' and 'glass', 10-anonymous imprecise rules perform best. We observe that the number of rules are most at $K = 10$ in those datasets. We note that the number of rules does not increase monotonously because, as K increases, the K-anonymity becomes stronger condition while the size of B at $\langle 5 \rangle$ of the proposed method increases. The more the size of B, the more imprecise rules are induced at $\langle 6 \rangle$. Indeed, as k increases, the average number of rules increases in datasets 'dermatology' and 'zoo' while it decreases in datasets 'hayes-roth', 'iris' and 'wine'. In other datasets, it attains the largest number at $k = 10$. In general, except 'hayes-roth', the classifier composed of the rules induced by the proposed method preserves the classification accuracy of the conventional classification while the induced rules improve the anonymity.

More investigation of the proposed rule anonymization is found in [17].

5 Concluding Remarks

In this chapter, we described the rough set approaches to decision tables based on the lower approximations of unions of k decision classes instead of lower approximations of single decision classes. We demonstrated that significantly different results are obtained by the selection of k. In attribute reduction and importance, the selection of k depends on to what extent of imprecision is meaningful/allowable in object classification. On the other hand, in rule induction, k can be selected about $p/2$ (a half of the number of decision classes), because of the classification accuracy of the classifier. The imprecise rules have applied to privacy protection when the publication of rules is requested.

In attribute reduction, bigger k preserves more information about classification. Therefore, we may guess that bigger k is better. On the other hand, in rule induction, we showed that k around a half of the number of decision classes seems good. This results are obtained when the proposed approach is applied to consistent decision tables. Therefore, the analysis of inconsistent decision tables and selection of the best performed k are conceivable for future topics. Moreover, the proposed imprecise rule induction method can be improved so as to reduce the number of induced rules without big deterioration of classification accuracy. The applications and improvements of the proposed approaches as well as the comparison with other multi-class rule mining methods are other future topics.

Acknowledgements This work was partially supported by JSPS KAKENHI Grant Number 26350423. This chapter is the extended version of [13] with new data and discussions.

References

1. Błaszczyński, J., Greco, S., Słowiński, R., Szelg, M.: Monotonic variable consistency rough set approaches. Int. J. Approx. Reason. **50**(7), 979–999 (2009)
2. Dubois, D., Prade, H.: Rough fuzzy sets and fuzzy rough sets. Int. J. General Syst. **17**, 191–209 (1990)

3. Flores, V., et al. (eds.): Rough Sets: International Joint Conference, IJCRS 2016, Proceedings, LNAI 9920. Springer, Cham (2016)
4. Grabisch, M.: Set Functions. Games and Capacities in Decision Making. Springer, Swizerland (2016)
5. Greco, S., Matarazzo, B., Słowiński, R.: Fuzzy measure technique for rough set analysis. In: Proceedings of EUFIT'98, pp. 99–103 (1998)
6. Greco, S., Matarazzo, B., Słowiński, R.: Rough sets theory for multi-criteria decision analysis. Eur. J. Oper. Res. **129**(1), 1–7 (2001)
7. Greco, S., Matarazzo, B., Słowiński, R.: Decision rule approach. In: Figueira, J., Greco, S., Ehrgott, M. (eds.) Multiple Criteria Decision Analysis, pp. 507–561. Springer, New York (2005)
8. Grzymala-Busse, J.W.: LERS–a system for learning from examples based on rough sets. In: Słowiński, R. (ed.) Intelligent Decision Support: Handbook of Application and Advances of Rough Set Theory, pp. 3–18. Kluwer Academic Publishers, Dordrecht (1992)
9. Grzymala-Busse, J.W.: MLEM2—discretization during rule induction. In: Proceedings of the IIPWM2003, pp. 499–508 (2003)
10. Hamakawa, T., Inuiguchi, M.: On the utility of imprecise rules induced by MLEM2 in classification. In: Kudo Y., Tsumoto, S. (Eds.) Proceedings of 2014 IEEE International Conference on Granular Computing (GrC), pp. 76–81 (2014)
11. Inuiguchi, M.: Approximation-oriented fuzzy rough set approaches. Fundamenta Informaticae **142**(1–4), 21–51 (2015)
12. Inuiguchi, M.: Variety of rough set based attribute importance. Proc. SCIS-ISIS **2016**, 548–551 (2016)
13. Inuiguchi, M.: Rough set approaches to imprecise modeling. In: Flores, V., et al. (eds.) Rough Sets: International Joint Conference, IJCRS 2016, Proceedings, LNAI 9920, pp. 54–64. Springer, Cham (2016)
14. Inuiguchi, M.: Attribute importance degrees corresponding to several kinds of attribute reduction in the setting of the classical rough sets. In: Torra, V., Dahlbom, A., Narukawa, Y. (Eds.) Fuzzy Sets, Rough Sets, Multisets and Clustering. Springer (in press)
15. Inuiguchi, M., Hamakawa, T.: The utilities of imprecise rules and redundant rules for classifiers. In: Huynh, V.-N., Denoeux, T., Tran, D.H., Le, A.C., Pham, S.B. (Eds.) Knowledge and Systems Engineering: Proceedings of the Fifth International Conference KSE 2013, vol. 2, AISC 245, pp. 45–56. Springer, Cham (2013)
16. Inuiguchi, M., Hamakawa, T., Ubukata, S.: Utilization of imprecise rules induced by MLEM2 algorithm. In: Proceedings of the 10th Workshop on Uncertainty Processing (WUPES'15), pp. 73–84 (2015)
17. Inuiguchi, M., Hamakawa, T., Ubukata, S.: Imprecise rules for data privacy. In: Ciucci, D., Wang, G., Mitra, S., Wu, W.-Z. (eds.) Rough Sets and Knowledge Technology 10th International Conference, RSKT 2015, LNCS 9436, pp. 129–139. Springer, Cham (2015)
18. Inuiguchi, M., Kusunoki, Y., Inoue, M.: Rule induction considering implication relations between conclusions. Ind. Eng. Manage. Syst. Int. J. **10**(1), 66–74 (2011)
19. Inuiguchi, M., Tsurumi, M.: Measures based on upper approximations of rough sets for analysis of attribute importance and interaction. Int. J. Innov. Comput. Inf. Control **2**(1), 1–12 (2006)
20. Inoue, K. (ed.): Application of Rough Sets to Kansei Engineering. Kaibundo, Tokyo (2009) (in Japanese)
21. Klir, G.J.: Where do we stand on measures of uncertainty, ambiguity, fuzziness, and the like? Fuzzy Sets Syst. **24**(2), 141–160 (1987)
22. Kryszkiewicz, M.: Rough set approach to incomplete information systems. Inf. Sci. **112**, 39–49 (1998)
23. Murofushi, T., Soneda, S.: Techniques for reading fuzzy measures (iii): interaction index. In Proceedings of the 9th Fuzzy Systems Symposium, pp. 693–696 (1993) (in Japanese)
24. Murofushi, T., Sugeno, M.: A theory of fuzzy measures: representations, the Choquet integral and null sets. J. Math. Anal. Appl. **159**, 532–549 (1991)
25. Pawłak, Z.: Rough sets. Int. J. Comput. Inf. Sci. **11**(5), 341–356 (1982)

26. Pawłak, Z.: Rough Sets: Theoretical Aspects of Reasoning About Data. Kluwer Academic Publishing, Dordrecht (1991)
27. Pawłak, Z., Skowron, A.: Rough sets and Boolean reasoning. Inf. Sci. **177**(1), 41–73 (2007)
28. Polkowski, L., Skowron, A. (eds.): Rough Sets in Knowledge Discovery 2: Applications. Case Studies and Software Systems. Physica-Verlag, Heidelberg (2010)
29. Shapley, L.S.: A value for *n*-person games. In: Kuhn, H.W., Tucker, A.W. (eds.) Contributions to the Theory of Games II, pp. 307–317. Princeton, Princeton University Press (1953)
30. Ślęzak, D.: Various approaches to reasoning with frequency based decision reducts: a survey. In: Polkowski, L., Tsumoto, S., Lin, T.Y. (eds.) Rough Set Methods and Applications, pp. 235–285. Physica-Verlag, Heidelberg (2000)
31. Ślęzak, D.: On generalized decision functions: Reducts, networks and ensembles. In: Yao, Y., Hu, Q., Yu, H., Grzymala-Busse, J.W. (eds.) Rough Sets, Fuzzy Sets, Data Mining, and Granular Computing, 15th International Conference, RSFDGrC 2015, LNCS 9437, pp. 13–23. Springer, Cham (2015)
32. Skowron, A., Rauser, C.M.: The discernibility matrix and function in information systems. In: Słowiński, R. (ed.) Intelligent Decision Support: Handbook of Application and Advances of Rough Set Theory, pp. 331–362. Kluwer Academic Publishers, Dordrecht (1992)
33. Słowiński, R., Vanderpooten, D.: A generalized definition of rough approximations based on similarity. IEEE Trans. Data Knowl. Eng. **12**(2), 331–336 (2000)
34. Sweeney, L.: K-anonymity: a model for protecting privacy. Int. J. Uncertain. Fuzziness. Knowl.-Based Syst. **10**(5), 557–570 (2002)
35. UCI Machine Learning Repository. http://archive.ics.uci.edu/ml/
36. Yao, Y.Y., Zhao, Y.: Attribute reduction in decision theoretic rough set models. Inf. Sci. **178**(17), 3356–3373 (2008)
37. Ziarko, W.: Variable precision rough sets model. J. Comput. Syst. Sci. **46**(1), 39–59 (1993)

Pawlak's Many Valued Information System, Non-deterministic Information System, and a Proposal of New Topics on Information Incompleteness Toward the Actual Application

Hiroshi Sakai, Michinori Nakata and Yiyu Yao

Abstract This chapter considers Pawlak's Many Valued Information System (MVIS), Non-deterministic Information System (NIS), and related new topics on information incompleteness toward the actual application. Pawlak proposed rough sets, which were originally defined in a standard table, however his research in non-standard tables like MVIS and NIS is also seen. Since rough sets have been known to many researchers deeply and several software tools have been proposed until now, it will be necessary to advance from this research on a standard table to research on MVIS and NIS, especially in regards to NIS. In this chapter, previous research is surveyed and new topics toward the actual application of NIS are proposed, namely data mining under various types of uncertainty, rough set-based estimation of an actual value, machine learning by rule generation, information dilution, and an application to privacy-preserving questionnaire, in NIS. Such new topics will further extend the role of Pawlak's rough sets.

1 Introduction

Rough sets proposed by Pawlak have been known to many researchers, and the concept on a discernibility relation is applied to several research areas [5, 9, 12–14, 17, 23, 30, 31, 40, 41, 44, 49, 50]. We briefly survey the history of rough sets and non-deterministic information at first.

H. Sakai (✉)
Department of Mathematical Science, Kyushu Institute of Technology, Sensui 1-1, Tobata, Kitakyushu 804-8550, Japan
e-mail: sakai@mns.kyutech.ac.jp

M. Nakata
Faculty of Management and Information Science, Josai International University, Gumyo, Togane, Chiba 283-0002, Japan
e-mail: nakatam@ieee.org

Y. Yao
Department of Computer Science, University of Regina, Regina, Saskatchewan S4S 0A2, Canada
e-mail: yyao@cs.uregina.ca

G. Wang et al. (eds.), *Thriving Rough Sets*, Studies in Computational Intelligence 708, DOI 10.1007/978-3-319-54966-8_9

In the 1970s, a mathematical framework of information retrieval for a standard table [20] and relational algebra [6] were investigated. Based on research of the past, a *Deterministic Information System* (DIS) ψ is usually considered for specifying a standard table [26–29, 41]:

$$\psi = (OB, AT, \{VAL_A \mid A \in AT\}, f), \quad (1)$$

where OB is a finite set whose elements are called *objects*, AT is a finite set whose elements are called *attributes*, VAL_A is a finite set whose elements are called *attribute values* for an attribute $A \in AT$, and f is a mapping below:

$$f : OB \times AT \rightarrow \cup_{A \in AT} VAL_A. \quad (2)$$

At the beginning of the 1980s, rough sets seem to be defined with respect to question-answering [20] and relational algebra [6], and Pawlak also dealt with question-answering and relational algebra in non-standard tables like *Many Valued Information System* (MVIS) and *Non-deterministic Information System* (NIS). Tables 1 and 2 are examples of MVIS [27] and NIS [24]. The keyword 'nondeterministic information' [24] is used by Orłowska and Pawlak, and many valued information [27] by Pawlak. The attribute values are enumerated in Table 1.

In MVIS and NIS, each attribute value is given as a set, and this set is mathematically defined by a mapping g:

$$g : OB \times AT \rightarrow 2^{(\cup_{A \in AT} VAL_A)}. \quad (3)$$

Table 1 An example of many valued information system [27]. Each possible color value is enumerated

OB	*color*
x_1	*blue*
x_2	*blue, red*
x_3	*blue*
x_4	*blue, red*
x_5	*blue, red, green*
x_6	*green*

Table 2 An example of nondeterministic information system [24]. Each attribute value is a set of possible values

OB	a_1	a_2
D_1	$\{v_1, v_3\}$	$\{u_1, u_2, u_3\}$
D_2	$\{v_2, v_5\}$	$\{u_1\}$
D_3	$\{v_1, v_3, v_4\}$	$\{u_1, u_2\}$
D_4	$\{v_1\}$	$\{u_1, u_2\}$
D_5	$\{v_1, v_3\}$	$\{u_1\}$
D_6	$\{v_5\}$	$\{u_1\}$

Table 3 An example of Lipski's incomplete information database [18]. The age of x_2 is one of 52, 53, 54, 55, and 56 years old, which is non-deterministic information

OB	*age*	*dept#*	*hireyear*	*salary*
x_1	$[60, 70]$	$\{1, \cdots, 5\}$	$\{70, \cdots, 75\}$	$\{10000\}$
x_2	$[52, 56]$	$\{2\}$	$\{72, \cdots, 76\}$	$(0, 20000]$
x_3	$\{30\}$	$\{3\}$	$\{70, 71\}$	$(0, \infty)$
x_4	$(0, \infty)$	$\{2, 3\}$	$\{70, \cdots, 74\}$	$\{22000\}$
x_5	$\{32\}$	$\{4\}$	$\{75\}$	$(0, \infty)$

For example, MVIS handles such information as that Tom can speak both English and French by $g(Tom, Language) = \{English, French\}$ [25]. We employ the conjunctive interpretation for this description. On the other hand, NIS seems to be defined in order to handle information incompleteness in tables, and we employ the disjunctive interpretation (more correctly the exclusive disjunctive interpretation). For example, we see $g(Tom, Age) = \{24, 25, 26\}$ that Tom's age is not certain but his age is one of 24, 25, and 26. Based on tables with a mapping g, we can formally consider information incompleteness in DIS.

A framework of Lipski's incomplete information databases is known well [18, 19]. Table 3 is cited from [18]. In his framework, the purpose is to realize an actual question-answering system based on possible world semantics [15]. Lipski proposed a set of axioms for the equivalent query transformation and a normal form query, and proved the soundness (a transformed query becomes a normal form query) and the completeness (a normal form query can be transformed from a query by the set of axioms) of the set of axioms. This set of axioms is equal to the system S4 in modal logic [18, 19]. By using this transformation, it is possible to handle a normal form query for any query. This causes the simplification of the query evaluation procedure, and reduces the execution time.

Generally, if we employ possible world semantics, it will be necessary to consider some algorithms for reducing the execution time, because there may be a huge number of possible worlds. In rule generation described in the subsequent section, the number of possible worlds may exceed 10^{100}. We follow Lipski's way of thinking, and consider rule generation based on possible world semantics later.

In the 1990s, the research trend seemed to move from question-answering to data mining, and rough sets seemed to be employed as the mathematical framework on data mining. Now, we enumerate research on information incompleteness during this decade. The complexity on incomplete information and the theoretical aspect including logic were investigated by Demri and Orłowska [7]. Then, the *LERS* system was implemented by Grzymała-Busse [10–12]. We understand that the *LERS* system employs a covering method for a target set, and rules are obtained as a side effect of the covering. Furthermore, rules in incomplete information systems were defined by Kryszkiewicz, and reduction algorithms were shown [16, 17]. In two interesting pieces of research, missing values are employed instead of non-deterministic information, and rules are defined based on some assumptions about the missing

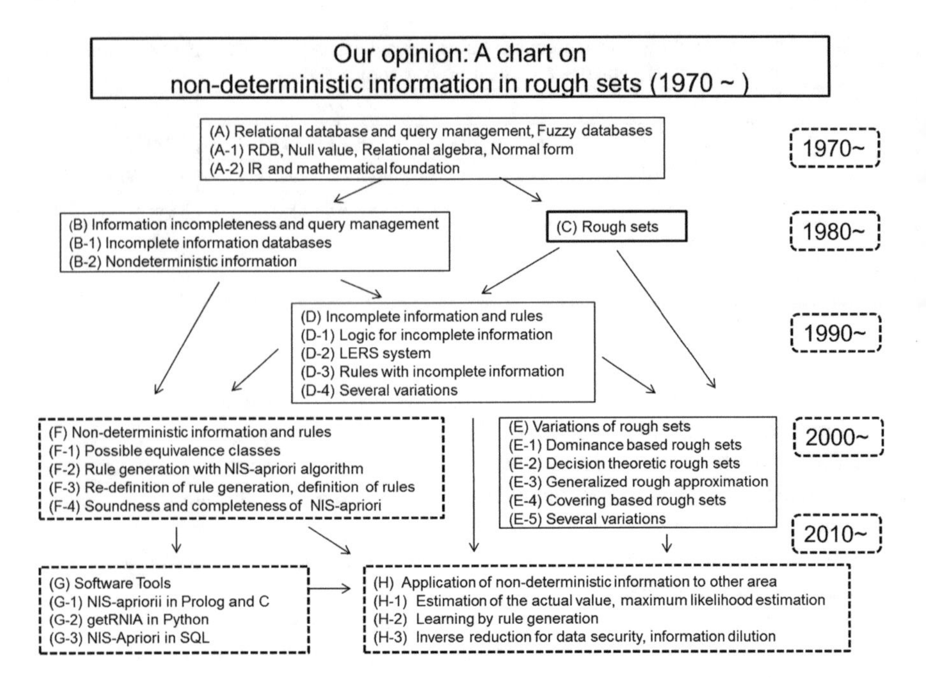

Fig. 1 A chart on non-deterministic information in rough sets in our opinion. The blocks with broken line show our work. This chart is a revised version in [35]

values. However, we think that non-deterministic information describes more general information than a missing value does, because it includes every missing value as a special case. For example non-deterministic information such that 'Tom's age is one of 24, 25, and 26' is expressed by $g(Tom, Age) = \{24, 25, 26\}$. On the other hand, if Tom's age is missing, we express it by $g(Tom, Age) = \{0, 1, 2, \cdots, 120\}$ by non-deterministic information. In rule generation in NIS, we face with the problem that the number of possible worlds becomes very huge, and we recently showed a solution for this problem [33, 36].

In the 2000s, we have several variations of rough sets, for example *dominance-based rough sets* [9], *generalized rough approximations* [5], *covering-based rough sets* [50], etc. We have also coped with NIS, and surveyed about rough sets and non-deterministic information [35]. Figure 1 shows our opinion. We are investigating the issues in the blocks with broken lines.

Here, we have to remark that each author has his major part with respect to NIS and information incompleteness, for example the first and the second authors' *Rough Non-deterministic Information Analysis* (RNIA) [32, 33, 36], the second author's information tables containing possibilistic values, *Generalized Discernibility Relation* (GDR), and *Twofold Rough Approximations* [21–23], and the third author's *Three Way Decision* (TWD), *Multi-Granular Rough Sets* (MGRS), *Decision Theoretic Rough Sets*, etc. [31, 46–49]. Basically, RNIA is research toward realizing the application software tools, and the second and the third authors' work is research for establishing a new mathematical framework on rough sets. Relying on interdisciplinary research the authors cooperate to extend the role of Pawlak's rough sets in more depth. In this chapter, the application of RNIA is especially focused on, and new topics on information incompleteness are proposed toward the actual application.

This chapter is organized as follows: Sect. 2 surveys RNIA and some theoretical frameworks. Section 3 proposes new topics on information incompleteness, and Sect. 4 concludes this chapter.

2 Background

This section at first simply surveys RNIA toward the actual application, then describes the theoretical frameworks of GDR, TWD, and MGRS.

2.1 Background of RNIA

1. In RNIA, we handle NIS like Table 2 and Fig. 2. We obtain DIS by replacing each set in NIS with a value in the set. We call such DIS a *derived DIS*, and we see there is the actual DIS in all derived DISs. In Fig. 2, there are 24 ($=2^3 \times 3$) derived DISs. The definition of derived DISs is coming from the definition of the *extension* by Lipski [18, 19]. Lipski introduced modal logic into incomplete information databases by using extensions, and axiomatized the equivalent query transformation for question-answering.
2. A pair $[A_i, val_i]$ of attribute A_i and its attribute value val_i is called a *descriptor* in a table, like $[color, red]$ and $[size, m]$ in Fig. 2. For a decision attribute Dec, an implication $\tau : \wedge_i [A_i, val_i] \Rightarrow [Dec, val]$ with an appropriate property is called a *rule* in a table. Namely, any rule consists of descriptors in a table.

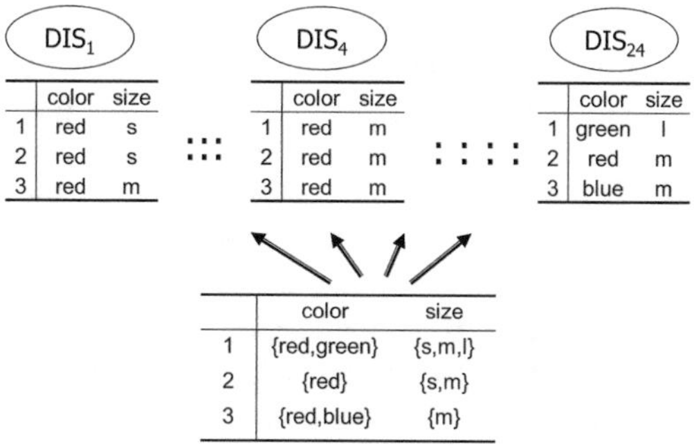

DIS_1

	color	size
1	red	s
2	red	s
3	red	m

DIS_4

	color	size
1	red	m
2	red	m
3	red	m

DIS_{24}

	color	size
1	green	l
2	red	m
3	blue	m

	color	size
1	{red,green}	{s,m,l}
2	{red}	{s,m}
3	{red,blue}	{m}

Fig. 2 NIS and 24 derived DISs [36]

3. In rule generation, we employ the usual definition of a rule in DIS [14, 28, 43, 44], and extended it to a certain rule and a possible rule in NIS below [32, 33]:
   (Rule in DIS) An implication τ is a *rule*, if τ satisfies $support(\tau) \geq \alpha$ and $accuracy(\tau) \geq \beta$ in DIS for the given α and β.
   (Certain rule in NIS) An implication τ is a *certain rule*, if τ is a rule in each derived DIS for the given α and β.
   (Possible rule in NIS) An implication τ is a *possible rule*, if τ is a rule in at least one derived DIS for the given α and β.
4. If τ is a certain rule, we can conclude τ is also a rule in the unknown actual DIS. This property is also described in Lipski's incomplete information databases. Let us consider Fig. 2. For an implication $\tau : [color, red] \Rightarrow [size, m]$, threshold values $\alpha = 0.3$, and $\beta = 0.5$, $support(\tau) = 3/3 > 0.3$ and $accuracy(\tau) = 3/3 > 0.5$ in DIS_4. This means τ is a rule in DIS_4 and τ is a possible rule in NIS. However, $support(\tau) = 1/3 > 0.3$ and $accuracy(\tau) = 1/3 < 0.5$ in DIS_1, so τ is not a rule in DIS_1. Thus, it is concluded that τ is not a certain rule in NIS.
5. Here, we give an example on non-deterministic information and missing values. In Tables 4 and 5, there is no information incompleteness except Tom's age. Since we generally see that ? may become any possible value, so we may have an implication $[age, senior] \Rightarrow [salary, low]$ from Tom's tuple. This contradicts the implication $\tau : [age, senior] \Rightarrow [salary, high]$ from Mary's tuple. Therefore, τ is not a certain rule in Table 4. However in Table 5, there are two derived DISs, and τ is consistent in each derived DIS. Therefore, τ is a certain rule in Table 5.
6. The definition of rules in NIS seems to be natural, and it follows possible world semantics. However, the number of all derived DISs increases exponentially for the number of non-deterministic values. For example, there are more than 10^{100} derived DISs for Mammographic data set and Hepatitis data set in UCI machine learning repository [8]. So, it will be hard to examine whether τ is a certain rule or not, if the trivial algorithm is employed (we sequentially pick up every derived DIS, and examine whether τ is a rule or not).
7. For this computational problem, we proved some mathematical properties [33, 36], and added these properties to the *Apriori* algorithm [3, 4]. (The Apriori algorithm originally handles transaction data sets, however we can consider the Apriori algorithm by identifying each item in the transaction data with a descriptor in a table.) We named this new algorithm the *NIS-Apriori* algorithm, which does not depend upon the number of derived DISs. By using this algorithm, we provided a solution for escaping from the computational problem. The details of the NIS-Apriori algorithm are in [33, 36, 40].
8. This NIS-Apriori algorithm preserves a logical property, namely it is sound and complete for the defined rules [37]. If we fix a decision attribute, and two threshold values α and β, the whole set of all certain rules and possible rules is fixed. For this fixed set, the NIS-Apriori algorithm has the next property:
   (Soundness) Each implication generated by the NIS-Apriori algorithm is either a certain rule or a possible rule. Any other implication is not generated.
   (Completeness) Each certain rule and possible rule is obtained by the NIS-Apriori algorithm.

Table 4 An example of DIS with a missing value

OB	*age*	*salary*
Tom	?	*low*
Mary	*senior*	*high*

Table 5 An example of NIS

OB	*age*	*salary*
Tom	$\{young, middle\}$	$\{low\}$
Mary	$\{senior\}$	$\{high\}$

In logic, this is seen as the most important property, and this property assures the validity of the NIS-Apriori algorithm. In Lipski's framework [18, 19], the axiomatized query transformation system is also sound and complete.

9. The analysis on the computational complexity of the NIS-Apriori algorithm is still in progress. This algorithm consists of two parts, the certain rule generation part and the possible rule generation part, and the Apriori algorithm is applied to each part. The calculation on the minimum support, the minimum accuracy, the maximum support, and the maximum accuracy for each implication τ does not depend upon the number of all derived DISs, and we can calculate them in polynomial time [33, 36]. Therefore, we figure out that the computational complexity of the NIS-Apriori algorithm is more than twice the complexity of the Apriori algorithm. However, we can escape from the exponential order problem by using the NIS-Apriori algorithm. We actually obtained certain rules and possible rules from tables with more than 10^{100} derived DISs. It will be hard to obtain them by using the trivial method.
10. The set of implications obtained by the NIS-Apriori algorithm is equal to the whole set of all certain rules and possible rules. This means that it is enough for us to apply NIS-Apriori algorithm for obtaining rules in NIS. We have implemented some software tools by using this algorithm. On the web page [40], the files for details and the execution log files are uploaded.

2.2 *Background of Theoretical Framework on Rough Sets*

Now, we refer to our theoretical framework on rough sets. The framework named Three Way Decision (TWD) [47] is the most common concept on information incompleteness. For handling information incompleteness, we often depend upon the modal concept like certainty and possibility, the minimum and the maximum, or an optimistic view and a pessimistic view. Certain rules and possible rules in RNIA also belong to the framework in TWD.

In NIS, we usually have a huge number of possible tables. For example in Fig. 2, we have 24 derived DISs, and we have different equivalence relations. In Multi-Granular Rough Sets (MGRS) [31, 49], the property on different equivalence rela-

Table 6 An exemplary table with containing possibilistic information by a possibility distribution. Each attribute value is given as a set of pairs of a value and it possibility

OB	*P*	*Q*
x_1	$\{(a, 1.0)\}$	$\{(w, 1.0), (z, 0.6)\}$
x_2	$\{(a, 1.0), (b, 0.8)\}$	$\{(w, 0.4), (x, 1.0)\}$
x_3	$\{(b, 1.0)\}$	$\{(x, 1)\}$

tions is clarified, and an *optimistic interpretation* (causing the best decision) and a *pessimistic interpretation* (causing the worst decision) are proposed by the third author. In rule generation in RNIA, we employ these concepts, namely a certain rule is defined by an implication τ satisfying the constraint in the pessimistic interpretation for τ. A possible rule is defined by an implication τ satisfying the constraint in the optimistic interpretation for τ.

We are also focusing on a variation of NIS, for example a table containing with possibilistic information by a possibility distribution in Table 6. The second author proposes rough sets in such tables, which we simply call research on Generalized Discernibility Relation (GDR). We need to apply these theoretical results to the actual application of rough sets.

3 New Topics on Information Incompleteness

In the previous research, we focused on how we obtain certain and possible rules in NIS. Since we gave a solution for rule generation in NIS, we think that it will be able to cope with next new topics in Fig. 3, which will extend the role of rough sets and information incompleteness. We sequentially describe the new topics based on Fig. 3.

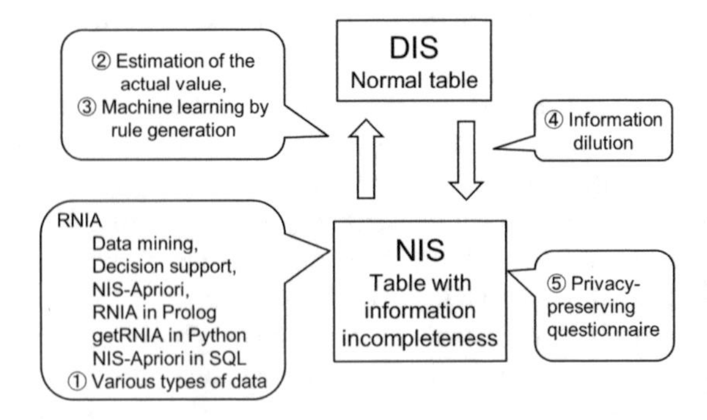

Fig. 3 An overview of new topics

3.1 Handling Various Types of Uncertain Data

This subsection considers rule generation from various types of uncertain data in Fig. 3.

In NIS, we employ variations $inf([A, val])$ and $sup([A, val])$ of an equivalence class $eq([A, val])$ for each descriptor $[A, val]$ below [32, 33]:

$$\begin{array}{l} (1)\ inf([A, val]) = \{x \in OB \mid g(x, A) = \{val\}\}, \\ (2)\ inf(\wedge_i[A_i, val_i]) = \cap_i\ inf([A_i, val_i]), \\ (3)\ sup([A, val]) = \{x \in OB \mid val \in g(x, A)\}, \\ (4)\ sup(\wedge_i[A_i, val_i]) = \cap_i\ sup([A_i, val_i]), \\ \text{Here, } g \text{ is a mapping in formula (3).} \end{array} \quad (4)$$

In DIS, $inf([A, val]) = sup([A, val])$ holds, however $inf([A, val]) \subseteq sup([A, val])$ generally holds in NIS. For example in Fig. 2, $inf([color, red]) = \{2\}$ and $sup([color, red]) = \{1, 2, 3\}$ hold. The difference in $sup([A, val]) \setminus inf([A, val])$ means a set of objects which may be changeable.

For an implication $\tau : \wedge_i[A_i, val_i] \Rightarrow [Dec, val]$, we consider sets $inf(\wedge_i[A_i, val_i])$, $sup(\wedge_i[A_i, val_i])$, $inf([Dec, val])$, and $sup([Dec, val])$. Then, $OUTACC$ in formula (5) shows a set of objects which may have the same condition of τ and different decisions of τ.

$$OUTACC = [sup(\wedge_i[A_i, val_i]) \setminus inf(\wedge_i[A_i, val_i])] \setminus inf([Dec, val]) \quad (5)$$

To reduce the accuracy value, we manipulate each object in $OUTACC$. On the other hand, $INACC$ in formula (6) shows a set of an object which may have the same condition of τ and the same decision of τ.

$$INACC = [sup(\wedge_i[A_i, val_i]) \setminus inf(\wedge_i[A_i, val_i])] \cap sup([Dec, val]) \quad (6)$$

For increasing the accuracy value, we manipulate each object in $INACC$ [33, 36]. In NIS, these sets $inf([A, val])$ and $sup([A, val])$ take an important role instead of the equivalence class $eq([A, val])$ in DIS.

Referring to Fig. 4, the block with the broken line shows the procedure in the NIS-Apriori algorithm. For a given NIS Φ, we obtain descriptors occurring in Φ and two sets *inf* and *sup* for each descriptor by formula (4). Namely, the descriptors and two sets *inf* and *sup* are essential in rule generation. Therefore, if we define the descriptors and two sets *inf* and *sup* in any type of data sets, we can apply the NIS-Apriori algorithm to these data sets.

In Fig. 5, if we obtain descriptors and two sets *inf* and *sup*, even in the different tables and plotted data, we are able to consider rules based on the different tables. Let us consider tables T_1 and T_2. Since the set of objects in T_1 and the set of objects in T_2 may not be the same, we generate rules based on a set $OB(T_1 \cap T_2)$ of all objects, which occur in T_1 and T_2. We can also consider a set $OB(T_1 \cap T_2 \cap \cdots \cap T_n)$ for tables

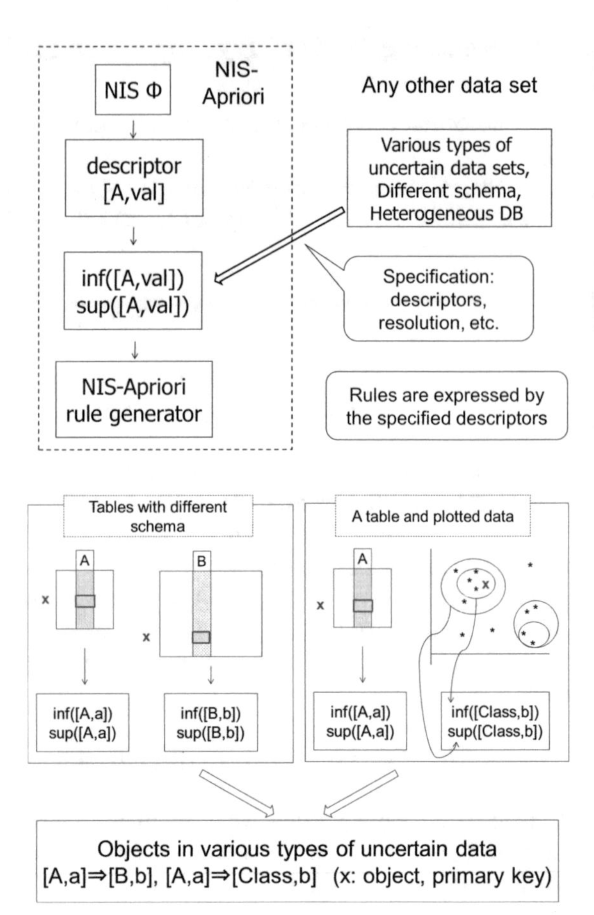

Fig. 4 Inf and sup information in NIS-Apriori algorithm

Fig. 5 Unified Apriori-based rule generation for various types of uncertain data

T_1, T_2, ⋯, T_n. Each element in $OB(T_1 \cap T_2 \cap \cdots \cap T_n)$ takes the role of the primary key. In plotted data, probably each cluster center will take the role of a descriptor. Using this concept, we are able to handle not only table data sets but also any types of uncertain data.

3.2 Rough Set-Based Estimation of an Actual Value in NIS

This section considers rough set-based estimation of an actual value in NIS. As for estimation, we know statistical estimation, and we focus on the strategy of the *maximum likelihood estimation* (MLE) [1]. The idea is that 'A sample value is obtained, because its probability is high'. In MLE, at first we obtain a set of sample values, and we estimate some parameters in a distribution function so as to maximize the probability of the set of sample values.

Based on MLE, we propose rough set-based estimation. Even though we need some details for this proposal, we show an example in the next subsection. Thus, here it will be described in brief. In certain rule $\tau_{certain}$ generation, a derived DIS Ψ_{min} (Ψ_{min} may not be unique) which causes $\tau_{certain}$ is implicitly fixed. For $\tau_{certain}$, we also obtain a derived DIS $\Psi_{max,\tau_{certain}}$ ($\Psi_{max,\tau_{certain}}$ may not be unique). In $\Psi_{max,\tau_{certain}}$, $\tau_{certain}$ is obtained as a possible rule [33, 36]. Since $\tau_{certain}$ is the most reliable rule, we should estimate each attribute value so as to cause $\tau_{certain}$ as many as possible. This follows the strategy by MLE. Namely, we estimate $\Psi_{max,\tau_{certain}}$ from NIS based on $\tau_{certain}$. We name this estimation *Rough set-based estimation of actual values.*

It will also be possible to consider constraints like the equivalence classes, data dependency, and consistency. In each constraint, we have a set M_γ (γ: constraint) of estimated DISs, and we estimate the actual DIS as an element of $\cap_\gamma M_\gamma$. In this case, the estimation will be dealt with as the constraint satisfaction problem [45].

3.3 Machine Learning by Rule Generation

This subsection extends the estimation concept in the previous subsection to *Machine Learning by Rule Generation* (MLRG) as seen in Fig. 6. We have just proposed this framework [38], and we show an example in this subsection.

Even though Table 7 is a toy example, it will be easy to know the framework of MLRG. In Table 7, there are 144 ($=2^4 \times 3^2$) derived DISs. At first, we fix the thresholds α to 0.3 and β to 0.6, and the NIS-Apriori algorithm generates every certain rule τ satisfying $support(\tau) \geq 0.3$ and $accuracy(\tau) \geq 0.6$ in each of the 144 derived DISs. The following is a part of the log file.

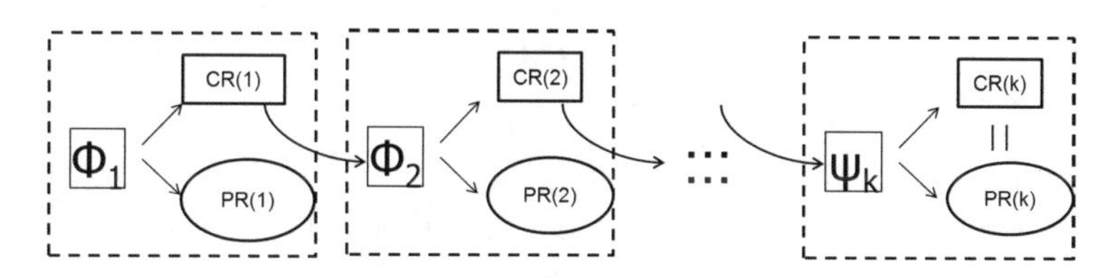

CR(i): A set of certain rules in NIS Φi,
PR(i): A set of possible rules in NIS Φi,
CR(1)⊂CR(2)⊂ : : : ⊂CR(k)=PR(k)⊂PR(k-1)⊂ : : : ⊂PR(1),

DIS ψ_k: Estimated DIS from Φ1,
CR(k)=PR(k): Learned rules from Φ1.

Fig. 6 An overview of machine learning by rule generation [38]

Table 7 An exemplary NIS

object	*age*	*dept.*	*smoker*	*salary*
$x1$	$\{young\}$	$\{first\}$	$\{yes\}$	$\{low\}$
$x2$	$\{young, senior\}$	$\{first, second, third\}$	$\{yes\}$	$\{low\}$
$x3$	$\{senior\}$	$\{second\}$	$\{yes, no\}$	$\{high\}$
$x4$	$\{young, senior\}$	$\{second\}$	$\{no\}$	$\{high\}$
$x5$	$\{young\}$	$\{first, second, third\}$	$\{yes, no\}$	$\{high\}$
$x6$	$\{senior\}$	$\{third\}$	$\{no\}$	$\{high\}$

Table 8 NIS in step 2

object	*age*	*dept.*	*smoker*	*salary*
$x1$	$\{young\}$	$\{first\}$	$\{yes\}$	$\{low\}$
$x2$	$\{young\}$	$\{first, third\}$	$\{yes\}$	$\{low\}$
$x3$	$\{senior\}$	$\{second\}$	$\{no\}$	$\{high\}$
$x4$	$\{senior\}$	$\{second\}$	$\{no\}$	$\{high\}$
$x5$	$\{young\}$	$\{second\}$	$\{no\}$	$\{high\}$
$x6$	$\{senior\}$	$\{third\}$	$\{no\}$	$\{high\}$

```
File=salary2016.rs Support=0.3, Accuracy=0.6
===== 1st STEP =======================================
===== Lower System ===================================
    [4][age,senior]=>[salary,high](0.333,0.667) Objects:[3,6]
    [6][dept,second]=>[salary,high](0.333,0.667) Objects:[3,4]
    [10][smoker,no]=>[salary,high](0.333,1.000) Objects:[4,6]
The Rest Candidates: [[[3,1],[4,1]]]
(Lower System Terminated)
```

Based on the execution log, we see a certain rule τ_{10} : [*smoker*, *no*] $\Rightarrow$ [*salary*, *high*] is the most reliable implication, because the pair (0.330,1.000) of values is better. In order to create τ_{10} as many as possible based on the MLE strategy, we estimate *no* from both $g(x3, smoker) = \{yes, no\}$ and $g(x5, smoker) = \{yes, no\}$. Then, we consider a certain rule τ_4 : [*age*, *senior*] $\Rightarrow$ [*salary*, *high*]. In order to create τ_4 as many as possible, we estimate *senior* from $g(x4, age) = \{young, senior\}$. On the other hand, we estimate *young* from $g(x2, age) = \{young, senior\}$ so as not to make a contradiction. Like this we obtain a new NIS in Step 2 (Table 8).

For Table 8, we fix the thresholds α to 0.1 and β to 0.4, and the NIS-Apriori algorithm generates certain rules. The following is a part of the log file.

```
File=salary2016(STEP2).rs Support=0.1, Accuracy=0.4
===== 1st STEP =======================================
===== Lower System ===================================
```

Table 9 An estimated DIS from NIS

object	*age*	*dept.*	*smoker*	*salary*
*x*1	*young*	*first*	*yes*	*low*
*x*2	*young*	*first*	*yes*	*low*
*x*3	*senior*	*second*	*no*	*high*
*x*4	*senior*	*second*	*no*	*high*
*x*5	*young*	*second*	*no*	*high*
*x*6	*senior*	*third*	*no*	*high*

```
    [1][age,young]=>[salary,low](0.333,0.667) Objects:[1,2]
    [4][age,senior]=>[salary,high](0.500,1.000) Objects:[3,4,6]
    [5][dept,first]=>[salary,low](0.167,1.000) Objects:[1]
    [8][dept,second]=>[salary,high](0.500,1.000) Objects:[3,4,5]
    [10][dept,third]=>[salary,high](0.167,0.500) Objects:[6]
    [11][smoker,yes]=>[salary,low](0.333,1.000) Objects:[1,2]
    [14][smoker,no]=>[salary,high](0.667,1.000) Objects:[3,4,5,6]
The Rest Candidates: [[[1,1],[4,2]]]
(Lower System Terminated)
```

Based on the execution log, we focus on τ_5 with (0.167,1.000) and τ_{10} with (0.167,0.500). Since τ_5 has a better minimum accuracy value, we try to fix attribute values so as to generate τ_5 as many as possible. Thus, we estimate *first* from $g(x2, dept) = \{first, third\}$, and we estimate one DIS in Table 9 from 144 DISs.

In MLRG, NIS recognizes certain rules by itself after rule generation. NIS tries to create certain rules as many as possible. In this process, non-deterministic information is fixed to one value, and NIS repeats this process by reducing the threshold values α and β sequentially. In order to obtain new certain rules, it will be necessary to reduce these threshold values. The selection of the threshold values will strongly cause the result by MLRG, and we need to consider what selection of the threshold values is proper in MLRG.

We have proposed MLRG, whose concept comes from the maximum likelihood estimation. Even though it will be impossible to know the actual value for non-deterministic information [18, 19], we think that MLRG will give us a plausible estimation.

3.4 *Information Dilution*

We consider generating NIS from DIS intentionally in Fig. 3. We add noisy information to DIS, and generate NIS for hiding the actual information. NIS Φ is seen as a *diluted* DIS ψ, and we can hide the actual values in ψ by using Φ. We name this

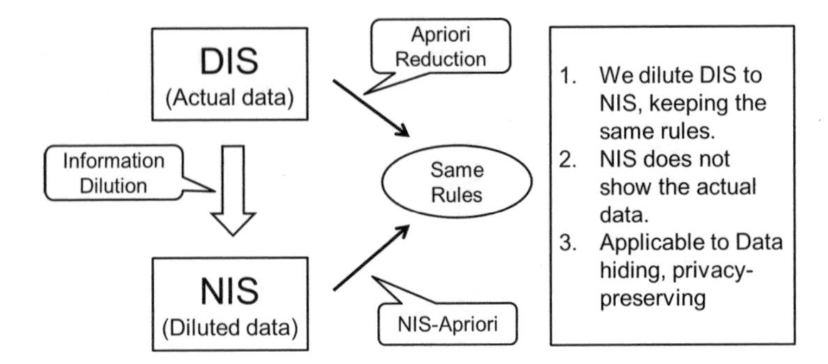

Fig. 7 Formalization of information dilution with constraint cited from [34]

Table 10 An exemplary DIS Ψ

OB	*A*	*B*	*C*	*D*
1	3	1	1	1
2	2	1	1	1
3	1	1	1	2
4	3	1	2	1
5	3	1	1	1
6	2	2	2	2
7	1	2	1	2
8	2	2	2	2

method of hiding *Information Dilution* by non-deterministic information as seen in Fig. 7. We are going to apply this idea to data hiding and privacy-preserving [2].

Reduction is one of the most important concepts in rough sets, and we remove any redundant information from each table by reduction. Pawlak's original reduction is defined by preserving the consistency of a data set [28], and this is extended to reduction with the preservation of lower and upper approximations [13], etc. On the other hand, in dilution we add some noisy information to some attribute values. We may say that dilution is inverse-reduction, and we consider a constraint that each rule in DIS is obtained as a possible rule in NIS in Fig. 7.

Here, we consider rules defined by $support(\tau) > 0$ and $accuracy(\tau) = 1.0$ in DIS Ψ. Although we omit the details of the dilution algorithm, the set of rules in Ψ (Table 10) is equal to the set of possible rules in Φ (Table 11). Namely, Φ is a diluted NIS from DIS Ψ with preserving the same rules. The details of this example are in [34]. As for dilution, even though we have shown such examples, further research is now in progress. Information dilution will also be the next topic in rough sets (Fig. 7).

Table 11 NIS Φ diluted from Ψ

OB	A	B	C	D
1	$\{3\}$	$\{1,2\}$	$\{1,2\}$	$\{1\}$
2	$\{2\}$	$\{1\}$	$\{1\}$	$\{1\}$
3	$\{1,2,3\}$	$\{1\}$	$\{1\}$	$\{2\}$
4	$\{1,2,3\}$	$\{1\}$	$\{2\}$	$\{1\}$
5	$\{1,2,3\}$	$\{1,2\}$	$\{1,2\}$	$\{1,2\}$
6	$\{2\}$	$\{1,2\}$	$\{2\}$	$\{2\}$
7	$\{1\}$	$\{1,2\}$	$\{1,2\}$	$\{2\}$
8	$\{1,2,3\}$	$\{2\}$	$\{1,2\}$	$\{2\}$

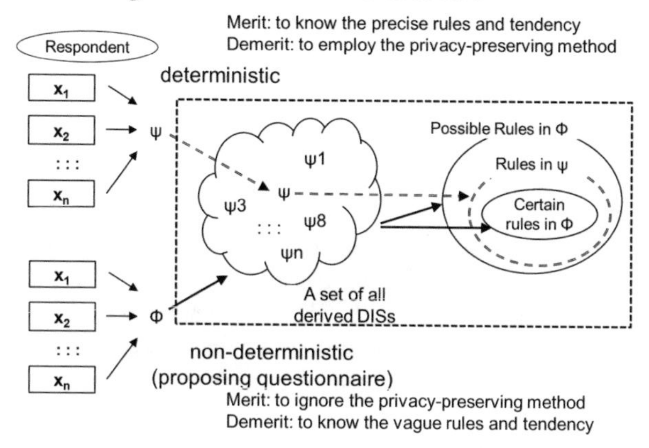

Fig. 8 A chart on a privacy-preserving questionnaire

3.5 Application to a Privacy-Preserving Questionnaire

Let us consider rule generation from DIS and NIS in Fig. 8. We have implicitly coped with the block with the broken lines, namely DIS and NIS are given. For given NIS, we investigated how we generate rules in order to know the tendency and the property of a data set.

Now, we consider the application of NIS to the privacy preserving questionnaire, which consists of three-choice question or multiple-choice question. In questionnaire, we may respond personal information, which the organizer of the questionnaire should deal with safely. However, there are several cases of information leaks, so it is necessary for the organizer to pay attention to such questionnaire.

If we answer 'either A or B' instead of the actual choice A, we intentionally dilutes our choice. This will be the similar concept on the 2-anonymity [2], and non-deterministic information will be desirable for preserving each respondent's informa-

tion. Since we can generate rules from NIS now, we can consider this new questionnaire in Fig. 8 [39]. In the usual case, the actual table Ψ is obtained, and the organizer acquires the tendency of all respondents by using rules in Ψ. On the other hand in our proposal, Ψ is diluted to one DIS in all derived DISs from Φ, and the organizer knows the tendency by using certain and possible rules. For knowing the tendency of respondents, Ψ will be more suitable than Φ. However, Φ is more suitable than Ψ for information security. Like this non-deterministic information will take an important role in information security, and we propose the next questionnaire.
(A privacy-preserving questionnaire) Usually, each respondent answers one choice from the multiple-choice question. However, each respondent may answer non-deterministic information like 'either A or B' from the multiple choices, if the question is inconvenient for him. It is possible to know the tendency of all respondents by using certain rules and possible rules in Φ. Even though rules in Φ may not be clearer than rules in Ψ, Φ is more suitable than Ψ for data security.

4 Concluding Remarks

This chapter surveyed rough sets and non-deterministic information, and considered the topics related to non-deterministic information toward the actual application. Recently, we realized the software tool NIS-Apriori in SQL [40, 42], which is implemented by the procedures in SQL. The environment for analyzing NIS is getting better, so we will be able to cope with proposing topics on NIS. We think such topics on NIS will further extend the role of Pawlak's rough set.

Acknowledgements The first author is grateful to Prof. Shusaku Tsumoto for his comments on the privacy-preserving questionnaire. He also thanks Prof. Dominik Ślęzak for his guidance on SQL. This work is supported by JSPS (Japan Society for the Promotion of Science) KAKENHI Grant Number 26330277.

References

1. Aldrich, J.: R. A. Fisher and the making of maximum likelihood 1912–1922. Stat. Sci. **12**(3), 162–176 (1997)
2. Aggarwal, C., Yu, S.: A general survey of privacy-preserving data mining models and algorithms. In: Aggarwal, C., Yu, S. (eds.) Privacy-Preserving Data Mining, Models and Algorithms, pp. 11–52. Springer (2008)
3. Agrawal, R., Srikant, R.: Fast algorithms for mining association rules in large databases. In: Bocca, J.B., Jarke, M., Zaniolo, C (eds.) Proceedings of VLDB'94, pp. 487–499. Morgan Kaufmann (1994)
4. Agrawal, R., Mannila, H., Srikant, R., Toivonen, H., Verkamo, A. I.: Fast discovery of association rules. In: Advances in Knowledge Discovery and Data Mining, pp. 307–328. AAAI/MIT Press (1996)
5. Ciucci, D., Flaminio, T.: Generalized rough approximations in PI 1/2. Int. J. Approximate Reasoning **48**(2), 544–558 (2008)

6. Codd, E.F.: A relational model of data for large shared data banks. Commun. ACM **13**(6), 377–387 (1970)
7. Demri, S., Orłowska, E.: Incomplete Information: Structure, Inference. Complexity. An EATCS Series, Springer, Monographs in Theoretical Computer Science (2002)
8. Frank, A., Asuncion, A.: UCI machine learning repository. Irvine, CA: University of California, School of Information and Computer Science. http://mlearn.ics.uci.edu/MLRepository.html (2010)
9. Greco, S., Matarazzo, B., Słowiński, R.: Granular computing and data mining for ordered data: The dominance-based rough set approach. In: R.A. Meyers (ed.) Encyclopedia of Complexity and Systems Science, pp. 4283–4305. Springer (2009)
10. Grzymała-Busse, J. W., Werbrouck, P.: On the best search method in the LEM1 and LEM2 algorithms. In: E. Orłowska (ed.) Incomplete Information: Rough Set Analysis, Studies in Fuzziness and Soft Computing, vol. 13, pp. 75–91. Springer (1998)
11. Grzymała-Busse, J.W.: Data with missing attribute values: Generalization of indiscernibility relation and rule induction. Trans. Rough Sets **1**, 78–95 (2004)
12. Grzymała-Busse, J., Rząsa, W.: A local version of the MLEM2 algorithm for rule induction. Fundamenta Informaticae **100**, 99–116 (2010)
13. Inuiguchi, M., Yoshioka, Y., Kusunoki, Y.: Variable-precision dominance-based rough set approach and attribute reduction. Int. J. Approximate Reasoning **50**(8), 1199–1214 (2009)
14. Komorowski, J., Pawłak, Z., Polkowski, L., Skowron, A.: Rough sets: a tutorial, In: Pal, S. K., Skowron, A. (eds.) Rough Fuzzy Hybridization: A New Method for Decision Making, pp. 3–98. Springer (1999)
15. Kripke, S.A.: Semantical considerations on modal logic. Acta Philosophica Fennica **16**, 83–94 (1963)
16. Kryszkiewicz, M.: Rough set approach to incomplete information systems. Inf. Sci. **112**(1–4), 39–49 (1998)
17. Kryszkiewicz, M.: Rules in incomplete information systems. Inf. Sci. **113**(3–4), 271–292 (1999)
18. Lipski, W.: On semantic issues connected with incomplete information databases. ACM Trans Database Syst. **4**(3), 262–296 (1979)
19. Lipski, W.: On databases with incomplete information. J. ACM **28**(1), 41–70 (1981)
20. Marek, W., Pawłak, Z.: Information storage and retrieval systems: Mathematical foundations. Theor. Comput. Sci. **1**(4), 331–354 (1976)
21. Nakata, M., Sakai, H.: Lower and upper approximations in data tables containing possibilistic information. Trans. Rough Sets **7**, 170–189 (2007)
22. Nakata, M., Sakai, H.: Applying rough sets to information tables containing possibilistic values. Trans. Comput. Sci. **2**, 180–204 (2008)
23. Nakata, M., Sakai, H.: Twofold rough approximations under incomplete information. Int. J. Gen. Syst. **42**(6), 546–571 (2013)
24. Orłowska, E., Pawłak, Z.: Representation of nondeterministic information. Theoret. Comput. Sci. **29**(1–2), 27–39 (1984)
25. Orłowska, E.: Introduction: What you always wanted to know about rough sets, In: Orłowska, E (ed.) Incomplete Information: Rough Set Analysis, Studies in Fuzziness and Soft Computing, vol. 13, pp. 1–20. Springer (1998)
26. Pawłak, Z.: Information systems theoretical foundations. Inf. Syst. **6**(3), 205–218 (1981)
27. Pawłak, Z.: Systemy Informacyjne: Podstawy Teoretyczne (In Polish), p. 186. WNT Press (1983)
28. Pawłak, Z.: Rough Sets: Theoretical Aspects of Reasoning about Data, p. 229. Kluwer Academic Publishers (1991)
29. Pawłak, Z.: Some issues on rough sets. Trans. Rough Sets **1**, 1–58 (2004)
30. Polkowski, L., Skowron, A. (Eds.): Rough sets in knowledge discovery 1: Methodology and applications. In: Studies in Fuzziness and Soft Computing, vol. 18, p. 576. Springer (1998)
31. Qian, Y.H., Liang, J.Y., Yao, Y.Y., Dang, C.Y.: MGRS: A multi-granulation rough set. Inf. Sci. **180**, 949–970 (2010)

32. Sakai, H., Okuma, A.: Basic algorithms and tools for rough non-deterministic information analysis. Trans. Rough Sets **1**, 209–231 (2004)
33. Sakai, H., Ishibashi, R., Koba, K., Nakata, M.: Rules and apriori algorithm in non-deterministic information systems. Trans. Rough Sets **9**, 328–350 (2008)
34. Sakai, H., Wu, M., Yamaguchi, N., Nakata, M.: Rough set-based information dilution by non-deterministic information. In: Ciucci, Davide, et al. (eds.) Proceedings of RSFDGrC2013, vol. 8170, pp. 55–66. Springer, LNCS (2013)
35. Sakai, H., Wu, M., Yamaguchi, N., Nakata, M.: Non-deterministic information in rough sets: A survey and perspective. In: Pawan Lingras et al. (eds.) Proceedings of RSKT2013 vol. 8171, pp. 7–15. LNCS, Springer (2013)
36. Sakai, H., Wu, M., Nakata, M.: Apriori-based rule generation in incomplete information databases and non-deterministic information systems. Fundamenta Informaticae **130**(3), 343–376 (2014)
37. Sakai, H., Wu M.: The completeness of NIS-Apriori algorithm and a software tool getRNIA. In: Mori, M. (ed.) Proceedings of International Conference on AAI2014, pp. 115–121. IEEE (2014)
38. Sakai, H., Liu, C.: A consideration on learning by rule generation from tables with missing values. In: Mine, T. (ed.) Proceedings of International Conference on AAI2015, pp. 183–188. IEEE (2015)
39. Sakai, H., Liu, C., Nakata, M., Tsumoto, S.: A proposal of the privacy-preserving questionnaire by non-deterministic information and its analysis. In: Proceedings of IEEE International Conference on Big Data, pp. 1956–1965 (2016)
40. Sakai, H.: Execution logs by RNIA software tools. http://www.mns.kyutech.ac.jp/~sakai/RNIA (2016)
41. Skowron, A., Rauszer, C.: The discernibility matrices and functions in information systems. In: Słowiński, R. (ed.) Intelligent Decision Support—Handbook of Advances and Applications of the Rough Set Theory, pp. 331–362. Kluwer Academic Publishers (1992)
42. Ślęzak, D., Sakai, H.: Automatic extraction of decision rules from non-deterministic data systems: Theoretical foundations and SQL-based implementation. In: Ślęzak, D., Kim, T.H., Zhang, Y., Ma, J., Chung, K.I. (eds.) Database Theory and Application, Communications in Computer and Information Science, vol. 64, pp. 151–162. Springer (2009)
43. Tsumoto, S.: Knowledge discovery in clinical databases and evaluation of discovered knowledge in outpatient clinic. Inf. Sci. **124**(1–4), 125–137 (2000)
44. Tsumoto, S.: Automated extraction of hierarchical decision rules from clinical databases using rough set model. Expert Syst. Appl. **24**, 189–197 (2003)
45. Wikipedia: Constraint satisfaction problem. https://en.wikipedia.org/wiki/Constraint_satisfaction_problem
46. Yao, Y.Y.: A note on definability and approximations. Trans. Rough Sets **7**, 274–282 (2007)
47. Yao, Y.Y.: Three-way decisions with probabilistic rough sets. Inf. Sci. **180**, 314–353 (2010)
48. Yao, Y.Y.: Two sides of the theory of rough sets. Knowl. Based Syst. **80**, 67–77 (2015)
49. Yao, Y.Y., She, Y.: Rough set models in multigranulation spaces. Inform. Sci. **327**, 40–56 (2016)
50. Zhu, W.: Topological approaches to covering rough sets. Inform. Sci. **177**(6), 1499–1508 (2007)

Part III
Rough Set Theory

From Information Systems to Interactive Information Systems

Andrzej Skowron and Soma Dutta

Abstract In this chapter we propose a departure from classical notion of information systems. We propose to bring in the background of agent's interaction with physical reality in arriving at a specific information system. The proposals for generalizing the notion of information systems are made from two aspects. In the first aspect, we talk about incorporating relational structures over the value sets from where objects assumes values with respect to a set of attributes. In the second aspect, we introduce interaction with physical reality within the formal definition of information systems, and call them as interactive information systems.

1 Introduction

Professor Zdzisław Pawlak published several papers [8–14, 16, 18–24, 26, 27] as well as a book (in Polish) [25] on information systems (see Figs. 1, 2, 3 and 4). The first definition of information systems, as proposed by him, appeared in [18, 19].

An information system was defined as a tuple consisting of a finite set of objects and a set of attributes defined over the set of objects with values in attribute value sets. More formally, an information system is a tuple

$$IS = (U, \mathscr{A}, \{f_a : U \rightarrow V_a\}_{a \in \mathscr{A}}), \quad (1)$$

A. Skowron (✉)
Institute of Mathematics, University of Warsaw, Banacha 2, 02-097 Warsaw, Poland
e-mail: skowron@mimuw.edu.pl

A. Skowron
Systems Research Institute, Polish Academy of Sciences, Newelska 6,
01-447 Warsaw, Poland

S. Dutta
Vistula University, Stokłosy 3, 02-787 Warsaw, Poland
e-mail: somadutta9@gmail.com

S. Dutta
Faculty of Mathematics, Informatics, and Mechanics, University of Warsaw,
Warsaw, Poland

G. Wang et al. (eds.), *Thriving Rough Sets*, Studies in Computational
Intelligence 708, DOI 10.1007/978-3-319-54966-8_10

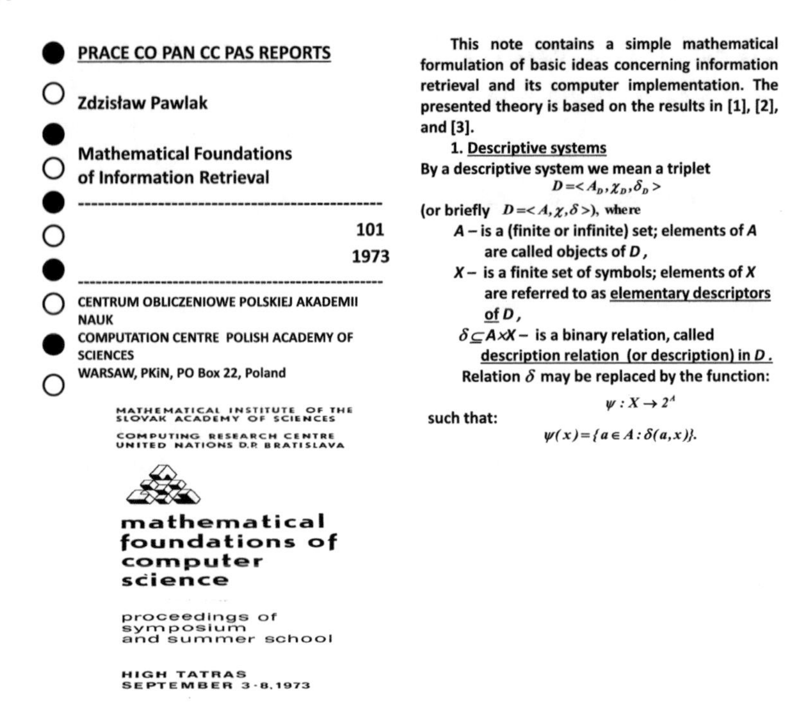

PRACE CO PAN CC PAS REPORTS

Zdzisław Pawlak

Mathematical Foundations of Information Retrieval

101
1973

CENTRUM OBLICZENIOWE POLSKIEJ AKADEMII NAUK
COMPUTATION CENTRE POLISH ACADEMY OF SCIENCES
WARSAW, PKiN, PO Box 22, Poland

This note contains a simple mathematical formulation of basic ideas concerning information retrieval and its computer implementation. The presented theory is based on the results in [1], [2], and [3].

1. Descriptive systems

By a descriptive system we mean a triplet

$$D = < A_D, \chi_D, \delta_D >$$

(or briefly $D = < A, \chi, \delta >$), where

A – is a (finite or infinite) set; elements of A are called objects of D,

X – is a finite set of symbols; elements of X are referred to as elementary descriptors of D,

$\delta \subseteq A \times X$ – is a binary relation, called description relation (or description) in D.

Relation δ may be replaced by the function:

$$\psi : X \to 2^A$$

such that:

$$\psi(x) = \{a \in A : \delta(a, x)\}.$$

MATHEMATICAL INSTITUTE OF THE SLOVAK ACADEMY OF SCIENCES
COMPUTING RESEARCH CENTRE UNITED NATIONS D.P. BRATISLAVA

mathematical foundations of computer science

proceedings of symposium and summer school

HIGH TATRAS
SEPTEMBER 3-8, 1973

Fig. 1 The first papers on information systems by Zdzisław Pawlak [18, 19]

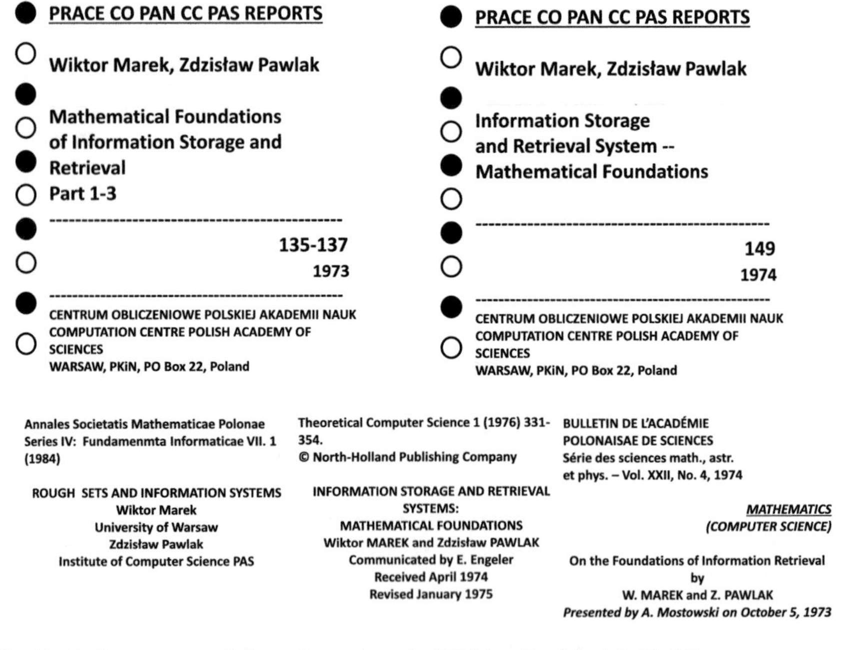

PRACE CO PAN CC PAS REPORTS

Wiktor Marek, Zdzisław Pawlak

Mathematical Foundations of Information Storage and Retrieval Part 1-3

135-137
1973

CENTRUM OBLICZENIOWE POLSKIEJ AKADEMII NAUK
COMPUTATION CENTRE POLISH ACADEMY OF SCIENCES
WARSAW, PKiN, PO Box 22, Poland

PRACE CO PAN CC PAS REPORTS

Wiktor Marek, Zdzisław Pawlak

Information Storage and Retrieval System -- Mathematical Foundations

149
1974

CENTRUM OBLICZENIOWE POLSKIEJ AKADEMII NAUK
COMPUTATION CENTRE POLISH ACADEMY OF SCIENCES
WARSAW, PKiN, PO Box 22, Poland

Annales Societatis Mathematicae Polonae
Series IV: Fundamenmta Informaticae VII. 1 (1984)

ROUGH SETS AND INFORMATION SYSTEMS
Wiktor Marek
University of Warsaw
Zdzisław Pawlak
Institute of Computer Science PAS

Theoretical Computer Science 1 (1976) 331-354.
© North-Holland Publishing Company

INFORMATION STORAGE AND RETRIEVAL SYSTEMS: MATHEMATICAL FOUNDATIONS
Wiktor MAREK and Zdzisław PAWLAK
Communicated by E. Engeler
Received April 1974
Revised January 1975

BULLETIN DE L'ACADÉMIE POLONAISAE DE SCIENCES
Série des sciences math., astr. et phys. – Vol. XXII, No. 4, 1974

MATHEMATICS (COMPUTER SCIENCE)

On the Foundations of Information Retrieval
by
W. MAREK and Z. PAWLAK
Presented by A. Mostowski on October 5, 1973

Fig. 2 Further papers on information systems by Zdzisław Pawlak et al. [8–14]

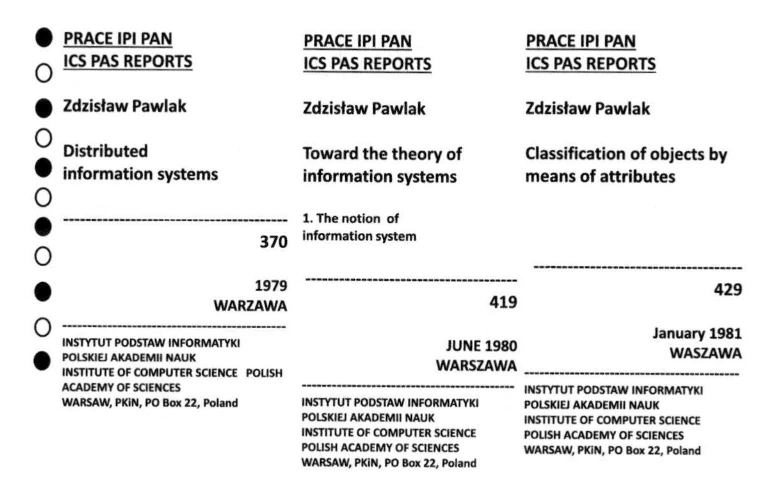
PRACE IPI PAN
ICS PAS REPORTS

Zdzisław Pawlak

Distributed information systems

370

1979
WARZAWA

INSTYTUT PODSTAW INFORMATYKI
POLSKIEJ AKADEMII NAUK
INSTITUTE OF COMPUTER SCIENCE POLISH ACADEMY OF SCIENCES
WARSAW, PKiN, PO Box 22, Poland

PRACE IPI PAN
ICS PAS REPORTS

Zdzisław Pawlak

Toward the theory of information systems

1. The notion of information system

419

JUNE 1980
WARSZAWA

INSTYTUT PODSTAW INFORMATYKI
POLSKIEJ AKADEMII NAUK
INSTITUTE OF COMPUTER SCIENCE
POLISH ACADEMY OF SCIENCES
WARSAW, PKiN, PO Box 22, Poland

PRACE IPI PAN
ICS PAS REPORTS

Zdzisław Pawlak

Classification of objects by means of attributes

429

January 1981
WASZAWA

INSTYTUT PODSTAW INFORMATYKI
POLSKIEJ AKADEMII NAUK
INSTITUTE OF COMPUTER SCIENCE
POLISH ACADEMY OF SCIENCES
WARSAW, PKiN, PO Box 22, Poland

Fig. 3 Further papers on information systems by Zdzisław Pawlak [21–23] (CONT)

Zdzisław Pawlak

SYSTEMY INFORMACYJNE
Podstawy teoretyczne

WYDAWNICTWA NAUKOWO-TECHNICZNE • WARSZAWA 1983

Fig. 4 Book (in Polish) dedicated to information systems by Zdzisław Pawlak [25]

where

- U is a finite set of objects,
- $\mathscr{A}$ is a finite set of attributes,
- any attribute a from $\mathscr{A}$ can be characterized as a function f_a from U to a value set V_a corresponding to a.

In the mentioned papers and in the book, Pawlak investigated different kinds of information systems such as deterministic, nondeterministic, information systems with missing values, probabilistic, stochastic as well as distributed. From the point of view of rough sets, information systems are used for constructive definition of indiscernibility relation. Then the indiscernibility or similarity classes can be tuned to relevant ones (in order to get relevant indiscernibility classes, also called elementary granules), e.g., by selecting or extracting relevant attributes. On the basis of information systems (as data sets) data models are induced using different methods, in particular based on rough sets.

In this chapter, we propose some generalizations of information systems.

First we will consider a bit more general definition of the value sets for attributes. In particular, together with the value set V_a for any attribute a, we will also consider a relational structure $\mathscr{R}_a$ over V_a. These $\mathscr{R}_a$'s are not restricted to the case of the relational structure consisting of only the equality relation on V_a's, as it was originally considered by Pawlak. More general cases can include a linear order over V_a as well as more complex relations with arity greater than 2 [34]. Together with the relational structure $\mathscr{R}_a$ we consider a language $\mathscr{L}_a$ of formulas defining (under a given interpretation over $\mathscr{R}_a$) subsets of V_a. It is to be noted, that such formulas can be obtained from formulas with many free variables by substituting a constant for each of them except one. Some relevant formulas from this set of formulas become useful as they can play a role in inducing data models. For example, one can consider an attribute with real values and formalize a discretization problem. In this case for any real-valued attribute a we can consider a set of formulas $\{x \in [c, d]\}$, where x is a free variable corresponding to the attribute a taking real values and c, d are constants defining an interval. Then we search for a minimal set of such formulas discerning in the optimal by decisions labeling the attribute values and defining the partition of the real numbers [15]. Another example may be related to the dominance rough set approach (see, e.g., [3, 35]), where linear orders are considered on attribute value sets.

In this chapter, we also introduce a network of information systems over such generalized information systems. This is done analogous to the notion of information flow approach proposed by Barwise and Seligman [1, 32, 33]. However, first we consider different kinds of aggregation of relational structures corresponding to attributes from a given set of attributes A. Then we define a set of formulas $\mathscr{L}$ which can be interpreted over such relational structures. In this context, one may introduce relations with many arguments. Discovery of such kinds of relevant relations, based on purpose, is the task of relational learning [2].

Our final stage of generalization of information systems concerns of interactions of information systems with the environment. This issue is strongly related to the discussed interactive granular computations (see e.g., [4–7, 28–31]), where information systems are treated as open objects, which are continuously evolving based on the interactions with the environment. This extension can be used as a basis for developing Perception Based Computing (PBC) [17, 36] and for developing the foundations of Interactive Granular Computing (IGrC) [4–7, 28–31].

The chapter is structured as follows. In Sect. 2, we first discuss the roles of relational structures over the value sets corresponding to attributes of an information system. We present different examples to elucidate the fact that aggregation of such relational structures plays an important role in representation and granulation of data of an information system, which often contains huge and scattered data. Section 3 introduces the notion of interactive information system as a generalization of the notion of *IS* (cf. Eq. (1)) presented at the beginning of this chapter. In the last section, as concluding remarks, we add some discussion regarding incorporation of some other finer aspects of interactive information systems.

2 Role of Relational Structures in Aggregation of Information Systems

Depending on purpose we need to gather information of different nature, such as images of some object as well as quantitative values for some features of the same object, together in order to make an overall understanding about the object. So, values corresponding to different features as well as the intra-relational structures among the values become important. The aim of this section is to present different kinds of aggregation of relational structures, which we need to perform in order to aggregate information collected from, and for, different perspectives.

The chapter is organized so that, in one aspect we would talk about relations over the value sets of the attributes of an information system, and in another aspect we also would like to address the issue of the relational objects lying in the real world, about which we only able to gather some information through some attributes and their values. This aspect of real world will be discussed in the next section where we propose to introduce interaction with physical reality in the process of obtaining an information system. A physical object o, being in a complex relation with other objects in the real physical world, sometimes cannot be directly accessed. We sometimes identify the object with some of its images or with some of its parts or components, and try to gauge information about the object with respect to some parameters. One possible way of measuring the real state of an object through some other state is proposed through the notion of complex granule in [4–7, 28–31]. Here we will address this introducing a notion of infomorphism in the line of [ ], and call that *interaction with physical reality*. In this section, we only stick to the relations among the values of attributes using which we learn about objects in the physical world. Let us start with some examples in order to make the issue more lucid.

Let $\mathscr{A}_{rect} = \{a, b, c\}$ be a set of attributes representing respectively *length*, *breadth*, and *angle between two sides* of a rectangle. Clearly, a and b are of the same nature and can assume values from the same set, say $V_a = [0, 300]$ in some unit of length, and be endowed with the same relation $\leq$. Let us call the relational structure over the values for the attribute a as $\mathscr{R}_a = (V_a, \leq)$, which is same as $\mathscr{R}_b$ too, in this context. Let $V_c = [0°, 180°]$ and $\mathscr{R}_c = (V_c, =)$. Now we can construct a language $\mathscr{L}_{a,b}$ (cf. Table 1).

Table 1 Language $\mathscr{L}_{a,b}$

Variable: x_1
Constants: any value from V_a
Function symbol: a, b
Relational symbol: $\leq$
Terms: (i) Variable and constants are terms (ii) $a(x_1)$ and $b(x_1)$ are terms
Examples of wffs: $b(x_1) \leq a(x_1)$ is an atomic wff

Table 2 Language $\mathscr{L}_c$

Variable: x_1
Constants: any value from V_c
Function symbol: c
Relational symbol: =
Terms: (i) Variable and constants are terms (ii) $c(x_1)$ is a term
Examples of wffs: $c(x_1) = 90°$ is an atomic wff

In particular, we may call $b(x_1) \leq a(x_1)$, which represents *breadth of* x_1 *is less or equal to the length of* x_1, as $\phi_{11}(x_1)$. Considering that the variable x_1 is ranging over a set of objects, say $\mathscr{O}$, values from $\mathscr{R}_a = (V_a, \leq)$ can be assigned to $a(x_1), b(x_1)$, and thus the semantics of $\mathscr{L}_{a,b}$ can be given over the relational structure $\mathscr{R}_a$. In the similar way we can have the language $\mathscr{L}_c$, semantics of which can be given over the relational structure $\mathscr{R}_c$ (cf. Table 2).

Before passing on to the next table for $\mathscr{L}_c$, it is to be noted that, the values of terms $a(x_1), b(x_1)$, belonging to V_a and V_b respectively, are obtained by some agent *ag* observing a complex granule (c-granule, for short) [4–7, 28–31] grounded on a configuration of physical objects. Relations among the parts of the configuration can be perceived partially by the c-granule through $a(x_1), b(x_1)$. Some objects in the configuration have states which may be directly measurable, and those can be encoded by elements of V_a and V_b. They can be be treated as values, e.g., $a(x_1), b(x_1)$ of the example, under the assumption that they represent states of one distinguished object o in the configuration. They considered as a current value of x_1, identified by some mean with o. However, the states of o may not be directly measurable. Information about not directly measurable states may be obtained using relevant interactions with physical objects pointed by the c-granule, and making it possible to transmit information about such states and encode it using measurable states. In this chapter we represent interactions of agents with the physical reality using infomorphisms [1].

Like previous case, here also we can assume $\phi_{12}(x_1)$ as the formula $c(x_1) = 90°$, which represents *angles between two sides of* x_1 is 90°. So, we have two relational structures, namely $\mathscr{R}_a = (V_a, \leq)$ and $\mathscr{R}_c = (V_c, =)$, on which respectively the formulas $\phi_{11}(x_1)$ and $\phi_{12}(x_1)$ are interpreted with respect to the domain of interpretation of x_1, which can be considered as a set of objects. The value of the term $c(x_1)$ is obtained in an analogous way as mentioned before for $a(x_1), b(x_1)$. Now, the question arises how can we combine these two relational structures to gather information about whether an object is rectangle or not. Here, as the attributes a, b, c are relevant for the same sort of objects, we may simply extend the language combining all the components of $\mathscr{L}_{a,b}$ and $\mathscr{L}_c$ together (cf. Table 3).

In $\mathscr{L}_{rect}$ instead of relational symbols $\leq$ and $=$, one can also consider a new three-place relational symbol r_1^3 such that $r_1^3(a(x_1), b(x_1), c(x_1))$ holds for some object from the domain of x_1 if $b(x_1) \leq a(x_1)$ (*i.e.*, $\phi_{11}(x_1)$) is true over $\mathscr{R}_a$ and $c(x_1) = 90°$ (*i.e.*, $\phi_{12}(x_1)$) is true over $\mathscr{R}_c$. So, assuming a set of objects as the domain of interpretation

Table 3 Language $\mathscr{L}_{rect}$: combination of $\mathscr{L}_{a,b}$ and $\mathscr{L}_c$

Variable: x_1
Constants: any value from $V_a \cup V_c$
Function symbol: a, b, c
Relational symbol: $\leq, =$
Terms: (i) Variable and constants are terms (ii) $a(x_1)$, $b(x_1)$, $c(x_1)$ are terms
Examples of wffs: $b(x_1) \leq a(x_1)$, $c(x_1) = 90°$ are atomic wffs

Table 4 Language $\mathscr{L}_{tri}$: combination of $\mathscr{L}_d$ and $\mathscr{L}_e$

Variable: x_2
Constants: any value from $V_d \cup V_e$
Function symbol: d, e
Relational symbol: $\leq, \preceq$
Terms: (i) Variable and constants are terms (ii) $d(x_2)$, $e(x_2)$ are terms
Examples of wffs: $d(x_2) = 1$, $e(x_2) = 180°$ are atomic wffs.

for x_1, this new language $\mathscr{L}_{rect}$ can be interpreted over the combined relational structure $\mathscr{R}_{rect} = (V_{rect}, \{\leq, =\})$, where $V_{rect} = V_a \cup V_c$. We may call $r_1^3(a(x_1), b(x_1), c(x_1))$ as $\phi_{13}(x_1)$.

Let us consider another context where the attributes are relevant for a triangle-shaped object. So, we consider $\mathscr{A}_{tri} = \{d, e\}$, where d stands for *three-sided* and e stands for *sum of the angles*. Again the relational structures suitable for the values of the attributes are respectively $\mathscr{R}_d = (\{0, 1\}, \leq)$ and $\mathscr{R}_e = ([0°, 180°], \preceq)$. It is to be noted that $\preceq$ is the same relation as that of the real numbers (*i.e.*, $\leq$). We use different symbol in order to emphasize that the values relevant for d and e are of different types. Now as shown in the previous case, we can construct different languages over the different relations from $\mathscr{R}_d$ and $\mathscr{R}_e$, and combining them together we can have the language $\mathscr{L}_{tri}$ (cf. Table 4).

In this context too, in $\mathscr{L}_{tri}$, instead of two relation symbols $\leq, \preceq$, one can take a two-place relation symbol r_1^2 such that for some object from the domain of x_2, $r_1^2(d(x_2), e(x_2))$ holds if with respect to that object $d(x_2) = 1$ and $e(x_2) = 180°$ are true over $\mathscr{R}_{tri} = (V_{tri}, \{\leq, \preceq\})$ where $V_{tri} = V_d \cup V_e$. As above, $r_1^2(d(x_2), e(x_2))$ may be called $\phi_{23}(x_2)$, where the values of $d(x_2)$ and $e(x_2)$ are obtained in an analogous way as before.

In the above two cases we have obtained the extended relational structures $\mathscr{R}_{rect}$ and $\mathscr{R}_{tri}$ by combining the respective relational structures for each attribute from $\mathscr{A}_{rect}$ and $\mathscr{A}_{tri}$. In some context, we need to gather information about objects whose domain consists of tuples of elements of different natures. As an example we can consider a situation where we need to collect information about objects which are prisms with rectangular bases and triangular faces. So, we need to have a language over $\mathscr{A} = \mathscr{A}_{rect} \cup \mathscr{A}_{tri}$, and contrary to the earlier cases of combining languages here

Table 5 Language $\mathscr{L}_{prism}$: aggregation of $\mathscr{L}_{rect}$ and $\mathscr{L}_{tri}$

Variable: $x = (y, z)$
Constants: any value from $V_{rect} \cup V_{tri}$, x_1, x_2, and $\phi_{13}(x_1)$, $\phi_{23}(x_2)$
Relational symbol: r_{11}, r_{21}
Terms: (i) Variable and constants are terms
Wffs: $r_{11}((x_1, x_2), \phi_{13}(x_1))$, $r_{21}((x_1, x_2), \phi_{23}(x_2))$ are atomic wffs

we need to aggregate information of two different languages $\mathscr{L}_{rect}$ and $\mathscr{L}_{tri}$ with different domains of concern focusing on different parts of an object. In this context, we would construct the language $\mathscr{L}_{prism}$ one level above the languages $\mathscr{L}_{rect}$ and $\mathscr{L}_{tri}$, and the variables, constants, and wffs of those languages will be referred to as constant symbols of the language of $\mathscr{L}_{prism}$ (cf. Table 5).

Here we have introduced a pair of variables (y, z) to represent a single object with respectively first component for the base and the second component for the face, and (y, z) is assumed to range over a set of objects of the form $o = (o_b, o_f)$ where o_b's are taken from the domain of interpretation of $\mathscr{L}_{rect}$ (i.e., objects on which x_1 ranges), and o_f's from that of $\mathscr{L}_{tri}$ (i.e., objects on which x_2 ranges). So, $r_{11}((x_1, x_2), \phi_{13}(x_1))$ is introduced to represent that an object, characterized by the pair of components base and face (x_1, x_2), has the property of a prism with rectangular base. On the other hand, $r_{21}((x_1, x_2), \phi_{23}(x_2))$ is introduced to represent that (x_1, x_2) is a prism with triangular face. Let us call $r_{11}((x_1, x_2), \phi_{13}(x_1)) = \alpha$ and $r_{21}((x_1, x_2), \phi_{23}(x_2)) = \beta$. Then with respect to the set of objects of the form (o_b, o_f)'s, α and β are true over $\mathscr{R}_{\mathscr{A}}$ for the subsets of objects given by $\{(o_b, o_f) = o : o_b \in ||\phi_{13}(x_1)||_{\mathscr{R}_{\mathscr{A}_{rect}}}\}$ and $\{(o_b, o_f) = o : o_f \in ||\phi_{23}(x_2)||_{\mathscr{R}_{\mathscr{A}_{tri}}}\}$ respectively.

Let us now come to the discussion of how these relational structures over the values for attributes and the respective languages help in the representation of different purposes of information systems.

Let us assume that we have a technical set-up to abstract out images of some parts of the objects appearing in front of a system. Two cameras are set up in a way that any object appearing to the system through a specified way can have their images recorded in the database of the system through some way of measurements. Let the first camera be able to capture the image of the base of the object and the second camera be able to capture the face of the object. So, there are two information systems, namely $IS_{rect} = (B, \mathscr{A}_{rect}, \{\mathscr{R}_a\}_{a \in \mathscr{A}_{rect}}, \{f_a : B \mapsto V_a\}_{a \in \mathscr{A}_{rect}})$ where $\mathscr{R}_a = (V_a, \leq) = \mathscr{R}_b$ and $\mathscr{R}_c = (V_c, =)$, and $IS_{tri} = (F, \mathscr{A}_{tri}, \{\mathscr{R}_d\}_{d \in \mathscr{A}_{tri}}, \{f_d : F \mapsto V_d\}_{d \in \mathscr{A}_{tri}})$ with $\mathscr{R}_d = (V_d, \leq)$ and $\mathscr{R}_e = (V_e, \preceq)$. At the first level we may need to gather information from both IS_{rect} and IS_{tri} in a single information system, say IS_{prism}, in a way that a copy of each of IS_{rect} and IS_{tri} is available. So, we construct a sum of information systems as $IS_{prism} = (B \times F, \mathscr{A}_{prism}, \mathscr{R}_{prism}, \{f_a : B \times F \mapsto V_{prism}\}_{a \in \mathscr{A}_{prism}})$ where $\mathscr{A}_{prism} = (\{1\} \times \mathscr{A}_{rect}) \cup (\{2\} \times \mathscr{A}_{tri})$, $V_{prism} = V_{rect} \cup V_{tri}$, and the relational structure $\mathscr{R}_{prism} = (V_{prism}, \{r\}_{r \in \mathscr{R}_{rect} \cup \mathscr{R}_{tri}})$.

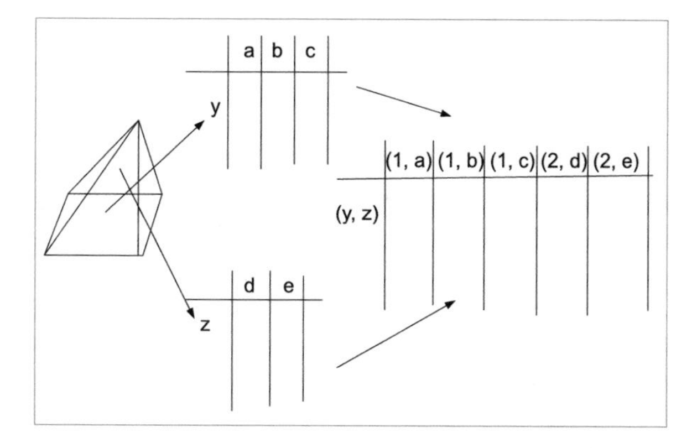

Fig. 5 IS_{prism}: sum of the information systems IS_{rect} and IS_{tri}

Table 6 Sum of information systems

	$(1, a)$	$(1, b)$	$(1, c)$	$(2, d)$	$(2, e)$
(o_b, o_f)	$f_a(o_b)$	$f_b(o_b)$	$f_c(o_b)$	$f_d(o_f)$	$f_d(o_f)$
$\vdots$	$\vdots$	$\vdots$	$\vdots$	$\vdots$	$\vdots$
(o'_b, o'_f)	$f_a(o'_b)$	$f_b(o'_b)$	$f_c(o'_b)$	$f_d(o'_f)$	$f_d(o'_f)$
$\vdots$	$\vdots$	$\vdots$	$\vdots$	$\vdots$	$\vdots$

Language over the relational structure comes into play when we want to have an information system with an added constraint. Let us assume that from all possible three-dimensional objects from $B \times F$, in the above information system IS_{prism}, we are interested in the chunk which has objects with rectangular bases and triangular faces. We can construct an information system imposing a constraint from the language $\mathscr{L}_{prism}$, and thus have the *constraint-based information system*, viz., $CIS_{prism} = ((B \times F) \cap (||\alpha||_{\mathscr{R}_{\mathscr{A}}} \cap ||\beta||_{\mathscr{R}_{\mathscr{A}}}), \mathscr{A}_{prism}, \mathscr{R}_{prism}, \{f_a : B \times F \mapsto V_{prism}\}_{a \in \mathscr{A}_{prism}})$.

Simultaneous consideration of each relational structure included in $\mathscr{R}_{prism} = (V_{prism}, \{r\}_{r \in \mathscr{R}_{rect} \cup \mathscr{R}_{tri}})$ becomes useful when looking at values for one object, say (o_b, o_f) one needs to predict about another object, say (o'_b, o'_f). Let us consider the following situation where we have the information aggregated from IS_{rect} and IS_{tri} in the information system IS_{prism} (cf. Fig. 5 and Table 6).

Let us assume that the values corresponding to each attribute for $o = (o_b, o_f)$ are known. A new object $o' = (o'_b, o'_f)$ appears to the system for which some of the values are missing or because of some technical error measurements of values are not precise. So, one may need to check how value corresponding to each attribute of o is related to the respective value of the other object o'. Let $(v^1_o, v^2_o, v^3_o, v^4_o, v^5_o)$ be the tuple of values corresponding to the attributes (a, b, c, d, e) for object o, and $(v^1_{o'}, v^2_{o'}, v^3_{o'}, v^4_{o'}, v^5_{o'})$ be that of o'. Now on $\Pi_{a \in \mathscr{A}} V_a$, the cartesian product of the value sets corresponding to each attributes of $\mathscr{A}$, we can define a relation r such that $r((v^1_o, v^2_o, v^3_o, v^4_o, v^5_o), (v^1_{o'}, v^2_{o'}, v^3_{o'}, v^4_{o'}, v^5_{o'}))$ if and only if the following relations hold among their components.

(i) $r_{1,2}((v^1_o, v^2_o), (v^1_{o'}, v^2_{o'}))$ iff $v^1_o \leq v^2_o$ and $v^1_{o'} \leq v^2_{o'}$,
(ii) $r_3(v^3_o, v^3_{o'})$ iff $|v^3_o - v^3_{o'}| \leq \epsilon$ for some $\epsilon > 0$,
(iii) $r_4(v^4_o, v^4_{o'})$ iff $v^4_o = v^4_{o'}$, and
(iv) $r_5(v^5_o, v^5_{o'})$ iff $|v^5_o - v^5_{o'}| \leq \delta$ for some $\delta > 0$.

So, based on the relational structure $\mathscr{R}_{\mathscr{A}} = (\Pi_{a\in\mathscr{A}} V_a, r)$ we can have the information system $(B \times F,\ \Pi_{a\in\mathscr{A}} a,\ \mathscr{R}_{\mathscr{A}}, f : B \times F \mapsto \Pi_{a\in\mathscr{A}} V_a)$, which will allow us to cluster objects together, satisfying the relation r.

So, in this section we observe different types of aggregation of relational structures, starting from simple combination of all value sets and their relations in a single aggregated relational structure to cartesian product of value sets and some new relations defined over the relations of the component value sets. We also notice that how based on the requirement of presentation of data in a form of a table, aggregation of those relational structures, and languages interpreted over them become useful. This gives a hint that we need to depart from the classical way of presenting an information system as IS $= (U, \mathscr{A}, \{f_a : U \mapsto V_a\}_{a\in\mathscr{A}})$ to $(U, \mathscr{A}, \{\mathscr{R}_a\}_{a\in\mathscr{A}}, \{f_a : U \mapsto V_a\}_{a\in\mathscr{A}})$, where $\mathscr{R}_a = (V_a, \{r_{ai}\}_{i=1}^k)$.

In the next section, we would present another aspect of generalizing the classical notion of information systems bringing in a component of interaction of an information system, via the respective agent, with the physical reality, and letting the information system to evolve with time.

3 Interactive Information Systems

Information system (cf. Eq. (1)) allow us to present the information about the real world phenomenon in the form a table with object satisfying certain properties to certain degrees. Moreover, as discussed in the previous section, through information systems we can also present how two objects are related based on the interrelation among the values/degrees they obtain for different parameters/attributes/properties. But presentation of the reality through an information system is subjective to the agent's perspective towards viewing, perceiving the reality, which can change with time. At some time t an agent can manage to access some parts of a real object or phenomenon through some process of interactions with the real physical world and abstract out some relevant information, which then can be presented in the form of an information system. That this interaction with reality, based on the factors of time and accessibility, plays a great role in the presented form of an information system cannot be ignored. Two information systems approximating the same real phenomenon may yield quite different views. Thus, incorporating the process of interactions, through which one obtains a particular information system, may help in understanding the background of the presented data. In this regard, below we propose a notion of interactive information system.

As presented the definition of *IS* in Sect. 1, we start with an information system $IS_t = (\mathscr{O}_t, \mathscr{A}_t, \{f_a : U_t \mapsto V_a\}_{a\in\mathscr{A}_t})$ at time t, and trace back to the interactions with the physical reality causing the formation of such an IS_t. The idea is to incorporate those interactions with reality in the mathematical model of *IS* and have an interactive information system, which we may call *ISAPR*, an *information system approximating the physical reality*. Let at time t, there be an existing physical reality which is nothing but a complex network of objects and their interrelations.

1. We assume *physical reality at time t*, denoted as PR_t, as $PR_t = (U'_t, \mathscr{R})$, where U'_t is the set of real objects at time t, and $\mathscr{R}$ is a family of relations of different arities on U'_t. The time point t can be considered as the time related to the local clock of the agent involved in the process of interaction with the physical reality.
2. A part of PR_t is a subrelational structure $(U_\#, \mathscr{R}_\#)$ where $U_\# \subseteq U'_t$ and $\mathscr{R}_\#$ $(\subseteq \mathscr{R})$ is a subfamily of relations over $U_\#$.
3. The set of all subrelational structures at time t is denoted by S_{PR_t}. It is to be noted that for any subset of U'_t endowed with the set of relations from $\mathscr{R}$, restricted to that subset, is a member of S_{PR_t}. So, $S_{PR_t} = \{(U_\#, \mathscr{R}_\#) : U_\# \in P(U'_t), \mathscr{R}_\# = \mathscr{R}_{|U_\#}\}$.
4. At time t, by some means, we can access some information about some parts of the reality. As an instance we can consider a real tree as an object of the physical world, and some houses, park surrounding it as a description of a relational structure among objects in the reality. This we may call $(U_\#, \mathscr{R}_\#)$, a fragment of the real world. Now when an agent captures some images of the tree using a camera, which is another physical object, some states of the real tree are recorded. The real object tree may then be identified with those states or images in the agent's information system. So, we introduce a function $AR_t : S_{PR_t} \mapsto P(S^*)$, where $\mathscr{O}_t \subseteq S^*$, and call it a *function accessing reality at time t*. S^* may be interpreted as a set of states, which can be accessible by some means, and $\mathscr{O}_t$ is the set of states which is possible to access at time t. The role of the camera here is like a tool, which mathematically can be thought of as the function AR_t, through which we access the reality. If instead of a standard camera, one uses a high-resolution camera, then the same fragment of a real physical world may be accessed better than before. So, change of AR_t may give different perspectives of the same physical object.
5. The pair (S_{PR_t}, AR_t) can also be viewed as an information system. For any pair $(U_\#, \mathscr{R}_\#) \in S_{PR_t}$, the first component of the pair can be considered as object and the second can be considered as the set of relations characterizing the object in the physical reality. Given such a pair $(U_\#, \mathscr{R}_\#)$, the function AR_t, which may be a tool (like camera) to interact with the physical reality, basically selects out a set of states, say $\{s_1, s_2, \dots, s_m\}$ $(\subseteq S^*)$ representing the object $(U_\#, \mathscr{R}_\#)$. That is, we can visualize the information system as follows (Table 7).

Table 7 Information system corresponding to (S_{PR_t}, AR_t)

	AR_t	...
$(U_\#, \mathscr{R}_\#)$	$\{s_1, s_2, \dots, s_m\}$	...
⋮		

6. When we fix such a function AR_t, i.e., the mean by which a part of the real world is accessed, we can identify different parts of the real world, say $(U_\#, \mathscr{R}_\#)$ with the respective set of states $\{s_1, s_2, \ldots, s_m\}$ obtained through the function.
   Let us call $\text{States}_{AR_t}(S_{PR_t}) = \bigcup_{(U_\#, \mathscr{R}_\#) \in S_{PR_t}} AR_t(U_\#, \mathscr{R}_\#)$. So, we may consider the physical reality with respect to a specific accessibility function AR_t as an information system $PR_{AR_t} = (S_{PR_t}, States_{AR_t}, \models_{PR_t})$, where $(U_\#, \mathscr{R}_\#) \models_{PR_t} s$ iff $s \in AR_t(U_\#, \mathscr{R}_\#) \cap \mathscr{O}_t$.
7. On agent's side there is $EIS_t = (P(S^*), \mathscr{A}_t, \models_{IS_t})$, which is grounded on the information system $IS_t = (\mathscr{O}_t, \mathscr{A}_t, \{f_a : \mathscr{O}_t \mapsto V_a\}_{a \in \mathscr{A}_t})$. In EIS_t agent also considers states belonging to $S^* - \mathscr{O}_t$, which are potentially measurable with respect to some parameters, but not measured at time t. So, the satisfaction relation is defined with respect to the information available at IS_t, and given by $\{s_1, s_2, \ldots, s_l\} \models_{IS_t} a$ iff for all $i = 1, 2, \ldots, l, f_a(s_i) \in V_a$.
8. Now an *interaction of the agent with* PR_t, the physical reality at time t, can be presented as an infomorphism from $EIS_t = (P(S^*), \mathscr{A}_t, \models_{PR_t})$, an extension of IS_t to $PR_{AR_t} = (S_{PR_t}, States_{AR_t}(S_{PR_t}), \models_{PR_t})$ following the sense of Barwise and Seligman [1].
   The infomorphism from $(P(S^*), \mathscr{A}_t, \models_{PR_t})$ to $(S_{PR_t}, States_{AR_t}(S_{PR_t}), \models_{PR_t})$ is defined as follows. An infomorphism

$$I_t : (P(S^*), \mathscr{A}_t, \models_{IS_t}) \rightleftarrows (S_{PR_t}, States_{AR_t}(S_{PR_t}), \models_{PR_t})$$

   consists of a pair of functions $(\hat{I}_t, \check{I}_t)$ where for any $a \in \mathscr{A}_t, \hat{I}_t(a) \in States_{AR_t}(S_{PR_t})$ and for any $(U_\#, \mathscr{R}_\#) \in S_{PR_t}, \check{I}_t(U_\#, \mathscr{R}_\#) \in P(S^*)$, and $(U_\#, \mathscr{R}_\#) \models_{PR_t} \hat{I}_t(a)$ iff $\check{I}_t(U_\#, \mathscr{R}_\#) \models_{IS_t} a$.
9. An *ISAPR* at time point t, denoted as $ISAPR_t$, is represented by the tuple (PR_{AR_t}, I_t, IS_t). In the classical sense of information system, $ISAPR_t$ is also an information system consisting of a single object PR_{AR_t} representing the information at the physical world with respect to the accessibility function AR_t, a single parameter I_t representing a specific interaction, and an outcome of the interaction with the physical reality viz., $IS_t = (\mathscr{O}_t, \mathscr{A}_t, \{f_a : \mathscr{O}_t \mapsto V_a\}_{a \in \mathscr{A}_t})$ (cf. Table 8).
   Here we use the same symbol I_t considering an interaction as a parameter. Thus there is a function f_{I_t} corresponding to the parameter I_t such that $f_{I_t}(PR_{AR_t}) = IS_t$ if $I_t : (P(S^*), \mathscr{A}_t, \models_{IS_t}) \rightleftarrows (S_{PR_t}, States_{AR_t}(S_{PR_t}), \models_{PR_t})$.
10. An *interactive information system approximating the physical reality*, denoted as *IISAPR*, represents an information system of the following kind. (Fig. 6)

Table 8 An information system as an outcome of an interaction with the physical reality

	I_t
PR_{AR_t}	$(\mathscr{O}_t, \mathscr{A}_t, \{f_a : \mathscr{O}_t \mapsto V_a\}_{a \in \mathscr{A}_t})$

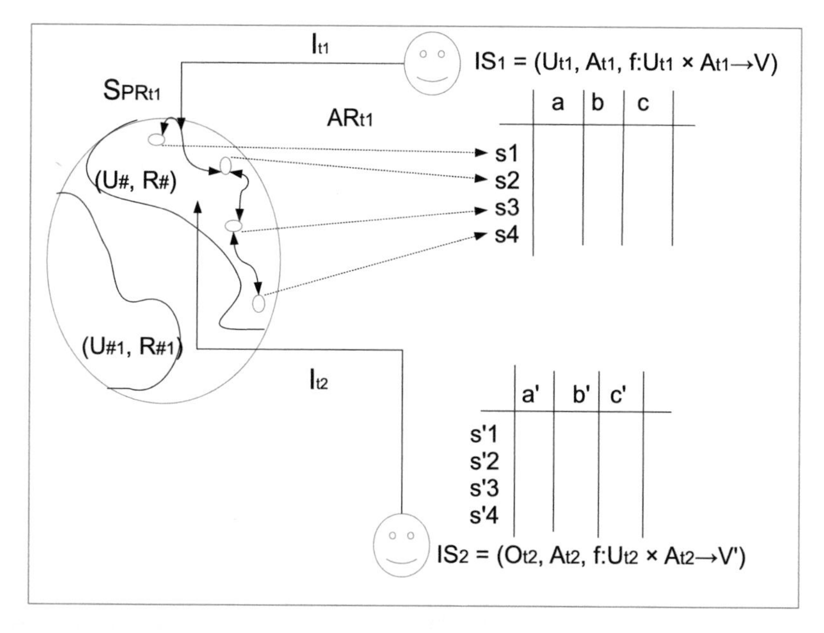

Fig. 6 Interaction between agent and physical reality through an infomorphism from the agent's information system to the physical reality

Table 9 An interactive information system approximating the physical reality

	I_{j_1}	I_{j_2}	...
$PR_{AR_{t_1}}$	$(\mathcal{O}_{t_1}, \mathcal{A}_{t_1}, \{f_a : \mathcal{O}_{t_1} \mapsto V_a\}_{a\in\mathcal{A}_{t_1}})$	$(\mathcal{O}^1_{t_1}, \mathcal{A}^1_{t_1}, \{f_a : U^1_{t_1} \mapsto V_a\}_{a\in\mathcal{A}^1_{t_1}})$	...
$PR_{AR^1_{t_2}}$	$(\mathcal{O}'_{t_2}, \mathcal{A}_{t_2}, \{f_a : \mathcal{O}'_{t_2} \mapsto V_a\}_{a\in\mathcal{A}_{t_2}})$	...	...
⋮	⋮	⋮	⋮

$$IISAPR = (\{PR_{AR_t}\}_{AR_t\in A_f, t\in T}, \{I_j\}_{j\in J}, \{f_{I_j} : \{PR_{AR_t}\}_{AR_t\in A_f, t\in T} \mapsto \{IS_l\}_{l\in L}\}_{j\in J}),$$

where $\{PR_{AR_t}\}_{AR_t\in A_f, t\in T}$ is a family of fragments of reality indexed by both time $t \in T$ and possible accessibility functions $AR_t \in A_f$, $\{I_j\}_{j\in J}$ is a family of possible interactions of agents with the physical world, and corresponding to each interaction I_j, f_{I_j} is a function assigning a unique information system from $\{IS_l\}_{l\in L}$ to each of $\{PR_{AR_t}\}_{AR_t\in A_f, t\in T}$. That is, we have Table 9 for *IISAPR*.

So, the whole process of arriving at relevant information systems with the passage of time and different interactions may be visualized through the picture presented in Fig. 7. There can be different cases when the time factor is considered. At some point

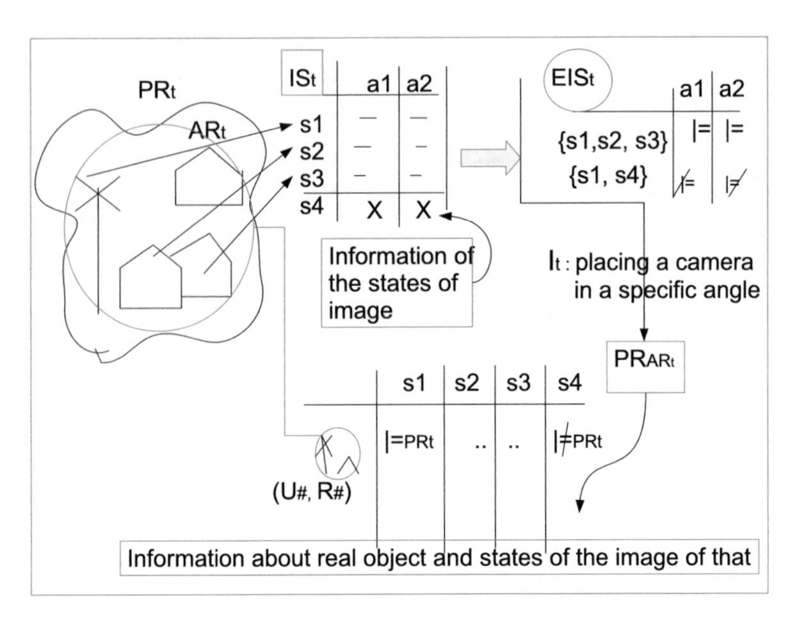

Fig. 7 Example: the process of interaction between agent and physical reality and obtaining an information system

of time t' ($> t$), $PR_{t'}$ may remain the same as PR_t, but with time the agent can manage to access more or something different than before. That is, $AR_{t'}$ may change; that is $PR_{AR_{t'}}$ may become different from PR_{AR_t}, and one can arrive at a different set of states $\{s'_1, s'_2, \ldots, s'_n\}$ from the same fragment of the physical world $(U_\#, \mathscr{R}_\#)$. On the other hand, with time the physical reality itself can change, and we may have $PR_{t'}$ different from PR_t (Fig. 6).

It is to be noted that apart from the time point t there are two more factors, namely the *function accessing the reality at time t* (AR_t) and the *interaction of agent with reality at time t* (I_t). Let us take an attempt to explain the role of the different components we have introduced in the model for an interactive information system.

Let there be a tree surrounded by a number of houses and a park, which may be considered as a fragment of reality $(U_\#, \mathscr{R}_\#)$. An agent can gather some information about $(U_\#, \mathscr{R}_\#)$ from an image of the fragment of the physical world taken by some particular camera. So, the camera works as a tool for accessing the information about the real physical world. That is, mathematically the functionality of such a tool or mode of accessing reality is availed by a function AR_t. Let through the camera the agent become able to get a good overview of the tree and two houses close to the tree, which are respectively captured by some states s_1, s_2, s_3 of the image in terms of brightness of colours (a_1), pixel-points (a_2) etc. Let us assume that some other state s_4, representing another object in the vicinity of the tree, appeared blurred in the image. So, though s_4 seems to be a possible measurable state, is not measured properly at time t. So, following the terminologies used above $s_4 \in S^* - \mathscr{O}_t$. The agent manages to get an image of $(U_{\#}, \mathscr{R}_{\#})$ as there was a specific interaction I_t between the agent and the reality at time t, which might be considered as *placing a camera in a specific angle*. At some further point of time t' the agent may initiate a different interaction $I_{t'}$ by changing the angle of the same camera, or replacing the earlier camera by a high-resolution camera. In the former case, interaction $I_{t'}$ changes but the tool for accessing the reality $AR_{t'}$ remains the same as AR_t. On the other hand, in the latter case, both $I_{t'}$ and $AR_{t'}$ change.

4 Concluding Remarks

As concluding remarks, let us present some important aspects, which can come up naturally from the proposed idea of interactive information systems, as future issues of investigation.

- One is, how can we assure that the set of states, about which we learn through AR_t, depicts the relational structure present in $(U_{\#}, \mathscr{R}_{\#})$ properly. In this context, we need to concentrate on the relations over the sets of values *i.e.*, the range sets of the functions AR_t as well as f_a's for each $a \in \mathscr{A}_t$. To illustrate the point we can think about the example of image of a real object captured by a camera. The expectation to AR_t is that when applied on a pair $(U_{\#}, \mathscr{R}_{\#})$ as an outcome it should produce a set of states $\{s_1, s_2, \ldots, s_m\}$ which together represents a prototypical image of the real object $(U_{\#}, \mathscr{R}_{\#})$ preserving the relations among the real parts of the object. So, we may consider that if two objects $o_i, o_j \in U_{\#}$ are such that for some $r_{\#} \in \mathscr{R}_{\#}$, $r_{\#}(o_i, o_j)$, then the relation should also be preserved somehow under the transformation of $(U_{\#}, \mathscr{R}_{\#})$ to $AR_t(U_{\#}, \mathscr{R}_{\#}) = \{s_1, s_2, \ldots, s_m\}$. So, the relational structures as well as the languages having interpretation over them would come into play in the context of interactive information systems.
- The above point leads us towards another important aspect. The question is, what happens if for two object $o_i, o_j \in U_{\#}$, $r_{\#}(o_i, o_j)$ holds for some $r_{\#} \in \mathscr{R}_{\#}$ but the relation is not preserved among the states which represent $(U_{\#}, \mathscr{R}_{\#})$ under the accessibility function AR_t. In the context of an human agent, it is quite natural that such a situation would generate some action to initiate new interaction with the physical reality. So, satisfaction or dissatisfaction of some interrelations between the relational structures lying in the physical object, and that of in its representation with respect to the the agent's information system may generate some typical actions. These actions in turn, with the progress of time, generates new interaction and keeps on modifying the agent's information system characterizing some fragment of the physical world. So, for each interactive information system approximating the physical reality at some time point t, viz., $IISAPR_t$, we also need to count a set of decision attributes consisting of actions, say A_c, so that depending on the level of accuracy of an agent's information system in characterizing a fragment of the physical world which action to be taken can be determined.

Acknowledgements This work was partially supported by the Polish National Science Centre (NCN) grants DEC-2011/01/D /ST6/06981, DEC-2013/09/B/ST6/01568 as well as by the Polish National Centre for Research and Development (NCBiR) under the grant DZP/RID-I-44 / 8 /NCBR/2016.

References

1. Barwise, J., Seligman, J.: Information Flow: The Logic of Distributed Systems. Cambridge University Press, Cambridge, UK (1997)
2. Getoor, L., Taskar, B. (eds.): Introduction to Statistical Relational Learning. The MIT Press (2007)
3. Greco, S., Matarazzo, B., Słowiński, R.: Granular computing and data mining for ordered data: The dominance-based rough set approach. In: Encyclopedia of Complexity and Systems Science, pp. 4283–4305. Springer, Heidelberg (2009)
4. Jankowski, A., Skowron, A., Dutta, S.: Toward problem solving support based on big data and domain knowledge: interactive granular computing and adaptive judgement. In: Japkowicz, N., Stefanowski, J. (eds.) Big Data Analysis: New Algorithms for a New Society, Series Big Data, vol. 16, pp. 44–90. Springer, Heidelberg (2015)
5. Jankowski, A., Skowron, A., Swiniarski, R.: Interactive complex granules. Fundamenta Informaticae **133**(2–3), 181–196 (2014)
6. Jankowski, A., Skowron, A., Swiniarski, R.W.: Interactive computations: toward risk management in interactive intelligent systems. In: Maji, P., Ghosh, A., Murty, M.N., Ghosh, K., Pal, S.K. (eds.) Proceedings of the 5th International Conference on Pattern Recognition and Machine Intelligence (PReMI 2013), Kolkata, India, December 10–14, 2013. Lecture Notes in Computer Science, vol. 8251, pp. 1–12. Springer, Heidleberg (2013)
7. Jankowski, A., Skowron, A., Swiniarski, R.W.: Perspectives on uncertainty and risk in rough sets and interactive rough-granular computing. Fundamenta Informaticae **129**(1–2), 69–84 (2014)
8. Marek, W., Pawłak, Z.: Mathematical foundations of information storage and retrieval. part 1. In: CC PAS Reports 135/73, pp. 1–10. Computation Center Polish Academy of Sciences (CC PAS), Warsaw, Poland (1973)
9. Marek, W., Pawłak, Z.: Mathematical foundations of information storage and retrieval. part 2. In: CC PAS Reports 136/73, pp. 1–11. Computation Center Polish Academy of Sciences (CC PAS), Warsaw, Poland (1973)
10. Marek, W., Pawłak, Z.: Mathematical foundations of information storage and retrieval. part 3. In: CC PAS Reports 137/73, pp. 1–8. Computation Center Polish Academy of Sciences (CC PAS), Warsaw, Poland (1973)
11. Marek, W., Pawłak, Z.: Information storage and retrieval system—mathematical foundations. In: CC PAS Reports 149/74, pp. 1–54. Computation Center Polish Academy of Sciences (CC PAS), Warsaw, Poland (1974)
12. Marek, W., Pawłak, Z.: On the foundation of information retrieval. Bull. Pol. Acad. Sci. Math. Sci. **22**(4), 447–452 (1974)
13. Marek, W., Pawłak, Z.: Information storage and retrieval, mathematical foundations. Theor. Comput. Sci. **1**, 331–354 (1976)
14. Marek, W., Pawłak, Z.: Rough sets and information systems. Fundamenta Informaticae **7**(1), 105–115 (1984)
15. Nguyen, H.S.: Approximate boolean reasoning: foundations and applications in data mining. Trans. Rough Sets: J. Subline **5**, 344–523. LNCS 4100. Springer, Heidelberg (2006)
16. Orłowska, E., Pawłak, Z.: Representation of nondeterministic information. Theor. Comput. Sci. **29**, 27–39 (1984)

17. Ortiz Jr., C.L.: Why we need a physically embodied Turing test and what it might look like. AI Mag. **37**, 55–62 (2016)
18. Pawłak, Z.: Mathematical foundation of information retrieval. In: Proceedings of the International Symposium and Summer School on Mathematical Foundations of Computer Science, pp. 135–136. Mathematical Institute of the Slovak Academy of Sciences, Strbske Pleso, High Tatras, Czechoslovakia (1973)
19. Pawłak, Z.: Mathematical foundations of information retrieval. In: CC PAS Reports 101/73, pp. 1–8. Computation Center Polish Academy of Sciences (CC PAS), Warsaw, Poland (1973)
20. Pawłak, Z.: Information systems. In: ICS PAS Reports 338/78, pp. 1–15. Institute of Computer Science Polish Academy of Sciences (ICS PAS), Warsaw, Poland (1978)
21. Pawłak, Z.: Distribued information systems. In: ICS PAS Reports 370/79, pp. 1–28. Institute of Computer Science Polish Academy of Sciences (ICS PAS), Warsaw, Poland (1979)
22. Pawłak, Z.: Toward the theory of information systems. the notion of information system. In: ICS PAS Reports 419/80, pp. 1–35. Institute of Computer Science Polish Academy of Sciences (ICS PAS), Warsaw, Poland (1980)
23. Pawłak, Z.: Classification of objects by means of attributes. In: ICS PAS Reports 429/81, pp. 1–20. Institute of Computer Science Polish Academy of Sciences (ICS PAS), Warsaw, Poland (1981)
24. Pawłak, Z.: Information systems theoretical foundations. Inf. Syst. **6**(3), 205–218 (1981)
25. Pawłak, Z.: Systemy Informacyjne. Podstawy Teoretyczne (Information Systems. Theoretical Foundations). Wydawnictwa Naukowo-Techniczne, Warsaw, Poland (1983). (book in Polish)
26. Pawłak, Z.: Rough sets and information systems. Podstawy Sterowania **18**, 175–200 (1988)
27. Pawłak, Z.: Information systems and decision tables a rough set perspective. Archiwum Informatyki Teoretycznej i Stosowanej **2**(3–4), 139–166 (1990)
28. Skowron, A., Jankowski, A.: Interactive computations: toward risk management in interactive intelligent systems. Nat. Comput. **15**(3) (2016)
29. Skowron, A., Jankowski, A.: Toward W2T foundations: interactive granular computing and adaptive judgement. In: Zhong, N., Ma, J., Liu, J., Huang, R., Tao, X. (eds.) Wisdom Web of Things (W2T), pp. 47–71. Springer, Heidelberg (2016)
30. Skowron, A., Jankowski, A., Wasilewski, P.: Interactive computational systems. In: Popowa-Zeugmann, L. (ed.) Proceedings of the Workshop on Concurrency, Specification, and Programming (CS&P 2012), Humboldt University, Berlin, September 26–28, 2012, Informatik-Bericht, vol. 225, pp. 358–369. Humboldt-Universität, Berlin (2012)
31. Skowron, A., Jankowski, A., Wasilewski, P.: Risk management and interactive computational systems. J. Adv. Math. Appl. **1**, 61–73 (2012)
32. Skowron, A., Stepaniuk, J.: Hierarchical modelling in searching for complex patterns: constrained sums of information systems. J Exp. Theor. Artif. Intell. **17**, 83–102 (2005)
33. Skowron, A., Stepaniuk, J., Peters, J.F.: Rough sets and infomorphisms: towards approximation of relations in distributed environments. Fundamenta Informaticae **54**(1–2), 263–277 (2003)
34. Skowron, A., Szczuka, M.: Toward interactive computations: a rough-granular approach. In: Koronacki, J., Raś, Z., Wierzchoń, S., Kacprzyk, J. (eds.) Advances in Machine Learning II: Dedicated to the Memory of Professor Ryszard S. Michalski, Studies in Computational Intelligence, vol. 263, pp. 23–42. Springer, Heidelberg (2009)
35. Słowiński, R., Greco, S., Matarazzo, B.: Rough set methodology for decision aiding. In: Springer Handbook of Computational Intelligence, pp. 349–370. Springer, Heidelberg (2015)
36. Zadeh, L.A.: A new direction in AI: toward a computational theory of perceptions. AI Mag. **22**(1), 73–84 (2001)

Back to the Beginnings: Pawlak's Definitions of the Terms *Information System* and *Rough Set*

Davide Ciucci

Abstract There are several notions and terms in Rough Set Theory not having a crystal clear definition. I discuss here two basic ones: Rough Set and Information System. The discussion will be lead by the two founding papers by Z. Pawlak. We will see that the term Information System has a narrow sense (the most used one in the rough set community) and a wide one (the world wide common use of information system and also Pawlak's one). In case of the term Rough Set, several definitions are possible, none of them without problems. Some other minor issues related to Pawlak's papers will be highlighted.

1 Introduction

After 35 years of rough set research, we tend to give for granted some basic terms such as "information system" and "rough set". However, information system has a different and much wider meaning in the computing science community with respect to the intended meaning of the Rough Set Theory (RST) community. On the other hand, the term rough set is overloaded inside the RST community itself, as widely discussed by M. Chakraborty [3, 4]. Moreover, once faced with the simple question "what is a rough set?", whatever definition we choose, we are forced to a (to some extent) long explanation in order to give a precise answer.

Surprisingly, while reading the two founding papers by Z. Pawlak, we see that the term Information System was used in a more conscious way and strictly related to the standard way of using it. On the other hand, the difficulty of defining a Rough Set is perhaps intrinsic to its notion. In [18], a formal definition arrives only in section five, using several notions defined in the previous sections. We also notice that the definition given in [18] is one of the list given in [3]. More precisely, a rough set is defined as a family of subsets of the universe which are roughly equal, that is whose lower and upper approximations coincide.

D. Ciucci (✉)
University of Milano–Bicocca, Milano, Italy
e-mail: ciucci@disco.unimib.it

G. Wang et al. (eds.), *Thriving Rough Sets*, Studies in Computational Intelligence 708, DOI 10.1007/978-3-319-54966-8_11

In this work, I will discuss the meaning of the term Information System and the difficulty to give a problem-free definition of a Rough Set. Besides these two terminological problems, other issues arising from a reading of the two founding papers will be pointed out and in particular the relationship of RST with other disciplines such as Database Theory and Artificial Intelligence. The structure of the chapter will reflect this objective: in Sect. 2, two definitions of information system will be given, whereas in Sect. 3, I will recall the possible definitions of a rough set and show some problems in these definitions. Finally, some conclusions will close the chapter.

2 Information System

An Information System is defined according to Encyclopædia Britannica[1] as

> an integrated set of components for collecting, storing, and processing data and for providing information, knowledge, and digital products. [...] The main components of information systems are computer hardware and software, telecommunications, databases and data warehouses, human resources, and procedures.

On the other hand, in Rough Set Theory the meaning of Information System is confined to data representation and in particular to a simple mathematical modelization of data. It relies on the formal definition given by Pawlak [17]:

Definition 1 An *Information System* is a tuple $S = (X, A, V, \rho)$ where X is a finite set of objects, A is a finite set of attributes, $V = \cup_{a \in A} V_a$ where v_a is the set of values of attribute a and $|V_a| > 1$, ρ is a function mapping objects and attributes to values, $\rho : X \times A \mapsto V$.

Remark 1 As a side remark, we notice that attribute values are kept distinguished in Definition 1. Sometimes, in definitions of Information Systems, we can find simply V instead of V_a. Clearly, the original formulation Definition 1 is more informative.

If we compare Definition 1 with the worldwide point of view, the former only gives a partial view of what an Information System is. However, with respect to Pawlak's work, this is only a part of the story. First of all, let me notice that this is just one of the names used by Pawlak himself to denote this concept. Indeed, in his publications we can find at least the following names:

- Descriptive System [16]
- Information storage and retrieval system [14]
- *Information System* [17]
- Knowledge representation system [19]

However, the one that is most used in rough set literature is exactly Information System. For instance in RSDS[2] we can find 98 papers with "Information system"

---

[1] https://global.britannica.com/topic/information-system

[2] Rough Set Database System, http://rsds.univ.rzeszow.pl/

in title or keyword, 12 with "Knowledge representation system" and none using the other two terms.

2.1 A Complex Notion

Besides the fact that different alternative names exist, it is more important to notice that if we consider the whole paper [17], we see that the notion of Information System has a more wide meaning and it is perfectly in line with the development of the area at that time. Pawlak explicitly refers to the standard notion of Information System and his proposal tries to unify the two research lines of relational database and data language, by considering, at the same time, data and language: "Syntax and semantics of a *query language* is introduced". Indeed, the above Definition 1 comes together with a formally defined query language. Moreover, implementation issues and distributed Information Systems are also discussed. In the sequel of the section we review these points, in order to highlight how complex was the notion of Information System described in [17] with respect to the simple Definition 1.

Query language

The query language introduced by Pawlak is a purely logical one whose syntax and semantics *depend on a given Information System*. Indeed, the language is built up from the constants 0, 1 and so called *descriptors*, that is attribute-value pairs, i.e., elements of $A \times V$. Terms are combination of descriptors and formulas combination of terms. Of course, semantics assigns to each formula a set of objects of a given Information System. There are two kinds of query that a user can perform, namely *Terms*, that return a set of objects and *Formulas*, that return a logical value yes/no.

Example 1 Let us consider the Information System in Table 1.

A "term query" is for instance `(SEX=Female) + (AGE = 20-30)`, where + means OR. Clearly, the answer to this query is the set of objects $\{x_2, x_3, x_4, x_6\}$ being x_2, x_6 females and x_3, x_4 aged 20–30. On the other hand the "formula query" `(SEX = female) = (EMPLOYED = yes)` has answer *no* since the set of females does not coincide with set of employed people.

Table 1 Example of information system

X	Sex	Employed	Age
x_1	Male	No	30–40
x_2	Female	No	30–40
x_3	Male	Yes	20–30
x_4	Male	Yes	20–30
x_5	Male	Yes	30–40
x_6	Female	Yes	30–40

Other important notions (which will be useful later) in the query language are *elementary terms*: a conjunction of basic descriptors, and *normal form*: a disjunction of elementary terms.

So, (Pawlak) Information System is an approach explicitly antagonistic to the standard relational model and links and differences with other database models are stressed in his paper. In particular, a fundamental difference regards the way functional dependencies work. In the relational model are defined independently from data and they represent a constraint on data. On the other hand, in (what will be) Rough Set Theory they are extracted from data, thus they are not universally valid, but can change according with data changes.

Remarks on Implementation

Besides mathematical foundations, Pawlak also discusses implementation details for an "efficient method of information retrieval". The basic idea is to keep in memory the *elementary terms* and to transform a query in *normal form*. Of course some problems must be solved, Pawlak lists them and also proposes a solution. Namely:

- There are too many elementary terms in order to be stored. To overcome this problem some heuristics are proposed, for instance considering only some of the attributes, in order to reduce the number of the elementary terms.
- How to (efficiently) find the elementary terms. Pawlak refers to the solution proposed in [14].
- How to efficiently transform a query in normal form. With respect to this issue, standard methods already existed.

Distributed System

Further, the problem of separate (called distributed) systems and how to retrieve information in this case is discussed.

The problem is that we have n systems $S_1, \dots, S_n$ each with its own logical language L_i, we have to combine them in a unique system S with a unique language L and understand if it is possible to give a global answer to a query by combining local answers. The issues to be addressed are:

- combine different systems. To this purpose three *connection* operations are proposed: a general one (a sort of full outer join in SQL); an attribute connection, when attributes are the same (so it corresponds to a SQL union); an object connection, when objects are the same (a sort of SQL inner join).
- give access to a local user to other sub-system (this issue is not addressed in his paper).

Moreover, the concept of local *lower and upper approximations* to a query are formally defined. Thus, the idea of *approximations*, which will then be developed in Rough Set Theory, already appears here.

We can conclude that for Pawlak, Information Systems seem to be something more complex than the only Definition 1. So, considering the terminology, my advice is to avoid the term Information System in the narrow sense of Definition 1 and use an alternative name such as *information table* [26], *attribute-value system* [28], *knowledge representation system, descriptive system*, etc.

Further, whereas Pawlak's work was settled in the context of databases and Information Systems (in the wider sense), now there is a gap between rough sets and Databases/Information Systems with, of course, some few exceptions, notably the approximate query approach by Infobright Inc.[3] The rough set community should investigate if this gap can be reduced.

3 Rough Sets

If we are faced with the simple question "what is a rough set?", the answer is not so simple and linear. First of all, it is a complex notion, secondly, even if we consider the classical case, several possible definitions exist.

At least four definitions of (equivalence-based) rough set can be given (see [3]), divided in two categories: *rough* used as an adjective (Definition 2 below), identifying a particular classical set or *rough set* as a noun defining a new category of sets (Definitions 3–5 below). They are (L is the lower and U the upper approximation):

Definition 2 A set A is rough iff its boundary $U(A)\setminus L(A)$ is empty;

Definition 3 A rough-set is a pair $(L(A), U(A))$ (or equivalently $(L(A), U(A)^c)$);

Definition 4 A rough-set is a pair (D_1, D_2) where D_1, D_2 are definable and $D_1 \subseteq D_2$;

Definition 5 A rough-set is an equivalence class $[A]_{\equiv}$ with respect to rough equality.

The last definition is the one given by Pawlak in [18] and it belongs to the second category: a rough set is a family of subsets that are equivalent with respect to the *rough equality* relation, that is X and Y are equivalent if their lower and upper approximations coincide.

Example 2 Let us consider the data Table 1 and the partition $\pi = \{ \{x_1, x_2\}, \{x_3, x_4\}, \{x_5, x_6\} \}$ given by the two attributes Employees and Age. Then, rough sets are for instance the set containing the emptyset: $\{\emptyset\}$ and the families $\{\{x_1\}, \{x_2\}\}$; $\{\{x_1, x_2\}\}$; $\{\{x_1, x_3\}, \{x_2, x_4\}\}$; ...

In order to arrive at this definition, we thus need an approximation space, a definition of the approximations and of the rough equality: a rich and complex notion. In the light of the actual researches on rough sets, it seems that this complexity is intrinsic to the model. Rough sets share this feature with other models, such as Formal Concept Analysis. If we are asked what a formal concept is, we need to define at first a formal context, then the derivation operations which finally permits to introduce formal concepts [10]. On the other hand, we notice that this is different from Fuzzy Set Theory. Indeed, usually a fuzzy subset f of X is defined through its membership function[4]: $f : X \mapsto [0, 1]$.

---

[3] https://infobright.com/.

[4] A slightly more complex but mathematically equivalent definition can be given inside classical set theory [24].

We also notice that Definition 5 is not the most commonly used definition. Indeed, as a rough set the lower-upper approximation pair (Definition 3) is often meant.

3.1 Is There an Optimal/Correct Definition?

Given the fact that there are several definitions, we can wonder if there is some which is best according to some criteria. I argue that all of them suffer from some problem. First of all, I do not tend to consider Definition 4 as a good definition of a rough set. Indeed, it eliminates the idea of approximations. The pair (D_1, D_2) with $D_1 \subseteq D_2$, is more similar to what I call an *orthopair* [7], the only difference being the requirement that D_1, D_2 are definable sets. The notion of orthopair is more generic than the rough set one, with several other applications. On the other hand, this definition is the only one that avoids the language problem we are going to discuss.

Let us consider all the other three definitions. They all suffer from what we can call the *two level* or *language* problem. The two levels in question are objects and granules. Each level has its own language and we can reason inside each level with no problems. However, with the above rough set definitions we mix the two levels. This has some not desirable consequences, for instance:

- it is not possible to capture "the variability in the interpretation of a vague concept" (Chakraborty, [3]);
- there are difficulties in defining (and interpreting) the operations on approximations.

The first point has been discussed by M. Chakraborty in [3], basically he points out that there are different interpretations of a vague concept and the above definitions are not able to capture them all.

The second problem is detailed in [3, 4, 8] and I am going to recall it in the next paragraph.

Language problem

Let us consider the rough set Definition 4. We want to show the two-level problem by defining the standard intersection operation on rough sets:

$$(L(A), U(A)) \sqcap (L(B), U(B)) = (L(A) \cap L(B), U(A) \cap L(B)). \tag{1}$$

In order to show that $\sqcap$ is a well defined operation on rough sets, we have to find a set C such that $r(C) = r(A) \sqcap r(B)$ (where $\forall H \subseteq X$, $r(H) = (L(H), U(H))$, since in general, $C \neq A \cap B$. The problem has been solved independentely by different authors [1, 2, 11]. In particular, let us consider the definition given in [1]:

$$C = (A \cap B) \cup ((A \cap U(B)) \cap (U(A \cap B)^c)). \tag{2}$$

and show how it works on an example.

Example 3 Let us consider the data Table 1 and the two subsets $X_1 = \{x_1, x_2, x_6\}$ and $X_2 = \{x_3, x_5\}$. If we consider the two attributes Employees and Age we obtain the partition $\pi_1 = \{\ \{x_1, x_2\}, \{x_3, x_4\}, \{x_5, x_6\}\ \}$. Hence, the approximation of X_1 and X_2 are respectively:

$$r(X_1) = (\{x_1, x_2\}, \{x_1, x_2, x_5, x_6\}) \quad r(X_2) = (\emptyset, \{x_3, x_4, x_5, x_6\})$$

According to Eq. 2, the set giving the intersection is $C = \{x_6\}$, in effect, $r(C) = (\emptyset, \{x_5, x_6\}) = r(\{x_1, x_2, x_6\}) \sqcap r(\{x_3, x_5\})$. However, this is not the unique solution, indeed also $r(\{x_5\}) = (\emptyset, \{x_5, x_6\})$. Clearly this is due to the granulation of the universe, where different objects, x_5, x_6 in the example, (the first level) are glued into a unique granule (the second level). Moreover, if we change partition, by using the attributes Sex and Age, we obtain $\pi_2 = \{\ \{x_1, x_5\}, \{x_3, x_4\}, \{x_2, x_6\}\ \}$. The approximations are thus,

$$r(X_1) = (\{x_2, x_6\}, \{x_1, x_2, x_5, x_6\}\}) \quad r(X_2) = (\emptyset, \{x_1, x_3, x_4, x_5\})$$

and the intersection $C' = \{x_1\}$ with $r(C') = r(\{x_1, x_2, x_6\}) \sqcap r(\{x_3, x_5\}) = (\emptyset, \{x_1, x_5\})$. So summarizing, changing or not partition, the intersection of two sets A and B can be very different. We got in the example, $\{x_3\}$, $\{x_5\}$ or $\{x_1\}$.

We have seen in the previous example that the result of an operation strongly depends on the partition (which of course represent our available knowledge), through the two approximations L, U and finally, in case of a data table, on the attributes. Moreover, the solution is not unique even inside the same partition. So, the two languages we are implicitely using are

- The language of sets, representing the extension of the set;
- The language of attributes (or approximations, granules), representing the intension of the set.

In order to compute an aggregation operator, we can operate on the language of attributes but then we are not able to interpret the result on sets.

As a further issue let us consider that if we want to compute $A \cap A^c$, that is we set $B = A^c$ in Eq. 2, then we generally get that $C \neq \emptyset$, which may sound odd since the intersection of a set with its complement should be empty. Finally, we notice that the problem does not concern only the intersection as defined above. In [8] we considered 14 intersections and 14 implications and showed that they all depend on the approximations L and U, hence the problem is present in all these operations.

An open issue

In conclusion, the question if there does exist a mathematical problem-free definition of a rough set is still open. In order to find a new definition, Chakraborty proposes what we can summarize with a slogan as *try to move vagueness inside the language*. Indeed, given a propositional language representing rough sets, the interpretation should not be a function but a relation[5] such that a rough set would be a collection of sets $(A_1, A_2, \ldots)$ with the same lower and upper approximations, but not necessarily all such subsets. However, Chakraborty also suggests to use as intersection between two collections $(A_1, A_2, \ldots)$ and $(B_1, B_2, \ldots)$ the operation given in equation (2), so the language problem could rise once more.

Whatever the definition, I think it has to have some characteristics in order to be considered a good one. First of all, whether intended as a classical or a new concept of set, it should be defined in a way such that *the extension is known, the intension is uncertain*. That is, given a set $S = \{o_1, o_2, \ldots\}$ it is not possible to exactly describe it with the available knowledge (relation, attributes,...). Moreover, the set S *can be approximated by a lower and an upper approximation*.

Clearly, these characterizations have to be casted in a specific domain, by giving formal definitions. So, several ways to proceed and several problems open here. For instance: what to define at first, approximations or definable sets? Which properties (duality, monotonicity, etc.) should the approximations satisfy? These topics have already been discussed (see for instance [6, 9, 13, 27]) but not in the light of finding a new problem-free definition, so more reflections are needed.

3.2 Other Issues

There are other points worth considering in the paper [18], here I list two of them.

Non standard examples

We are used to consider as a paradigmatic example of rough sets a data table, introduce an equivalence relation and then compute the approximations. A peculiarity of [18] is that the first two given examples on rough approximations are not given in an Information System context. Indeed, the first one is on real number approximations, with the idea of a use in measurement theory. Roughly speaking a real number r is approximated by the lower and upper integer. The second one is on approximation of formal languages, with a possible use in "speech recognition, pattern matching, fault tolerant computer, etc.". Let $\mathcal{L} \subseteq V^*$ be a language on the vocabulary V and (V^*, R) an approximation space (R to be defined), then a language $\mathcal{L}$ is recognizable if $L(\mathcal{L}) = U(\mathcal{L})$ otherwise it can be approximated.

These aspects have been partly investigated also later [15, 23], but perhaps to a few extent. Hence, the rough-set community could take care of these preliminary ideas and develop them. More generally, rough sets have been mostly successfully applied to data analysis, thus using approximation spaces originated from data tables. Some applications using other kinds of approximation spaces, as the two mentioned above, can be also found in the literature, for instance approximation spaces based

[5] Let me notice that this is done for instance in many valued paraconsistent logics where a proposition can map to 0, 1, *both* or none of the two values [22].

on graphs (see [5] and its references), matroids [25], relations [12], ... The challenge is to make this research line grow.

Artificial Intelligence

Finally, we notice that in the paper [18] there are several references to Artificial Intelligence. Quoting from the introduction : "The rough set concept can be of some importance, primarily in some branches of artificial intelligence" and "we are primarily aiming at laying mathematical foundations for artificial intelligence". His interest for artificial intelligence was also evident in the following years, at least in two papers [20, 21] he explicitly connected the two disciplines pointing out applications of rough sets to AI related problems. Nowadays, however, the two communities seem to be at a great extent not connected.[6] In my opinion, it would be worth to increase these connections and attract the attention of AI researches to Rough Set Theory and vice versa. Indeed, rough set theory is still applied in several branches of artificial intelligence, such as knowledge representation and reasoning, machine learning, cognitive science, decision support systems, conflict analysis, etc.

4 Conclusion

In this note, I put forward a terminological discussion on *Information Systems* and *Rough Sets* in the light of the two founding papers by Z. Pawlak.

We have seen that the notion of Information System, as presently used in many rough set papers, is much more restricted than the one used by other communities and by Pawlak himself in his founding paper. This of course represents a communication problem which could isolate rough set research(ers). My suggestion is to avoid to use the term Information System in the narrow sense and substitute it with one of the existing alternatives.

In case of rough sets, several possible definitions are available and I highlighted the fact that there are problems with any definition, in particular a language or two-level problem. It is a task for future research to find a problem-free definition.

Finally, other ideas for the future are to increase the connections with Information System, Database and Artificial Intelligence research areas and people and to apply rough set ideas in a general setting of approximation spaces, not necessarily based on an information table.

[6] For instance, recently, we had to cancel the Rough Set Theory workshop at ECAI 2016 (European Conference on Artificial Intelligence) due to insufficient participants.

References

1. Banerjee, M., Chakraborty, M.: Rough sets through algebraic logic. Fundamenta Informaticae **28**, 211–221 (1996)
2. Bonikowski, Z.: A certain conception of the calculus of rough sets. Notre Dame J. Formal Logic **33**(3), 412–421 (1992)
3. Chakraborty, M.K.: On some issues in the foundation of rough sets: the problem of definition. Fundamenta Informaticae (2016) (To appear)
4. Chakraborty, M.K., Banerjee, M.: Rough sets: some foundational issues. Fundamenta Informaticae **127**(1–4), 1–15 (2013)
5. Chiaselotti, G., Ciucci, D., Gentile, T.: Simple graphs in granular computing. Inf. Sci. **340–341**, 279–304 (2016)
6. Ciucci, D.: Approximation algebra and framework. Fundam. Info. **94**(2), 147–161 (2009)
7. Ciucci, D.: Orthopairs and granular computing. Granular Comput. **1**, 159–170 (2016)
8. Ciucci, D., Dubois, D.: Three-valued logics, uncertainty management and rough sets. Trans. Rough Sets **XVII**, 1–32 (2014)
9. Ciucci, D., Mihálydeák, T., Csajbók, Z.E.: On definability and approximations in partial approximation spaces. In: Miao, D., Pedrycz, W., Slezak, D., Peters, G., Hu, Q., Wang, R. (eds.) Rough Sets and Knowledge Technology—9th International Conference, RSKT 2014. Proceedings of Lecture Notes in Computer Science, vol. 8818, pp. 15–26. Shanghai, China, 24–26 Oct 2014
10. Ganter, B., Wille, R.: Formal Concept Analysis: Mathematical Foundations. Springer-Verlag, Berlin Heidelberg (1999)
11. Gehrke, M., Walker, E.: On the structure of rough sets. Bull. Pol. Acad. Sci. (Math.) **40**, 235–245 (1992)
12. Janicki, R.: Approximations of arbitrary binary relations by partial orders: classical and rough set models. Trans. Rough Sets **13**, 17–38 (2011)
13. Mani, A.: Dialectics of counting and the mathematics of vagueness. Trans. Rough Sets **15**, 122–180 (2012)
14. Marek, V.W., Pawlak, Z.: Information storage and retrieval systems: mathematical foundations. Theor. Comput. Sci. **1**(4), 331–354 (1976)
15. Orlowska, E., Pawlak, Z.: Measurement and indiscernibility. Bull. Pol. Acad. Sci. Math. **32**, 617–624 (1984)
16. Pawlak, Z.: Mathematical foundation of information retrieval. In: Mathematical Foundations of Computer Science: Proceedings of Symposium and Summer School, Strbské Pleso, High Tatras, pp. 135–136. Mathematical Institute of the Slovak Academy of Sciences, Czechoslovakia, 3–8 Sept 1973
17. Pawlak, Z.: Information systems theoretical foundation. Info. Syst. **6**, 205–218 (1981)
18. Pawlak, Z.: Rough sets. Int. J. Comput. Info. Sci. **11**, 341–356 (1982)
19. Pawlak, Z.: On discernibility of objects in knowledge representation systems. Bull. Pol. Acad. Sci., Math. Sci. **32**, 613–615 (1984)
20. Pawlak, Z.: Rough sets and some problems of artificial intelligence. Tech. Rep. ICS PAS Reports 565/85, Institute of Computer Science Polish Academy of Sciences (ICS PAS), Warsaw, Poland (1985)
21. Pawlak, Z.: AI and intelligent industrial applications: the rough set perspective. Cybern. Syst. **31**(3), 227–252 (2000)
22. Priest, G., Tanaka, K., Weber, Z.: Paraconsistent logic. In: Zalta, E.N. (ed.) The Stanford Encyclopedia of Philosophy. Spring 2015 edn. (2015)
23. Păun, G., Polkowski, L., Skowron, A.: Rough set approximations of languages. Fundamenta Informaticae **32**(2), 149–162 (1997)
24. Ramík, J., Vlach, M.: A non-controversial definition of fuzzy sets. Trans. Rough Sets **II**(3135), 201–207 (2004)
25. Wang, S., Zhu, Q., Zhu, W., Min, F.: Rough set characterization for 2-circuit matroid. Fundam. Inform. **129**(4), 377–393 (2014)

26. Yao, J.T., Yao, Y.Y.: Induction of classification rules by granular computing. In: Alpigini, J.J., Peters, J.F., Skowron, A., Zhong, N. (eds.) Rough Sets and Current Trends in Computing, Third International Conference, RSCTC 2002. Proceedings of Lecture Notes in Computer Science vol. 2475, pp. 331–338. Malvern, PA, USA, 14–16 October 2002
27. Yao, Y.: A note on definability and approximations. Trans. Rough Sets **7**, 274–282 (2007)
28. Ziarko, W., Shan, N.: A method for computing all maximally general rules in attribute-value systems. Comput. Intell. **12**, 223–234 (1996)

Knowledge and Consequence in AC Semantics for General Rough Sets

A. Mani

Abstract Antichain based semantics for general rough sets was introduced recently by the present author. In her paper two different semantics, one for general rough sets and another for general approximation spaces over quasi-equivalence relations, were developed. These semantics are improved and studied further from a lateral algebraic logic and an algebraic logic perspective in this research. The framework of granular operator spaces is also generalized. The main results concern the structure of the algebras, deductive systems and the algebraic logic approach. The epistemological aspects of the semantics is also studied in this chapter in some depth and revolve around nature of knowledge representation, Peircean triadic semiotics and temporal aspects of parthood. Examples have been constructed to illustrate various aspects of the theory and applications to human reasoning contexts that fall beyond information systems.

1 Introduction

It is well known that sets of rough objects (in various senses) are quasi or partially orderable. Specifically in classical or Pawlak rough sets [31], the set of roughly equivalent sets has a Quasi Boolean order on it while the set of rough and crisp objects is Boolean ordered. In the classical semantic domain or classical meta level, associated with general rough sets, the set of crisp and rough objects is quasi or partially orderable. Under minimal assumptions on the nature of these objects, many orders with rough ontology can be associated—these necessarily have to do with concepts of discernibility. Concepts of rough objects, in these contexts, depend additionally on approximation operators and granulations used. These were part of the motivations of the development of the concept of granular operator spaces by the present author in [26].

A. Mani (✉)
Department of Pure Mathematics, University of Calcutta, 9/1B, Jatin Bagchi Road, Kolkata (calcutta) 700029, India
e-mail: a.mani.cms@gmail.com
URL: http://www.logicamani.in

G. Wang et al. (eds.), *Thriving Rough Sets*, Studies in Computational Intelligence 708, DOI 10.1007/978-3-319-54966-8_12

In quasi or partially ordered sets, sets of mutually incomparable elements are called *antichains* (for basics see [9, 11, 15]). The possibility of using antichains of rough objects for a possible semantics was mentioned in [23, 24, 27] by the present author and developed in [26]. The semantics is applicable for a large class of operator based rough sets including specific cases of RYS [21], other general approaches like [7, 12, 13] and all specific cases of relation based and cover based rough set approaches. In [7], negation like operators are assumed in general and these are not definable operations relative the order related operations/relation. A key problem in many of the latter types of approaches is of closure of possible aggregation and commonality operations [14, 27, 44, 45].

In the present paper, the semantics of [26] is improved and developed further from an algebraic logic point of view (based on defining ternary deduction terms), the concept of knowledge in the settings is also explored in some depth and related interpretations are offered. The basic framework of granular operator spaces used in [26] is generalized in this paper as most of the mathematical parts carry over. The semantics of [26], as improved in the present paper by way of ternary terms, is very general, open ended, extendable and optimal for lateral studies. Most of it applies to *general granular operator spaces*, introduced in a separate paper by the present author. In the same framework, the machinery for isolation of deductive systems is developed and studied from a purely algebraic logic point of view. New results on representation of roughness related objects are also developed. Last but not least, the concept of knowledge as considered in [21, 22, 27, 32] is recast in very different terms for describing the knowledge associated with representation of data by maximal antichains. These representations are also examined for compatibility with triadic semiotics (that is not necessarily faithful to Peirce's ideas) for integration with ontology. Philosophical questions relating to perdurantism and endurantism are also solved in some directions. Illustrative examples that demonstrate applicability to *human reasoning contexts involving approximations but no reasonable data tables* have also been constructed in this chapter.

In the next subsection, relevant background is presented. Granular operator spaces are generalized and nature of parthood is explained in the next section. In the third section, the essential algebraic logic approach used is outlined. In the next section, possible operations on sets of maximal antichains derived from granular operator spaces are considered, AC-algebras are defined and their generation is studied. Representation of antichains derived from the context are also improved and earlier examples are refined. The algebras of quasi-equivalential rough sets formed by related procedures is presented to illustrate key aspects of the semantics in the fifth and sixth sections. Ternary deduction terms in the context of the AC-algebra are explored next and various results are proved. The connections with epistemology and knowledge forms the following section. Further directions are provided in Sect. 9.

Background

Let K be any set and l, u be lower and upper approximation operators on $\mathcal{K} \subseteq \wp(K)$ that satisfy monotonicity and $(\forall a \subseteq K)\, a \subseteq a^u$. An element $a \in \mathcal{K}$ will be said to be *lower definite* (resp. *upper definite*) if and only if $a^l = a$ (resp. $a^u = a$) and *definite*, when it is both lower and upper definite. In this chapter, the operators will be on $\wp(K)$ and not on a proper subset thereof. For possible concepts of rough objects [21, 26] may be consulted. Finiteness of K and granular operator spaces, defined below, will be assumed (though not always essential) unless indicated otherwise.

Let K be any set and l, u be lower and upper approximation operators on $\mathcal{K} \subseteq \wp(K)$ that satisfy monotonicity and $(\forall a \subseteq K)\, a \subseteq a^u$. An element $a \in \mathcal{K}$ will be said to be *lower definite* (resp. *upper definite*) if and only if $a^l = a$ (resp. $a^u = a$) and *definite*, when it is both lower and upper definite. The following are some of the the possible concepts of rough objects considered in the literature, (and all considerations will be restricted as indicated in the next definition):

- A non definite subset of K, that is x is a rough object if and only if $x^l \neq x^u$.
- Any pair of definite subsets of the form (a, b) satisfying $a \subseteq b$.
- Any pair of subsets of the form (x^l, x^u).
- Sets in an interval of the form (x^l, x^u).
- Sets in an interval of the form (a, b) satisfying $a \subseteq b$ and a, b being definite subsets.
- A non-definite element in a RYS, that is an x satisfying $\neg \mathbf{P} x^u x^l$ (x may be a subset and both upper and lower case letters may be used for them).
- An interval of the form, (a, b) satisfying $a \subseteq B$ and a, b being definite subsets.

Set framework with operators will be used as all considerations will require quasi orders in an essential way. The evolution of the operators need not be induced by a cover or a relation (corresponding to cover or relation based systems respectively), but these would be special cases. The generalization to some rough Y-systems RYS (see [21] for definitions), will of course be possible as a result.

Definition 1 A *Granular Operator Space* [26] S will be a structure of the form $S = \langle \underline{S}, \mathcal{G}, l, u \rangle$ with $\underline{S}$ being a set, $\mathcal{G}$ an *admissible granulation* (defined below) over S and l, u being operators $: \wp(\underline{S}) \longmapsto \wp(\underline{S})$ satisfying the following ($\underline{S}$ will be replaced with S if clear from the context. Lower and upper case alphabets will both be used for subsets):

$$a^l \subseteq a \,\&\, a^{ll} = a^l \,\&\, a^u \subset a^{uu}$$
$$(a \subseteq b \longrightarrow a^l \subseteq b^l \,\&\, a^u \subseteq b^u)$$
$$\emptyset^l = \emptyset \,\&\, \emptyset^u = \emptyset \,\&\, \underline{S}^l \subseteq S \,\&\, \underline{S}^u \subseteq S.$$

Here, Admissible granulations are granulations $\mathcal{G}$ that satisfy the following three conditions (Relative RYS [21], $\mathbf{P} = \subseteq$, $\mathbb{P} = \subset$) and t is a term operation formed from set operations):

$$(\forall a \exists b_1, \ldots b_r \in \mathcal{G})\, t(b_1, b_2, \ldots b_r) = a^l$$
$$\text{and } (\forall a)\, (\exists b_1, \ldots b_r \in \mathcal{G})\, t(b_1, b_2, \ldots b_r) = a^u, \quad \text{(Weak RA, WRA)}$$
$$(\forall b \in \mathcal{G})(\forall a \in \wp(\underline{S}))\, (b \subseteq a \longrightarrow b \subseteq (a^l)), \text{ (Lower Stability, LS)}$$
$$(\forall a, b \in \mathcal{G})(\exists z \in \wp(\underline{S}))\, a \subset z,\, b \subset z \,\&\, z^l = z^u = z, \quad \text{(Full Underlap, FU)}$$

In the present context, these conditions mean that every approximation is somehow representable by granules, that granules are lower definite, and that all pairs of distinct granules are contained in definite objects.

On $\wp(\underline{S})$, the relation $\sqsubset$ is defined by

$$A \sqsubset B \text{ if and only if } A^l \subseteq B^l \,\&\, A^u \subseteq B^u. \tag{1}$$

The rough equality relation on $\wp(\underline{S})$ is defined via $A \approx B$ if and only if $A \sqsubset B$ & $B \sqsubset A$.

Regarding the quotient $\underline{S}| \approx$ as a subset of $\wp(\underline{S})$, the order $\Subset$ will be defined as per

$$\alpha \Subset \beta \text{ if and only if } \alpha^l \subseteq \beta^l \,\&\, \alpha^u \subseteq \beta^u. \tag{2}$$

Here α^l is being interpreted as the lower approximation of α and so on. $\Subset$ will be referred to as the *basic rough order*.

Definition 2 By a *roughly consistent object* will be meant a set of subsets of $\underline{S}$ of the form $H = \{A; (\forall B \in H)\, A^l = B^l, A^u = B^u\}$. The set of all roughly consistent objects is partially ordered by the inclusion relation. Relative this maximal roughly consistent objects will be referred to as *rough objects*. By *definite rough objects*, will be meant rough objects of the form H that satisfy

$$(\forall A \in H)\, A^{ll} = A^l \,\&\, A^{uu} = A^u. \tag{3}$$

Proposition 1 *$\Subset$ is a bounded partial order on $\underline{S}| \approx$.*

Proof Reflexivity is obvious. If $\alpha \Subset \beta$ and $\beta \Subset \alpha$, then it follows that $\alpha^l = \beta^l$ and $\alpha^u = \beta^u$ and so antisymmetry holds.

If $\alpha \Subset \beta$, $\beta \Subset \gamma$, then the transitivity of set inclusion induces transitivity of $\Subset$. The poset is bounded by $0 = (\emptyset, \emptyset)$ and $1 = (S^l, S^u)$. Note that 1 need not coincide with (S, S). □

Theorem 1 *Some known results relating to antichains and lattices are the following:*

1. *Let X be a partially ordered set with longest chains of length r, then X can be partitioned into k number of antichains implies $r \leq k$.*
2. *If X is a finite poset with k elements in its largest antichain, then a chain decomposition of X must contain at least k chains.*
3. *The poset $AC_m(X)$ of all maximum sized antichains of a poset X is a distributive lattice.*
4. *For every finite distributive lattice L and every chain decomposition C of J_L (the set of join irreducible elements of L), there is a poset X_C such that $L \cong AC_m(X_C)$.*

Proof Proofs of the first three of the assertions can be found in in [9, 17] for example. Many proofs of results related to Dilworth's theorems are known in the literature and some discussion can be found in [17] (pp. 126–135).

1. To prove the first, start from a chain decomposition and recursively extract the minimal elements from it to form r number of antichains.
2. This is proved by induction on the size of X across many possibilities.
3. See [9, 17] for details.
4. In [15], the last connection between chain decompositions and representation by antichains reveals important gaps—there are other posets X that satisfy $L \cong AC_m(X)$. Further the restriction to posets is too strong and can be relaxed in many ways [39].

□

If R is a binary relation on a set $\underline{S}$, then the neighborhood generated by an $x \in \underline{S}$ will be

$$[x] = \{y \, : \, Ryx\}$$

A binary relation R on a set $\underline{S}$ is said to be a *Quasi-Equivalence* if and only if it satisfies:

$$(\forall x, y)\,([x] = [y] \leftrightarrow Rxy \,\&\, Ryx).$$

It is useful in algebras when it behaves as a good factor relation [2]. But the condition is of interest in rough sets by itself. *Note that Rxy* is a compact form of $(x, y) \in R$.

2 General Granular Operator Spaces (GSP)

Definition 3 A *General Granular Operator Space* (GSP) S shall be a structure of the form $S = \langle \underline{S}, \mathcal{G}, l, u, \mathbf{P} \rangle$ with $\underline{S}$ being a set, $\mathcal{G}$ an *admissible granulation* (defined below) over S, l, u being operators $: \wp(\underline{S}) \longmapsto \wp(\underline{S})$ and $\mathbf{P}$ being a definable binary generalized transitive predicate (for parthood) on $\wp(\underline{S})$ satisfying the following: ($\underline{S}$ will be replaced with S if clear from the context. Lower and upper case alphabets will both be used for subsets):

$$a^l \subseteq a \,\&\, a^{ll} = a^l \,\&\, a^u \subset a^{uu}$$
$$(a \subseteq b \longrightarrow a^l \subseteq b^l \,\&\, a^u \subseteq b^u)$$
$$\emptyset^l = \emptyset \,\&\, \emptyset^u = \emptyset \,\&\, \underline{S}^l \subseteq S \,\&\, \underline{S}^u \subseteq S.$$

Here, the generalized transitivity can be any proper nontrivial generalization of parthood (see [23]) and Admissible granulations are granulations $\mathcal{G}$ that satisfy the following three conditions (In the granular operator space of [26], $\mathbf{P} = \subseteq$, $\mathbb{P} = \subset$ only in that definition), $\mathbb{P}$ is proper parthood (defined via $\mathbb{P}ab$ iff $\mathbf{P}ab \,\&\, \neg \mathbf{P}ba$) and t is a term operation formed from set operations:

$$(\forall x \exists y_1, \ldots y_r \in \mathcal{G})\, t(y_1, y_2, \ldots y_r) = x^l$$
$$\text{and } (\forall x)\, (\exists y_1, \ldots y_r \in \mathcal{G})\, t(y_1, y_2, \ldots y_r) = x^u, \quad \text{(Weak RA, WRA)}$$
$$(\forall y \in \mathcal{G})(\forall x \in \wp(\underline{S}))\, (\mathbf{P}yx \longrightarrow \mathbf{P}yx^l), \text{ (Lower Stability, LS)}$$
$$(\forall x, y \in \mathcal{G})(\exists z \in \wp(\underline{S}))\, \mathbb{P}xz, \,\&\, \mathbb{P}yz \,\&\, z^l = z^u = z, \quad \text{(Full Underlap, FU)}$$

On $\wp(\underline{S})$, if the parthood relation $\mathbf{P}$ is defined via a formula Φ as per

$$\mathbf{P}ab \text{ if and only if } \Phi(a, b), \tag{4}$$

then the Φ-rough equality would be defined via

$$a \approx_{\Phi} b \text{ if and only if } \mathbf{P}ab \,\&\, \mathbf{P}ba. \tag{5}$$

In a granular operator space, $\mathbf{P}$ is the same as $\sqsubset$ and is defined by

$$a \sqsubset b \text{ if and only if } a^l \subseteq b^l \,\&\, a^u \subseteq b^u. \tag{6}$$

The rough equality relation on $\wp(\underline{S})$ is defined via $a \approx b$ if and only if $a \sqsubset b \,\&\, b \sqsubset a$. Regarding the quotient $\underline{S}| \approx$ as a subset of $\wp(\underline{S})$, the order $\Subset$ will be defined as per

$$\alpha \Subset \beta \text{ if and only if } \Phi(\alpha, \beta) \tag{7}$$

Here $\Phi(\alpha, \beta)$ is an abbreviation for $(\forall a \in \alpha, b \in \beta)\Phi(a, b)$. $\Subset$ will be referred to as the *basic rough order*.

Definition 4 By a *roughly consistent object* will be meant a set of subsets of $\underline{S}$ of the form $H = \{A; (\forall B \in H)\, A^l = B^l, A^u = B^u\}$. The set of all roughly consistent objects is partially ordered by the inclusion relation. Relative this maximal roughly consistent objects will be referred to as *rough objects*. By *definite rough objects*, will be meant rough objects of the form H that satisfy

$$(\forall A \in H)\, A^{ll} = A^l \,\&\, A^{uu} = A^u. \tag{8}$$

Other concepts of rough objects will also be used in this chapter.

Proposition 2 *When S is a granular operator space, $\Subset$ is a bounded partial order on $\underline{S}| \approx$. More generally it is a bounded quasi order.*

Proof The proof of the first part is in [26], the second part is provable analogously.

2.1 Parthood and Frameworks

Many of the philosophical issues relating to mereology take more specific forms in the context of rough sets in general and in the GSP framework. The axioms of parthood that can be seen to be not universally satisfied in all rough contexts include the following:

$$\mathbf{P}ab \,\&\, \mathbf{P}bc \longrightarrow \mathbf{P}ac \qquad \text{(Transitivity)}$$
$$(\mathbf{P}ab \leftrightarrow \mathbf{P}ba) \longrightarrow a = b \qquad \text{(Extensionality)}$$
$$(\mathbf{P}ab \,\&\, \mathbf{P}ba \longrightarrow a = b) \qquad \text{(Antisymmetry)}$$

This affords many distinct concepts of *proper parthoods* $\mathbb{P}$:

$$\mathbb{P}ab \text{ if and only if } \mathbf{P}ab \,\&\, a \neq b \qquad \text{(PP1)}$$
$$\mathbb{P}ab \text{ if and only if } \mathbf{P}ab \,\&\, \neg\mathbf{P}ba \qquad \text{(PP2)}$$
$$\mathbb{P}ab \longrightarrow (\exists z)\mathbf{P}zb \,\&\, (\forall w)\neg(\mathbf{P}wa \,\&\, \mathbf{P}wz) \qquad \text{(WS)}$$

PP1 does not follow from PP2 without antisymmetry and WS (weak supplementation) is a kind of proper parthood. All this affords a mereological approach with much variation to abstract rough sets.

3 Deductive Systems

In this section, key aspects of the approach to ternary deductive systems in [4, 5] are presented. These are intended as natural generalizations of the concepts of ideals and filters and classes of congruences that can serve as subsets or subalgebras closed under consequence operations or relations (also see [10]).

Definition 5 Let $\mathbb{S} = \langle S, \Sigma \rangle$ be an algebra, then the set of term functions over it will be denoted by $\mathbf{T}^{\Sigma}(\mathbb{S})$ and the set of r-ary term functions by $\mathbf{T}^{\Sigma}_{r}(\mathbb{S})$. Further let

$$g \in \mathbf{T}^{\Sigma}_{1}(\mathbb{S}),\ z \in S,\ \tau \subset \mathbf{T}^{\Sigma}_{3}(\mathbb{S}), \qquad (0)$$
$$g(z) \in \Delta \subset S, \qquad (1)$$
$$(\forall t \in \tau)(a \in \Delta \,\&\, t(a, b, z) \in \Delta \longrightarrow b \in \Delta), \qquad (2)$$
$$(\forall t \in \tau)(b \in \Delta \longrightarrow t(g(z), b, z) \in \Delta), \qquad (3)$$

then Δ is a $(g, z)-$ τ-*deductive system* of $\mathbb{S}$. If further for each k-ary operation $f \in \Sigma$ and ternary $p \in \tau$

$$(\forall a_i, b_i \in S)(\&_{i=1}^{k} p(a_i, b_i, z) \in \Delta \longrightarrow p(f(a_1, \ldots, a_k), f(b_1, \ldots, b_k), x) \in \Delta), \qquad (9)$$

then Δ is said to be compatible.

τ is said to be a g-*difference system* for $\mathbb{S}$ if τ is finite and the condition

$$(\forall t \in \tau)t(a, b, c) = g(c) \text{ if and only if } a = b \text{ holds.} \qquad (10)$$

A variety $\mathcal{V}$ of algebras is regular with respect to a unary term g if and only if for each $S \in \mathcal{V}$,

$$(\forall b \in S)(\forall \sigma, \rho \in con(S))([g(b)]_\sigma = [g(b)]_\rho \longrightarrow \sigma = \rho). \tag{11}$$

It should be noted that in the above τ is usually taken to be a finite subset and a variety has a g-difference system if and only if it is regular with respect to g.

Proposition 3 *In the above definition, it is provable that*

$$(\forall t \in \tau)(t(g(z), b, z) \in \Delta \longrightarrow b \in \Delta). \tag{12}$$

Definition 6 In the context of Definition 5, $\Theta_{Delta,z}$ shall be a relation induced on S by τ as per the following

$$(a, b) \in \Theta_{\Delta,z} \text{ if and only if } (\forall t \in \tau)\, t(a, b, z) \in \Delta. \tag{13}$$

Proposition 4 *In the context of Definition 6, $\Delta = [g(z)]_{\Theta_{\Delta,z}}$.*

Proposition 5 *Let $\tau \subset T_3^\Sigma(\mathbb{S})$ with the algebra $\mathbb{S} = \langle S, \Sigma \rangle$, $v \in T_1^\Sigma(\mathbb{S})$, $e \in S$, $K \subseteq S$ and let $\Theta_{K,e}$ be induced by τ. If $\Theta_{K,e}$ is a reflexive and transitive relation such that $K = [v(e)]_{Theta_{K,e}}$, then K is a (v, e)- τ-deductive system of $\mathbb{S}$.*

Theorem 2 *Let h is a unary term of a variety $\mathcal{V}$ and τ a h-difference system for $\mathcal{V}$. If $\mathbb{S} \in \mathcal{V} m\, \Theta \in Con(\mathbb{S})$, $z \in S$ and $\Delta = [h(z)]_\Theta$, then $\Theta_{\Delta,z} = \Theta$ and Δ is a compatible (h, z)-τ-deductive system of $\mathbb{S}$.*

The converse holds in the following sense:

Theorem 3 *If h is a unary term of a variety $\mathcal{V}$, τ is a h-difference system in it, $\mathbb{S} \in \mathcal{V}$, $z \in S$ and if Δ is a compatible (h, z)-τ-deductive system of $\mathbb{S}$, then $\Theta_{\Delta,z} \in Con(\mathbb{S})$ and $\Delta = [g(z)]_{\Theta_{\Delta,z}}$.*

When $\mathcal{V}$ is regular relative h, then $\mathcal{V}$ has a h-difference system relative τ and for each $\mathbb{S} \in \mathcal{V}$, $z \in S$ and $\Delta \subset S$, $\Delta = [h(z)]$ if and only if Δ is a (h, z)- τ-deductive system of $\mathbb{S}$.

In each case below, $\{t\}$ is a h-difference system ($x \oplus y = ((x \wedge y^*)^* \wedge (x^* \wedge y)^*)^*$):

$h(z) = z \;\&\; t(a, b, c) = a - b + c$	(Variety of Groups)
$h(z) = z \;\&\; t(a, b, c) = a \oplus b \oplus c$	(Variety of Boolean Algebras)
$h(z) = z^{**} \;\&\; t(a, b, c) = (a + b) + c$	(Variety of p-Semilattices)

4 Anti Chains for Representation

In this section, the main algebraic semantics of [26] is summarized, extended to *AC*-algebras and relative properties are studied. It is also proved that the number of maximal antichains required to generate the AC-algebra is rather small.

Definition 7 $\mathbb{A}, \mathbb{B} \in \underline{S}| \approx$, will be said to be *simply independent* (in symbols $\Xi(\mathbb{A}, \mathbb{B})$)if and only if

$$\neg(\mathbb{A} \Subset \mathbb{B}) \text{ and } \neg(\mathbb{B} \Subset \mathbb{A}). \tag{14}$$

A subset $\alpha \subseteq \underline{S}| \approx$ will be said to be *simply independent* if and only if

$$(\forall \mathbb{A}, \mathbb{B} \in \alpha)\, \Xi(\mathbb{A}, \mathbb{B}) \vee (\mathbb{A} = \mathbb{B}). \tag{15}$$

The set of all simply independent subsets shall be denoted by $\mathcal{SY}(S)$.

A *maximal simply independent subset*, shall be a simply independent subset that is not properly contained in any other simply independent subset. The set of maximal simply independent subsets will be denoted by $\mathcal{SY}_m(S)$. On the set $\mathcal{SY}_m(S)$, $\ll$ will be the relation defined by

$$\alpha \ll \beta \text{ if and only if } (\forall \mathbb{A} \in \alpha)(\exists \mathbb{B} \in \beta)\, \mathbb{A} \Subset \mathbb{B}. \tag{16}$$

Theorem 4 *$\langle \mathcal{SY}_m(S), \ll \rangle$ is a distributive lattice.*

Analogous to the above, it is possible to define essentially the same order on the set of maximal antichains of $\underline{S}| \approx$ denoted by $\mathfrak{S}$ with the $\Subset$ order. This order will be denoted by $\lessdot$ - this may also be seen to be induced by maximal ideals.

Theorem 5 *If $\alpha = \{\mathbb{A}_1, \mathbb{A}_2, \ldots, \mathbb{A}_n, \ldots\} \in \mathfrak{S}$, and if L is defined by*

$$L(\alpha) = \{\mathbb{B}_1, \mathbb{B}_2, \ldots, \mathbb{B}_n, \ldots\} \tag{17}$$

with $X \in \mathbb{B}_i$ if and only if $X^l = \mathbb{A}_i^{ll} = \mathbb{B}_i^l$ and $X^u = \mathbb{A}_i^{lu} = \mathbb{B}_i^u$, then L is a partial operation in general.

Proof The operation is partial because $L(\alpha)$ may not always be a maximal antichain. This can happen in general in which the properties $A^{ll} \subset A^l$ and/or $A^{ul} \subset A$ hold for some elements. The former possibility is not possible by our assumptions, but the latter is scenario is permitted.

Specifically this can happen in bitten rough sets when the bitten upper approximation [19] operator is used in conjunction with the lower approximation. But many more examples are known in the literature (see [21]). □

Definition 8 Let $\chi(\alpha \cap \beta) = \{\xi;\ \xi \text{ is a maximal antichain } \& \ \alpha \cap \beta \subseteq \xi\}$ be the set of all possible extensions of $\alpha \cap \beta$. The function $\delta : \mathfrak{S}^2 \longmapsto \mathfrak{S}$ corresponding to *extension under cognitive dissonance* will be defined as per $\delta(\alpha, \beta) \in \chi(\alpha \cap \beta)$ and (LST means *maximal subject to*)

$$\delta(\alpha, \beta) = \begin{cases} \xi, & \text{if } \xi \cap \beta \text{ is a maximum subject to } \xi \neq \beta \text{ and } \xi \text{is unique,} \\ \xi, & \text{if } \xi \cap \beta \,\&\, \xi \cap \alpha \text{ are LST } \xi \neq \beta, \alpha \text{ and } \xi \text{is unique,} \\ \beta, & \text{if } \xi \cap \beta \,\&\, \xi \cap \alpha \text{ are LST } \&\, \xi \neq \beta, \alpha \text{ but } \xi \text{ is not unique,} \\ \beta, & \text{if } \chi(\alpha \cap \beta) = \{\alpha, \beta\}. \end{cases} \tag{18}$$

Definition 9 In the context of the above definition, the function $\varrho : \mathfrak{S}^2 \longmapsto \mathfrak{S}$ corresponding to *radical extension* will be defined as per $\varrho(\alpha, \beta) \in \chi(\alpha \cap \beta)$ and (MST means *minimal subject to*)

$$\varrho(\alpha, \beta) = \begin{cases} \xi, & \text{if } \xi \cap \beta \text{ is a minimum under } \xi \neq \beta \text{ and } \xi \text{ is unique,} \\ \xi, & \text{if } \xi \cap \beta \,\&\, \xi \cap \alpha \text{ are MST } \xi \neq \beta, \alpha \text{ and } \xi \text{is unique,} \\ \alpha, & \text{if } (\exists \xi)\, \xi \cap \beta \,\&\, \xi \cap \alpha \text{ are MST } \xi \neq \beta, \alpha \,\&\, \xi \text{ is not unique,} \\ \alpha, & \text{if } \chi(\alpha \cap \beta) = \{\alpha, \beta\}. \end{cases} \tag{19}$$

Theorem 6 *The operations ϱ, δ satisfy all of the following:*

$$\varrho, \delta \text{ are groupoidal operations,} \quad (1)$$

$$\varrho(\alpha, \alpha) = \alpha \quad (2)$$

$$\delta(\alpha, \alpha) = \alpha \quad (3)$$

$$\delta(\alpha, \beta) \cap \beta \subseteq \delta(\delta(\alpha, \beta), \beta) \cap \beta \quad (4)$$

$$\delta(\delta(\alpha, \beta), \beta) = \delta(\alpha, \beta) \quad (5)$$

$$\varrho(\varrho(\alpha, \beta), \beta) \cap \beta \subseteq \varrho(\alpha, \beta) \cap \beta. \quad (6)$$

Proof 1. Obviously ϱ, δ are closed as the cases in their definition cover all possibilities. So they are groupoid operations. Associativity can be easily shown to fail through counterexamples.
2. Idempotence follows from definition.
3. Idempotence follows from definition.

For the rest, note that by definition, $\alpha \cap \beta \subseteq \delta(\alpha, \beta)$ holds. The intersection with β of $\delta(\alpha, \beta)$ is a subset of $\delta(\delta(\alpha, \beta), \beta) \cap \beta$ by recursion. □

In general, a number of possibilities (potential non-implications) like the following are satisfied by the algebra: $\alpha \lessdot \beta \,\&\, \alpha \lessdot \gamma \nrightarrow \alpha \lessdot \delta(\beta, \gamma)$. Given better properties of l and u, interesting operators can be induced on maximal antichains towards improving the properties of ϱ and δ. The key constraint hindering the definition of total l, u induced operations can be avoided in the following way:

Definition 10 In the context of Theorem 5, operations $\Box, \Diamond$ can be defined as follows:

- Given $\alpha = \{\mathbb{A}_1, \mathbb{A}_2, \ldots, \mathbb{A}_n, \ldots\} \in \mathfrak{S}$, form the set $\gamma(\alpha) = \{\mathbb{A}_1^l, \mathbb{A}_2^l, \ldots, \mathbb{A}_n^l, \ldots\}$. If this is an antichain, then α would be said to be *lower pure*.
- Form the set of all relatively maximal antichains $\gamma_+(\alpha)$ from $\gamma(\alpha)$.
- Form all maximal antichains $\gamma_*(\alpha)$ containing elements of $\gamma_+(\alpha)$ and set $\Box(\alpha) = \wedge\gamma_*(\alpha)$.
- For $\Diamond$, set $\pi(\alpha) = \{\mathbb{A}_1^u, \mathbb{A}_2^u, \ldots, \mathbb{A}_n^u, \ldots\}$. If this is an antichain, then α would be said to be *upper pure*.
- Form the set of all relatively maximal antichains $\pi_+(\alpha)$ from $\pi(\alpha)$.
- Form all maximal antichains $\pi_*(\alpha)$ containing elements of $\pi_+(\alpha)$ and set $\Diamond(\alpha) = \vee\pi_*(\alpha)$.

Theorem 7 *In the context of the above definition, the following hold:*

$$\alpha \lessdot \beta \longrightarrow \Box(\alpha) \lessdot \Box(\beta) \,\&\, \Diamond(\alpha) \lessdot \Diamond(\beta)$$
$$\Box(\alpha) \lessdot \alpha \lessdot \Diamond(\alpha),\ \Box(0) = 0 \,\&\, \Diamond(1) = 1$$

Based on the above properties, the following algebra can be defined.

Definition 11 By a *Concrete AC* algebra (AC -algebra) will be meant an algebra of the form $\left\langle \mathfrak{S}, \varrho, \delta, \vee, \wedge, \Box, \Diamond, 0, 1 \right\rangle$ associated with a granular operator space S satisfying all of the following:

- $\langle \mathfrak{S}, \vee, \wedge \rangle$ is a bounded distributive lattice derived from a granular operator space as in the above.
- $\varrho, \delta, \Box, \Diamond$ are as defined above.

The following concepts of ideals and filters are of interest as deductive systems in a natural sense and relate to ideas of rough consequence (detailed investigation will appear separately).

Definition 12 By a *LD-ideal* (resp. LD-filter) K of an AC-algebra $\mathfrak{S}$ will be meant a lattice ideal (resp. filter) that satisfies:

$$(\forall \alpha \in K)\, \Box(\alpha), \Diamond(\alpha) \in K \tag{20}$$

By a *VE-ideal* (resp. VE-filter) K of an AC-algebra $\mathfrak{S}$ will be meant a lattice ideal (resp. filter) that satisfies:

$$(\forall \xi \in \mathfrak{S})(\forall \alpha \in K)\, \varrho(\xi, \alpha), \delta(\xi, \alpha) \in K \tag{21}$$

Proposition 6 *Every VE filter is closed under* $\Diamond$

4.1 Generating AC-Algebras

Now it will be shown below that specific subsets of AC-algebras suffice to generate the algebra itself and that the axioms satisfied by the granulation affect the generation process and properties of AC-algebras and forgetful variants thereof.

An element $x \in \mathfrak{S}$ will be said to be *meet irreducible* (resp. join irreducible) if and only if $\wedge\{x_i\} = x \longrightarrow (\exists i)\, x_i = x$ (resp. $\vee\{x_i\} = x \longrightarrow (\exists i)\, x_i = x$). Let $W(S)$, $J(S)$ be the set of meet and join irreducible elements of $\mathfrak{S}$ and let $l(\mathfrak{S})$ be the length of the distributive lattice.

Theorem 8 *All of the following hold:*

- $(\mathfrak{S}, \vee, \wedge, 0, 1)$ *is a isomorphic to the lattice of principal ideals of the poset of join irreducibles.*
- $l(\mathfrak{S}) = \#(J(S)) = \#(W(S))$.
- $J(S)$ *is not necessarily the set of sets of maximal antichains of granules in general.*
- *When* $\mathcal{G}$ *satisfies mereological atomicity that is* $(\forall a \in \mathcal{G})(\forall b \in S)(\mathbf{P}ba,\ a^l = a^u = a \longrightarrow a = b)$, *and all approximations are unions of granules, then elements of* $J(S)$ *are proper subsets of* $\mathcal{G}$.
- *In the previous context,* $W(S)$ *must necessarily consist of two subsets of* S *that are definite and are not parts of each other.*

Proof • The first assertion is a well known.

- Since the lattice is distributive and finite, its length must be equal to the number of elements in $J(S)$ and $W(S)$. For a proof see [29].
- Under the minimal assumptions on $\mathcal{G}$, it is possible for definite elements to be properly included in granules as in esoteric or prototransitive rough sets [18, 23]. These provide the required counterexamples.
- The rest of the assertions follows from the nature of maximal antichains and the constructive nature of approximations. □

Theorem 9 *In the context of the previous theorem if* $R(\Diamond)$, $R(\Box)$ *are the ranges of the operations* $\Diamond, \Box$ *respectively, then these have a induced lattice order on them. Denoting the associated lattice operations by* $\curlyvee, \curlywedge$ *on* $R(\Diamond)$, *it can be shown that*

- $R(\Diamond)$ *can be reconstructed from* $J(R(\Diamond)) \cup W(R(\Diamond))$.
- $R(\Box)$ *can be reconstructed from* $J(R(\Box)) \cup W(R(\Box))$.
- *When* $\mathcal{G}$ *satisfies mereological atomicity and absolute crispness (i.e.* $(\forall x \in \mathcal{G})\, x^l = x^u = x$*), then* $R(\Diamond)$ *are lattices which are constructible from two sets* A, C *(with* $A = \{\mathcal{G} \cup \{g_1 \cup g_2\}^u \setminus \{g_1, g_2\};\ g_1, g_2 \in \mathcal{G}\}$ *and* C *being the set of two element maximal antichains formed by sets that are upper approximations of other sets).*

Proof It is clear that $R(\Diamond)$ is a lattice in the induced order with $J(R(\Diamond))$ and $W(R(\Diamond))$ being the partially ordered sets of join and meet irreducible elements respectively. This holds because the lattice is finite.

The reconstruction of the lattice can be done through the following steps:

- Let $Z = J(R(\Diamond)) \cup W(R(\Diamond))$. This is a partially ordered set in the order induced from $R(\Diamond)$.
- For $b \in J(R(\Diamond))$ and $a \in W(R(\Diamond))$, let $b \prec a$ if and only if $a \neq b$ in $R(\Diamond)$.
- On the new poset Z with $\prec$, form sets including elements of $W(R(\Diamond))$ connected to it.
- The set of union of all such sets including empty set ordered by inclusion would be isomorphic to the original lattice [29].
- Under additional assumptions on $\mathcal{G}$, the structure of Z can be described further.

When the granulation satisfies the properties of crispness and mereological atomicity, then $A = J(R(\Diamond))$ and $C = W(R(\Diamond))$. So the third part holds as well. □

The results motivate this concept of purity: A maximal antichain will be said to *pure* if and only if it is both lower and upper pure.

4.2 Enhancing the Anti Chain Based Representation

An integration of the orders on sets of maximal antichains or antichains and the representation of rough objects and possible orders among them leads to interesting multiple orders on the resulting structure. A major problem is that of defining the orders or partials thereof in the first place among the various possibilities.

Definition 13 By the *rough interpretation of an antichain* will be meant the sequence of pairs obtained by substituting objects in the rough domain in place of objects in the classical perspective. Thus if $\alpha = \{a_1, a_2 \ldots, a_p\}$ is a antichain, then its rough interpretation would be ($\pi(a_i) = (a_i^l, a_i^u)$ for each i)

$$\underline{\alpha} = \{\pi(a_1), \pi(a_2), \ldots, \pi(a_p)\}. \tag{22}$$

Proposition 7 *It is possible that some rough objects are not representable by maximal antichains.*

Proof Suppose the objects represented by the pairs (a, b) and (e, f) are such that $a = e$ and $b \subset f$, then it is clear that any maximal antichain containing (e, f) cannot contain any element from $\{x : x^l = a \,\&\, x^u = b\}$. This situation can happen, for example, in the models of proto transitive rough sets [24, 27]. Concrete counterexamples can be found in the same paper. □

Definition 14 A set of maximal antichains V will be said to be *fluent* if and only if $(\forall x)(\exists \alpha \in V)(\exists (a, b) \in \alpha)\, x^l = a \,\&\, x^u = b$.

It will be said to be *well fluent* if and only if it is fluent and no proper subset of it is fluent.

A related problem is of finding conditions on $\mathcal{G}$, that ensure that V is fluent.

Table 1 Successor neighborhoods

Objects E	a	b	c	e	f
Neighborhoods [E]	$\{a\}$	$\{a, b, e\}$	$\{c, e\}$	$\{c, f\}$	$\{e\}$

4.3 Extended Abstract Example-1

The following example is intended to illustrate some aspects of the intricacies of the inverse problem situation where anti chains may be described. It is done within the relation based paradigm and the assumption that objects are completely determined by their properties.

Let $\S = \{a, b, c, e, f\}$ and let R be a binary relation on it defined via

$$R = \{(a, a), (b, b), (c, c), (a, b), \\ (c, e), (e, f), (e, c), (f, e), (e, b)\}$$

If the formula for successor neighborhoods is

$$[x] = \{z ; Rzx\},$$

then the table for successor neighborhoods would be as in Table 1.

Using the definitions

$$x^l = \bigcup_{[z]\subseteq x} [z] \;\&\; x^u = \bigcup_{[z]\cap x\neq\emptyset} [z],$$

the approximations and rough objects that follow are in Table 2 (strings of letters of the form *abe* are intended as abbreviation for the subset $\{a, b, e\}$ and $\lrcorner$ is for, among subsets).

Under the rough inclusion order, the bounded lattice of rough objects in Fig. 1 (arrows point towards smaller elements) is the result:

From this ordered structure, maximal antichains can be evaluated by standard algorithms or by a differential process of looking at elements, their order ideals (and order filters) and maximal antichains that they can possibly form. In the figure, for example, elements in the order ideals of 69 cannot form antichains with it. This computation is targeted at representation in terms of relatively exact objects. The direct computation that is likely to come first before representation in practice is presented after the Table 3 in which some of the maximal antichains are computed by representation:

$\{60, 54, 69, 72\}$ is a maximal antichain because no more elements can be added to the set without violating incomparability. Note that the singletons $\{0\}$ and $\{1\}$ are also maximal antichains by definition. A diagram of the associated distributive lattice will not be attempted because of the number of elements.

Table 2 Approximations and rough objects

Rough object x	z^l	z^u	RO identifier
$\{a \lrcorner b \lrcorner ab\}$	$\{a\}$	$\{abe\}$	$\{3\}$
$\{ae \lrcorner abe\}$	$\{a\}$	$\{abce\}$	$\{6\}$
$\{e \lrcorner be\}$	$\{e\}$	$\{abec\}$	$\{9\}$
$\{c\}$	$\{\emptyset\}$	$\{cef\}$	$\{15\}$
$\{f\}$	$\{\emptyset\}$	$\{cf\}$	$\{24\}$
$\{cf\}$	$\{cf\}$	$\{cef\}$	$\{27\}$
$\{bc \lrcorner bf\}$	$\{\emptyset\}$	$\{S\}$	$\{30\}$
$\{ac \lrcorner af \lrcorner abc \lrcorner abf\}$	$\{a\}$	$\{S\}$	$\{33\}$
$\{aef\}$	$\{ae\}$	$\{S\}$	$\{36\}$
$\{ef \lrcorner bef\}$	$\{e\}$	$\{S\}$	$\{42\}$
$\{ec \lrcorner bce\}$	$\{ec\}$	$\{S\}$	$\{45\}$
$\{bcf\}$	$\{fc\}$	$\{S\}$	$\{51\}$
$\{abef\}$	$\{abe\}$	$\{S\}$	$\{54\}$
$\{ace\}$	$\{ace\}$	$\{S\}$	$\{60\}$
$\{acf\}$	$\{acf\}$	$\{S\}$	$\{63\}$
$\{ecf \lrcorner bcef\}$	$\{cef\}$	$\{S\}$	$\{69\}$
$\{abcf\}$	$\{abcf\}$	$\{S\}$	$\{72\}$
$\{abce\}$	$\{abcf\}$	$\{S\}$	$\{78\}$

Table 3 Maximal antichains

Rough object Z	Antichains including Z (differential)
78	$\{78, 69, 72\}$
60	$\{60, 54, 69, 72\}, \{60, 54, 69, 63\}, \{60, 54, 51\}, \{60, 54, 27\}$
54	$\{54, 45, 72\}$
72	$\{72, 45, 36\}, \{36, 69, 72\}, \{42, 72\}, \{9, 72\}$
69	$\{36, 69, 63\}, \{69, 33, 42\}, \{69, 6\}, \{69, 3\}$
42	$\{42, 33, 51\}, \{42, 33, 27\}, \{42, 6, 27\}, \{42, 6, 51\}, \{42, 63\}$
36	$\{36, 45, 63\}, \{36, 51, 45\}, \{36, 27, 45\}$
33	$\{45, 33, 51\}, \{45, 33, 27\}$
6	$\{9, 6, 15\}, \{9, 6, 27\}, \{9, 6, 51\}, \{9, 6, 24\}$
9	$\{9, 3, 15\}, \{9, 3, 24\}, \{9, 3, 27\}, \{9, 3, 51\}, \{9, 63\}$

4.3.1 Comparative Computations

In practice, the above table corresponds to only one aspect of information obtained from information systems. The scope of the anti chain based is intended to be beyond that including the inverse problem [21]. The empirical aspect is explained in this part.

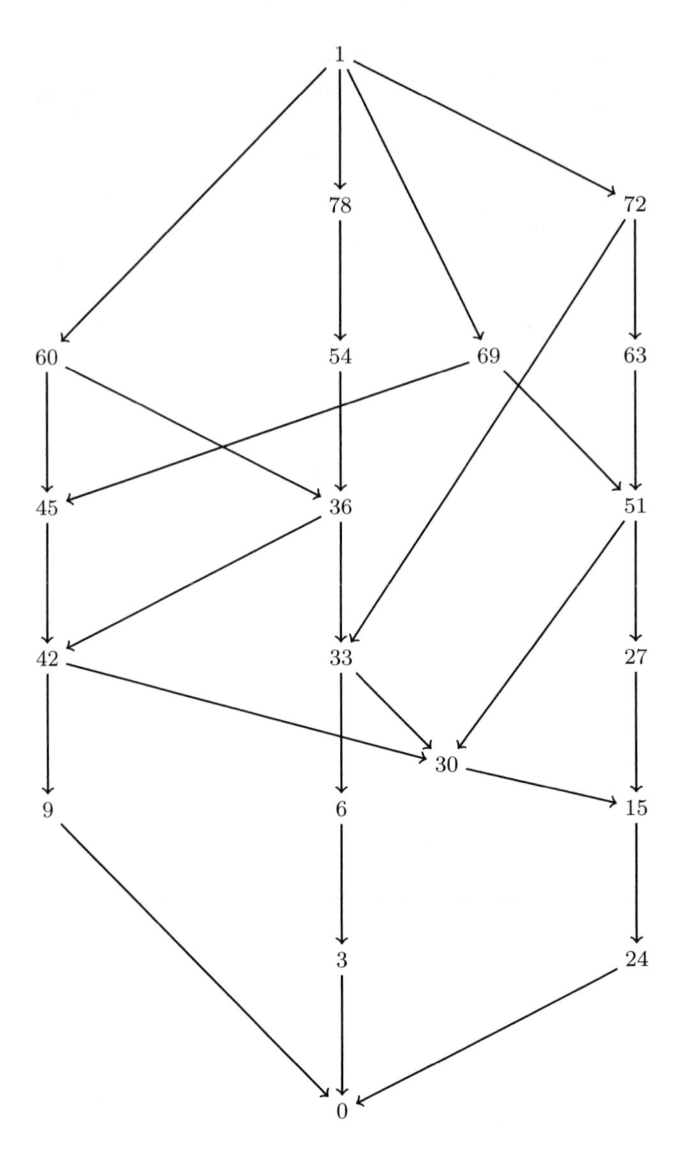

Fig. 1 Lattice of rough objects

Antichains are formed from $\wp(S)$ or subsets of it with some implicit temporal order (because of the order in which elements are accessed). If the elements of $\wp(S)$ are accessed in lexicographic order, and the sequence is decomposed by rough object discernibility alone, then it would have the following form ($\lceil, \rceil$ being group boundaries):

$$\{\lceil\{a\}\rceil, \lceil\{b\}\,\{c\}, \{e\}, \{f\}\rceil, \lceil\{a,b\}, \{a,c\}, \{a,e\}\rceil, \lceil\{a,f\}, \{b,c\}, \{b,e\}\rceil, \lceil\{b,f\}, \{c,e\}, \{c,f\}, \{e,f\}, \{a,b,c\}, \{a,b,e\}\}\rceil, \ldots\}$$

If these are refined by rough inclusion, then a decomposition into antichains would have the following form ($\lceil, \rceil$ now serve as determiners of antichain boundaries)

$$\{\lceil\{a\}\rceil, \lceil\{b\}\,\{c\}, \{e\}\rceil, \lceil\{f\}, \{a,b\}\rceil, \lceil\{a,c\}\rceil, \lceil\{a,e\}\rceil, \lceil\{a,f\}\rceil, \lceil\{b,c\}, \{b,e\}\rceil, \lceil\{b,f\}\rceil, \lceil\{c,e\}, \{c,f\}\rceil, \lceil\{e,f\}, \{a,b,c\}\rceil, \lceil\{a,b,e\}\}\rceil, \ldots\}$$

Implicit in all this is that the agent can perceive

- rough approximations,
- rough inclusion,
- rough equality and

have good intuitive algorithms for arriving at maximal antichains. In the brute force approach, the agent would need as much as $\frac{2^{\#(\wp(S))}!}{2}$ orders for obtaining all maximal antichains. The number of computations can be sharply reduced by the table of rough objects and known algorithms in the absence of intuitive algorithms.

A reading of the above sequence of antichains in terms of approximations (the compact notation introduced earlier is used) is

$$\{\lceil(a, abe)\rceil, \lceil(a, abe), (\emptyset, cef), (e, abec)\rceil, \lceil(\emptyset, cf), (a, abe)\rceil, \lceil(a, S)\rceil, \lceil(a, abec)\rceil, \lceil(a, S)\rceil, \lceil(\emptyset, S), (e, abec)\rceil, \lceil(\emptyset, S)\rceil, \lceil(ec, S), (cf, cef)\rceil, \lceil(e, S), (a, S)\rceil, \lceil(a, abec)\rceil, \ldots\}$$

Relative the order structure this reads as

$$\{\lceil 3\rceil, \lceil 3, 15, 9\rceil, \lceil 24, 3\rceil, \lceil 33\rceil, \lceil 6\rceil, \lceil 33\rceil, \lceil 30, 9\rceil, \lceil 30\rceil, \lceil 45, 27\rceil, \lceil 42, 33\rceil, \lceil 6\rceil, \ldots\}$$

4.4 Example: Micro-Fossils and Descriptively Remote Sets

This is a somewhat extended version of the example mentioned by the present author in [26]. In the case study on numeric visual data including micro-fossils with the help of nearness and remoteness granules in [34], the difference between granules and approximations is very fluid as the precision level of the former can be varied.

The data set consists of values of probe functions that extract pixel data from images of micro-fossils trapped inside other media like amethyst crystals.

The idea of remoteness granules is relative a fixed set of nearness granules formed from upper approximations—so the approach is about reasoning with sets of objects which in turn arise from tolerance relations on a set. In [34], antichains of rough objects are not used, but the computations can be extended to form maximal antichains at different levels of precision towards working out the best antichains from the point of view of classification.

Let X be an object space consisting of representation of some patterns and $\Phi : X \longmapsto \mathbb{R}^n$ be a *probe function*, defined by

$$\Phi(x) = (\phi_1(x), \phi_2(x), \dots, \phi_n(x)), \tag{23}$$

where $\phi_i(x)$ is intended as a measurement of the ith component in the feature space $\Im(\Phi)$. The concept of descriptive intersection of sets permits migration from classical ideas of proximity to ones based on descriptions. A subset $A \subseteq X$'s descriptive intersection with subset $B \subseteq X$ is defined by

$$A \cap_{\Phi} B = \{x \in A \cup B : \Phi(x) \in \Phi(A) \,\&\, \Phi(x) \in \Phi(B)\} \tag{24}$$

A is then *descriptively near* B if and only if their descriptive intersection is nonempty. Peter's version of proximity π_{Φ} is defined by

$$A \pi_{\Phi} B \leftrightarrow \Phi(A) \cap \Phi(B) \neq \emptyset \tag{25}$$

In [8], weaker implications for defining *descriptive nearness* are considered :

$$A \cap_{\Phi} B \neq \emptyset \rightarrow A \delta_{\Phi} B. \tag{26}$$

Specifically, if δ is a proximity on R^n, then a descriptive proximity δ_{Φ} is definable via

$$A \delta_{\Phi} B \leftrightarrow \Phi(A) \delta \Phi(B). \tag{27}$$

All these are again approachable from an anti-chain perspective.

4.5 *Example: Beyond Data Tables*

In this example subjective data is cast in terms of rough language for the purpose of understanding appropriate frameworks and solving context related problems.

Suppose agent X wants to complete a task and this task is likely to involve the use of a number of tools. X thinks tool-1 suffices for the task that a tool-2 is not suited for the purpose and that tool-3 is better suited than tool-1 for the same task. X also believes that tool-4 is as suitable as tool-1 for the task and that tool-5 provides more than what is necessary for the task. X thinks similarly about other tools but not much is known about the consistency of the information. X has a large repository of tools and limited knowledge about tools and their suitability for different purposes, and at the same time X might be knowing more about difficulty of tasks that in turn require better tools of different kinds.

Suppose also that similar heuristics are available about other similar tasks.

The reasoning of the agent in the situation can be recast in terms of lower, upper approximations and generalized equality and questions of interest include those relating to the agent's understanding of the features of tools, their appropriate usage contexts and whether the person thinks rationally.

To see this it should be noted that the key predicates in the context are as below:

- suffices for can be read as *includes potential lower approximation of* a right tool for the task.
- is not suited for can be read as *is neither a lower or upper approximation of* any of the right tools for the task.
- better suited than can be read as *potential rough inclusion*,
- is as suitable as can be read as *potential rough equality* and
- provides more than what is necessary for is for *upper approximation* of a right tool for the task.

If *table rationality* is the process of reasoning by information tables and approximations, then when does X's reasoning become table rational?

This problem fits in easily with the antichain perspective, but not the information table approach because the latter requires extra information about properties.

5 Quasi Equivalential Rough Sets and More

Entire semantics of various general rough set approaches can be recast in the antichain based perspective. For example, prototransitive rough sets [27] can be dealt with the same way. A finer characterization of the same will appear separately. Quasi Equivalential rough set were considered also as an example of the approach in [26]. As this can serve as an important example, the part is repeated below.

One of the most interesting type of granulation $\mathcal{G}$ in relational RST is one that satisfies

$$(\forall x, y)\,(\phi(x) = \phi(y) \leftrightarrow Rxy \,\&\, Ryx), \tag{28}$$

where $\phi(x)$ is the granule generated by $x \in \underline{S}$. This granular axiom says that if x is left-similar to y and y is left-similar to x, then the elements left similar to either of x and y must be the same. R is being read as *left-similarity* because it is directional and has effect similar to tolerances on neighborhood granules.

Reflexivity is not assumed as the present author's intention is to isolate the effect of the axiom alone.

For example, it is possible to find quasi equivalences that do not satisfy other properties from contexts relating to numeric measures. Let S be a set of variables such that Rxy if and only if $x \approx \kappa y \,\&\, y \approx \kappa' x \,\&\, \kappa, \kappa' \in (0.9, 1.1)$ for some interpretation of $\approx$.

Definition 15 By a *Quasi-Equivalential Approximation Space* will be meant a pair of the form $S = \langle \underline{S}, R \rangle$ with R being a quasi equivalence. For an arbitrary subset $A \in \wp(S)$, the following can be defined:

$$(\forall x \in \underline{S})\, [x] = \{y\,;\, y \in \underline{S} \,\&\, Ryx\}.$$

$$A^l = \bigcup\{[x]\,;\, [x] \subseteq A \,\&\, x \in \underline{S}\} \,\&\, A^u = \bigcup\{[x]\,;\, [x] \cap A \neq \emptyset \,\&\, x \in \underline{S}\}$$

$$A^{l_o} = \bigcup\{[x]\,;\, [x] \subseteq A \& x \in A\} \,\&\, A^{u_o} = \bigcup\{[x]\,;\, [x] \cap A \neq \emptyset \,\&\, x \in A\}$$

$$A^L = \{x\,;\, \emptyset \neq [x] \subseteq A \,\&\, x \in \underline{S}\} \,\&\, A^U = \{x\,;\, [x] \cap A \neq \emptyset \,\&\, x \in \underline{S}\}$$

$$A^{L_o} = \{x\,;\, [x] \subseteq A \,\&\, x \in A\} \,\&\, A^{U_o} = \{x\,;\, [x] \cap A \neq \emptyset \vee x \in A\}.$$

$$A^{L_1} = \{x\,;\, [x] \subseteq A \,\&\, x \in \underline{S}\} \,\&\, A^U = \{x\,;\, [x] \cap A \neq \emptyset \,\&\, x \in \underline{S}\}.$$

Note the requirement of non-emptiness of $[x]$ in the definition of A^L, but it is not necessary in that of A^{L_o}.

Theorem 10 *The following properties hold:*

1. *All of the approximations are distinct in general.*
2. $(\forall A \in \wp(S))\, A^{L_o} \subseteq A^{l_o} \subseteq A^l \subseteq A$ *and* $A^{L_o} \subseteq A^L$.
3. $(\forall A \in \wp(S))\, A^{l_o l} = A^{l_o} \,\&\, A^{ll_o} \subseteq A^{l_o} \,\&\, A^{l_o l_o} \subseteq A^{l_o}$
4. $(\forall A \in \wp(S))\, A^u = A^{ul} \subseteq A^{uu}$, *but it is possible that* $A \nsubseteq A^u$
5. *It is possible that* $A^L \nsubseteq A$ *and* $A \nsubseteq A^U$, *but* $(\forall A \in \wp(S))\, A^L \subseteq A^U$ *holds. In general* A^L *would not be comparable with* A^l *and similarly for* A^U *and* A^u.
6. $(\forall A \in \wp(S))\, A^{L_o L_o} \subseteq A^{L_o} \subseteq A \subseteq A^{U_o} \subseteq A^{U_o U_o}$. *Further* $A^U \subseteq A^{U_o}$.

Clearly the operators l, u are granular approximations, but the latter is controversial as an upper approximation operator. The point-wise approximations L, U are more problematic—not only do they fail to satisfy representability in terms of neighborhood granules, but the lower approximation fails inclusion.

Example 1 (General)

$$\text{Let } \underline{S} = \{a, b, c, e, f, k, h, q\} \tag{29}$$

and let R be a binary relation on it defined via

$$R = \{(a, a), (b, a), (c, a), (f, a),$$
$$(k, k), (e, h), (f, c), (k, h)$$
$$(b, b), (c, b), (f, b), (a, b), (c, e), (e, q)\}.$$

The neighborhood granules $\mathcal{G}$ are then

$$[a] = \{a, b, c, f\} = [b], [c] = \{f\}, [e] = \{c\},$$
$$[k] = \{k\}, [h] = \{k, e\}, [f] = \emptyset \,\&\, [q] = \{e\}.$$

So R is a quasi-equivalence relation.

If $A = \{a, k, q, f\}$, then

$$A^l = \{k, f\},\ A^u = \{a, b, c, f, k, e\},\ A^{uu} = \{a, b, c, f, k, e, h\}$$
$$A^{l_o} = \{k\},\ A^{u_o} = \{a, b, c, k, f\}.$$
$$A^L = \{k, f\},\ A^U = \{a, b, c, k, h, q\}.$$
$$A^{L_o} = \{q, k, f\},\ A^{U_o} = \{a, k, q, f, b, c, h\}.$$
$$A^{L_1} = \{k, c, f\},\ A^U = \{a, b, c, k, h, q\}.$$

$$\text{Note that } A^{L_1} \not\subseteq A \,\&\, A^{L_1} \not\subseteq A^U \,\&\, A \not\subseteq A^U. \tag{30}$$

6 Semantics of QE-Rough Sets

In this section a semantics of quasi-equivalential rough sets (QE-rough sets), using antichains generated from rough objects, is developed. Interestingly the properties of the approximation operators of QE-rough sets fall short of those of granular operator space. Denoting the set of maximal antichains of rough objects by $\mathfrak{S}$ and carrying over the operations $\ll, \varrho, \delta$, the following algebra can be defined.

Definition 16 A *maximal simply independent algebra* Q of quasi equivalential rough sets shall be an algebra of the form

$$Q = \langle \mathfrak{S}, \ll, \varrho, \delta \rangle \tag{31}$$

defined as in Sect. 4 with the approximation operators being l, u uniformly in all constructions and definitions.

Theorem 11 *Maximal simply independent algebras are well defined.*

Proof None of the steps in the definition of the maximal antichains, or the operations ϱ or δ are problematic because of the properties of the operators l, u. □

The above theorem suggests that it would be better to try and define more specific operations to improve the uniqueness aspect of the semantics or at least the properties of ϱ, δ. It is clearly easier to work with antichains as opposed to maximal antichains as more number of suitable operations are closed over the set of antichains as opposed to those over the set of maximal antichains.

Definition 17 Let $\mathfrak{K}$ be the set of antichains of rough objects of S then the following operations $\mathfrak{L}$, $\mathfrak{U}$ and extensions of others can be defined:

- Let $\alpha = \{\mathbb{A}_1, \mathbb{A}_2, \ldots, \mathbb{A}_n, \ldots\} \in \mathfrak{K}$ with $\mathbb{A}_i$ being rough objects; the lower and upper approximation of any subset in $\mathbb{A}_i$ will be denoted by $\mathbb{A}_i^l$ and $\mathbb{A}_i^u$ respectively.
- Define $\mathfrak{L}(\alpha) = \{\mathbb{A}_1^l, \mathbb{A}_2^l, \ldots, \mathbb{A}_r^l, \ldots\}$ with duplicates being dropped
- Define $\mathfrak{U}(\alpha) = \{\mathbb{A}_1^u, \mathbb{A}_2^u, \ldots, \mathbb{A}_r^u, \ldots\}$ with duplicates being dropped
- Define

$$\mu(\alpha) = \begin{cases} \alpha & \text{if } \alpha \in \mathfrak{S} \\ \text{undefined,} & \text{else.} \end{cases} \tag{32}$$

- Partial operations ϱ^*, δ^* corresponding to ϱ, δ can also be defined as follows: Define

$$\varrho^*(\alpha, \beta) = \begin{cases} \varrho(\alpha, \beta) & \text{if } \alpha, \beta \in \mathfrak{S} \\ \text{undefined,} & \text{else.} \end{cases} \tag{33}$$

$$\delta^*(\alpha, \beta) = \begin{cases} \delta(\alpha, \beta) & \text{if } \alpha, \beta \in \mathfrak{S} \\ \text{undefined,} & \text{else.} \end{cases} \tag{34}$$

The resulting partial algebra $\mathfrak{K} = \langle \mathfrak{K}, \mu, \vee, \wedge, \varrho^*, \delta^*, \mathfrak{L}, \mathfrak{U}, 0 \rangle$ will be said to be a *simply independent QE algebra.*

Theorem 12 *Simply independent QE algebras are well defined and satisfy the following:*

- $\mathfrak{L}(\alpha) \vee \alpha = \alpha.$
- $\mathfrak{U}(\alpha) \vee \alpha = \mathfrak{U}(\alpha).$

7 Ternary Deduction Terms

Since AC-algebras are distributive lattices with additional operations, a natural strategy should be to consider terms similar to Boolean algebras and p-Semilattices. For isolating deductive systems in the sense of Sect. 3, a strategy can be through complementation-like operations. This motivates the following definition:

Definition 18 In a AC-algebra $\mathfrak{S}$, if an antichain $\alpha = (X_1, X_2, \ldots, X_n)$, then some possible general complements on the schema

$$\alpha^c = \mathfrak{H}(X_1^c, X_2^c, \ldots, X_n^c)$$

are as follows:

$$X_i^* = \{w;\ (\forall a \in X_i)\, \neg \mathbf{P}aw \,\&\, \neg \mathbf{P}wa\} \tag{Class A}$$
$$X_i^\# = \{w;\ (\forall a \in X_i)\, \neg a^l = w^l \text{ or } \neg a^u = w^u\} \tag{Light}$$
$$X_i^\flat = \{w;\ (\forall a \in X_i)\, \neg a^l = w^l \text{ or } \neg a^{uu} = w^{uu}\} \tag{UU}$$

$\mathfrak{H}$ is intended to signify the maximal antichain containing the set if that is definable.

Note that under additional assumptions (similarity spaces), the light complementation is similar to the preclusivity operation in [3] for Quasi BZ-lattice or Heyting-Wajsburg semantics and variants.

The above operations on α are partial in general, but can be made total with the help of an additional order on α and the following procedure:

1. Let $\alpha = \{X_1, X_2, \ldots, X_n\}$ be a finite sequence,
2. Form α^c and split into longest ACs in sequence,
3. Form maximal ACs containing each AC in sequence
4. Join resulting maximal ACs.

Proposition 8 *Every general complement defined by the above procedure is well defined.*

Proof
- Suppose $\{X_1^c, X_2^c\}, \{X_3^c, \ldots X_n^c\}$ form antichains, but $\{X_1^c, X_2^c, X_3^c\}$ is not an antichain.
- Then form the maximal antichains $\eta_1, \ldots, \eta_p$ containing either of the two antichains.
- The join of this finite set of maximal antichains is uniquely defined. By induction, it follows that the operations are well defined. □

7.1 Translations

As per the approach of Sect. 3, possible definitions of translations are as follows:

Definition 19 A *translation* in a AC-algebra $\mathfrak{S}$ is a $\sigma : \mathfrak{S} \longmapsto \mathfrak{S}$ that is defined in one of the following ways (for a fixed $a \in \mathfrak{S}$):

$$\sigma_\theta(x) = \theta(a, x) \; ; \theta \in \{\vee, \wedge, \varrho, \delta\}$$

$$\sigma_\mu(x) = \mu(x) \; ; \mu \in \{\Box, \Diamond\}$$

$$\sigma_t(x) = (x \oplus a) \oplus b \text{ for fixed a, b}$$

$$\sigma_{t+}(x) = (a \oplus b) \oplus x \text{ for fixed a, b}$$

Theorem 13

$$\sigma_\vee(0) = a = \sigma_\vee(a \, ; \sigma_\vee(1) = 1$$

$$Ran(\sigma_\vee) \textit{ is the principal filter generated by a}$$

$$Ran(\sigma_\wedge) \textit{ is the principal ideal generated by a}$$

$$x \lessdot w \longrightarrow \sigma_\vee(x) \lessdot \sigma_\vee(w) \,\&\, \sigma_\wedge(x) \lessdot \sigma_\wedge(w)$$

Proof
- Let $\mathbb{F}(a)$ be the principal lattice filter generated by a.

- If $a \lessdot w$, then $a \vee w = w = \sigma_{\vee}(w)$. So $w \in Ran(\sigma_{\vee})$.
- $\sigma_{\vee}(x) \wedge \sigma_{\vee}(w) = (a \vee x) \wedge (a \vee w) = a \vee (x \wedge w) = \sigma_{\vee}(x \wedge w)$.
- So if $x, w \in Ran(\sigma_{\vee})$, then $x \wedge w, x \vee w \in Ran(\sigma_{\vee})$
- Similarly it is provable that $Ran(\sigma_{\wedge})$ is the principal ideal generated by a.

□

7.2 Ternary Terms and Deductive Systems

Possible ternary terms that can cohere with the assumptions of the semantics include the following $t(a, b, z) = a \wedge b \wedge z$, $t(a, b, z) = a \oplus b \oplus z$ ($\oplus$ being as indicated earlier) and $t(a, b, z) = \Box(a \wedge b) \wedge z$. These have admissible deductive systems associated. Further under some conditions on granularity, the distributive lattice structure associated with $\mathfrak{S}$ becomes pseudo complemented.

Theorem 14 *If $t(a, b, z) = a \wedge b \wedge z$, $\tau = \{t\}$, $z \in H$, $h(x) = x$ $\sigma(x) = x \wedge z$ and if H is a ternary τ-deduction system at z, then it suffices that H be an filter.*

Proof All of the following must hold:

- If $a \in H$, $t(z, a, z) = a \wedge z \in H$
- If $t(a, b, z) \in H$, then $t(\sigma(a), \sigma(b), z) = t(a, b, z) \in H$
- If $a, t(a, b, z) \in H$ then $t(a, b, z) = (a \wedge z) \wedge b \in H$. But H is a filter, so $b \in H$. □

Theorem 15 *If $t(a, b, z) = (a \vee (\Box b)) \wedge z$, $\tau = \{t\}$, $z \in H$, $h(x) = x$ $\sigma(x) = x \wedge z$ and if H is a ternary τ-deduction system at z, then it suffices that H be a principal LD-filter generated by z.*

Proof All of the following must hold:

- If $a \in H$, $t(z, a, z) = (z \vee (\Box a)) \wedge z \in H$ because $(z \vee (\Box a)) \in H$.
- If $t(a, b, z) \in H$, then $t(\sigma(a), \sigma(b), z) = t((a \wedge z), (b \wedge z), z) = ((a \wedge z) \vee \Box(b \wedge z)) \wedge z \in H$
- If $a, t(a, b, z) \in H$ then $t(a, b, z) = (a \vee \Box(b)) \wedge z = (a \wedge z) \vee (\Box(b) \wedge z) \in H$. But H is a LD-filter, so $a \vee \Box(b) \in H$. This implies $\Box(b) \in H$, which in turn yields $b \in H$. □

In the above two theorems, the conditions on H can be weakened considerably. The converse questions are also of interest.

The existence of pseudo complements can also help in defining ternary terms that determine deductive systems (or subsets closed under consequence). In general, pseudo complementation $\circledast$ is a partial unary operation on $\mathfrak{S}$ that is defined by $x^{\circledast} = \max\{a\,;\, a \wedge x = 0\}$ (if the greatest element exists).

There is no one answer to the question of existence as it depends on the granularity assumptions of representation and stability of granules. The following result guarantees pseudo complementation (in the literature, there is no universal approach—it has always been the case that in some case they exist):

Theorem 16 *In the context of AC-algebras, if the granulation satisfies mereological atomicity and absolute crispness, then a pseudo complementation is definable.*

Proof Under the conditions on the granulation, it is possible to form the rough interpretation of each antichain. Moreover the granules can be moved in every case to construct the pseudo complement. The inductive steps in this proof have been omitted. □

8 Relation to Knowledge Interpretation

In Pawlak's concept of knowledge in classical RST [30, 32], if $\underline{S}$ is a set of attributes and P an indiscernibility relation on it, then sets of the form A^l and A^u represent clear and definite concepts (the semantic domain associated is the rough semantic domain). Extension of this to other kinds of RST have also been considered in [20, 22, 24, 27] by the present author. In [20], the concept of knowledge advanced by her is that of union of pairwise independent granules (in set context corresponding to empty intersection) correspond to clear concepts. This granular condition is desirable in other situations too, but may not correspond to the approximations of interest. In real life, clear concepts whose parts may not have clear relation between themselves are too common. If all of the granules are not definite objects, then analogous concepts of knowledge may be graded or typed based on the properties satisfied by them [24, 27]. *Then again the semantic domains in which these are being considered can be varied and so knowledge is relative.* Some examples of granular knowledge axioms are as follows:

1. Individual granules are atomic units of knowledge.
2. If collections of granules combine subject to a concept of mutual independence, then the result would be a concept of knowledge. The 'result' may be a single entity or a collection of granules depending on how one understands the concept of *fusion* in the underlying mereology. In set theoretic (ZF) setting the fusion operation reduces to set-theoretic union and so would result in a single entity.
3. Maximal collections of granules subject to a concept of mutual independence are admissible concepts of knowledge.
4. Parts common to subcollections of maximal collections are also knowledge.
5. All stable concepts of knowledge consistency should reduce to correspondences between granular components of knowledges. Two knowledges are *consistent* if and only if the granules generating the two have 'reasonable' correspondence.
6. Knowledge A is consistent with knowledge B if and only if the granules generating knowledge B are part of some of the granules generating A.

An antichain of rough objects is essentially a set of *some-sense mutually distinct rough concepts* relative that interpretation. Maximal antichains naturally correspond to represented knowledge that can be handled in a clear way in a context involving vagueness. The stress here should be on possible operations both within and over them. It is fairly clear that better the axioms satisfied by a concept of granular knowledge, better will be the nature of possible operations over sets of *some-sense mutually distinct rough concepts*.

From decision making perspectives, antichains of rough objects correspond to forming representative partitions of the whole and semantics relate to relation between different sets of representatives.

8.1 Knowledge Representation

In Sect. 4.2, the developed representation has the following features:

- Every object in a antichain is representable by a pair of objects (a, b) that are respectively of the form x^l and z^u.
- Some of these objects might be of the form (a, a) under the restriction $a = a^l = a^u$
- The above means that antichains can be written in terms of objects that are approximations of other objects or themselves.
- At another level, the concepts of rough objects mentioned in the background section suggest classification of the possible concepts of knowledge.
- The representation is perceivable in a rough semantic domain and this will be referred to as the *AC-representable rough domain* ACR.
- If *definable rough objects* are those rough objects representable in the form (a, b) with a, b being definite objects, then these together with definite objects may not correspond to maximal antichains in the classical semantic domain—the point is that some of the non crisp objects may fail to get represented under the constraints. The semantic domain associated with the definable rough objects with the representation and crisp objects will be referred to as the *strict rough domain* (SRD).

The above motivates the following definition sequence

Definition 20 All of the following constitute the basic knowledge structure in the context of AC-semantics:

- A *Proper Knowledge Sequence* in ACR corresponds to the representation of any of the maximal antichains.
- An *Abstract Proper Knowledge Sequence* in ACR corresponds to the representation of possible maximal antichains. These may be realized in particular models.
- A *Knowledge Sequence* in SRD corresponds to the relatively maximal antichains formed by sequences of definable rough objects and definite objects.
- Definable rough objects.
- Representation of rough and crisp objects.

More complex objects formed by antichains are also of interest. The important thing about the idea of knowledge sequences is the explicit admission of temporality and the relation to all of the information available in the context. This is considered next.

8.2 Perdurantism and Endurantism

The *endurantist* position is that objects persist over time by being *wholly present* at all times of their existence. Endurantism is also known as *Three dimensionalism*. The *perdurantist* position (or four dimensionalism),in contrast, is that objects persist over time by having *temporal parts* in addition to their spatial parts at all times of their existence. Some references are [16, 38, 42, 43]. In the context of rough sets, though temporality has been investigated, these concepts do not figure explicitly in earlier work in both the present author's work and the rough membership function based approach [35].

Classical extensional mereology CEM is seen by most endurantists and perdurantists as a reasonable framework for handling real objects with material existence. In the framework, such objects may be viewed as mereological aggregations including sum and fusion. But when it comes to concepts as in rough sets, such an approach need not suffice (see [21]). Given the assumptions of CEM, it can be seen that maximal antichains have temporal parts and by modification of temporal parts their identity changes. Again they can be seen as the same information with irrelevant temporal parts—but in doing this the semantic domain needs to be changed. This means that the positions of perdurantism and endurantism need to rely on choice of semantic domain for their validation. From the way the algebraic semantics has been constructed, it is clear that any two distinct maximal antichain have distinct temporal components. So the following meta theorem follows:

Theorem 17 *The semantic domain associated with the the AC-algebras is perdurantist (or four dimensional).*

But the argument does not stop here. It can be argued that two distinct maximal antichains

- are insufficient references to the same knowledge U (say).
- U is guaranteed to exist by generalized granular operator spaces.
- U is not affected by the *so-called temporal parts* and in the classical semantic domain maximal antichains correspond to distinct knowledge.

All this confirms the present author's position that the two positions happen because of choice of semantic domains—the main problem is then of constructing/describing consistent domains.

8.3 Patterns of Triads

Pierce's approach to semiotics does not provide a consistent perspective for ontology in general. This is often seen as a failure of the enterprise [1]. But attempts have been made to adopt aspects of the semiotics to get to new insights in the structure of knowledge and consequence. In this section, parts of how these might play with the concept of knowledge afforded by rough sets and the antichain based knowledge representation is examined. Admittedly the developed approach may differ substantially with Pierce's ideas and in particular on assumptions relating to existence of fixed ontologies.

The first thing to be noted is that the concept of knowledge in general rough sets must be viewed in a deconstructive perspective rather than in the perspective of aggregation. The deconstructive process in the triadic perspective may not always yield parts that correspond to exact knowledge and so a few redefinitions of the ontology are introduced. Multiple sources of knowledge can also be handled within the framework with scope for handling conflict resolution by algebraic strategies especially when the graphical take forever.

If *knowledge* is viewed as the result of a *process of knowing* an object or phenomena then the process in question may be about placing all of the parts or the whole in a logical space of reasons that refers to justification and the justified. This kind of knowledge will be referred to as *pre-rational knowledge* and if the understanding of *pre-rational* is something less than rational, then the machinery of general rough sets affords a specific instance of this definition of pre-rational knowledge. This definition may be seen as a variation of the definition of knowledge in [36] (passage 36) wherein the part-whole distinctions are missing. Unless quasi orders or preference orders is imposed on granulations or attributes or the mereological *is necessary part of* is specified between subsets of granules and objects, the concept of *rationality* in approximations is difficult to ensure scientific rationality.

In Peirce's view all forms of reality exist in Signs. A sign is a spatiotemporal reality that may be material or conceptual. These exist in relation to other signs and is determined by its Object, and determines Interpretant through the mediating role of representations. Somewhat controversially all this can be assumed to establish the nature of the sign as a triad of interactions or relations. Thus in [33] pp. 272–273 it is stated that

> A Sign, (or Representamen), is a First which stands in such a genuine triadic relation to a Second, called its Object, as to be capable of determining a Third, called its Interpretant, to assume the same triadic relation to its Object in which it stands itself to the same Object.

The semiotic process of Object-Representation; Representation-Representation; Representation-Interpretant is then a triad of three relations, correlated by the mediation of the Representation. Some authors [40] reject the singularity of *a triadic relation* because it contributes to setting up a kind of closed and isolated process, while others believe that it is but one of the possibilities.

Signs are definable through six relations (called semiosic predicates) and interact as functions that act as mediation between input and output. The triad may be read as Object-Mediation/Representation-Interpretant in Peirce's sense and being irreducible. One scheme described in [37, 40], consists of ten basic signs that can be seen as classes with further subdivisions. This approach requires further ontology for a reasonable ascription of thresholds for concepts of reasonable/rational approximations.

The triadic approach again refers some fragment of rationality because of the way in which sign-vehicles [1] refer parts of attributes possessed—for example all attributes associated with red traffic lights are not used by drivers for the intended interpretation. The main features of sign vehicles in the triadic approach are classified by the signified by virtue of qualities, existential facts, or conventions respectively into *qualisigns, sinsigns, and legisigns* respectively.

Explicit integration with rough theoretic ideas of knowledge is however hindered by the ontological commitments of the semiosis and is an interesting problem—Will the triadic approach yield good thresholds of rationality in approximations?

Omitting even the ten sign classification, knowledge can be associated with ontological types along the following simple lines—these can in turn be applied to knowledge and its parts.

- Simplified Firstness [f]: these directly refer possible qualities or primitives that interact or become combined in multiple ways to form real objects or entities.
- Simplified Secondness [s]: these directly refer real objects/entities.
- Simplified Thirdness [t]: these directly refer general principles, rules, laws, methodologies and categories.

The resulting triads may not be easily applicable for handling questions relating to rational approximations and concepts, but permits contextual reasoning to some degree of rigor. An example ontological assignment may be like the following [41]:

```
Entities-[f]
.[SingleEntities-[f]
...[Objects-[f]
...States-[s]
...Events-[t]]
.PartOfEntities-[s]
...[Members-[f]
...Parts-[s]
...FunctionalComponents-[t]]
.ComplexEntities-[t]
...[CollectiveStuff-[f]
...MixedStuff-[s]
...CompoundEntities-[t]]
```

In the ontology, it is claimed that the process of investigation of knowledge by way of thirdness is fractal in nature. In the present author's perspective it is obvious that the claim is provably false at sufficient depth and is not a mathematical one. How deep are fractals formed on three symbols?

The components of maximal antichains and antichains in general rough sets can also be perceived in the triadic systems of reasoning and in the above mentioned ontological scheme. For closely related considerations on ontology and parthood the reader is referred to [25].

9 Concluding Remarks

In this research, general approaches to semantics of rough sets using antichains of rough objects and maximal antichains have been developed further. Specifically the operations used in [26] have been improved. The semantics is shown to be valid for a very large class of general rough set theories. This has been possible mainly because the objects of study have been taken to be antichains of rough objects as opposed to plain rough objects.

The problem of finding deductive systems in the context of antichain based semantics for general rough sets has also been explored in considerable detail and key results have been proved by the present author. The lateral approach used by her is justified by the wide variety of possible concepts of rough consequence in the general setting.

In forthcoming papers including [28], the framework of granular operator spaces has been expanded with definable parthood relations and semantics has been considered through counting strategies. All this will be explored in greater detail in future work.

The concept of knowledge afforded by the antichain based approach is explored in much detail and contrasted with concepts of knowledge studied in earlier papers by the present author. New concepts of fullness of knowledge have also been isolated by her and it is shown (in a sense) that the knowledge afforded by antichains is four dimensional. Last but not least methods of integrating Peirce's triadic approach to semiotics is shown to be possible.

This research also motivates the following:

- Further study of specific rough sets from the perspective of antichains.
- Research into connections with the rough membership function based semantics of [6] and extensions by the present author in a forthcoming paper. This is justified by advances in concepts of so-called cut-sets in antichains.
- Research into computational aspects as the theory is well developed for antichains. The abstract example illustrates parts of this aspect in particular.
- Study of consequence and special ideals afforded by the semantics and
- Research into ontologies indicated by the triadic approach.

Acknowledgements The present author would like to thank the referees for useful comments that helped in improving the readability of this research chapter.

References

1. Atkin, A.: Peirce's theory of signs. In: Zalta, E.N. (ed.) The Stanford Encyclopedia of Philosophy, summer 2013 edn. (2013)
2. Bosnjak, I., Madarasz, R.: On some classes of good quotient relations. Novisad J. Math. **32**(2), 131–140 (2002)

3. Cattaneo, G., Ciucci, D.: Algebras for rough sets and fuzzy logics. In: Transactions on Rough Sets 2 (LNCS 3100), pp. 208–252 (2004)
4. Chajda, I.: Ternary Deductive Systems. In: Pinus, A.G., Ponomaryov, K.N. (eds.) Algebra i Teoria Modelej 3, pp. 14–18. Novosibirsk STU (1998)
5. Chajda, I.: Generalized deductive systems in subregular varieties. Math. Bohemica **128**(3), 319–324 (2003)
6. Chakraborty, M.K.: Membership function based rough set. Inf. Sci. **55**, 402–411 (2014)
7. Ciucci, D.: Approximation algebra and framework. Fundam. Inf. **94**, 147–161 (2009)
8. Concilio, A.D., Guadagni, C., Peters, J., Ramanna, S.: Descriptive proximities I: properties and interplay between classical proximities and overlap, pp. 1–12 (2016). arXiv:1609.06246v1
9. Davey, B.A., Priestley, H.A.: Introduction to Lattices and Order, 2nd edn. Cambridge University Press (2002)
10. Font, J.M., Jansana, R.: A general algebraic semantics for sentential logics. In: Association of Symbolic Logic, vol. 7 (2009)
11. Gratzer, G.: General Lattice Theory. Birkhauser (1998)
12. Iwinski, T.B.: Rough orders and rough concepts. Bull. Pol. Acad. Sci (Math.) **3–4**, 187–192 (1988)
13. Jarvinen, J.: Lattice theory for rough sets. In: Peters, J.F. et al. (eds.) Transactions on Rough Sets VI, vol. LNCS 4374, pp. 400–498. Springer (2007)
14. Jarvinen, J., Pagliani, P., Radeleczki, S.: Information completeness in Nelson algebras of rough sets induced by quasiorders. Studia Logica pp. 1–20 (2012)
15. Koh, K.: On the lattice of maximum-sized antichains of a finite poset. Algebra Univers. **17**, 73–86 (1983)
16. Koslicki, K.: Towards a Neo-Aristotelian Mereology. dialectica **61**(1), 127–159 (2007)
17. Kung, J.P.S., Rota, G.C., Yan, C.H.: Combinatorics-The Rota Way. Cambridge University Press (2009)
18. Mani, A.: Esoteric rough set theory-algebraic semantics of a generalized VPRS and VPRFS. In: Skowron, A., Peters, J.F. (eds.) Transactions on Rough Sets, LNCS 5084, vol. VIII, pp. 182–231. Springer (2008)
19. Mani, A.: Algebraic semantics of similarity-based bitten rough set theory. Fundam. Inf. **97** (1–2), 177–197 (2009)
20. Mani, A.: Choice inclusive general rough semantics. Inf. Sci. **181**(6), 1097–1115 (2011). doi:10.1016/j.ins.2010.11.016
21. Mani, A.: Dialectics of counting and the mathematics of vagueness. In: Peters, J.F., Skowron A. (eds.) Transactions on Rough Sets, LNCS 7255, vol. XV, pp. 122–180. Springer (2012)
22. Mani, A.: Towards logics of some rough perspectives of knowledge. In: Suraj, Z., Skowron A. (eds.) Intelligent Systems Reference Library dedicated to the memory of Prof. Pawlak ISRL 43, pp. 419–444. Springer (2013)
23. Mani, A.: Approximation dialectics of proto-transitive rough sets. In: Chakraborty, M.K., Skowron, A., Kar, S. (eds.) Facets of Uncertainties and Applications, Springer Proceedings in Mathematics and Statistics, vol. 125, pp. 99–109. Springer (2013–15)
24. Mani, A.: Algebraic semantics of proto-transitive rough sets, 1st edn. arXiv:1410.0572 (2014)
25. Mani, A.: Ontology, rough Y-systems and dependence. Int. J. Comput. Sci. Appl. **11**(2), 114–136 (2014). Special Issue of IJCSA on Computational Intelligence
26. Mani, A.: Antichain based semantics for rough sets. In: Ciucci, D., Wang, G., Mitra, S., Wu, W. (eds.) RSKT 2015, pp. 319–330. Springer (2015)
27. Mani, A.: Algebraic semantics of proto-transitive rough sets. In: Peters, J.F., Skowron, A. (eds.) Transactions on Rough Sets LNCS 10020, vol. XX, pp. 51–108. Springer (2016)
28. Mani, A.: Pure rough mereology and counting. In: WIECON, 2016, pp. 1–8. IEEXPlore (2016)
29. Markowsky, G.: Representations of posets and lattices by sets. Algebra Univers. **11**, 173–192 (1980)
30. Orlowska, E., Pawlak, Z.: Logical foundations of knowledge representation—reports of the computing centre. Technical Report, Polish Academy of Sciences (1984)

31. Pagliani, P., Chakraborty, M.: A Geometry of Approximation: Rough Set Theory: Logic. Algebra and Topology of Conceptual Patterns. Springer, Berlin (2008)
32. Pawlak, Z.: Rough Sets: Theoretical Aspects of Reasoning About Data. Kluwer Academic Publishers, Dodrecht (1991)
33. Peirce, C.S.: The Essential Peirce, vol. 2. Indiana University Press (1998)
34. Peters, J., Skowron, A., Stepaniuk, J.: Nearness of visual objects—application of rough sets in proximity spaces. Fundam. Inf. **128**, 159–176 (2013)
35. Polkowski, L.: Approximate Reasoning by Parts. Springer (2011)
36. Sellars, W.: Empiricism and the philosophy of mind. In: Feigl, H., Scriven, M. (eds.) Minnesota Studies in the Philosophy of Science, vol. 1, pp. 253–329. University of Minnesota Press, (1956)
37. Short, T.L.: Peirce's Theory of Signs. Cambridge University Press (2007)
38. Sider, T.: Four-Dimensionalism: An Ontology of Persistence and Time. Clarendon Press (2001)
39. Siggers, M.: On the representation of finite distributive lattices, pp. 1–17 (2014) arXiv:1412.0011
40. Taborsky, E.: The methodolgy of semiotic morphology: an introduction. Seed **2**, 5–26 (2005)
41. Team, C.: KKO upper structure . Technical Report (2016). URL http://github.com/Cognonto
42. Thomson, J.J.: Parthood and identity across time. J. Philos. **80**, 201–220 (1983)
43. Thomson, J.J.: The statue and the clay. Noûs **32**, 149–173 (1998)
44. Yao, Y.Y., Yao, B.X.: Covering based rough set approximations. Inf. Sci. **200**, 91–107 (2012)
45. Zhang, X., Dai, J., Yu, Y.: On the union and intersection operations of rough sets based on various approximation spaces. Inf. Sci. **292**, 214–229 (2015)

Measuring Soft Roughness of Soft Rough Sets Induced by Covering

Amr Zakaria

Abstract In this chapter, important properties of soft rough sets induced by soft covering have been studied and different examples are mentioned. A measure of soft roughness has been introduced via soft covering approximation. Further, integral properties of the measure have been discussed and an example to show the prominence of the measure has been presented. A New approach of soft rough approximation space has been presented via a measure of soft roughness. Moreover, the concepts of soft lower and soft upper approximations via soft roughness have been mentioned. Finally, essential properties of this new approach have been elaborated.

1 Introduction

Rough set theory, introduced by Pawlak [16], is an expansion of set theory for the study of knowledge classification [7, 19] and rule learning [20, 21]. Moreover, the theory has been successfully applied in machine learning [1], data mining [2], decision making support and analysis [12, 17, 18], and expert system [23]. Other important research about rough set theory and its generalizations can be found in [4, 8–11, 25–27, 30].

In 1999, Molodtsov [13] suggested the unprecedented concept of *soft set theory*, which furnishes a fully new approach for modeling vagueness and uncertainty. Soft set theory has a prosperous potential for applications in various directions, few of which were shown by Molodtsov in [13]. After Molodtsov's work, some different applications of soft sets were studied in Chen [3]. Further theoretical sides of soft sets were explored by Maji et al. [14]. Also the same authors [15] presented the definition of a fuzzy soft set. Some results are acquired about the relevance between rough sets and soft sets in [5, 22, 28, 29].

A. Zakaria (✉)
Institute of Mathematics, University of Debrecen, H-4002 Debrecen Pf. 400, Hungary
e-mail: amr.zakaria@edu.asu.edu.eg

Permanent address:
A. Zakaria
Department of Mathematics, Faculty of Education, Ain Shams University, Cairo 11341, Egypt

G. Wang et al. (eds.), *Thriving Rough Sets*, Studies in Computational Intelligence 708, DOI 10.1007/978-3-319-54966-8_13

In this chapter, integral properties of soft rough sets induced by *soft covering* have been investigated. This chapter is organized as follows: Sect. 2 has a collection of all basic definitions and notions for further study. The aim of Sect. 3 is to introduce essential properties of soft rough set induced by soft covering. Moreover, some illustration examples have been mentioned. In Sect. 4, the definition of a measure of soft roughness via soft covering has been introduced. Many properties of this new definition have been discussed and an example in order to show the importance of the measure has been mentioned. In Sect. 5, a new approach of soft rough sets via a measure of soft roughness has been discussed. Furthermore, the soft lower and soft upper approximations of this new approach have been defined. Finally, the properties of this new approach of soft rough sets are presented.

2 Preliminaries

In this section, some fundamental concepts and properties of soft sets and covering-based soft rough sets have been mentioned.

Definition 1 ([13]) Let U be a nonempty set, E be a set of parameters, and $P(U)$ denotes the power set of U. A pair (F, E) is said to be a *soft set* over U, where F is a mapping given by $F : E \to P(U)$. The family of all these soft sets on (U, E) denoted by $P(U)^E$.

Definition 2 ([14]) Let $F, G \in P(U)^E$. Then F is *soft subset* of G, denoted by $F \tilde{\subseteq} G$, if $F(e) \subseteq G(e)$ for all $e \in E$.

Definition 3 ([14]) Two soft subsets F and G over a nonempty set U are said to be *soft equal* if F is a soft subset of G and G is a soft subset of F.

Definition 4 ([14]) A soft set (F, E) over U is said to be a *null soft set*, denoted by $\tilde{\emptyset}$, if $F(e) = \emptyset$ for all $e \in E$.

Definition 5 ([14]) A soft set (F, E) over U is said to be an *absolute soft set*, denoted by $\tilde{U}$, if $F(e) = U$ for all $e \in E$.

Definition 6 ([5]) The intersection of two soft sets (F, A) and (G, B) over a nonempty set U is a soft set (H, D), denoted by $(F, A) \tilde{\cap} (G, B) = (H, D)$, where $D = A \cap B$ and $H(e) = F(e) \cap G(e)$ for all $e \in D$.

Definition 7 ([14]) The union of two soft sets (F, A) and (G, B) over a nonempty set U is a soft set (H, D), denoted by $(F, A) \tilde{\cup} (G, B) = (H, D)$, where $D = A \cup B$ and for all $e \in D$,

$$H(e) = \begin{cases} F(e) & \text{if } e \in A - B \\ G(e) & \text{if } e \in B - A \\ F(e) \cup G(e) & \text{if } e \in A \cap B \end{cases}$$

Definition 8 ([5]) Let (F, E) be a soft set over U, $A \subseteq E$ and $\mathscr{C} = \{F(e) : e \in A\}$. $\mathscr{C}$ is said to be a *soft covering* of U if $F(e) \neq \emptyset$ for all $e \in A$ and $\cup_{e \in A} F(e) = U$. The ordered pair $\langle U, \mathscr{C} \rangle$ is called *soft covering approximation space.*

Definition 9 ([6]) Let $\langle U, \mathscr{C} \rangle$ be a soft covering approximation space and $u \in U$. The *soft minimal description of* u, denoted by $Md(u)$, is defined as follows:

$$Md(u) := \{K \in \mathscr{C} : u \in K \wedge (\forall C \in \mathscr{C} \wedge u \in C \subseteq K) \Rightarrow K = C\}.$$

Definition 10 ([24]) Let $\langle U, \mathscr{C} \rangle$ be a soft covering approximation space. For $X \subseteq U$, *soft covering lower approximation* and *soft covering upper approximation* are, respectively, defined as

$$\underline{\mathscr{C}}(X) := \cup_{e \in E} \{F(e) : F(e) \subseteq X\}, \tag{1}$$

$$\overline{\mathscr{C}}(X) := \cup \{Md(x) : x \in X\}. \tag{2}$$

3 Extended Properties of Soft Rough Sets Induced by Covering

Definition 11 ([6]) Let $\langle U, \mathscr{C} \rangle$ be a soft covering approximation space. $N(u) := \cap \{K \in \mathscr{C} : u \in K\}$ is called the *neighborhood of an element* $u \in U$.

Definition 12 ([6]) Let $\langle U, \mathscr{C} \rangle$ be a soft covering approximation space. For any $X \subseteq U$, the *soft lower approximation* and *soft upper approximation* of X are defined, respectively, as:

$$X_* := \{u \in U : N(u) \subseteq X\}, \tag{3}$$

$$X^* := \{u \in U : N(u) \cap X \neq \emptyset\}. \tag{4}$$

The pair (X_*, X^*) is referred to as the *soft rough set* of X.

Definition 13 Let $\langle U, \mathscr{C} \rangle$ be a soft covering approximation space. For $X \subseteq U$, the *soft boundary*, *soft positive*, *soft negative regions* and *accuracy measure* of X are defined, respectively, as:

$$BND_{\mathscr{C}}(X) := X^* - X_*, \tag{5}$$

$$POS_{\mathscr{C}}(X) := X_*, \tag{6}$$

$$NEG_{\mathscr{C}}(X) := U - X^*, \tag{7}$$

Table 1 Tabular representation of the soft set G_1

	u_1	u_2	u_3	u_4	u_5
e_1	1	1	0	0	0
e_2	0	1	1	0	0
e_3	0	0	1	1	0
e_4	0	0	0	1	1

$$\nu_{\mathscr{C}}(X) := \frac{|X_*|}{|X^*|}. \tag{8}$$

It may be noted that $0 \leq \nu_{\mathscr{C}}(X) \leq 1$.

Example 1 Let $U = \{u_1, u_2, u_3, u_4, u_5\}$, $E = \{e_1, e_2, e_3, e_4\}$. $G_1 = (F, E)$ be a soft set over U given by Table 1. It is clear that $\mathscr{C} = \{F(e_1), F(e_2), F(e_3), F(e_4)\}$ is a covering of U. Hence $\langle U, \mathscr{C} \rangle$ is soft covering approximation space. One easily sees that $Md(u_1) = \{F(e_1)\}$, $Md(u_2) = \{F(e_1), F(e_2)\}$, $Md(u_3) = \{F(e_2), F(e_3)\}$, $Md(u_4) = \{F(e_3), F(e_4)\}$ and $Md(u_5) = \{F(e_4)\}$. In addition, $N(u_1) = \{u_1, u_2\}$, $N(u_2) = \{u_2\}$, $N(u_3) = \{u_3\}$, $N(u_4) = \{u_4\}$ and $N(u_5) = \{u_4, u_5\}$. For $X = \{u_2\}$, the covering-based approximations of X, according to (1) and (2), are $\underline{\mathscr{C}}(X) = \emptyset$ and $\overline{\mathscr{C}}(X) = \{u_1, u_2, u_3\}$. On the other hand, the covering-based approximations of X, according to (3) and (4), are $X_* = \{u_2\}$ and $X^* = \{u_1, u_2\}$. Therefore, the two approaches are different.

Lemma 1 *Let $\langle U, \mathscr{C} \rangle$ be a soft covering approximation space and $u, v \in U$ such that $v \in N(u)$. Then $N(v) \subseteq N(u)$.*

Proof Let $w \in N(v) = \cap\{K \in \mathscr{C} : v \in K\}$. That is, w belongs to all elements of $\mathscr{C}$ containing v, but v belongs to all element of $\mathscr{C}$ containing u. Hence w belongs to all elements of $\mathscr{C}$ containing u. Consequently, $w \in N(u)$. Thus $N(v) \subseteq N(u)$. □

Proposition 1 *Let $\langle U, \mathscr{C} \rangle$ be a soft covering approximation space and $u \in U$. Then $(N(u))_* = N(u)$.*

Proof Let $v \in N(u)$. Hence Lemma 1 implies that $N(v) \subseteq N(u)$. That is to say $v \in (N(u))_*$. Hence $N(u) \subseteq (N(u))_*$. On the other hand, let $v \in (N(u))_*$. It follows that $N(v) \subseteq N(u)$. Since $v \in N(v)$, hence $v \in N(v) \subseteq N(u)$. Then $(N(u))_* \subseteq N(u)$. Consequently, $(N(u))_* = N(u)$. □

Theorem 1 *Let $\langle U, \mathscr{C} \rangle$ be a soft covering approximation space. Then the soft lower approximation, defined in (3), satisfies the following properties:*

(L_1) $X_* \subseteq X$,
(L_2) $(X^c)_* = (X^*)^c$,
(L_3) $\emptyset_* = \emptyset$,
(L_4) $U_* = U$,

(L_5) $(X \cap Y)_* = X_* \cap Y_*$,
(L_6) $X \subseteq Y \Rightarrow X_* \subseteq Y_*$,
(L_7) $X_* \cup Y_* \subseteq (X \cup Y)_*$,
(L_8) $(X_*)_* = X_*$.

Proof (L_1) Let $u \in X_*$. It follows that $N(u) \subseteq X$. Since $u \in N(u)$, this implies that $u \in X$. Thus $X_* \subseteq X$.
(L_2) Let $u \in (X^c)_*$, then

$$\begin{aligned} u \in (X^c)_* &\Leftrightarrow N(u) \subseteq X^c \\ &\Leftrightarrow N(u) \cap X = \emptyset \\ &\Leftrightarrow u \notin X^* \\ &\Leftrightarrow u \in (X^*)^c. \end{aligned}$$

(L_3) Part (L_1) in this theorem implies that $\emptyset_* \subseteq \emptyset$. Since $\emptyset \subseteq \emptyset_*$. Hence $\emptyset_* = \emptyset$.
(L_4) It is clear that $U_* = \{u \in U : N(u) \subseteq U\} = U$.
(L_5) For any $u \in (X \cap Y)_*$, the following implications are hold:

$$\begin{aligned} u \in (X \cap Y)_* &\Leftrightarrow N(u) \subseteq X \cap Y \\ &\Leftrightarrow N(u) \subseteq X \, and \, N(u) \subseteq Y \\ &\Leftrightarrow u \in X_* \, and \, u \in Y_* \\ &\Leftrightarrow u \in X_* \cap Y_*. \end{aligned}$$

(L_6) Let $u \in X_*$. Hence $N(u) \subseteq X$. Since $X \subseteq Y$. It follows that $N(u) \subseteq Y$. Therefore, $u \in Y_*$. Hence $X_* \subseteq Y_*$.
(L_7) The result follows directly from (L_6).
(L_8) Since $X_* \subseteq X$, then (L_6) implies that $(X_*)_* \subseteq X_*$. On the other hand, let $u \in X_*$. Hence $N(u) \subseteq X$. (L_6) implies that $(N(u))_* \subseteq X_*$. This result, combined with Proposition 1, implies that $N(u) \subseteq X_*$. Thus $u \in (X_*)_*$, and hence $X_* \subseteq (X_*)_*$. As a consequence, $(X_*)_* = X_*$. □

Theorem 2 *Let $\langle U, \mathscr{C} \rangle$ be a soft covering approximation space. Then the soft upper approximation, defined in (4), satisfies the following properties:*

(U_1) $X \subseteq X^*$,
(U_2) $(X^c)^* = (X_*)^c$,
(U_3) $\emptyset^* = \emptyset$,
(U_4) $U^* = U$,
(U_5) $(X \cup Y)^* = X^* \cup Y^*$,
(U_6) $X \subseteq Y \Rightarrow X^* \subseteq Y^*$,
(U_7) $(X \cap Y)^* \subseteq X^* \cap Y^*$,
(U_8) $(X^*)^* = X^*$.

Proof The proof is similar to that of Theorem 1. □

Table 2 Tabular representation of the soft set G_2

	u_1	u_2	u_3	u_4
e_1	1	0	0	1
e_2	0	1	0	1
e_3	1	0	1	0

Corollary 1 *Let $\langle U, \mathscr{C} \rangle$ be a soft covering approximation space. Then the soft lower approximation, defined in (3), satisfies Kuratowski's axioms and induces a topology on U called $\tau_{\mathscr{C}}$ given by*

$$\tau_{\mathscr{C}} := \{X \subseteq U : X_* = X\}. \tag{9}$$

Proof The proof is a direct consequence of Theorem 1. □

The following example shows that the equality does not hold in (L_7) of Theorem 1 and (U_7) of Theorem 2, in general.

Example 2 Let $U = \{u_1, u_2, u_3, u_4\}$, $E = \{e_1, e_2, e_3, e_4, e_5\}$ and $A = \{e_1, e_2, e_3\} \subseteq E$. $G_2 = (F, A)$ be a soft set over U given by Table 2. It is clear that $\mathscr{C} = \{F(e_1), F(e_2), F(e_3)\}$ is soft covering of U. Hence $\langle U, \mathscr{C} \rangle$ is soft covering approximation space. Obviously, $N(u_1) = \{u_1\}$, $N(u_2) = \{u_2, u_4\}$, $N(u_3) = \{u_1, u_3\}$ and $N(u_4) = \{u_4\}$. For $X = \{u_1, u_2\}$ and $Y = \{u_2, u_3, u_4\}$, then $X_* = \{u_1\}$, $Y_* = \{u_2, u_4\}$ and $(X \cup Y)_* = U$. Hence $(X \cup Y)_* \neq X_* \cup Y_*$. Also, $X^* = \{u_1, u_2, u_3\}$, $Y^* = \{u_2, u_3, u_4\}$ and $(X \cap Y)^* = \{u_2\}$. Thus $(X \cap Y)^* \neq X^* \cap Y^*$.

4 A Measure of Soft Roughness via Soft Covering

4.1 Soft Rough Membership Function

Definition 14 Let $\langle U, \mathscr{C} \rangle$ be a soft covering approximation space and $X \subseteq U$. The *soft rough membership function* with respect to $\mathscr{C}$

$$\mu_X^{\mathscr{C}} : U \to [0, 1]$$

is defined by

$$\mu_X^{\mathscr{C}}(u) := \frac{| N(u) \cap X |}{| N(u) |} \qquad u \in U.$$

It should be noted that $0 \leq \mu_X^{\mathscr{C}}(u) \leq 1$.

Theorem 3 *Let $\langle U, \mathscr{C}\rangle$ be a soft covering approximation space. Then the following assertions are hold:*

(i) $\mu_X^{\mathscr{C}}(u) = 1$ *if and only if* $u \in X_*$,
(ii) $\mu_X^{\mathscr{C}}(u) = 0$ *if and only if* $u \in NEG_{\mathscr{C}}(X)$,
(iii) $0 < \mu_X^{\mathscr{C}}(u) < 1$ *if and only if* $u \in BND_{\mathscr{C}}(X)$,
(iv) $\mu_{\emptyset}^{\mathscr{C}}(u) = 0$ *and* $\mu_U^{\mathscr{C}}(u) = 1$ *for all* $u \in U$,
(v) $\mu_{X^c}^{\mathscr{C}}(u) = 1 - \mu_X^{\mathscr{C}}(u)$ *for all* $u \in U$,
(vi) *If* $X \subseteq Y$, *then* $\mu_X^{\mathscr{C}}(u) \leq \mu_Y^{\mathscr{C}}(u)$ *for all* $u \in U$,
(vii) $\mu_{X\cup Y}^{\mathscr{C}}(u) \geq \max\{\mu_X^{\mathscr{C}}(u), \mu_Y^{\mathscr{C}}(u)\}$ *for all* $u \in U$, *the equality holds if* $X \subseteq Y$ *or* $Y \subseteq X$,
(viii) $\mu_{X\cap Y}^{\mathscr{C}}(u) \leq \min\{\mu_X^{\mathscr{C}}(u), \mu_Y^{\mathscr{C}}(u)\}$ *for all* $u \in U$, *the equality holds if* $X \subseteq Y$ *or* $Y \subseteq X$.

Proof (i) By definitions, the following implications are directly hold

$$\mu_X^{\mathscr{C}}(u) = 1 \Leftrightarrow N(u) \subseteq X \Leftrightarrow u \in X_*.$$

(ii) By definitions, the following implications are follow:

$$\begin{aligned}\mu_X^{\mathscr{C}}(u) = 0 &\Leftrightarrow N(u) \cap X = \emptyset \\ &\Leftrightarrow u \notin X^* \\ &\Leftrightarrow u \in U - X^* \\ &\Leftrightarrow u \in NEG_{\mathscr{C}}(X).\end{aligned}$$

(iii) The result is a direct consequence of (i) and (ii).
(iv) It is clear that $\mu_{\emptyset}^{\mathscr{C}}(u) = \frac{|N(u)\cap\emptyset|}{|N(u)|} = \frac{|\emptyset|}{|N(u)|} = 0$. Also, $\mu_U^{\mathscr{C}}(u) = \frac{|N(u)\cap U|}{|N(u)|} = \frac{|N(u)|}{|N(u)|} = 1$.
(v) $\mu_X^{\mathscr{C}}(u) + \mu_{X^c}^{\mathscr{C}}(u) = \frac{|N(u)\cap X| + |N(u)\cap X^c|}{|N(u)|} = \frac{|N(u)|}{|N(u)|} = 1$. Thus $\mu_{X^c}^{\mathscr{C}}(u) = 1 - \mu_X^{\mathscr{C}}(u)$.
(vi) Let $u \in U$ and $X \subseteq Y$, then $| N(u) \cap X | \leq | N(u) \cap Y |$. Hence $\frac{|N(u)\cap X|}{|N(u)|} \leq \frac{|N(u)\cap Y|}{|N(u)|}$. That is to say $\mu_X^{\mathscr{C}}(u) \leq \mu_Y^{\mathscr{C}}(u)$.
(vii) For all $u \in U$,

$$\begin{aligned}\mu_{X\cup Y}^{\mathscr{C}}(u) &= \frac{| N(u) \cap (X \cup Y) |}{| N(u) |} \\ &= \frac{| (N(u) \cap X) \cup (N(u) \cap Y) |}{| N(u) |} \\ &\geq \frac{\max\{| (N(u) \cap X) |, | (N(u) \cap Y) |\}}{| N(u) |} \\ &= \max\{\frac{| (N(u) \cap X) |}{| N(u) |}, \frac{| (N(u) \cap Y) |}{| N(u) |}\} \\ &= \max\{\mu_X^{\mathscr{C}}(u), \mu_Y^{\mathscr{C}}(u)\}.\end{aligned}$$

Obviously, if $X \subseteq Y$ or $Y \subseteq X$. It follows that $\max\{\mu_X^{\mathscr{C}}(u), \mu_Y^{\mathscr{C}}(u)\} = \mu_Y^{\mathscr{C}}(u)$ or $\mu_X^{\mathscr{C}}(u)$, respectively. Thus $\mu_{X\cup Y}^{\mathscr{C}}(u) = \max\{\mu_X^{\mathscr{C}}(u), \mu_Y^{\mathscr{C}}(u)\}$.

(viii) For all $u \in U$,

$$\begin{aligned}
\mu_{X\cap Y}^{\mathscr{C}}(u) &= \frac{\mid N(u) \cap (X \cap Y) \mid}{\mid N(u) \mid} \\
&= \frac{\mid (N(u) \cap X) \cap (N(u) \cap Y) \mid}{\mid N(u) \mid} \\
&\leq \frac{\min\{\mid (N(u) \cap X) \mid, \mid (N(u) \cap Y) \mid\}}{\mid N(u) \mid} \\
&= \min\{\frac{\mid (N(u) \cap X) \mid}{\mid N(u) \mid}, \frac{\mid (N(u) \cap Y) \mid}{\mid N(u) \mid}\} \\
&= \min\{\mu_X^{\mathscr{C}}(u), \mu_Y^{\mathscr{C}}(u)\}.
\end{aligned}$$

Obviously, if $X \subseteq Y$ or $Y \subseteq X$. It follows that $\min\{\mu_X^{\mathscr{C}}(u), \mu_Y^{\mathscr{C}}(u)\} = \mu_X^{\mathscr{C}}(u)$ or $\mu_Y^{\mathscr{C}}(u)$, respectively. Thus $\mu_{X\cap Y}^{\mathscr{C}}(u) = \min\{\mu_X^{\mathscr{C}}(u), \mu_Y^{\mathscr{C}}(u)\}$. □

The following example shows that the equality does not hold in (vii) and (viii) of Theorem 3, in general.

Example 3 Consider Example 1. Let $X = \{u_1\}$ and $Y = \{u_2\}$. Hence $\mu_X^{\mathscr{C}}(u_1) = \frac{1}{2}$, $\mu_Y^{\mathscr{C}}(u_1) = \frac{1}{2}$, $\mu_{X\cup Y}^{\mathscr{C}}(u_1) = 1$ and $\mu_{X\cap Y}^{\mathscr{C}}(u_1) = 0$. Therefore, $\max\{\mu_X^{\mathscr{C}}(u_1), \mu_Y^{\mathscr{C}}(u_1)\} = \frac{1}{2}$ and $\min\{\mu_X^{\mathscr{C}}(u_1), \mu_Y^{\mathscr{C}}(u_1)\} = \frac{1}{2}$. Consequently, $\mu_{X\cup Y}^{\mathscr{C}}(u_1) \neq \max\{\mu_X^{\mathscr{C}}(u_1), \mu_Y^{\mathscr{C}}(u_1)\}$ and $\mu_{X\cap Y}^{\mathscr{C}}(u_1) \neq \min\{\mu_X^{\mathscr{C}}(u_1), \mu_Y^{\mathscr{C}}(u_1)\}$.

4.2 Example

Suppose U is the set of washing machines, E is the set of parameters, each parameter being a word or a sentence. A soft set $G_3 = (F, E)$ describes the attractiveness of the washing machines which Mr. X is going to buy. Suppose that there are six washing machines in the universe, given by

$$U = \{h_1, h_2, h_3, h_4, h_5, h_6\}$$

and $E := \{\textit{washing quality}, \textit{power use}, \textit{spin capacity}, \textit{noise}, \textit{design and ease of use}\}$ be the set of parameters. Then $G_3 = (F, E)$ be a soft set over U given by Table 3.

Hence

$$\mathscr{C} = \{F(\textit{washing quality}), F(\textit{power use}), F(\textit{spin capacity}), F(\textit{noise}), F(\textit{design and ease of use})\}$$

is soft covering of U and $\langle U, \mathscr{C}\rangle$ is soft covering approximation space. Therefore,

Table 3 Tabular representation of the soft set G_3

	h_1	h_2	h_3	h_4	h_5	h_6
Washing quality	1	0	1	0	0	0
Power use	1	1	1	0	0	0
Spin capacity	1	0	0	0	1	1
Noise	0	0	1	0	1	0
Design and ease of use	1	1	0	1	0	0

$$N(h_1) = \{h_1\} \qquad N(h_2) = \{h_1, h_2\}$$
$$N(h_3) = \{h_3\} \qquad N(h_4) = \{h_1, h_2, h_4\}$$
$$N(h_5) = \{h_5\} \qquad N(h_6) = \{h_1, h_5, h_6\}.$$

It is clear that washing quality $\mapsto \{h_1, h_3\}$, power use $\mapsto \{h_1, h_2, h_3\}$, spin capacity $\mapsto \{h_1, h_5, h_6\}$, noise $\mapsto \{h_3, h_5\}$ and design and ease of use $\mapsto \{h_1, h_2, h_4\}$. For any $h_i \in U, i = 1, \ldots, 6$, the degree of soft rough membership in X can be directly obtained as follows:

$$\mu^{\mathscr{C}}_{\{h_1,h_3\}}(h_1) = 1 \qquad \mu^{\mathscr{C}}_{\{h_1,h_3\}}(h_2) = \frac{1}{2}$$
$$\mu^{\mathscr{C}}_{\{h_1,h_3\}}(h_3) = 1 \qquad \mu^{\mathscr{C}}_{\{h_1,h_3\}}(h_4) = \frac{1}{3}$$
$$\mu^{\mathscr{C}}_{\{h_1,h_3\}}(h_5) = 0 \qquad \mu^{\mathscr{C}}_{\{h_1,h_3\}}(h_6) = \frac{1}{3}.$$

Similarly, all degrees of soft rough membership are represented in Table 4. It is clear that the degree of the fifth machine belonging to $\{washing\ quality\}$ is 0, which means that the washing quality of this machine is bad. The membership degree of the first or the third in $\{washing\ quality\}$ is 1. The second, the fourth and the sixth belong to $\{washing\ quality\}$ because the degrees of soft rough membership are $\frac{1}{2}$, $\frac{1}{3}$ and $\frac{1}{3}$, respectively. So, we can decide to which category a machine belongs according to the degrees of soft rough membership.

Table 4 Degrees of soft rough membership in different decisions

	Washing quality	Power use	Spin capacity	Noise	Design and ease of use
h_1	1	1	1	0	1
h_2	$\frac{1}{2}$	1	$\frac{1}{2}$	0	1
h_3	1	1	0	1	0
h_4	$\frac{1}{3}$	$\frac{2}{3}$	$\frac{1}{3}$	0	1
h_5	0	0	1	1	0
h_6	$\frac{1}{3}$	$\frac{1}{3}$	1	$\frac{1}{3}$	$\frac{1}{3}$

5 New Approach of Soft Rough Sets

In this section, a new approach of soft rough set via soft rough membership function has been introduced. Moreover, some properties of this new approach have been discussed.

Definition 15 Let $\langle U, \mathscr{C} \rangle$ be a soft covering approximation space and $X \subseteq U$. Then the *soft lower approximation* and *soft upper approximation* of X are defined, respectively, as:

$$\underline{X} := \{u \in U : \mu_X^{\mathscr{C}}(u) = 1\}, \tag{10}$$

$$\overline{X} := \{u \in U : \mu_X^{\mathscr{C}}(u) > 0\}. \tag{11}$$

The pair $(\underline{X}, \overline{X})$ is referred to as the *soft rough set* of X.

Theorem 4 *Let $\langle U, \mathscr{C} \rangle$ be a soft covering approximation space. Then the soft lower approximation, defined in (10), satisfies the following properties:*

(i) $\underline{X} = (\overline{X^c})^c$,
(ii) $\underline{X} \subseteq X$,
(iii) $\underline{\emptyset} = \emptyset$,
(iv) $\underline{U} = U$,
(v) $X \subseteq Y \Rightarrow \underline{X} \subseteq \underline{Y}$,
(vi) $\underline{X \cap Y} = \underline{X} \cap \underline{Y}$,
(vii) $\underline{X} \cup \underline{Y} \subseteq \underline{X \cup Y}$,
(viii) $\underline{\underline{X}} = \underline{X}$.

Proof (i)

$$\begin{aligned}
(\overline{X^c})^c &= \{u \in U : \mu_{X^c}^{\mathscr{C}}(u) > 0\}^c \\
&= \{u \in U : 1 - \mu_X^{\mathscr{C}}(u) > 0\}^c \\
&= \{u \in U : \mu_X^{\mathscr{C}}(u) < 1\}^c \\
&= \{u \in U : \mu_X^{\mathscr{C}}(u) \geq 1\} \\
&= \{u \in U : \mu_X^{\mathscr{C}}(u) = 1\} \\
&= \underline{X}.
\end{aligned}$$

(ii) Let $u \in \underline{X}$. Then $\mu_X^{\mathscr{C}}(u) = 1$. Hence (i) of Theorem 3 implies that $u \in X_*$. It follows that $N(u) \subseteq X$. Since $u \in N(u)$, hence $u \in N(u) \subseteq X$. Thus $\underline{X} \subseteq X$.
(iii) Part (iv) of Theorem 3 implies that $\mu_{\emptyset}^{\mathscr{C}}(u) = 0$ for all $u \in U$. Hence $\underline{\emptyset} = \emptyset$.
(iv) Part (iv) of Theorem 3 implies that $\mu_U^{\mathscr{C}}(u) = 1$ for all $u \in U$. Thus $\underline{U} = U$.

(v) Let $u \in \underline{X}$. Hence $\mu_X^{\mathscr{C}}(u) = 1$. Since $X \subseteq Y$, then (vi) of Theorem 3 implies $\mu_X^{\mathscr{C}}(u) \leq \mu_Y^{\mathscr{C}}(u)$. That is to say $\mu_Y^{\mathscr{C}}(u) \geq 1$, but any soft rough membership lies between 0 and 1. Consequently, $\mu_Y^{\mathscr{C}}(u) = 1$. Hence $u \in \underline{Y}$. Therefore, $\underline{X} \subseteq \underline{Y}$.

(vi) Let $u \in \underline{X} \cap \underline{Y}$. Hence $\mu_X^{\mathscr{C}}(u) = 1$ and $\mu_Y^{\mathscr{C}}(u) = 1$. Thus (i) of Theorem 3 implies $u \in X_*$ and $u \in Y_*$. This result combined with (L_5) of Theorem 1, implies $u \in (X \cap Y)_*$. Again, (i) of Theorem 3 implies $\mu_{X \cap Y}^{\mathscr{C}}(u) = 1$. Therefore, $u \in \underline{X \cap Y}$. Thus $\underline{X} \cap \underline{Y} \subseteq \underline{X \cap Y}$. The other inclusion is a direct consequence of (v) of this theorem. Hence $\underline{X \cap Y} = \underline{X} \cap \underline{Y}$.

(vii) The result is a direct consequence of (v) of this theorem.

(viii) It is sufficient to prove that $\underline{X} \subseteq \underline{\underline{X}}$ and the other inclusion follows directly by (ii) and (v) of this theorem. Let $u \in \underline{X}$. Hence $\mu_X^{\mathscr{C}}(u) = 1$. This result, combined with Theorem 3 part (i) and Theorem 1 (L_8), implies $u \in (X_*)_*$. Again, Theorem 3 part (i) implies $\mu_{\underline{X}}^{\mathscr{C}}(u) = 1$. Hence $u \in \underline{\underline{X}}$. Then the result. □

Theorem 5 *Let $\langle U, \mathscr{C} \rangle$ be a soft covering approximation space. Then the soft upper approximation, defined in (11), satisfies the following properties:*

(i) $\overline{X} = (\underline{X^c})^c$,
(ii) $X \subseteq \overline{X}$,
(iii) $\overline{\emptyset} = \emptyset$,
(iv) $\overline{U} = U$,
(v) $\overline{X \cup Y} = \overline{X} \cup \overline{Y}$,
(vi) $X \subseteq Y \Rightarrow \overline{X} \subseteq \overline{Y}$,
(vii) $\overline{X \cap Y} \subseteq \overline{X} \cap \overline{Y}$,
(viii) $\overline{\overline{X}} = \overline{X}$.

Proof The proof is similar to that of Theorem 4. □

The following example shows that the equality does not hold in (vii) of Theorems 4 and 5, in general.

Example 4 Consider Example 2. For $X = \{u_1, u_2\}$ and $Y = \{u_3\}$, then $\underline{X} = \{u_1\}$, $\underline{Y} = \emptyset$ and $\underline{X \cup Y} = \{u_1, u_3\}$. Hence $\underline{X \cup Y} \neq \underline{X} \cup \underline{Y}$. Also, $\overline{X} = \{u_1, u_2, u_3\}$, $\overline{Y} = \{u_3\}$ and $\overline{X \cap Y} = \emptyset$. Thus $\overline{X \cap Y} \neq \overline{X} \cap \overline{Y}$.

Corollary 2 *Let $\langle U, \mathscr{C} \rangle$ be a soft covering approximation space. Then the soft lower approximation, defined in (10), satisfies Kuratowski's axioms and induces a topology on U called $\tau_{\mathscr{C}}$ given by*

$$\tau_{\mathscr{C}} := \{X \subseteq U : \underline{X} = X\}. \tag{12}$$

Proof The proof is a direct consequence of Theorem 4. □

6 Conclusion

In this chapter, extended properties of soft rough sets induced via soft covering have been introduced. A measure of soft roughness via soft covering has been presented. Moreover, essential examples to show the significance of soft rough sets and a measure of soft roughness induced via soft covering have been studied. Moreover, properties of a measure of soft roughness have been discussed. A new approach of soft rough sets via soft rough membership has been introduced. Finally, many properties of this new approach are mentioned and an illustration examples have been presented.

Acknowledgements The author is grateful to the anonymous referee for a careful checking of the details and for helpful comments that improved this work.

References

1. Chmielewshi, M.R., Grzymala-Busse, J.W.: Global discretization of continuous attributes as preprocessing for machine learning. Int. J. Approx. Reason. **15**, 319–331 (1966)
2. Chan, C.C.: A rough set approach to attribute generalization in data mining. J. Inf. Sci. **107**, 169–176 (1998)
3. Chen, D.: The parametrization reduction of soft sets and its applications. Comput. Math. Appl. **49**, 757–763 (2005)
4. El-Sheikh, S.A., Zakaria, A.: Note on rough multiset and its multiset topology. Ann. Fuzzy Math. Inf. **10**, 235–238 (2015)
5. Feng, F., Li, C., Davvaz, B., Ali, M.I.: Soft sets combined with fuzzy sets and rough sets: a tentative approach. Soft Comput. **14**, 899–911 (2010)
6. Feng, F., Liu, X., Fotea, V.L., Jun, Y.B.: Soft sets and soft rough sets. Inf. Sci. **181**, 1125–1137 (2011)
7. Korvin, A.d., McKeegan, C., Kleyle, R.: Knowledge acquisition using rough sets when membership values are fuzzy sets. Intell. Fuzzy Syst. **6**, 237–244 (1998)
8. Kandil, A., Yakout, M., Zakaria, A.: On bipreordered approximation spaces. J. Life Sci. **8**, 505–509 (2011)
9. Kandil, A., Yakout, M., Zakaria, A.: Generalized rough sets via ideals. Ann. Fuzzy Math. Inf. **5**, 525–532 (2013)
10. Kandil, A., Yakout, M.M., Zakaria, A.: New approaches of rough sets via ideals. In: Handbook of Research on Generalized and Hybrid Set Structures and Applications for Soft Computing. IGI Global, pp. 247–264 (2016). doi:10.4018/978-1-4666-9798-0. (Ch. 012, Web. 23 Mar. 2016)
11. Kong, Q., Wei, Z.: Covering-based fuzzy rough sets. Intell. Fuzzy Syst. **29**, 2405–2411 (2015)
12. Mcsherry, D.: Knowledge discovery by inspection. Decis. Support Syst. **21**, 43–47 (1997)
13. Molodtsov, D.A.: Soft set theory first results. Comput. Math. Appl. **37**, 19–31 (1999)
14. Maji, P.K., Biswas, R., Roy, A.R.: Soft set theory. Comput. Math. Appl. **45**, 555–562 (2003)
15. Maji, P.K., Biswas, R., Roy, A.R.: Fuzzy soft set theory. J. Fuzzy Math. **3**, 589–602 (2001)
16. Pawlak, Z.: Rough sets. Int. J. Comput. Inf. Sci. **11**, 341–356 (1982)
17. Pawlak, Z.: Rough set approach to knowledge-based decision support. Eur. J. Oper. Res. **99**, 48–57 (1997)
18. Pomerol, J.C.: Artificial intelligence and human decision making. Eur. J. Oper. Res. **99**, 3–25 (1997)
19. Skowron, A., Stepaniuk, J., Swiniarski, R.: Modeling rough granular computing based on approximation spaces. Inf. Sci. **184**, 20–43 (2012)

20. Tsumoto, S.: Accuracy and coverage in rough set rule induction. Rough Sets and Current Trends in Computing, pp. 373–380 (2002)
21. Wang, X., Tsang, E.C., Zhao, S., Chen, D., Yeung, D.S.: Learning fuzzy rules from fuzzy samples based on rough set technique. Inf. Sci. **177**, 4493–4514 (2007)
22. Xiao, Z., Zou, Y.: A comparative study of soft sets with fuzzy sets and rough sets. Intell. Fuzzy Syst. **27**, 425–434 (2014)
23. Yahia, M.E., Mahmod, R., Sulaiman, N., Ahamad, F.: Rough neural expert systems. Expert Syst. Appl. **18**, 87–99 (2000)
24. Yüksel, F., Ergül, Z.G., Tozlu, N.: Soft covering based rough sets and their application. Sci. World J. **2014**, Article ID 970893, 9 (2014)
25. Zhu, W., Wang, F.: Reduction and axiomization of covering generalized rough sets. Inf. Sci. **152**, 217–230 (2003)
26. Zhu, W., Wang, F.: On three types of covering rough sets. IEEE Trans. Knowl. Data Eng. **19**, 1131–1144 (2007)
27. Zhu, W.: Relationship between generalized rough sets based on binary relation and covering. Inf. Sci. **179**, 210–225 (2009)
28. Zhan, J., Davvaz, B.: A kind of new rough set: rough soft sets and rough soft rings. Intell. Fuzzy Syst. **30**, 475–483 (2015)
29. Zhang, G., Li, Z., Qin, B.: A method for multi-attribute decision making applying soft rough sets. Intell. Fuzzy Syst. **30**, 1803–1815 (2016)
30. Zakaria, A., John, S.J., El-Sheikh, S.A.: Generalized rough multiset via multiset ideals. Intell. Fuzzy Syst. **30**, 1791–1802 (2016)

Rough Search of Vague Knowledge

Edward Bryniarski and Anna Bryniarska

Abstract This chapter presents the theoretical basis of the vague knowledge search algorithmization of a rough method. It introduces some data granulation method which aggregates this data as rough sets of data or ways to search this data in the semantic networks. As a result of this method is the possibility of the rough sets description, analogically to sets in the classical theory of sets. We try to answer the question how the agent searching some knowledge can conceive the search of vague knowledge in the semantic networks: (1) if it can, accordingly to the semantic and the conceiving rules, describe the relationships between nodes in this semantic network which are identified as ways of searching knowledge, (2) if it can approximate sets of this ways by using ways described by some concepts and roles.

1 Introduction

This chapter is a continuation of the paper [3]. In [3] it is shown how rough pragmatic description logic *RPDL* defines accurately how to interpret formulas describing vague knowledge: an incomplete information, unclear, unprecise, ambiguously expressed knowledge [2, 20, 21], within the structure of rough sets, which is determined by a semantic network. This network is determined in some pragmatic information system, in which agents make use of only some distinguished data types. And this system is determined by a pragmatic system of knowledge representation. It is in this system that, for the distinguished data types, the vague knowledge is determined. Therefore, it is right to say that formulas of the *RPDL* refer to vague knowledge. Since, in practice, man most often communicates

---

E. Bryniarski (✉)
Institute of Mathematics and Informatics, Opole University, Opole, Poland
e-mail: edlog@math.uni.opole.pl

A. Bryniarska
Institute of Control and Computer Engineering, Opole University of Technology, Opole, Poland
e-mail: a.bryniarska@po.opole.pl

G. Wang et al. (eds.), *Thriving Rough Sets*, Studies in Computational Intelligence 708, DOI 10.1007/978-3-319-54966-8_14

with other people, he passes to them data representing vague knowledge, thus, he conceives his utterances in approximation, approximating their sense. This means that people communicating with one another apply the *RPDL*.

Accordingly to the quoted paper [3], the procedure of the agent searching vague knowledge can take in the following steps:

1. Establishing a pragmatic system of knowledge representation.
2. Determining the set of data types that will be available to the agent.
3. Checking whether an information system in Pawlak's sense can be determined for the available set of types of knowledge.

Due to the dynamic of language communication, the relations of using a data copy can be vague and data classification, which leads to recognition of data types, is not precise. Then, realization of this procedure is not possible and should be preceded by using methods of fuzzy sets theory [2, 20, 22–24]: fuzzification and defuzzification which leads to sharpening knowledge. It is necessary to define pragmatic system of the knowledge representation. There may also be used the evolutionary methods or other of machine-learning theory methods. However, the effective method in information retrieval is the method of data granulation [16–19]. Data granulation is an important paradigm of modeling and processing with uncertainty. Then the information granules are the main mathematical constructions in the context of granular calculations. The information granules relate to the description of objects and are mathematical models which describe aggregated data about these objects. By using Fuzzy Description Logic in the Information Retrieval [6], data from these aggregates are functionally and structurally connected with each other. For this purpose, we use similarities of language criteria of the aggregated data. The data granulation method for describing the best possible data in tables from the information system [7], lets us check, if for available attributes, can be applied the generalized information systems in Pawlak's sense.

4. After establishing the general information system, we create the semantic network which corresponds to this system.
5. Formulate the language of description logic and provide the syntax of TBox terms, ABox assertions and RBox axioms. For the defined semantic network we determine the general information system. For the determined system we define the set-theoretical structure of rough sets. We establish procedures of interpretation of the language of descriptive logic in the structure of rough sets. We distinguish primitive axioms. We distinguish a set of rules conceiving the axioms.
6. We determine the base of the knowledge *Ab, Tb, Rb*, where the sets *Ab*, *Tb*, *Rb* are finite sets of expressions (descriptions of nodes of a semantic network), which are possible to be computer-processed, respectively: (1) assertions, about which the agent **knows**, (2) concepts, which the agent **has knowledge of** and also (3) axioms, which the agent **conceives by means of the rules of conceiving**. Since we say about a human being who is an agent availing himself of the base of knowledge that he knows something, has knowledge about

something and conceives something, hence by replacing this human being by an AI agent, we can say the same about this very agent. We will call such an AI agent—a **pragmatic agent of an AI** and we will say that he is one who **knows something about** holding of assertions, **knowing** some concepts and **conceiving** axioms.

However, in that quoted paper [3] is no answer to the question if this agent conceives vague knowledge search in the semantic networks. In other words,

7. if it can, accordingly to the semantic and the conceiving rules, describe the relationships between nodes in this semantic network which are identified as ways of searching knowledge,
8. if it can approximate sets of this ways by using ways described by concepts and roles from knowledge base ⟨Ab, Tb, Rb⟩.

Answers for these questions are the main motivation of this chapter. In Sects. 2 and 3 of this work are repeated, presented in the paper [3], the theory of approximation in the information systems and description of the pragmatic rough description logic *PRDL*. Moreover, next three sections present the conceptual apparatus about the new algorithmization issues, which are pointed out in the steps 7, 8 of the described above algorithm of searching vague knowledge.

2 Approximation in Information Systems

The main goal of the information systems is to compare data from different states of searching knowledge (gathering information from these states accordingly to this system attributes). We can ask question: which states are equally described by specific types of expressions (attributes), or in other words are they *undistinguishable* in this description.

Definition 1 (*general information systems* [4, 11]) Let Σ is an ordered system

$$\Sigma = \langle U, A, A_1, A_2, \ldots, A_n, \{V_a\}_{a \in A} \rangle$$

where U is a finite set and A is set of all *attributes*, A_k ($k = 1, 2, \ldots, n$) is the set of k-ary attributes understood as k-ary functions, V_a is the set of all codes of the attribute $a \in A$.

If for any $a \in A$ and $a \in A_k$, $a\colon U^k \to V_a$, then Σ is called the *general information system.*

In an analogous way, as for Pawlak's information system, we can define indiscernibility of information states in the general information system $\Sigma = \langle U, A, A_1, A_2, \ldots, A_n, \{V_a\}_{a \in A} \rangle$.

Definition 2 (*indiscernibility of states in an information system* [4]) Let $U_{gen} = U \cup U^2 \cup \cdots \cup U^n$. The relation $\approx \subseteq U_{gen} \times U_{gen}$ such that for any x, $y \in U_{gen}$:

$x \approx y$ iff there is an $k \in \{1, 2, \ldots, n\}$ such that $x, y \in U^k$ and *any* $a \in A_k$, $a(x) = a(y)$, is called the *indiscernibility relation* in the system Σ. The family $C = \{[s]_\approx : s \in U_{gen}\}$ of the equivalence classes of the relation $\approx$ is a partition of the set U_{gen}. Such a family of equivalence classes of the relations $\approx$ is determined uniquely in every information system. For this family of sets of information states, one can determine—in a unique manner—all sets $X = \cup B$, for $B \subseteq C$. We will say about such sets that they are *deterministic* in an information system. Is it possible to characterize, by means of exact sets, those that are not exact?

According to the rough set theory of Pawlak [11–14], in any general information system $S = \langle U, A, A_1, A_2, \ldots, A_n, \{V_a\}_{a \in A} \rangle$ (including also a non-deterministic information system) the operation of *lower approximation* $C^-: 2^{Ugen} \to 2^{Ugen}$, as well as that of *upper approximation* $C^+: 2^{Ugen} \to 2^{Ugen}$ for set $X \subseteq \cup C \subseteq U_{gen} = U \cup U^2 \cup \ldots \cup U^n$ can be determined (if S is a set, then $2^S = \{X: X \subseteq S\}$). Let us accept the notation $C = \{[s]_\approx : s \in U_{gen}\}$. For this reason we use the letter 'C' in operations C^-, C^+. A system $\langle U_{gen}, C \rangle$ is called an *approximation space* of subsets of U_{gen} [8].

Definition 3 (*Pawlak* [11–14]) For any $X \subseteq U_{gen}$,

$$C^-(X) = \{x \in U_{gen}: [x]_\approx \subseteq X\}$$
$$C^+(X) = \{x \in U_{gen}: [x]_\approx \cap X \neq \emptyset\}$$
$$Bn(X) = C^+(X) \backslash C^-(X)$$

where the relations $\approx$ is a discernibility relation in the general information system Σ. The operation Bn is called the *boundary operation*.

Fact 1. [8]

For any $X \subseteq U_{gen}$,

1. $C^-(X) = \cup \{K \in C: K \subseteq X\}$,
2. $C^+(X) = \cup \{K \in C: K \cap X \neq \emptyset\}$,
3. $C^-(X) \subseteq C^+(X)$,
4. sets $C^-(X)$ and $C^+(X)$ are exact,
5. if X is exact, then $C^-(X) = C^+(X) = X$.

Definition 4 (*indiscernibility of sets*) Any sets $X, Y \subseteq U_{gen}$ are *indiscernible,* which we write down as follows: $X \sim Y$ iff $C^-(X) = C^-(Y)$ and $C^+(X) = C^+(Y)$.

The relation $\sim$ is an equivalence relation. We denote equivalence classes $[X]_\sim$ of this relation with the representative X by X_C. We denote $\varnothing_C$ by 0_C and $(U_{gen})_C$ by 1_C.

Definition 5 We call equivalence classes of the relations $\sim$ —*rough sets* in the information system Σ.

Definition 6 (*an element of a rough set* [8, 9]) For any $X, Y \subseteq U_{gen}$, $X \in_C Y_C$ iff $X \neq \varnothing$ and there is such $x \in U_{gen}$, such that $X \subseteq [x]_\approx$, $C^-(X) \subseteq C^-(Y)$ and $[x]_\approx \subseteq C^+(Y)$.

We call the relation $\in_C$—a *rough membership relation*. We read the expression $X \in_C Y_C$: X is a rough element of the rough set Y_C.

Using the relation $\in_C$, one can define inclusion of rough sets.

Definition 7 For any $X, Y \subseteq U_{gen}$,

$$X_C \subseteq_C Y_C \text{ iff for every } Z \subseteq U_{gen}, \text{ if } Z \in_C X_C, \text{ then } Z \in_C Y_C.$$

Theorem 1

1. $X_C \subseteq_C Y_C$ *iff* $C^-(X) \subseteq C^-(Y)$ *and* $C^+(X) \subseteq C^+(Y)$,
2. $X_C = Y_C$ *iff* $C^-(X) = C^-(Y)$ *and* $C^+(X) = C^+(Y)$,
3. $X_C = Y_C$ *iff for every* $Z \subseteq U_{gen}$, $Z \in_C X_C$ *iff* $Z \in_C Y_C$,
4. *It is not the case, that there is* $Z \subseteq U_{gen}$, $Z \in_C 0_C$,
5. *For every* $X \subseteq U_{gen}$, $X_C \subseteq_C 1_C$.

Bryniarski, in his works [8, 9], *defines the operations of addition* $\cup_C$, *multiplication* $\cap_C$, *substraction* $\setminus_C$ *and complement* $'^C$ *of rough sets in the family of rough sets.*

Definition 8 For any rough sets X_C, Y_C, for any $Z \subseteq U_{gen}$,

$$Z \in_C X_C \cup_C Y_C \text{ iff } Z \in_C X_C \text{ or } Z \in_C Y$$

$$Z \in_C X_C \cap_C Y_C \text{ iff } Z \in_C X_C \text{ and } Z \in_C Y$$

$$Z \in_C X_C \setminus_C Y_C \text{ iff } Z \in_C X_C \text{ and not } Z \in_C Y$$

$$Z \in_C (X_C)^{'C} \text{ iff } Z \in_C U_C \setminus_C X_C$$

Let us denote the family of all rough sets in the information system Σ by $R(\Sigma)$.

Theorem 2 [8, 9] *For any rough sets* X_C, Y_C, *there is the set* $Z \subseteq U_{gen}$ *such that* $X_C \cup_C Y_C = Z_C$ *or* $X_C \cap_C Y_C = Z_C$ *or* $X_C \setminus_C Y_C = Z_C$ *or* $(X_C)^{'C} = Z_C$.

Hence

Theorem 3 *The structure* $R(\sum) = \langle R(S), \cup_C, \cap_C, \backslash_C, {}^{'C} 0_C, 1_C\rangle$ *restricted to exact rough sets is homomorphic to a set-theoretical field of sets.*

Let us extend the structure $\mathbf{R}(\Sigma)$ *by relations of rough membership and inclusion of rough sets. Let us introduce, for any family A family* $A \subseteq R(\Sigma)$ *the generalized sum* $\cup_C A$:

$$X \in_C \cup_C A \text{ iff there is } Y_C \in A, \text{ such that } X \in_C Y_C.$$

Now, in a way analogous with the set-theoretical construction of approximation of sets (homomorphic with these constructions), one can provide the construction of approximation of rough sets in the approximation space $\langle (U_{gen})_C, \{K_C: K \in C\}\rangle$ of *rough sets.*

It is noted that $(U_{gen})_C = \cup_C\{K_C : K \in C\}$.

Definition 9 (*approximation of rough sets*) For any $X_C \in R(\Sigma)$,

$$F^-(X_C) = \cup_C \{K_C: K \in C, K_C \subseteq_C X_C\},$$

$$F^+(X_C) = \cup_C \{K_C: K \in C, K_C \cap X_C \neq 0_C\},$$

$$Fbn(X_C) = F^+(X_C)\backslash_C F^-(X_C).$$

We call these operations, respectively: *lower approximation, upper approximation* and *boundary approximation* of the rough set X_C.

Theorem 4 *For any* $X_C \in R(\sum)$,

1. $F^-(X_C) = (C^-(X))_C$,
2. $F^+(X_C) = (C^+(X))_C$,
3. $Fbn(X_C) = (Bn(X))_C$,
4. $F^-(X_C) = \{C^-(X)\}$,
5. $F^+(X_C) = \{C^+(X)\}$,
6. $Fbn(X_C) = \{Bn(X)\}$.

Let h: $2^{U_{gen}} \to \mathbf{R}(\mathbf{\Sigma})$ *be a function such that for any* $X \subseteq U_{gen}$, $h(X) = X_C$. *The function h is a homomorphism from the structure* $\langle 2^{U_{gen}}, C^-, C^+, Bn\rangle$ *to the structure* $\langle R(\sum), F^-, F^+, Fbn\rangle$, i.e.: for any $X, Y \subseteq U_{gen}$,

- if $X \subseteq Y$, then $h(X) \subseteq_C h(Y)$,
- $h({}^{C^-}X)) = F^-(h(X))$,

- $h(C^{+}(X)) = F^{+}(h(X))$,

- $h(BnX)) = Fbn(h(X))$,

The above theorem allows providing the properties of the upper and lower approximation operations, as well as boundary operation of rough sets as homomorphic to the standard properties.

Theorem 5 (cf. [15]) *For any X, Y* $\subseteq U_{gen}$,

1. $F^{-}(X_C) \subseteq_C F^{+}(X_C)$,
2. $F^{-}(X_C) \subseteq_C X_C \subseteq_C F^{+}(X_C)$,
3. $F^{-}(F^{-}(X_C)) = F^{-}(X_C)$,
4. $F^{+}(F^{-}(X_C)) = F^{-}(X_C)$,
5. $Fbn(F^{-}(X_C)) = 0_C$,
6. $F^{-}(F^{+}(X_C)) = F^{+}(X_C)$,
7. $F^{+}(F^{+}(X_C)) = F^{+}(X_C)$,
8. $Fbn(F^{+}(X_C)) = 0_C$,
9. $F^{-}(Fbn(X_C)) = Fbn(X_C)$
10. $F^{+}(Fbn(X_C)) = Fbn(X_C)$,
11. $Fbn(Fbn(X_C)) = 0_C$,
12. $F^{-}(X_C) \cup_C F^{-}(Y_C) \subseteq_C F^{-}(X_C \cup_C Y_C)$,
13. $F^{-}(X_C \cap_C Y_C) = F^{-}(X_C) \cap_C F^{-}(Y_C)$,
14. $F^{+}(X_C \cup_C Y_C) = F^{+}(X_C) \cup_C F^{+}(Y_C)$,
15. $F^{+}(X_C \cap_C Y_C) \subseteq_C F^{+}(X_C) \cap_C F^{+}(Y_C)$.
16. *If* X_C *is an exact rough set, then* $F^{-}(X_C) = F^{+}(X_C) = X_C$ *and* $Fbn(X_C) = 0_C$
17. $X_C \subseteq_C Y_C$ *iff* $F^{-}(X_C) \subseteq_C F^{-}(Y_C)$ *and* $F^{+}(X_C) \subseteq_C F^{+}(Y_C)$,
18. $X \subseteq_C Y_C$ iff $X_C \subseteq_C Y_C$, $F^{-}(X_C) = 0_C$ *and there is an exact rough set* K_C *such that* $F^{+}(X_C) = K_C$.

We will call the structure

$$F(\Sigma) = \langle R(\Sigma), F^{-}, F^{+}, Fbn, \cup_C, \cap_C, \setminus_C, {'}^{C}, 0_C, 1_C, \in_C, \subseteq_C \rangle$$

a set-theoretical structure of rough sets *determined by the information system* Σ.

3 Rough Description Logic

Accepting that we have an information system determined by a pragmatic system of knowledge representation [3], we can in this information system distinguish a set *IN* of *nodes*, i.e. a set of individual names of the described objects or pronouns which point individually to the objects being described. Also consider the set *AS* of the pairs: the value ds_k of n-argument attribute and the node or the *edge*, i.e. n-tuple $\langle s_1, s_2, \ldots, s_k \rangle$ of individual names of the described objects, for which this attribute has the given value ds_n. We obtain, then, another system of knowledge representation, other than the information system.

Definition 10 (*semantic network*) We call a *semantic network* the following ordered system:

$$SN = \langle IN, AS, \{DS_i\}_{i \in N,\, i < n+1} \rangle,$$

where the sets: *IN*—a set of nodes, DS_n, $n \in N$—sets of the descriptions, called sets of *descriptions of n-argument relations* and the set *AS* satisfy the following conditions:

$$AS \subseteq (DS_1 \times IN) \cup (DS_2 \times IN^2) \cup \cdots \cup (DS_n \times IN^n),$$

where $ds_i \in DS_i$ is a description of the relation R such that

$$\{ds_i\} \times R = \left(\{ds_i\} \times IN^i\right) \cap AS.$$

We call elements of the set *AS*—*assertions*. We call any relation R, when it is a one argument, satisfying the above equation—a *concept* (a notion), and when R is, at the least, a two-argument one—a *role*.

For example, the role of “filiation” that links the person bearing the name “John” with the one named “Charles”, the latter being the father to the former, leads to the assertion: filiation, John, Charles, which we also write down: *filiation(John, Charles)* or *(John, Charles): filiation*. We write the assertion expressed in the sentence “Eve is sitting between John and Charles” in the following way:

sitting_between(Eve, John, Charles) or *(Eve, John, Charles): sitting_between*.

Let us notice that when in the following three *(Eve, John, Charles)*, cyclically, we reverse the names, we will obtain the following three *(John, Charles, Eve)*, which is also an occurrence of a role, *e.g.* expressed in the sentence “John and Charles are sitting beside Eve”. We can write down this assertion as follows: *(John, Charles, Eve): are sitting_beside*. We will say about the role *sitting_beside* that it is *cyclically reverse* towards that of *sitting_between*. When the three *(Eve, John, Charles)*, which is an occurrence of the assertion *sitting_between*, is reduced by the first name, then the pair *(John, Charles)*, is also an occurrence of the assertion, *e.g.* one expressed in the sentence “someone is sitting between John and Charles”: *(John, Charles): someone_sitting_between*. We will say about this role that it is a reduction of that *sitting_between*.

For any description *ds* of the relation R (a concept or a role) there is the characteristic function$a\colon IN^i \to V_a \subseteq \{0, 1\}$, such that when $ds = ds_i$ and $\langle ds_i, x_1, x_2, \ldots, x_i \rangle \in AS$, then $a(x_1, x_2, \ldots, x_i) = 1$, and for $\langle ds_i, x_1, x_2, \ldots, x_i \rangle \notin AS$, $a(x_1, x_2, \ldots, x_i) = 0$. Let us denote the set of all such characteristic functions by A. Let $V_a = a(IN^i)$. Then the structure:

$$\Sigma(SN) = \langle U, A, \{V_a\}_{a \in A} \rangle,$$

where $U = IN$ $U = IN$, will be an information system.

Definition 11 We call the information system $\Sigma(SN)$—an *information system determined by the semantic network SN.*

In the context of research of *Semantic Web* (for example in the works [1, 2, 11]), knowledge representation in the semantic network is determined by two systems of representation: the terminology called ***TBox***, as well as a set of representation of assertion called ***ABox***. A semantic network can be extended with nodes which render the available knowledge about concepts or roles, and also extended with boundaries determining dependences between concepts or roles. Descriptions of these dependences are called *axioms*, and the system of representing this knowledge is called ***RBox***. In the presented research trend, the base of the knowledge represented in the semantic network, in the descriptive language *AL* (*attributive language*) of the logic *DL* (*Description Logic*), is defined as the following triple *Ab, Tb, Rb*, where the sets *Ab, Tb, Rb* are finite sets of expressions (descriptions of nodes of a semantic network) that can be computer-processed, respectively: assertions, concepts, axioms. Describing rough knowledge in information systems determined by the semantic network *SN*, we will apply the suitably reformulated and extended language *AL*. We will call this language—*rough pragmatic language* (***RPL***; a set of its expressions will be denoted by *RPL*). It will be the language of the proposed logic RPDL.

3.1 Syntax of the Language RPL

Let the data be some non-empty sets: individual variables, individual names, names of concepts, names of roles and symbols of modifying agents of concepts.

3.1.1 Syntax of Occurrences of Concepts and Roles

Occurrences of a concept are the symbols x, y, z, v …, x_1, y_1, …, variables and the symbols a, b, c, …, a_1, b_1 …, determining individual names. The *variables run over individual names*. Intuitively, they describe the nodes of the semantic web.

Occurrences of a role are tuples $(t_1, t2, \ldots, t_k)$ of occurrences of concepts.

Transposition of the occurrence $(t_1, t_2, \ldots, t_k)$ is tuples $(t_1, t_2, \ldots, t_k)^T = (t_k, t_{k-1}, \ldots, t_1)$.

3.1.2 Syntax of *TBox*

The following names belong to the set of names of concepts and roles:

9. T (Top)—universal concept and universal role,
10. ⊥ (Bottom)—empty concept and empty role.

Top includes all occurrences of concepts and roles, and *Bottom* includes knowledge about a lack of any occurrences of concepts and roles.

11. $\{t\}$—*singleton of the occurrences of t*, a concept determined unequivocally by an occurrence of concept t,
12. $\{(t_1, t_2, \ldots, t_k)\}$—a role being a *singleton of n-tuple of occurrences*.

Let A, B be the names of concepts, R be the name of a role, a m—the symbol of a *modifier*, then the following are concepts:

$\neg A$	*negation of a concept*—the expression denoting all the occurrences of concepts that are not occurrences of the concept A;
$A \wedge B$	*intersection (conjunction) of the concepts* A and B—the expression denoting all the occurrences of the concepts A and B;
$A \vee B$	*union (alternative) of the concepts* A and B—the expression denoting all the occurrences of the concept A or the concept B;
$A \setminus B$	*difference of the concepts* A and B—the expression denoting all the occurrences of the concept A, which are not an occurrence of the concept B;
$\exists A.R$	*existential quantification*—the concept, whose occurrences are those of the concept A remaining in the role R in the first place, at least once with appearances of some concepts related to the role R in the successive places of the role;
$\forall A.R$	*general quantification*—the concept, whose occurrences are those of the concept A, remaining in the role R in the first place, together with all the occurrences of some concepts related to the role R in the successive places of this role;
$m(A)$	*modification m of the concept A*—denoting a concept that is the concept C altered by the word m, *e.g.* m can have such occurrences as: very much, more, the most, or high, higher, the highest; in the approximated calculus, modifications are lower approximation or upper approximation, or the boundary—$m \in \{Upper, Lower, Boundary\}$,
$m(R)$	*modification m of the role R*—$m \in \{Upper, Lower, Boundary\}$,
R^{-1}	*the role cyclically reverse to the role R*;
$A.R$	*restriction of the occurrences of the role R by the occurrences of the concept A*—such a role that the occurrences of A remain in the role R in the first place,

	together with those of some concepts in the other places of the role R;
R^T	*transposition of the role R*—the occurrences of the role R^T are the transpositions of the occurrences R;
R^-	*reduction of the role R* by the first argument—is the role for at least a three-argument role R, and the concept for a two-argument role;
$A_1 \times A_2 \times \cdots \times A_n$	*the Cartesian product* of the concepts $A_1, A_2, \ldots, A_n$.

3.1.3 Syntax of *ABox*

For any variables x, y, individual names a, b, the names of the concept C and those of the role R, which is a two-argument one, assertions are denoted by means of expressions in the form "x: C", "a: C", "(x, y): R", "(a, y): R", "(x,b): R", "(a, b): R". Generally, for the n-argument role R, expressions of assertion take on the form $(t_1, t_2, \ldots, t_n) : R$, where t_i are any occurrences of the concepts. Inscriptions of the form $t_1 : A$, $(t_1, t_2, \ldots, t_k) : R$ are read as follows: *t_1 an occurrence of the concept A, n-th tuple $(t_1, t_2, \ldots, t_k)$ is an occurrence of the role R.*

3.1.4 Syntax of *RBox*

For any names of the concepts A, B, the names of the roles R_1, R_2, as well as expressions of the assertion α, β, *axioms* are expressions rendered in the following form:

$A \subseteq B$	inclusion of the concepts A, B,
$A = B$	identity of the concepts A, B,
$R_1 \subseteq R_2$	inclusion of the roles R_1, R_2,
$R_1 = R_2$	identity of the roles R_1, R_2,
$\alpha : -\beta$	Horn's clause for the assertion α, β; we read this in the following way: *if there holds an occurrence of the assertion β, then there holds an occurrence of the assertion α.*

3.2 *The Distinguished Axioms for RPDL*

Let us distinguish some selected axioms for *RPDL*, divided into three groups. For any names of concepts or roles A, B, C and R and any occurrences $t, t_1, t_2, \ldots, t_k$.

Ax.1. $T = \neg\bot$, $A \subseteq T$, $\bot \subseteq A$, $A \subseteq A$,
Ax.2. $A = A$, $R = R$,

Ax.3. $A \vee \bot = A,\ A \wedge T = A$,
Ax.4. $A \wedge \bot = \bot,\ A \vee T = T$,
Ax.5. $A \vee B = B \vee A,\ A \wedge B = B \wedge A$,
Ax.6. $(A \vee B) \wedge C = (A \wedge C) \wedge (B \wedge C), (A \wedge B) \vee C = (A \vee C) \wedge (B \vee C)$,
Ax.7. $A \vee \neg A \subseteq T,\ A \wedge \neg A = \bot$,
Ax.8. $\neg A \subseteq T \backslash A$,
Ax.9. $\forall C.R \subseteq \exists C.R$,
Ax.10. $t\!: T\!: -t\!: A$,
Ax.11. $t\!: \{t\}\!: -t\!: A$,
Ax.12. $(t_1, t_2, \ldots, t_k)\!: \{(t_1, t_2, \ldots, t_k)\}\!: -(t_1, t_2, \ldots, t_k)\!: R$,
Ax.13. $\exists\{t_1\}.R\!: -(t_1, t_2, \ldots, t_k)\!: R$,
Ax.14. $(t_1, t_2, \ldots, t_k)\!: R\!: -\forall\{t_1\}.R$,
Ax.15. $(t_2, t_3, \ldots, t_k, t_1)\!: R^{-1}\!: -(t_1, t_2, \ldots, t_k)\!: R$,
Ax.16. $(t_2, \ldots, t_k)\!: R^{-}\!: -(t_1, t_2, \ldots, t_k)\!: R$, dla $k > 2$,
Ax.17. $t_2\!: R^{-}\!: -(t_1, t_2)\!: R$,
Ax.18. $A^{-} = \bot$,
Ax.19. $Lower(A) \subseteq Upper(A)$,
Ax.20. $Boundary(A) \subseteq Lower(A)$,
Ax.21. $Lower(A) \subseteq A \subseteq Upper(A)$,
Ax.22. $Upper(Upper(A)) = Upper(A)$,
Ax.23. $Lower(Upper(A)) = Upper(A)$,
Ax.24. $Boundary(Upper(A)) = \bot$,
Ax.25. $Upper(Lower(A)) = Lower(A)$,
Ax.26. $Lower(Lower(A)) = Lower(A)$,
Ax.27. $Boundary(Lower(A)) = \bot$,
Ax.28. $Upper(Boundary(A)) = Boundary(A)$,
Ax.29. $Lower(Boundary(A)) = Boundary(A)$,
Ax.30. $Boundary(Boundary(A)) = \bot$,
Ax.31. $Lower(A) \vee Lower(B) \subseteq Lower(A \vee B)$,
Ax.32. $Lower(A \wedge B) = Lower(A) \wedge Lower(B)$,
Ax.33. $Upper(A \vee B) = Upper(A) \vee Upper(B)$,
Ax.34. $Upper(A \wedge B) \subseteq Upper(A) \wedge Upper(B)$,

The above axioms will be satisfied in a selected set-theoretical structure of rough sets

$$F(\sum(SN)) = \langle F, F^{-}, F^{+}, Fbn, \cup_C, \cap_C, \backslash_C, {}'^C, 0_C, 1_C, \in_C, \subseteq_C \rangle,$$

determined by the information system $\sum(SN) = \langle U, A, \{V_a\}_{a \in A} \rangle$. Where $U = IN$ for the semantic network $SN = \langle IN, AS, \{DS_i\}_{i \in N,\, i < n+1} \rangle$ under consideration, and $F = R(\Sigma(SN))$ is a set of all the rough sets determined in the information system $\Sigma(SN)$.

3.3 Semantics of the Language RPL

Let us determine interpretation $\boldsymbol{I} = (F, {}^I)$ of ***RPL*** language for which the interpretation function ^{I}: $RPL \rightarrow F$ (we write the values $I(E)$ as $E^{I)}$ satisfies the following conditions:

I1. any occurrences of concepts and roles are assigned elements of rough sets:

$$t^I \in_C U_C,$$

$$(t_1, t_2, \ldots, t_k)^I \in_C \{\langle x_1, x_2, \ldots, x_k\rangle\}_C,$$

$$\text{for } \langle x_1, x_2, \ldots, x_k\rangle \in U^k,\ t_1^I \in_C \{x_1\}_C,\ t_2^I \in_C \{x_2\}_C, \ldots, t_k^I \in_C \{x_k\}_C,$$

I2. names of the concepts A, including the singletons $\{t\}$, are assigned the following rough sets:

$$\{t\}^I = \{x\}_C, \text{ for some } x \in U,$$

$$\{(t_1, t_2, \ldots, t_k)\}^I = \{\langle x_1, x_2, \ldots, x_k\rangle\}_C,$$

$$\text{for } \langle x_1, x_2, \ldots, x_k\rangle \in U^k,\ t_1^I \in_C \{x_1\}_C,\ t_2^I \in_C \{x_2\}_C, \ldots, t_k^I \in_C \{x_k\}_C,\ A^I \in F,$$

I3. names of the role R are assigned the rough sets $R^I \in F$,
I4. the modifiers $m \in \{Upper, Lower, Boundary\}$ assign some functions $m^I : F \rightarrow F$, from the set $\{F^-, F^+, Fbn\}$:

$$Lower^I = F^-,$$
$$Upper^I = F^+,$$
$$Boundary^I = Fbn.$$

3.3.1 Semantics of the Concepts of *TBox* Language

For any names of the concepts A, B, the name of the role R and the modifier m

I5. $\mathrm{T}^{\mathrm{I}} = 1_C$,
I6. $\bot^{\mathrm{I}} = 0_C$,
I7. $(\neg A)^I = (A^I)'^C$,
I8. $(A \wedge B)^I = (A^I \cap_C B^I)$,
I9. $(A \vee B)^I = (A^I \cup_C B^I)$,
I10. $(A \backslash B)^I = (A^I \backslash_C B^I)$.

For $R^I \subseteq_C (U^k)_C$ where R is a k-argument role, there hold the following conditions of interpretation of concepts and roles:

I11. $(\exists A.R)^I = \cup_C\{\{t_1\}^I : ((t_1, t_2, \ldots, t_k)^I \in_C R^I \wedge t_1^I \in_C A^I)\}$,
I12. $(\forall A.R)^I = \cup_C\{\{t_1\}^I : ((t_1, t_2, \ldots, t_k)^I \in_C R^I \wedge t_1^I \in_C A^I)\}$,
I13. $(A.R)^I = \cup_C\{\{(t_1, t_2, \ldots, t_k)\}^I : ((t_1, t_2, \ldots, t_k)^I \in_C R^I \wedge t_1^I \in_C A^I)\}$,
I14. $(R^{-1})^I = \cup_C\{\{(t_1, t_2, \ldots, t_k)\}^I : (t_2, t_3, \ldots, t_k, t_1)^I \in_C R^I\}$,
I15. $(R^{-})^I = \cup_C\{\{(t_1, t_2, \ldots, t_k)\}^I :$ if for some t_{k+1}, $(t_{k+1}, t_1, t_2, \ldots, t_k)^I \in_C R^I\}$,
where the operations $\cup_C$, $\cup_C$ are generalized operations of addition and multiplication that are defined on subsets of the family F of rough sets. For an empty set the value of these generalized operations is an empty rough set 0_C.
I16. $(m(A))^I = m^I(A^I)$, for $m \in \{Upper, Lower, Boundary\}$,
I17. $(m(R))^I = m^I(R^I)$, for $m \in \{Upper, Lower, Boundary\}$,

3.3.2 Semantics of the Assertion of *ABox* Language

I18. For any occurrences t, t_1, t_2, …, t_k concepts or roles, the name of the concept C, as well as the name of the role R

$$(t: C)^I \text{iff} (x^I) \in_C C^I,$$

$$((t_1, t_2, \ldots, t_k): R)^I \text{iff} (t_1, t_2, \ldots, t_k)^I \in_C R^I.$$

3.3.3 Semantics of the *RBox* Axioms

I19. For any A, B the names of concepts or roles and any assertions α, β

$$(X \subseteq Y)^I \text{ iff } X^I \subseteq_C Y^I,$$
$$(X = Y)^I \text{ iff } X^I = Y^I,$$
$$(\alpha : -\beta)^I \text{ iff if } \beta^I \text{ then } \alpha^I.$$

We call the *rule of conceiving of axioms*—the expression in the form of α_1, α_2, …, α_k/β, for any axioms α_1, α_2,…,α_k, β. The rule is *adequate* for the given interpretation function ^{I}, if—from the fact that there hold the interpretations α_1^I, α_2^I, …, α_k^I—there follows the fact that the interpretation β^I holds.

3.3.4 Distinguished Adequate Rules of Conceiving

Rule 1. $A \subseteq B,\ B \subseteq C / A \subseteq C,$
Rule 2. $A \subseteq B,\ B \subseteq A / A = B,$
Rule 3. $A \subseteq B / (t{:}B){:}\ -(t{:}A),$
Rule 4. $R_1 \subseteq R_2 / ((t_1,\ t_2,\ \ldots, t_k){:}R_2){:}\ -((t_1,\ t_2,\ \ldots, t_k){:}R_1),$
Rule 5. $t_1{:}A_1,\ t_2{:}A_2,\ \ldots, t_k{:}A_k / (t_1,\ t_2,\ \ldots, t_k){:}A_1 \times A_2 \times \ldots \times A_n,$
Rule 6. $t_k{:}\ (\{t_{k-1}\}.(\ldots(\{t_2\}.(\{t_1\}.R)^-)^-)\ldots)^-)^- / (t_1,\ t_2,\ \ldots, t_k){:}R.$

The adequate rules of conceiving do not serve the purpose of proving theorems—they merely determine the logical relations between axioms. If agents of the pragmatic system of representation of knowledge apply these rules, it means that they *conceive*, in an appropriate manner, interpretations of axioms in the structure of rough sets.

4 Subject of Conceiving in the Logic *PRDL*

Let in logic *PRDL* the set $N = \{n_1,\ n_2,\ \ldots,\ n_k\}$ be a set of constant concepts occurrences, which correspond to the nodes in the semantic network *SN* for any the interpretation $\boldsymbol{I} = (F,\ ^I)$. We accept the following definitions, which derive important facts and theorems.

Definition 12 We call a *system of conceiving* by the semantic network, the ordered system:

$$Con(SN) = \langle N,\ C,\ R,\ N_0,\ \mathbf{C}_0,\ R_0 \rangle$$

which is relational system, in which:

- C_0 is a set of distinguished (*accepted on the basis of the* Definition 10) concepts in the *RPDL* logic, which have occurrences described by elements from the set N,
- R_0 a set of distinguished (*accepted on the basis of the* Definition 10) roles in the *RPDL* logic, which have occurrences described by elements from the set N,
- N_0 a choice set of distinguished occurrences of constants from the set N,
- $\boldsymbol{C}$ a choice set of concepts derived from the accepted concepts $\boldsymbol{C}_0$ and roles $\boldsymbol{R}_0$ and the set N_0 of nodes accordingly to the syntactic and the conceiving rules of the *RPDL* logic,
- $\boldsymbol{R}$ a choice set of roles derived from the accepted concepts $\boldsymbol{C}_0$ and roles $\boldsymbol{R}_0$ and the set N_0 of nodes accordingly to the syntactic and the conceiving rules of the *RPDL* logic.

Moreover, for any $x \in N$ is satisfied at least one of the following conditions:

1. x is an element of some set from the family $\boldsymbol{C}$,

2. x is an argument of some relations from the set $\boldsymbol{R}$,
3. x is an element of the set N_0.

Intuitively, thanks to knowledge of psychology, by *conceiving* in the semantic network

$$SN = \langle IN, AS, \{DS_i\}_{i \in N,\, i<n+1} \rangle,$$

We understand determining whether or not we can specify relationships between network nodes, accordingly to the syntactic and conceiving rules, by the concepts and roles occurrences defined in this network. In this sense, it is justified calling *conceiving ways* all ordered sets of occurrences which describe nodes of the network *SN*.

In this chapter we assume that $N_0 = N$.

Definition 13 A *conceiving way* in the *Con*(*SN*), (short: *a way*), we call any occurrences of concept or role:

$$(\alpha_1, \alpha_2, \ldots, \alpha_j), (k \geq j \geq 1)$$

such that for

$$\alpha_1, \alpha_2, \ldots, \alpha_j \in N.$$

A *part of conceiving way* $r = (\alpha_1, \alpha_2, \ldots, \alpha_j)$ we call any subsequence:

$$t = (\alpha_{i1}, \ldots, \alpha_{i2}) \text{ such that: } 1 \leq i_1 \leq i_2 \leq j \leq k.$$

The expression "the conceiving way t is part of the way r" we write as: $t \; \varepsilon \; r$.

We accept the following notation agreement: $(\alpha) = \alpha$.

Definition 14

a. *A conceiving subject* in the *Con*(*SN*), (short: *subject*), we call any subset M of the set of all conceiving ways N_{gen}.
b. *The conceiving subject M on the way t* (symbolic: $t \; \varepsilon \; M$) is defined as follows:

$$t\varepsilon M \Leftrightarrow \exists u \in M (t \varepsilon u).$$

c. *A closure* of the conceiving subject X:

$$X^{\#} = \{t: t\varepsilon X\}$$

d. *A minimalisation* of the conceiving subject X

$$X_{\#} = \cap \{Z: Z^{\#} = X^{\#}\}.$$

e. Let N^i be i-th Cartesian power of the set N, and number n the largest number of arguments, roles in the network SN. For the conceiving system $Con(SN) = \langle N, CR, N_0, C_0, R_0 \rangle$ we accept notation:

$$N_{\mathbf{gen}} = (\cup \{A: A \in C \cup R\})^{\#} \subseteq N \cup N^2 \cup \cdots \cup N^n.$$

N_{gen} is the closure of the set of all conceiving ways, in which are conceiving concepts and roles in the semantic network SN. We call it *a universe of conceiving*.

Fact 2.

1. $X \subseteq X^{\#}$,
2. $X^{\#} = X^{\#}$,
3. $\emptyset^{\#} = \emptyset$,
4. $(N_{gen})^{\#} = N_{gen}$
5. $(X^{\#})^{\#} = X$,
6. $X \subseteq Y \Rightarrow X^{\#} \subseteq Y^{\#}$,
7. $(X \cup Y)^{\#} = X^{\#} \cup Y^{\#}$,
8. $(X \cap Y)^{\#} = X^{\#} \cap Y^{\#}$,
9. $\forall r(r\varepsilon X \Leftrightarrow r\varepsilon Y) \Leftrightarrow X^{\#} = Y^{\#}$,
10. $r \in X^{\#} \Rightarrow \{r\}^{\#} \subseteq X^{\#}$,
11. $X^{\#} = Z^{\#} \wedge Y^{\#} = Z^{\#} \Rightarrow (X \cup Y)^{\#} = (X \cup Y)^{\#} = (X \cap Y)^{\#} = Z^{\#}$,
12. $(\cap \{Z: Z^{\#} = X^{\#}\})^{\#} = X^{\#}$,
13. $X^{\#} \cap Y^{\#} = \emptyset \Rightarrow (X \backslash Y)^{\#} = X^{\#} \backslash Y^{\#}$, where symbol '\' means the substractions of the sets.

Remark 1 Sometimes the dependency $(X \backslash Y)^{\#} = X^{\#} \backslash Y^{\#}$ is not satisfied; i.e. when $X = \{(c, a, b)\}$, $Y = \{(a, b, c)\}$, because $X^{\#} = \{a, b, c, (c, a), (a, b), (c, a, b)\}$, $Y^{\#} = \{a, b, c, (b, c), (a, b), (a, b, c)\}$, $X \backslash Y = X$ and $X^{\#} \backslash Y^{\#} = \{(c, a), (c, a, b)\}$, then $(X \backslash Y)^{\#} = X^{\#} \neq X^{\#} \backslash Y^{\#}$.

Statement 1

$$X^{\#} \cap Y^{\#} = \emptyset \Leftrightarrow hom(X^{\#}) \cap hom(Y^{\#}) = \emptyset, \text{ where } hom(A) = \{x \in U: (x)\varepsilon A\}.$$

Based on the closure of the conceiving subject.

Theorem 6 *The family of all closures of the conceiving subjects with operations: the sum and intersection of sets, defined in this family subjects* $\varnothing$ *and* N_{gen}, *is a complete distributive lattice with the zero* $\varnothing$ *and the unit* N_{gen}.

Analogically to the minimalisation of the conceiving subjects there are following conditions:

Fact 3.

1. $X^{\#} = Y^{\#} \Rightarrow X_{\#} \subseteq Y \subseteq X^{\#}$,
2. $X^{\#} = Y^{\#} \Rightarrow X_{\#} = Y_{\#}$,
3. $(X_{\#})_{\#} = X_{\#}$,
4. $(X_{\#})^{\#} = X^{\#}$,
5. $(X \cup Y)_{\#} = X_{\#} \cup Y_{\#}$,
6. $(X \cup Y)_{\#} = X_{\#} \cup Y_{\#}$,
7. $(X \cap Y)_{\#} = X_{\#} \cap Y_{\#}$.

Proof 6, 7: the Fact 2, 3 that:

$$(X \cup Y)_{\#} = (X^{\#} \cup Y^{\#})_{\#} = ((X_{\#})^{\#} \cup (Y_{\#})^{\#})_{\#} = ((X_{\#} \cup Y_{\#})^{\#})_{\#} = X_{\#} \cup Y_{\#},$$

$$(X \cap Y)_{\#} = (X^{\#} \cap Y^{\#})_{\#} = ((X_{\#})^{\#} \cap (Y_{\#})^{\#})_{\#} = ((X_{\#} \cap Y_{\#})^{\#})_{\#} = X_{\#} \cap Y_{\#}.$$

Therefore:

Theorem 7 *The family of all minimalisation of the conceiving subjects with operations: the sum and intersection of sets, defined in this family subjects* $\varnothing$ *and* N_{gen}, *is a complete distributive lattice with the zero* $\varnothing$ *and the unit* N_{gen}.

Example 1 Let the conceiving system of drawing a square (for the semantic network representing vague knowledge about drawing a square) is:

$$Con(SN) = \langle N, \mathbf{C}, \mathbf{R}, N_0, \mathbf{C}_0, R_0 \rangle,$$

where:

$N = \{w_1, w_2, w_3, w_4\}$—set of the constant concepts occurrences and the concepts—the ways of conceiving the vertices of the square,

$$C = \{N\}, R = \{R_1, R_2\}, N_0 = N, C_0 = \{N\}, R_0 = \{R_{01}, R_{02}, R_{03}, R_{04}\},$$
$$R_{01} = \{(w_1, w_2)\}, R_{02} = \{(w_2, w_3)\}, R_{03} = \{(w_3, w_4)\}, R_{04} = \{(w_4, w_1)\}\},$$

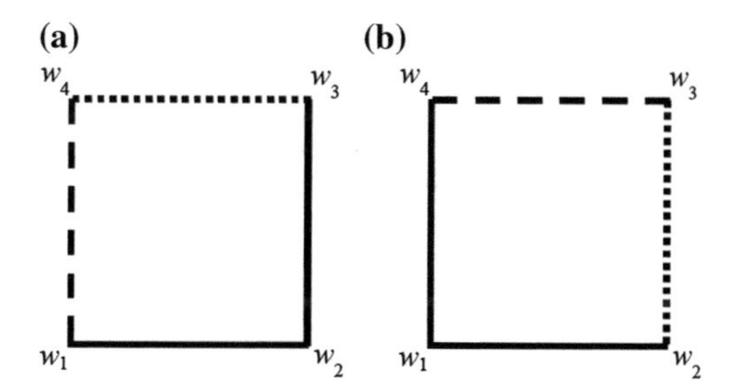

Fig. 1 Two different conceiving subjects of drawing *square*

$R_1 = \{(w_1, w_2, w_3), (w_4, w_1, w_2)\}$—the role of connecting with each other three vertices by line,

$R_2 = \{(w_2, w_3), (w_3, w_4), (w_4, w_1)\}$—the role of connecting with each other two vertices by line.

$M_1 = \{(w_1, w_2, w_3), (w_3, w_4), (w_4, w_1)\}$—the first conceiving subject of drawing the square (Fig. 1a):

(1) connect three vertices w_1, w_2, w_3,

(2) connect two vertices w_3, w_4

(3) connect two vertices w_4, w_1,

$M_2 = \{(w_4, w_1, w_2), (w_2, w_3), (w_3, w_4)\}$—the second conceiving subject of drawing the square (Fig. 1b). The closure of the conceiving subject M_1 and M_2:

$$M_1^{\#} = \{(w_1), (w_2), (w_3), (w_4), (w_1, w_2, w_3), (w_2, w_3), (w_3, w_4), (w_4, w_1)\},$$
$$M_2^{\#} = \{(w_1), (w_2), (w_3), (w_4), (w_4, w_1, w_2), (w_2, w_3), (w_3, w_4), (w_4, w_1)\}.$$

The minimalisation of the conceiving subject M_1 and M_2:

$$M_{1\#} = M_1, M_{2\#} = M_2.$$

5 Multistructures

Now we may introduce *the multistructure calculation.* We use concepts which were presented for the first time in papers [25, 26]. On this basis we formulate definition, easy to prove facts and theorems. Hence, we show that the family of multistructures is a complete distributive lattice with zero and unit.

Definition 15 The relation *Adq,* defined in the set of all conceiving subjects, is called a *relation of conceiving adequacy*, when for any $M_1, M_2 \subseteq N_{gen}$

$$M_1\,Adq\,M_2 \Leftrightarrow \forall r(r\varepsilon X \Leftrightarrow r\varepsilon Y)$$

The expression $M_1\,Adq\,M_2$ we read: the conceiving subjects M_1 and M_2 are conceived in the same ways.

Fact 4.

$$M_1\,Adq\,M_2 \Leftrightarrow M_1^{\#} = M_2^{\#}.$$

Fact 5.
A relation *Adq* is the equivalent relation in the set of all conceiving subjects.

Definition 16 [25, 26] An abstract class $[M]_{Adq}$ of the relation *Adq*, in which the subject $M \subseteq N_{gen}$ is represented, is called the *multistructure* defined by the subject M.

Fact 6.
For any two subjects M_1, M_2

1. $[M_1]_{Adq} = [M_2]_{Adq} \Leftrightarrow M_1^{\#} = M_2^{\#}$,
2. $M_1\,Adq\,M_2 \Leftrightarrow [M_1]_{Adq} = [M_2]_{Adq}$,
3. $[M_1]_{Adq} = [M_1^{\#}]_{Adq}$.

Definition 17 A conceiving way $x \in N_{gen}$ belongs to the *multistructure* $[M]_{Adq}$ iff $x\ \varepsilon\ M$. The fact that the way x belongs to the multistructure $[M]_{Adq}$ we write symbolically as:

$$x \in_m [M]_{Adq}.$$

Definition 18 Let $[M_1]_{Adq}$, $[M_2]_{Adq}$ be any multistructure.
$[M_1]_{Adq}$ *is included in* $[M_2]_{Adq}$ iff for any x, if $x \in_m [M_1]_{Adq}$, then $x \in_m [M_2]_{Adq}$.
The fact that the multistructure $[M_1]_{Adq}$ is included in the multistructure $[M_2]_{Adq}$ we write symbolically as:

$$[M_1]_{Adq} \subseteq_m [M_2]_{Adq}.$$

Fact 7.
For any two subjects M_1, M_2

$$[M_1]_{Adq} \subseteq_m [M_2]_{Adq} \Leftrightarrow M_1^{\#} \subseteq M_2^{\#}.$$

Theorem 8 For any two subjects M_1, M_2 and any way x

$$x \in_m [M_1^{\#} \cup M_2^{\#}]_{Adq} \Leftrightarrow x \in_m [M_1]_{Adq} \vee x \in_m [M_2]_{Adq}.$$

Definition 19 *A sum* $[M_1]_{Adq} \cup_m [M_2]_{Adq}$ of multistructures $[M_1]_{Adq}$ and $[M_2]_{Adq}$ is called the multistructure $[M_1^{\#} \cup M_2^{\#}]_{Adq}$.

Fact 8.

For any two subjects M_1, M_2 and any way x

$$x \in_m [M_1]_{Adq} \cup_m [M_2]_{Adq} \Leftrightarrow x \in_m \left[M_1^{\#}\right]_{Adq} \vee x \in_m \left[M_2^{\#}\right]_{Adq}.$$

Theorem 9 *For any two subjects M_1, M_2 and any way x*

$$x \in_m [M_1^{\#} \cap M_2^{\#}]_{Adq} \Leftrightarrow x \in_m \left[M_1^{\#}\right]_{Adq} \wedge x \in_m \left[M_2^{\#}\right]_{Adq}.$$

Definition 20 *A intersection* $[M_1]_{Adq} \cap_m [M_2]_{Adq}$ of multistructures $[M_1]_{Adq}$ and $[M_2]_{Adq}$ is called the multistructure $[M_1^{\#} \cap M_2^{\#}]_{Adq}$.

Fact 9.

For any two subjects M_1, M_2 and any way x

$$x \in_m [M_1]_{Adq} \cap_m [M_2]_{Adq} \Leftrightarrow x \in_m [M_1]_{Adq} \wedge x \in_m [M_2]_{Adq}.$$

Fact 10. (distributable)

For any subjects M_1, M_2, $M_3 \subseteq IN_{gen}$

$$[M_1]_{Adq} \cap_m ([M_2]_{Adq} \cup_m [M_3]_{Adq}) = ([M_1]_{Adq} \cap_m [M_2]_{Adq}) \cup_m ([M_1]_{Adq} \cap_m [M_3]_{Adq}).$$

Definition 21 Let $\boldsymbol{M} = \{[M]_{Adq} : M \subseteq N_{gen}\}$. A *generalized sum* of the multistructures family $\boldsymbol{M}$ is the multistructure:

$$\cup_m \boldsymbol{M} = [\cup \{M^{\#} : M \subseteq N_{gen} \wedge [M]_{Adq} \in \boldsymbol{M}\}]_{Adq}.$$

A *generalized intersection* of the multistructures family $\boldsymbol{M}$ is the multistructure:

$$\cap_m \boldsymbol{M} = [\cap \{M^{\#} : M \subseteq N_{gen} \wedge [M]_{Adq} \in \boldsymbol{M}\}]_{\mathrm{Adq}}.$$

Fact 11.

For any multistructures family $\boldsymbol{M}$:

1. $x \in_m \cup_m \boldsymbol{M} \Leftrightarrow$ exist such multistructure $[S]_{Adq} \in \boldsymbol{M}$, that $x \in_m [S]_{Adq}$,
2. $x \in_m \cap_m \boldsymbol{M} \Leftrightarrow$ for any multistructure $[S]_{Adq} \in \boldsymbol{M}$ is $x \in_m [S]_{Adq}$.

Definition 22

a. The multistructure $0_m = [\varnothing]_{Adq}$ we call the empty multistructure
b. The multistructure $1_m = \left[N_{gen}\right]_{Adq}$ we call the full multistructure.

Fact 12.

For any subject M

1. $[M]_{Adq} \cup_m 0_m' = [M]_{Adq}$,
2. $[M]_{Adq} \cap_m 0_m = 0_m$,
3. $M]_{Adq} \cup_m 1_m = 1_m$,
4. $[M]_{Adq} \cap_m 1_m = [M]_{Adq}$.

As a result of the given definitions, facts and theorems about multistructures we have:

Theorem 10 [25, 26] *The family of all multistructures with operations of the sum* $\cup_m$ *and the intersection* $\cap_m$, *defined on this family with the full* $\boldsymbol{1^m}$ *and empty* $\boldsymbol{0^m}$ *multistructure, is a full distributive lattice with the zero* $\boldsymbol{0^m}$ *and the unit* $\boldsymbol{1^m}$.

6 Approximation of Conceiving Subjects

Referring to work [5] which presents the theoretical apparatus about the *covering approximation space* $\langle U, C\rangle$ where U is nonempty set, C its finite covering ($U = \cup C$), we accept the following definition and theorems.

Definition 23 The *conceiving base* in the system *Con*(*SN*) is called the covering $\boldsymbol{B}$ of the set N_{gen} such that:

$$B = \{K: K = \left\{r, r^T\right\}^{\#} \wedge \exists A(A \in C \cup R \cup \{N_0\} \wedge r \in A)\},$$

where r^T is a transposition of the sequence r. The elements $K \in \boldsymbol{B}$ are called the *base conceiving subjects*.

The *approximation space* of the conceiving subjects is called an ordered pair $\langle N_{gen}, B\rangle$.

Remark 2 The base $\boldsymbol{B}$ conceiving subject $K = \{r, r^T\}^{\#}$ is determined by two conceiving ways which lead from one node to another of the semantic network, in one direction and a second direction (transposition).

Definition 24 (*a minimal description of the conceiving way*) In the approximation space $\langle N_{gen}, \mathbf{B}\rangle$, for any conceiving way $u \in N_{gen}$, *a minimal description* of this way is called a subset of the conceiving base:

$$Cl(r) = \{K \in B: r \in K \wedge \forall S \in B(u \in S \wedge S \subseteq K \Rightarrow K = S)\}.$$

The minimal description of the conceiving way r is a set of all minimal (in the sense of conclusion) base conceiving subjects (the subset of the base) for which is included the way r.

Definition 25 Let $M \subseteq N_{gen}$ be any conceiving subject.
The lower approximation of the subject M is a set:

$$B^{-}(M) = \{K \in B: K \subseteq M^{\#}\}.$$

The boundary approximation M is a set:

$$Bn(M) = \cup\{Cl(r): r \in M^{\#} \backslash \cup B^{-}(M)\}.$$

The upper approximation M is a set:

$$B^{+}(M) = B^{-}(M) \cup Bn(M).$$

Fact 13.
In the approximation space $\langle N_{gen}, B\rangle$, for any conceiving subject $X, Y \subseteq N_{gen}$

1. $B^{-}(\varnothing) = B^{+}(\varnothing) = \varnothing$,
2. $B^{-}(N_{gen}) = B^{+}(N_{gen}) = B$,
3. $X = \cup A \wedge A \subseteq B_{-} \Rightarrow (X) \subseteq = B^{+}(X)$,
4. $B^{-}(X) \subseteq B^{+}(X)$,
5. $X \subseteq Y \Rightarrow B^{-}(X) \subseteq B^{-}(Y) \wedge B^{+}(X) \subseteq B^{+}(Y)$,
6. $B^{-}(X) = B^{-}(X^{\#}) \wedge \mathrm{B}^{+}(X) = B^{+}(X^{\#})$, $Bn(X) = Bn(X^{\#})$.
   And for any $x \in N_{gen}$
7. $x \in_{Ngen} \Rightarrow \forall s \in_{Ngen}(s\, \varepsilon\, x \Rightarrow s\, \varepsilon \cup Cl(x))$,
8. $\forall K \in Cl(x) \forall s \in N_{gen}(s\, \varepsilon\, K \Rightarrow s\, \varepsilon \cup Cl(x))$,
9. $B^{-}(\{x\}) \neq \varnothing \Leftrightarrow \{x\} \in B$,
10. $\{x\} \in B \Leftrightarrow B^{-}(\{x\}) = \{\{x\}\}$,
11. $\{x\} \notin B \Rightarrow B^{-}(\{x\}) = \varnothing \wedge B^{+}(\{x\}) = Cl(x)$,
12. $x \in X^{\#} \Rightarrow x \in \cup B^{+}(X)$.

Remark 3 Based on Fact 13.6 of the lattice closures of conceiving subjects, the space approximation $\langle N_{gen}, B\rangle$ has all properties of the covering approximation space $\langle U, C\rangle$ described in the paper [5].

Theorem 11 *The base* ***B*** *of the approximation space* $\langle N_{gen}, B\rangle$ is *representative when it satisfies the condition:*

$$\forall K \in B \exists r \forall K \in S \in B(r \in S \Rightarrow K \subseteq S)$$

The subject $K \in \boldsymbol{B}$ has *representative* r_0, when

$$\forall S \in B(r_0 \in S \Rightarrow K \subseteq S)$$

Any approximation space $\langle N_{gen}, B \rangle$ *is representative and every, at least two-element set of the base* $\boldsymbol{B}$, *has at least two representative elements.*

Proof Let $K = \{(x_1, \ldots, x_j), (x_1, \ldots, x_j)^T\}^\# \in \boldsymbol{B}$, for some $(x_1, \ldots, x_j) \in \cup \boldsymbol{B}$. When the set $K \in \boldsymbol{B}$ is a one-element set, in other words: $K = \{(x)\}^\# = \{(x)\}$, $(x) \in \cup \boldsymbol{B}$, then from $(x) \in S$ appears that $K \subseteq S$.

For $K = \{(x_1, \ldots, x_j), (x_1, \ldots, x_j)^T\}^\#$, $(x_1, \ldots, x_j) \in N^j$, $j > 1$, it is assumed that $r = (x_1, \ldots, x_j)$. Let $r \in S = \{(y_1, \ldots, y_l), (y_1, \ldots, y_l)^T\}^\#$, then $r \, \varepsilon \, (y_1, \ldots, y_l)$ and $r^T \, \varepsilon$ $(y_1, \ldots, y_l)^T$ or $r^T \, \varepsilon \, (y_1, \ldots, y_l)$ and $r \, \varepsilon \, (y_1, \ldots, y_l)^T$, so $K = \{r, r^T\}^\# \subseteq S$.

Analogically is when we accept $r = (x_1, \ldots, x_j)^T$.

Furthermore, for $r_0 \in \{r, r^T\} \subseteq K$ is $\forall S \in \boldsymbol{B}(r_0 \in S \Rightarrow K \subseteq S)$, which means that the base subject K has two representatives.

Definition 26 The equivalence relation $\approx_B$ described as follows for any subjects X, $Y \subseteq N_{gen}$

$$X \approx_B Y \Leftrightarrow B^-(X) = B^-(Y) \wedge B^+(X) = B^+(Y)$$

is called a *indiscernibility relation of conceiving subject*. The expression "$X \approx_B Y$" we read: the subjects X, Y are undistinguishable. Two conceiving subjects are *indiscernible* in the established approximation space $\langle N_{gen}, B \rangle$ iff their representatives have identical lower and upper approximations.

Definition 27 In the approximate space of conceiving subjects $\langle N_{gen}, B \rangle$, *a rough conceiving subject* M_B appointed by the conceiving subject M, is called an abstract class with representative M of the equivalence relation $\approx_B$.

The set of all rough conceiving subjects will be denoted by $Rs(N_{gen}, \boldsymbol{B})$.

Definition 28 In the approximation space $\langle N_{gen}, B \rangle$ the relation $\subseteq_C$ is described for any subjects X, $Y \subseteq N_{gen}$ by model:

$$X_B \subseteq_B Y_B \Leftrightarrow B^-(X) \subseteq B^-(Y) \wedge B^+(X) \subseteq B^+(Y)$$

It is called the *inclusion relation of the rough conceiving subjects*.

Using Remark 3, we have

Theorem 12 [5] *Any approximation space* $\langle N_{gen}, B \rangle$ *is representative and every, at least two-element set of the base* $\boldsymbol{B}$, *has at least two representative elements if, and only if, for any family of rough subjects of conceiving in a* poset $\langle Rs(N_{gen}, B), \subseteq_B \rangle$ *there is a supremum and an infimum.*

From this, and Theorem 11 about the representative $\langle N_{gen}, B \rangle$, we obtain:

Fact 14.

The poset $\langle Rs(N_{gen}, B), \subseteq_B \rangle$ is a complete distributive lattice.

In the complete lattice $\langle Rs(N_{gen}, B), \subseteq_B \rangle$ are described general operations, such as a sum $\cup_B$ and an intersection $\cap_B$:

For family of sets $A \subseteq 2^{Ngen}$

$$\cup_B\{M_B: M \in A\} = \sup\{M_B: M \in A\} = X_B,$$

where $X \subseteq N_{gen}$ satisfies the conditions

$$\cup\{B^-(M): M \in A\} = B^-(X), \cup\{B^+(M): M \in A\} = B^+(X).$$

For infimum

$$\cap_B\{M_B: M \in A\} = inf\{M_B: M \in A\} = Y_B,$$

where $Y \subseteq N_{gen}$ satisfies the conditions

$$\cap\{B^-(M): M \in \mathrm{A}\} = B^-(X), \cap\{B^+(M): M \in A\} = B^+(X)$$

In particular, for two-argument operations, the sign $\cup_B$ for addition and $\cap_B$ for multiplication:

For any subjects of conceiving $X, Y \subseteq N_{gen}$

$$X_B \cup_B Y_B = \cup_B\{X_B, Y_B\}, X_B \cap_B Y_B = \cap_B\{X_B, Y_B\}.$$

Assuming the definition given in the paper [5]:

$$X \in_B Y_B \text{ iff } X \neq \varnothing \wedge X_B \subseteq_B Y_B \wedge \exists K \in B(B^+(X) = \{K\}).$$

We get all set theory interpretation of rough sets, for which algorithm was given in the second chapter of this chapter.

7 Further Research

In this chapter is shown the data granulation method for vague knowledge, which occurs in information systems and semantic networks. Finally, it is proposed to use granulation of data searching result in the semantic networks by using set theory (or lattice) operations, which are done on rough conceiving subjects. Realization of the vague knowledge search algorithm is finished, when are established the elements of the rough conceiving subject. Intuitively, finding set of all ways in the conceiving system, where represented knowledge is vague, is at the same time establishing (finding) this knowledge. Logic properties of these calculations remain to further research. Maybe, based on this research, we could formulate some *conceiving logic* [10].

References

1. Baader, F., Calvanese, D., McGuinness, D., Nardi, D., Patel-Schneider, P. (eds.): The Description Logic. Handbook Theory, Implementation and Application. Cambridge University Press, Cambridge (2003)
2. Bobillo, F., Straccia, U.: fuzzyDL: An expressive fuzzy description logic reasoner. In: Proceedings of IEEE International Conference on Fuzzy Systems FUZZ-IEEE 2008 (IEEEWorld Congress on Computational Intelligence), pp. 923–930 (2008)
3. Bonikowski, Z., Bryniarski, E., Wybraniec-Skardowska, U.: Rough pragmatic description logic. In: Skowron, A., Suraj, Z. (eds.) Rough Sets and intelligent Systems—Professor Zdzisław Pawlak in Memoriam, vol. 2, pp. 157–184. Springer, Berlin, Heidelberg, New York (2013). ISBN: 978-3-642-30340-1
4. Bonikowski, Z., Wybraniec-Skardowska, U.: Vagueness and Roughness. In: Peters, J.F., Skowron, A., Rybi´nski, H. (eds.) Transactions on Rough Sets IX. LNCS, vol. 5390, pp. 1–13. Springer, Heidelberg (2008)
5. Bonikowski, Z., Bryniarski, E., Wybraniec-Skardowska, U.: Extensions and intensions in the rough set theory. J. Inform. Sci. **107**, 149–167 (1998)
6. Bryniarska, A.: The paradox of the fuzzy disambiguation in the information retrieval. Int. J. Adv. Res. Artif. Intell. (IJARAI) **2**(9), 55–58 (2013)
7. Bryniarska, A.: The model of possible web data retrieval. In: Proceedings of 2nd IEEE International Conference on Cybernetics CYBCONF 2015, Gdynia, pp. 348–353 (2015). ISBN 978-1-4799-8320-9
8. Bryniarski, E.: A calculus of rough sets of the first order. Bull. Pol. Acad. Math. **37**, 109–136 (1989)
9. Bryniarski, E.: Formal conception of rough sets. Fund. Inf. **27**(2–3), 103–108 (1996)
10. Bryniarski, E.: The logic of conceiving of the sentential schemes. In: XIX Conference "Applications of Logic in Philosophy and the Foundations of Mathematics" Szklarska Poręba, Poland 5–9 May 2014/Abstract, University of Wrocław, 4–6 2014
11. Fanizzi, N., d'Amato, C., Esposito, F., Lukasiewicz, T.: Representing uncertain concepts in rough description logics via contextual indiscernibility relations. In: Bobillo, F., da Costa, P.C.G., d'Amato, C., et al. (eds.) Proceedings of 4th International Workshop on Uncertainty Reasoning for the Semantic Web (URSW 2008). CEUR-WS, vol. 423. CEUR-WS.org (2008). http://ceurs.org/Vol-423/paper7.pdf. Accessed 13 May 2011
12. Pawlak, Z.: Information systems—theoretical foundations. Inf. Syst. **6**, 205–218 (1981)
13. Pawlak, Z.: Rough sets. Int. J. Comput. Inf. Sci. **11**, 341–356 (1982)
14. Pawlak, Z.: Rough Sets: Theoretical Aspects of Reasoning About Data. Kluwer Academic Publishers, Dordrecht (1991)
15. Pawlak, Z.: Rough set elements. In: Polkowski, L., Skowron, A. (eds.) Rough Sets in Knowledge Discovery 1. Methodology and Applications, pp. 10–30. Springer, Heidelberg (1998)
16. Pedrycz, W., Skowron, A., Kreinovich, V.: Handbook of Granular Computing. Wiley, New York (2008)
17. Pedrycz, W.: Allocation of information granularity in optimization and decision-making models: towards building the foundations of granular computing. Eur. J. Oper. Res. **232**(1), 137–145 (2014). doi:10.1016/j.ejor.2012.03.038
18. Pedrycz, W., Bargiela, A.: An optimization of allocation of information granularity in the interpretation of data structures: toward granular fuzzy clustering. IEEE Trans. Syst. Man Cybern. Part B Cybern. **42**(3), 582–590 (2012). doi:10.1109/TSMCB.2011.2170067
19. Pedrycz, W.: Granular Computing: Analysis and Design of Intelligent Systems. Taylor & Francis Group, Abingdon (2013)
20. Polkowski, L.: Reasoning by Parts. An Outline of Rough Mereology, Warszawa (2011)
21. Skowron, A., Polkowski, L.: Rough mereology and analytical morphology. In: Orłowska, E. (ed.) Incomplete Information: Rough Set Analysis, pp. 399–437. Springer, Heidelberg (1996)

22. Skowron, A., Wasilewski, P.: An introduction to perception based computing. In: Kim, T.-h., Lee, Y.-h., Kang, B.-H., Ślęzak, D. (eds.) FGIT 2010. LNCS, vol. 6485, pp. 12–25. Springer, Heidelberg (2010)
23. Simou, N., Stoilos, G., Tzouvaras, V., Stamou, G., Kollias, S.: Storing and querying fuzzy knowledge in the semantic web. In: Bobillo, F., da Costa, P.C.G., d'Amato, C., et al. (eds.) Proceedngs 4th International Workshop on Uncertainty Reasoning for the Semantic Web URSW 2008, May 13. CEUR-WS, vol. 423, CEUR-WS.org (2008). http://ceur-ws.org/Vol-423/paper8.pdf. Accessed 13 May 2011
24. Simou, N., Mailis, T., Stoilos, G., Stamou, S.: Optimization techniques for fuzzy description logics. In: Description Logics. Proceedings 23rd International Workshop on Description Logics (DL 2010). CEUR-WS, vol. 573, CEUR-WS.org (2010). http://ceur-ws.org/Vol-573/paper_25.pdf. Accessed 13 May 2011
25. Waldmajer, J.: A representation of relational systems. In: Kłopotek, M.A., Wierzchoń, S.T., Trojanowski, K. (eds.) Inteligent Information Processing and Web Mining, June 2–5, Zakopane 2003, Proccedings of the International IIS: IIPWM'03 Conference, pp. 569–573. Springer, Berlin
26. Waldmajer, J.: Boolean algebra of multistructures, abstracts. In: Logic Colloqium'03, August 14–20, 2003, p. 151. University of Helsinki, Helsinki (2003)

Vagueness and Uncertainty: An F-Rough Set Perspective

Dayong Deng and Houkuan Huang

Abstract F-rough sets are the first dynamical rough set model for a family of information systems (decision systems). This chapter investigates vagueness and uncertainty from the viewpoints of F-rough sets. Some indexes, including two types of F-roughness, two types of F-membership-degree and F-dependence degree etc., are defined. Each of these indexes may be a set of number, not like other vague and uncertain indexes in Pawlak rough sets. These indexes extend those of Pawlak rough sets, and indicate vagueness and uncertainty in a family of information subsystems (decision subsystems). Moreover, these indexes themselves also include vagueness and uncertainty, namely, vagueness of vagueness and uncertainty of uncertainty. Further, we investigate some interesting properties of these indexes.

1 Introduction

Human beings have investigated vagueness and uncertainty for more than 100 years. Frege presented the concept of vagueness in 1904. Russell [25] investigated it further in 1923. Zadeh [36] proposed fuzzy set theory in 1965. Shafer [26] introduced the theory of evidence in 1976. Pawlak presented rough set theory in 1982 [19, 20]. Recently, type-2 fuzzy sets [18] and cloud model [15, 16] were proposed. These two theories can investigate not only vague and uncertain phenomena, but also vagueness of vagueness and uncertainty of uncertainty. However, these methods except for rough set theory have more or less subjectivity.

D. Deng (✉)
Xingzhi College, Zhejiang Normal University, Jinhua 321004, China
e-mail: dayongd@163.com

H. Huang
School of Computer and Information Technology, Beijing Jiaotong University,
Beijing 100044, China
e-mail: hkhuang@bjtu.edu.cn

G. Wang et al. (eds.), *Thriving Rough Sets*, Studies in Computational Intelligence 708, DOI 10.1007/978-3-319-54966-8_15

Rough set theory was introduced by Pawlak [19, 20, 22, 23], which can deal with imprecise, vague and incomplete information, and has no subjectivity. In order to deal with different kinds of data, many models of rough sets are extended, including various precision rough set model [1, 14, 41], rough fuzzy sets and fuzzy rough sets [13], bayesian rough set model [27–30], probabilistic rough set model [17, 24, 31, 32, 42, 43], rough sets based on accessible relation [5], covering rough set model [37–40] and decision-theoretic rough set model [32–35] etc. All of these models can investigate vagueness and uncertainty, but they can't investigate vagueness of vagueness or uncertainty of uncertainty, namely, can't investigate the change of vagueness or uncertainty.

The model of F-rough sets [7] is the first dynamical model of rough sets, which extends Pawlak rough sets from one information system (decision system) to a family of information systems (decision systems), and can combine with any models of rough sets. These information systems (decision systems) can be obtained from a large amount of data, increasing data and multi-source data. In [6, 7] Deng et al. used F-rough sets to obtain dynamic reducts [2, 3] and parallel reducts.

Moreover, F-rough sets can deal with vagueness of vagueness and uncertainty of uncertainty in vague and uncertain cases, and have been employed to detect concept drifting. In [8, 9] F-rough sets were employed to detect concept drifting, and various indexes of uncertainty and vagueness were proposed. In [10] the uncertainty and concept drifting in a single decision system were investigated from the view of F-rough sets, which extended concept drifting from time to space (or conditions), and defined the concept of cognition convergence. In [12] the loss of information for conditional reduction was presented, which consolidated the information theory basis for rough sets. In [11] double-level absolute reduction for multi-granulation rough sets were proposed, which obtained absolute reducts from heterogenous data.

In this chapter, the indexes of uncertainty and vagueness are investigated from the viewpoints of F-rough sets, which are different from those in Pawlak rough sets. In F-rough sets, every decision system (information system) indicates one situation, among which exists differences. Just like Pawlak rough sets, upper approximation and lower approximation are used to approximate one concept, but there may be differences for these upper approximations and lower approximations in F-rough sets. Moreover, these indexes, such as membership degree, roughness, will be significant differences between in Pawlak rough sets and in F-rough sets. In Pawlak rough sets, indexes of vagueness and uncertainty, which indicate vague and uncertain phenomenons, are fixed. However, in F-rough sets, indexes of vagueness and uncertainty are flexible, just like type-2 fuzzy sets.

The rest of this chapter is organized as follows. Section 2 reviews briefly vagueness and uncertainty in Pawlak rough sets. Section 3 refers to vagueness and uncertainty of the boundary region in F-rough sets. Section 4 defines two type of membership degree in F-rough sets, and investigates their properties. Section 5 defines two types of dependence degree in F-rough sets, and investigates their properties. Section 6 defines two types of roughness, and investigates their properties. At last, we draw a conclusion in Sect. 7.

2 Vagueness and Uncertainty in Pawlak Rough Sets

Pawlak rough set theory is to deal with imprecise, vague and inconsistent information in an information system (a decision system). There are some indexes to indicate vagueness and uncertainty, including upper approximation, lower approximation, roughness, membership function and dependence degree etc. [4, 21]. In the following paragraphs, we will briefly review some related knowledge about rough set theory and these vague and uncertain indexes.

An information system is a pair $IS = (U, A)$, where U is the universe of discourse with a finite number of objects (or entities), A is a set of attributes defined on U. Each $a \in A$ corresponds to the function $a : U \to V_a$, where V_a is called the value set of a. Elements of U are called situation, objects or rows, interpreted as, e.g., cases, states [20].

With any subset of attributes $B \subseteq A$, we associate the information set for any element $x \in U$ by

$$Inf_B(x) = \{(a, a(x)) : a \in B\}$$

An equivalence relation called B-indiscernible relation is defined by

$$IND(B) = \{(x, y) \in U \times U : Inf_B(x) = Inf_B(y)\}$$

Two elements x, y satisfying the relation $IND(B)$ are indiscernible by attributes from B. $[x]_B$ is referred to as the equivalence class of $IND(B)$ defined by x.

Suppose $IS = (U, A)$ is an information system, $B \subseteq A$ is a subset of attributes, and $X \subseteq U$ is a subset of discourse, the sets

$$\underline{B}(X) = \{x \in U : [x]_B \subseteq X\}$$

$$\overline{B}(X) = \{x \in U : [x]_B \cap X \neq \emptyset\}$$

are called B-lower approximation and B-upper approximation respectively. The lower approximation is also called positive region, denoted by $POS_B(X)$. Sometimes $\underline{B}(X)$ and $\overline{B}(X)$ are denoted by $\underline{B}(IS, X)$ and $\overline{B}(IS, X)$ respectively.

In a decision system $DS = (U, A, d)$, where $\{d\} \cap A = \emptyset$, the decision attribute d divides the universe U into parts, denoted by $U/\{d\} = \{Y_1, Y_2, ..., Y_p\}$, where Y_i is an equivalence class. The positive region is defined as

$$POS_B(d) = \bigcup_{Y_i \in U/\{d\}} \underline{B}(Y_i)$$

where $B \subseteq A$.

Sometimes the positive region $POS_B(d)$ is also denoted by $POS_B(DS, d)$ or POS (DS, B, d).

We always define $\partial(x) = \{d(y) : y \in [x]_A\}$ in a decision system $DS = (U, A, d)$, if $\forall_{x \in U}(|\partial(x)| = 1)$, the decision system $DS = (U, A, d)$ is called a consistent decision system, or else, it is called an inconsistent decision system.Where $|\bullet|$ denotes the cardinality of a set.

In rough set theory, membership degree is an important index to indicate a vague and uncertain degree which an element belongs to a set. The membership degree in rough set theory is defined as follows:

$$\mu_{B,X|DS}(x) = \frac{|X \cap [x]_B|}{|[x]_B|}$$

In a decision system $DS = (U, A, d)$ we will say d depends on $B \subseteq A$ to a degree $h(0 \leq h \leq 1)$, if

$$h = \gamma(B, DS) = \frac{|POS_B(d)|}{|U|}$$

The dependence degree is a very important index when we want to obtain a conditional reduct from a decision system and an information system.

In a broad sense, all of indexes in rough set theory can be called roughness (vagueness), but in the narrow sense roughness is a ratio which the lower approximation of a concept is divided by its upper approximation. Its mathematic formula is expressed as follows.

$$\alpha_{B,DS}(X) = \frac{|\underline{B}(X)|}{|\overline{B}(X)|}$$

In [21] Pawlak discussed vagueness and uncertainty in rough sets. We don't introduce them here in detail. In the following sections, we will discuss vagueness and uncertainty of lower approximation and upper approximation, membership degree, dependence degree and roughness from the viewpoints of F-rough sets respectively.

3 Vagueness and Uncertainty in the Boundary Region of F-Rough Sets

In F-rough sets [7], the domain of discourse is a family of decision subsystems F and the family of its corresponding information subsystems $FIS = \{IS_i\}(i = 1, 2, \ldots, n)$, where $IS_i = (U_i, A)$, $DT_i = (U_i, A, d) \in F$, $F \neq \emptyset \subseteq P(DS)$, $DS = (U, A, d)$.

Definition 1 Suppose X is a concept, and N is a situation. $X|N$ denotes the concept X in the situation N. In a family of information subsystems $FIS = \{IS_1, IS_2, \ldots, IS_n\}$, $IS_i \in FIS$, $X|IS_i = X \cap IS_i$, $X|FIS = \{X|IS_1, X|IS_2, \ldots, X|IS_n\}$. If it is not confused, the $X|N$ can be denoted by X for short.

Table 1 Decision subsystem DT_1

U_1	a	b	c	d
x_1	0	0	1	1
x_2	1	1	0	1
x_3	0	1	0	0
x_4	1	1	0	1

Table 2 Decision subsystem DT_2

U_2	a	b	c	d
y_1	0	1	0	0
y_2	1	1	0	1
y_3	1	1	0	1
y_4	0	1	0	0
y_5	1	2	0	0
y_6	1	2	0	1

Example 1 Suppose $F = \{DT_1, DT_2\}$ is a family of decision subsystems, corresponding to Tables 1 and 2 respectively, where a, b, c are condition attributes, d is a decision attribute.

For a concept $X = \{x : d(x) = 0\}$, it is different in the decision subsystems DT_1 and DT_2. $X|DT_1 = \{x : d(x) = 0\} \cap DT_1 = \{x_3\}$, $X|DT_2 = \{x : d(x) = 0\} \cap DT_2 = \{y_1, y_4, y_5\}$, $X|F = \{X|DT_1, X|DT_2\} = \{\{x_3\}, \{y_1, y_4, y_5\}\}$.

Suppose X is a concept in *FIS*, its lower approximation, upper approximation, boundary region and negative region are defined respectively in the following.

$$\underline{B}(FIS, X) = \{\underline{B}(IS_i, X) : IS_i \in FIS\} = \{\{x \in U_i : [x]_B \subseteq X\}\}$$

$$\overline{B}(FIS, X) = \{\overline{B}(IS_i, X) : IS_i \in FIS\} = \{\{x \in U_i : [x]_B \cap X \neq \phi\}\}$$

$$BND(FIS, X) = \{BND(IS_i, X) : IS_i \in FIS\} = \{\overline{B}(IS_i, X) - \underline{B}(IS_i, X) : IS_i \in FIS\}$$

$$NEG(FIS, X) = \{NEG(IS_i, X) : IS_i \in FIS\} = \{U_i - \overline{B}(IS_i, X) : IS_i \in FIS\}$$

The lower approximation, upper approximation, boundary region and negative region of X in *FIS* are the sets of corresponding regions of X in elements of *FIS*.

The pair $(\underline{B}(FIS, X), \overline{B}(FIS, X))$ is called F-rough sets. If $\underline{B}(FIS, X) = \overline{B}(FIS, X)$, then the pair $(\underline{B}(FIS, X), \overline{B}(FIS, X))$ is crisp.

From the definition of F-rough sets, the lower approximation, the upper approximation, the boundary region and the negative region of a concept in every element of F contain vagueness and uncertainty. Moreover, they are different in different element of F. This is to say, the lower approximation, the upper approximation, the boundary region and the negative region of a concept in F-rough sets contain vagueness of vagueness and uncertainty of uncertainty.

4 Uncertainty and Membership Function

Membership degree is an important index to measure vagueness and uncertainty in rough sets. In F-rough sets, we can also define its membership degree as follows:

Definition 2 For an element $x \in U$ in an information system $IS = (U, A)$, a subset of attributes $B \subseteq A$, a subset of universe $X \subseteq U$ and a family of information subsystems $FIS = \{IS_1, IS_2, \ldots, IS_n\}$, where $IS_i \subseteq IS$. the membership degree of $x \in X$ relative to FIS (F-membership-degree in short) is defined as

$$\mu_{B,X|FIS}(x) = \{\mu_{B,X|IS_1}(x), \mu_{B,X|IS_2}(x), \ldots, \mu_{B,X|IS_n}(x)\}$$

The average of membership degree of $x \in X$ relative to FIS is

$$\widetilde{\mu_{B,X|FIS}}(x) = \frac{\sum_{i=1}^{n} \mu_{B,X|IS_i}(x)}{n}$$

The maximum of membership degree of $x \in X$ relative to FIS is

$$\overline{\mu_{B,X|FIS}}(x) = \max_{IS_i \in FIS} \{\mu_{B,X|IS_i}(x)\}$$

The minimum of membership degree of $x \in X$ relative to FIS is

$$\underline{\mu_{B,X|FIS}}(x) = \min_{IS_i \in FIS} \{\mu_{B,X|IS_i}(x)\}$$

The variance of the membership degree of $x \in X$ relative to FIS is,

$$\delta_M^2 = \frac{\sum_{i=1}^{n} (\mu_{B,X|IS_i}(x) - \widetilde{\mu_{B,X|FIS}}(x))^2}{n}$$

and δ_M is the standard deviation of the membership degree $x \in X$ in FIS.

F-membership degree indicates the membership degree of an element x belonging to a concept X in different situations. It is a set which contains vagueness of vagueness and uncertainty of uncertainty. If the order of its elements should be fixed, it can be denoted as a vector by $\mu_{B,X|FIS}(x) =< \mu_{B,X|IS_1}(x), \mu_{B,X|IS_2}(x), \ldots, \mu_{B,X|IS_n}(x) >$.

Table 3 Decision sub-system DT_3

U_1	a	b	c	d
x_1	0	0	1	1
x_2	1	1	0	1
x_3	1	1	0	0
x_4	1	1	0	1

Table 4 Decision sub-system DT_4

U_2	a	b	c	d
y_1	0	1	0	0
y_2	1	1	0	1
y_3	1	1	0	1
y_4	1	1	0	0
y_5	1	1	0	0
y_6	1	2	0	1

F-membership degree can describe the changing of a membership degree and uncertainty of uncertainty,but if a number needs to be used to indicate the membership degree in F-rough sets, we can use the following definition of membership degree.

Definition 3 The total membership degree of $x \in X$ relative to FIS (F-total-membership-degree in short) is defined as

$$T\mu_{B,X|FIS}(x) = \frac{\sum_{X \subseteq IS_i \wedge IS_i \in FIS} |[x]_B \cap X|}{\sum_{[x]_B \subseteq IS_i \wedge IS_i \in FIS} |[x]_B|}$$

Both F-membership degree and F-total-membership-degree express the membership degree of an element x belonging to a concept X in F-rough sets, but they have some differences. F-membership degree focuses on uncertainty of uncertainty, and the changing of membership degrees in every element of F, while F-total-membership-degree emphasizes the total effectiveness of membership degrees in the family of information systems (decision systems) F. Both F-membership degree and F-total-membership-degree can be abbreviated as F-membership degree or membership degree if not confused.

Example 2 Assume that $F = \{DT_3, DT_4\}$ which shows in Tables 3 and 4, $B = \{a, b\}$, and that x_3 in DT_3 is the same as y_3 in DT_4. Then, the membership degree of $x_3 \in X = \{x : d(x) = 0\}$ can be calculated as follows:

$\mu_{B,X|DT_3}(x_3) = \frac{|[x_3]_B \cap X|}{|[x_3]_B|} = \frac{1}{3}$

$\mu_{B,X|DT_4}(x_3) = \frac{|[x_3]_B \cap X|}{|[x_3]_B|} = \frac{1}{2}$

So,

$\mu_{B,X|F}(x_3) = \{\mu_{B,X|DT_3}(x_3), \mu_{B,X|DT_4}(x_3)\} = \{\frac{1}{3}, \frac{1}{2}\}$

$\widetilde{\mu_{B,X|F}}(x_3) = \frac{\sum_{i=3}^{4} \mu_{B,X|DT_i}(x_3)}{2} = \frac{\frac{1}{3}+\frac{1}{2}}{2} = \frac{5}{12}$

$\overline{\mu_{B,X|F}}(x_3) = \max_{DT_i \in F}\{\mu_{B,X|DT_i}(x_3)\} = \frac{1}{2}$

$\underline{\mu_{B,X|F}}(x_3) = \min_{DT_i \in F}\{\mu_{B,X|DT_i}(x_3)\} = \frac{1}{3}$

$\delta_M^2 = \frac{\sum_{i=3}^{4}(\mu_{B,X|DT_i}(x_3) - \widetilde{\mu_{B,X|F}}(x_3))^2}{2} = \frac{1}{144}$

$T\mu_{B,X|F}(x_3) = \frac{\sum_{X \subseteq DT_i \wedge DT_i \in F} |[x_3]_B \cap X|}{\sum_{[x_3]_B \subseteq DT_i \wedge DT_i \in F} |[x_3]_B|} = \frac{3}{7}$

In the following paragraphs, we will investigate some properties of these two membership degrees in F-rough sets.

Proposition 1 *For an information system $IS = (U, A)$ and a family of its information subsystems $FIS = FIS_1 \cup FIS_2$, a concept $X \subseteq U$, $B \subseteq A$ and an element $x \in U$. The following propositions are true.*

(1) $\overline{\mu_{B,X|FIS}}(x) = \max\{\overline{\mu_{B,X|FIS_1}}(x), \overline{\mu_{B,X|FIS_2}}(x)\}$

(2) $\underline{\mu_{B,X|FIS}}(x) = \min\{\underline{\mu_{B,X|FIS_1}}(x), \underline{\mu_{B,X|FIS_2}}(x)\}$.

(3) $\min\{\widetilde{\mu_{B,X|FIS_1}}(x), \widetilde{\mu_{B,X|FIS_2}}(x)\} \leq \widetilde{\mu_{B,X|FIS}}(x) \leq \max\{\widetilde{\mu_{B,X|FIS_1}}(x), \widetilde{\mu_{B,X|FIS_2}}(x)\}$.

(4) $\min\{T\mu_{B,X|FIS_1}(x), T\mu_{B,X|FIS_2}(x)\} \leq T\mu_{B,X|FIS}(x) \leq \max\{T\mu_{B,X|FIS_1}(x), T\mu_{B,X|FIS_2}(x)\}$.

Proof It is easy to prove the above proposition.

Proposition 1 shows that both F-membership degree and F-total-membership-degree will be changed after two sets of information subsystems (decision subsystems) are united. The average of F-membership degree and the value of F-total-membership-degree will be a value between the two old corresponding ones. This means that both the average of F-membership degree and F-total-membership-degree will converge a stable number with the increment of data sets.

Proposition 2 *For an information system $IS = (U, A)$ and a family of its information subsystems $FIS = \{IS_1, IS_2, \ldots, IS_n\}$, a concept $X \subseteq U$, $B \subseteq A$ and an element $x \in U$, the following formulas are true.*

(1) $\underline{\mu_{B,X|FIS}}(x) = 1 \Leftrightarrow \forall_{IS_i \in FIS}(x \in \underline{B}(IS_i, X))$.

(2) $\underline{\mu_{B,X|FIS}}(x) > 0 \Leftrightarrow \forall_{IS_i \in FIS}(x \in \overline{B}(IS_i, X))$.

(3) $\overline{\overline{\mu_{B,X|FIS}}}(x) = 1 \Leftrightarrow \exists_{IS_i \in FIS}(x \in \underline{B}(IS_i, X))$.

(4) $\overline{\mu_{B,X|FIS}}(x) > 0 \Leftrightarrow \exists_{IS_i \in FIS}(x \in \overline{B}(IS_i, X))$.

(5) $\overline{\mu_{B,X|FIS}}(x) = 0 \Leftrightarrow \forall_{IS_i \in FIS}(x \in NEG(IS_i, X))$.

(6) $T\mu_{B,X|FIS}(x) = 1 \Leftrightarrow \forall_{IS_i \in FIS}(x \in \underline{B}(IS_i, X))$.

(7) $T\mu_{B,X|FIS}(x) = 0 \Leftrightarrow \forall_{IS_i \in FIS}(x \in NEG(IS_i, X))$.

(8) $T\mu_{B,X|FIS}(x) > 0 \Leftrightarrow \exists_{IS_i \in FIS}(x \in \overline{B}(IS_i, X))$.

Proof It is easy to prove the above propositions.

Proposition 2 shows: According to the value of F-membership degree or F-total-membership-degree, we can judge whether an element in the discourse is in the positive region, or the negative region,or the boundary region in a family of information subsystems (decision subsystems).

Proposition 3 *For an information system $IS = (U, A)$ and a family of its information subsystems $FIS = \{IS_1, IS_2, \dots, IS_n\}$, concepts $X, Y \subseteq U$, $B \subseteq A$ and an element $x \in U$, the following formulas are true.*

(1) $\overline{\mu_{B,(X \cup Y)|FIS}}(x) \geq \max\{\overline{\mu_{B,X|FIS}}(x), \overline{\mu_{B,Y|FIS}}(x)\}$.

(2) $\min\{\underline{\mu_{B,X|FIS}}(x), \underline{\mu_{B,Y|FIS}}(x)\} \geq \underline{\mu_{B,(X \cap Y)|FIS}}(x)$.

(3) $\widetilde{\mu_{B,(X \cup Y)|FIS}}(x) \geq \max\{\widetilde{\mu_{B,X|FIS}}(x), \widetilde{\mu_{B,Y|FIS}}(x)\}$.

(4) $\utilde{\mu_{B,(X \cap Y)|FIS}}(x) \leq \min\{\utilde{\mu_{B,X|FIS}}(x), \utilde{\mu_{B,Y|FIS}}(x)\}$.

(5) $T\mu_{B,X \cup Y|FIS}(x) \geq \max\{T\mu_{B,X|FIS}(x), T\mu_{B,Y|FIS}(x)\}$.

(6) $T\mu_{B,X \cap Y|FIS}(x) \leq \min\{T\mu_{B,X|FIS}(x), T\mu_{B,Y|FIS}(x)\}$.

Proof It is easy to prove these propositions according to their definitions.

According to Propositions 3, both the value of F-membership degree and F-total-membership-degree will increase with the increment of the extension of a concept, and will decrease with the decrement of the extension of a concept.

Both F-membership degree and F-total-membership-degree can be calculated strictly from a family of information systems (decision systems) F. F-membership degree focuses on the changing of membership degree in a family of information subsystems (decision subsystems), while F-total-membership-degree emphasizes the total effectiveness of membership degree. They preserve the advantages of both membership degree in rough sets and membership degree in fuzzy sets. They are both fixed and flexible.

5 Uncertainty and Dependence Degree

The dependence degree is a very important index in rough sets. It is usually used to obtain reducts and reason. In F-rough sets the index should also be defined. In [7] F-total-dependence-degree is defined and used to obtain parallel reducts. In the following paragraphs, we will define another dependence degree in F-rough sets, called F-dependence degree, and investigate properties of both F-dependence degree and F-total-dependence-degree.

Definition 4 In a decision system $DS = (U, A, d)$, a subset of attributes $B \subseteq A$, a family of decision subsystems $F = \{DT_1, DT_2, \dots, DT_n\}$, where $DT_i \subseteq DS$. the dependence degree relative to F(F-dependence degree in short) is defined as

$$\gamma(B, F) = \{\gamma(B, DT_1), \gamma(B, DT_2), \dots, \gamma(B, DT_n)\}$$

The average of dependence degree relative to F is

$$\widetilde{\gamma(B,F)} = \frac{\sum_{i=1}^{n} \gamma(B,DT_i)}{n}$$

The maximum of dependence degree relative to F is

$$\overline{\gamma(B,F)} = \max_{IS_i \in FIS} \{\gamma(B,DT_i)\}$$

The minimum of dependence degree relative to F is

$$\underline{\gamma(B,F)} = \min_{DT_i \in F} \{\gamma(B,DT_i)\}$$

The variance of the dependence degree relative to F is,

$$\delta_D^2 = \frac{\sum_{i=1}^{n} (\gamma(B,DT_i) - \widetilde{\gamma(B,F)})^2}{n}$$

and δ_D is the standard deviation of the membership degree $x \in X$ in F.

F-dependence degree is an index to indicate the changing of dependence degrees in a family of decision subsystems. If the elements of F-dependence degree should be fixed, it can be expressed as a vector $\gamma(B,F) =< \gamma(B,DT_1), \gamma(B,DT_2), \ldots, \gamma (B,DT_n) >$ instead. In [7] we defined another version of F-dependence degree, called F-total-dependence-degree, to express the dependence degree in a family of decision subsystems and use it to obtain parallel reducts from them. It is defined as follows.

Definition 5 ([7]) The total dependence degree relative to F (F-total-dependence-degree in short) is defined as

$$T\gamma(B,F) = \frac{\sum_{IS_i \in F} |POS(B,DT_i,d)|}{\sum_{DT_i \in F} |U_i|}$$

Both F-dependence degree and F-total-dependence-degree are to indicate the dependence degree in a family of decision subsystems. When we want to know the changing of dependence degree in different decision subsystems, F-dependence degree can be used, otherwise, F-total-dependence-degree can be used. F-dependence degree and F-total-dependence-degree can be abbreviated as F-dependence degree or dependence degree, if not confused. In the following, an example is given to explain the calculating process of F-dependence degree and F-total-dependence-degree.

Example 3 Assume that $B = \{a,b\} \subseteq A$ and $F = \{DT_1, DT_2\}$ which shows in Tables 1 and 2. Two types of F-dependence degree will be calculated as follows:

$\gamma(B,DT_1) = \frac{|POS(DT_1,B,d)|}{|U_1|} = 1$

$\gamma(B, DT_2) = \frac{|POS(DT_2,B,d)|}{|U_2|} = \frac{2}{3}$
So,
$\gamma(B, F) = \{\gamma(B, DT_1), \gamma(B, DT_2)\} = \{1, \frac{2}{3}\}$
$\widetilde{\gamma(B, F)} = \frac{\sum_{i=1}^{2} \gamma(B,DT_i)}{2} = \frac{5}{6}$
$\overline{\gamma(B, F)} = \max_{DT_i \in F}\{\gamma(B, DT_i)\} = 1$
$\underline{\gamma(B, F)} = \min_{DT_i \in F}\{\gamma(B, DT_i)\} = \frac{2}{3}$
$\delta_D^2 = \frac{\sum_{i=1}^{n}(\gamma(B,DT_i) - \widetilde{\gamma(B,F)})^2}{n} = \frac{1}{36}$
$T\gamma(B, F) = \frac{\sum_{DT_i \in F} |POS(B,DT_i,d)|}{\sum_{DT_i \in F} |U_i|} = \frac{4}{5}$

Some properties of F-dependence degree and F-total-dependence-degree are showed as follows.

Proposition 4 *Let* $\emptyset \neq B_1 \subseteq B_2 \subseteq A$*, then, the following expressions are true:*
(1) $\gamma(B_1, F) \leq \gamma(B_2, F)$ *(This means that every element in* $\gamma(B_2, F)$ *is not less than that* $\gamma(B_1, F)$*).*
(2) $T\gamma(B_1, F) \leq T\gamma(B_2, F)$*.*

Proof In every element DT_i of F, when the subset of condition attributes is increasing, say, $B_1 \to B_2$, the positive region $POS(DT_i, B_1, d) \subseteq POS(DT_i, B_2, d)$, and $\gamma(B_1, DT_i) \leq \gamma(B_2, DT_i)$. Thus, the above two expressions are true.

Proposition 4 shows that both F-dependence degree and F-total-dependence-degree will increase with the increment of the set of conditional attributes.

Proposition 5 *Let* $F = F_1 \cup F_2 \subseteq P(DS)$*, the following expressions are true:*
(1) $\overline{\gamma(B, F)} = \max\{\overline{\gamma(B, F_1)}, \overline{\gamma(B, F_2)}\}$.
(2) $\underline{\gamma(B, F)} = \min\{\underline{\gamma(B, F_1)}, \underline{\gamma(B, F_2)}\}$.
(3) $\min\{\widetilde{\gamma(B, F_1)}, \widetilde{\gamma(B, F_2)}\} \leq \widetilde{\gamma(B, F)} \leq \max\{\widetilde{\gamma(B, F_1)}, \widetilde{\gamma(B, F_2)}\}$.
(4) $\min\{T\gamma(B, F_1), T\gamma(B, F_2)\} \leq T\gamma(B, F) \leq \max\{T\gamma(B, F_1), T\gamma(B, F_2)\}$.

Proof According to their definitions and primary mathematical knowledge, the above expressions are true.

Proposition 5 shows: The F-dependence degree and the F-total-dependence-degree will change after two sets of decision subsystems are united. Especially, the average of F-dependence degree and the value of F-total-dependence-degree will be a value between the two corresponding old ones.

Proposition 6 *The following propositions are true.*
(1) $\underline{\gamma(B, F)} = 1 \Leftrightarrow T\gamma(B, F) = 1 \Leftrightarrow \forall_{DT_i \in F}(POS(B, DT_i, d) = U_i)$.
(2) $\overline{\gamma(B, F)} = 0 \Leftrightarrow T\gamma(B, F) = 0 \Leftrightarrow \forall_{DT_i \in F}(POS(B, DT_i, d) = \emptyset)$.
(3) $\overline{\gamma(B, F)} = 1 \Leftrightarrow \exists_{DT_i \in F}(POS(B, DT_i, d) = U_i)$.
(4) $\underline{\gamma(B, F)} = 0 \Leftrightarrow \exists_{DT_i \in F}(POS(B, DT_i, d) = \emptyset)$.

Proof According to their definitions, it is easy to get these results.

Proposition 6 shows we can know the positive region according to both the value of F-dependence degree and F-total-membership-degree in a family of decision subsystems.

F-dependence degree and F-total-dependence-degree indicate dependence degree from two ways, the first focuses on the changing and variation of dependence degree in elements of F, but the last emphasizes the total effectiveness of dependence degrees in all of elements of F. Both of them can be used to obtain parallel reducts.

6 Uncertainty and Roughness

Roughness is a unique index to indicate the ratio between lower approximation and upper approximation in rough set theory. In a family of information subsystems (decision subsystems), the indexes of roughness for all these information subsystems (decision subsystems) may be different. We will indicate them with two types of roughnesses as follows.

Definition 6 For a concept $X \subseteq U$ in an information system $IS = (U, A)$, a subset of attributes $B \subseteq A$, and a family of information subsystems $FIS = \{IS_1, IS_2, \ldots, IS_n\}$, where $IS_i \subseteq IS$. the roughness of $X \subseteq U$ relative to FIS (F-roughness in short) is defined as

$$\alpha_{B,FIS}(X) = \{\alpha_{B,IS_1}(X), \alpha_{B,IS_2}(X), \ldots, \alpha_{B,IS_n}(X)\}$$

The average of roughness of $X \subseteq U$ relative to FIS is

$$\widetilde{\alpha_{B,FIS}(X)} = \frac{\sum_{i=1}^{n} \alpha_{B,IS_i}(X)}{n}$$

The maximum of roughness of $X \subseteq U$ relative to FIS is

$$\overline{\alpha_{B,FIS}}(X) = \max_{IS_i \in FIS} \{\alpha_{B,IS_i}(X)\}$$

The minimum of roughness of $X \subseteq U$ relative to FIS is

$$\underline{\alpha_{B,FIS}}(X) = \min_{IS_i \in FIS} \{\alpha_{B,IS_i}(X)\}$$

The variance of the roughness degree of $X \subseteq U$ relative to FIS is,

$$\delta_R^2 = \frac{\sum_{i=1}^{n} (\alpha_{B,IS_i}(X) - \widetilde{\alpha_{B,FIS}(X)})^2}{n}$$

and δ_R is the standard deviation of the roughness degree X in FIS.

For $\alpha_{B,FIS}(X)$, if $\forall_{IS_i \in FIS}(\alpha_{B,IS_i}(X) = a), a \in [0,1]$, then it is called $\alpha_{B,FIS}(X) = a$. For $\alpha_{B_1,FIS}(X)$ and $\alpha_{B_2,FIS}(X)$, if $\forall_{IS_i \in FIS}(\alpha_{B_1,IS_i}(X) = \alpha_{B_2,IS_i}(X))$, then $\alpha_{B_1,FIS}(X)$ is equal to $\alpha_{B_2,FIS}(X)$, denoted by $\alpha_{B_1,FIS}(X) = \alpha_{B_2,FIS}(X)$. If $\forall_{IS_i \in FIS}(\alpha_{B_1,IS_i}(X) \leq \alpha_{B_2,IS_i}(X))$, then $\alpha_{B_1,FIS}(X)$ is not less than $\alpha_{B_2,FIS}(X)$, denoted by $\alpha_{B_1,FIS}(X) \leq \alpha_{B_2,FIS}(X)$.

Just like F-membership degree and F-dependence degree, F-roughness indicates the changing of roughnesses in different information subsystems (decision subsystems). If elements of F-roughness are required to be fixed, we also express it as a vector $\alpha_{B,FIS}(X) =< \alpha_{B,IS_1}(X), \alpha_{B,IS_2}(X), \ldots, \alpha_{B,IS_n}(X) >$. When we don't want to know the changing of roughnesses in different information subsystems (decision subsystems), a number, called F-total-roughness, is used. It is defined as follows.

Definition 7 The total roughness degree of $X \subseteq U$ relative to FIS (F-total-roughness-degree in short) is defined as

$$T\alpha_{B,FIS}(X) = \frac{\sum_{X \subseteq IS_i \wedge IS_i \in FIS} |\underline{B}(IS_i, X)|}{\sum_{X \subseteq IS_i \wedge IS_i \in FIS} |\overline{B}(IS_i, X)|}$$

Both F-roughness and F-total-roughness can indicate roughness in a family of information subsystems (decision subsystems). F-roughness focuses on the changing and variation of roughnesses in a family of information subsystems (decision subsystems), while F-total-roughness emphasizes the total effectiveness of roughnesses. Both of them can be abbreviated as F-roughness or roughness, if not confused. An example is given to show how to calculate F-roughness and F-total-roughness as follows.

Example 4 Assume that $X = \{x : d(x) = 0\}$, $F = \{DT_1, DT_2\}$ and $B = \{a, b\}$ in Tables 1 and 2.

$X|DT_1 = \{x : d(x) = 0\} \cap DT_1 = \{x_3\}$, $X|DT_2 = \{x : d(x) = 0\} \cap DT_2 = \{y_1, y_4, y_5\}$.

$\alpha_{B,DT_1}(X) = \frac{\underline{B}(X)}{\overline{B}(X)} = \frac{|\{x_3\}|}{|\{x_3\}|} = 1$, $\alpha_{B,DT_2}(X) = \frac{\underline{B}(X)}{\overline{B}(X)} = \frac{|\{y_1, y_4\}|}{|\{y_1, y_4, y_5, y_6\}|} = \frac{1}{2}$

$\alpha_{B,F}(X) = \{\alpha_{B,DT_1}(X), \alpha_{B,DT_2}(X)\} = \{1, \frac{1}{2}\}$

$\widetilde{\alpha_{B,F}(X)} = \frac{\sum_{i=1}^{2} \alpha_{B,DT_i}(X)}{2} = \frac{3}{4}$

$\overline{\alpha_{B,F}}(X) = \max_{DT_i \in F}\{\alpha_{B,DT_i}(X)\} = 1$

$\underline{\alpha_{B,F}}(X) = \min_{DT_i \in F}\{\alpha_{B,DT_i}(X)\} = \frac{1}{2}$

$\delta_R^2 = \frac{\sum_{i=1}^{2}(\alpha_{B,DT_i}(X) - \widetilde{\alpha_{B,F}}(X))^2}{2} = \frac{1}{16}$

$T\alpha_{B,F}(X) = \frac{\sum_{X \subseteq DT_i \wedge DT_i \in F} |\underline{B}(DT_i, X)|}{\sum_{X \subseteq DT_i \wedge DT_i \in F} |\overline{B}(DT_i, X)|} = \frac{3}{5}$

We will investigate some properties of F-roughness and F-total-rough in the following paragraphs.

Proposition 7 *In a family of information subsystems FIS, suppose that* $\emptyset \neq B_1 \subseteq B_2 \subseteq A$*, the following expressions are true.*

(1) $\alpha_{B_1,FIS}(X) \leq \alpha_{B_2,FIS}(X)$.
(2) $\widetilde{\alpha_{B_1,FIS}}(X) \leq \widetilde{\alpha_{B_2,FIS}}(X)$.
(3) $\overline{\alpha_{B_1,FIS}}(X) \leq \overline{\alpha_{B_2,FIS}}(X)$.
(4) $\underline{\alpha_{B_1,FIS}}(X) \leq \underline{\alpha_{B_2,FIS}}(X)$.
(5) $T\alpha_{B_1,FIS}(X) \leq T\alpha_{B_2,FIS}(X)$.

Proof According to their definitions, it is easy to prove the above expressions.

Proposition 7 shows that both F-roughness and F-total-roughness will increase with the increment of condition attributes.

Proposition 8 *Assume that both* FIS_1 *and* FIS_2 *are families of information subsystems and* $\emptyset \neq FIS_1 \subseteq FIS_2$*, then, the following expressions are true.*

(1) $\overline{\alpha_{B,FIS_1}}(X) \leq \overline{\alpha_{B,FIS_2}}(X)$.
(2) $\underline{\alpha_{B,FIS_2}}(X) \leq \underline{\alpha_{B,FIS_1}}(X)$.

Proof It is obvious that these expressions are true.

Proposition 8 shows that the maximum of F-roughness will increase with the increment of the set of information subsystems, while the minimum of F-roughness will decrease with the increment of the set of information subsystems.

Proposition 9 *Let* FIS_1 *and* FIS_2 *be families of information subsystems and* $\emptyset \neq FIS_1 \subseteq FIS_2$*,* $FIS = FIS_1 \cup FIS_2$*, then, the following expressions are true.*

(1) $\min\{\widetilde{\alpha_{B,FIS_1}}(X), \widetilde{\alpha_{B,FIS_2}}(X) \leq \widetilde{\alpha_{B,FIS}}(X) \leq \max\{\widetilde{\alpha_{B,FIS_1}}(X), \widetilde{\alpha_{B,FIS_2}}(X)\}$.
(2) $\overline{\alpha_{B,FIS}}(X) = \max\{\overline{\alpha_{B,FIS_1}}(X), \overline{\alpha_{B,FIS_2}}(X)\}$.
(3) $\underline{\alpha_{B,FIS}}(X) = \min\{\underline{\alpha_{B,FIS_1}}(X), \underline{\alpha_{B,FIS_2}}(X)\}$.
(4) $\min\{T\alpha_{B,FIS_1}(X), T\alpha_{B,FIS_2}(X)\} \leq T\alpha_{B,FIS}(X) \leq \max\{T\alpha_{B,FIS_1}(X), T\alpha_{B,FIS_2}(X)\}$.

Proof According to their definitions and primary mathematical knowledge, the above expressions are true.

According to Proposition 9, when two sets of information subsystems are united, the average of F-roughness and the F-total-roughness will be a value between the two old ones, and the maximum of F-roughness will be the maximum one in the union of two F-roughnesses, and the minimum of F-roughness will be the minimum one in the union of two F-roughnesses.

Proposition 10 *The following propositions are true.*

(1) $\alpha_{B,FIS}(X) = 1 \Leftrightarrow T\alpha_{B,FIS}(X) = 1 \Leftrightarrow \underline{B}(FIS, X) = \overline{B}(FIS, X)$.
(2) $\alpha_{B,FIS}(X) = 0 \Leftrightarrow T\alpha_{B,FIS}(X) = 0 \Leftrightarrow \forall_{IS_i \in FIS}(BND(IS_i, X) = X)$.
(3) $\overline{\alpha_{B,FIS}}(X) = 1 \Leftrightarrow \exists_{IS_i \in FIS}(\underline{B}(IS_i, X) = \overline{B}(IS_i, X))$.
(4) $\underline{\alpha_{B,FIS}}(X) = 0 \Leftrightarrow \exists_{IS_i \in FIS}(BND(IS_i, X) = X)$

Proof According to their definitions, it is easy to get the above results.

Proposition 10 shows that we can judge the relations among the upper approximation, the boundary region and the original concept according to the values of F-roughness and F-total-roughness in F-rough sets.

F-roughness and F-total-roughness are two types of roughness in F-rough sets. The first focuses on the changing and variation of roughness in every elements of F, while the last emphasizes the effectiveness of roughness in the family of information subsystems (decision subsystems) F.

7 Conclusion

In this chapter some indexes of vagueness and uncertainty are defined in F-rough sets, including F-membership-degree, F-total-membership-degree, F-dependence degree, F-total-dependence-degree, F-roughness and F-total roughness. These indexes may be a set of numbers or a vector, and can indicate vagueness of vagueness and uncertainty of uncertainty just like type-2 fuzzy sets and cloud model. After having defined these indexes, F-rough sets becomes the third tools to describe vagueness of vagueness and uncertainty of uncertainty in vague and uncertain phenomena. These indexes can be calculated strictly from a family of information subsystems (decision subsystems), but they are flexible and applicable. Moreover, This research provides not only a way to obtain vagueness of vagueness and uncertainty of uncertainty from data sets directly, but also a possibility to unify rough set theory and fuzzy set theory.

In the future we will investigate more properties of these indexes, and employ them to mine stream data and detect concept drift.

Acknowledgements The work is supported by National Natural Science Foundation of China (Nos 61473030), Zhejiang Provincial Natural Science Foundation of China (Nos LY15F020012) and Zhejiang Provincial Top Discipline of Cyber Security at Zhejiang Normal University.

References

1. An, A., Shan, N., Chan, C., et al.: Discovering rules for water demand prediction: an enhanced rough-set approach. Eng. Appl. Artif. Intell. **9**(6), 645–653 (1996)
2. Bazan, G.J.: A comparison of dynamic non-dynamic rough set methods for extracting laws from decision tables. In: Polkowski, L., Skowron, A. (eds.) Rough Sets in Knowledge Discovery 1: Methodology and Applications, pp. 321–365. Springer, Heidelberg (1998)
3. Bazan G.J., Nguyen H.S., Nguyen, S.H., et al.: Rough set algorithms in classification problem. In: Polkowski, L., Tsumot, S.O, Lin, T.Y. (eds.) Rough Set Methods and Applications, pp. 49–88. Springer, Heidelberg (2000)
4. Bello, R., Verdegay, J.L.: Rough sets in the soft computing environment. Inf. Sci. **212**(3), 1–14 (2012)

5. Deng, D.Y., Huang, H.K., Dong, H.B.: Rough sets based on accessible relation. J. Beijing Jiaotong Univ. **30**(5), 19–23 (2006). (in Chinese)
6. Deng, D.Y., Yan, D.X., Wang, J.Y.: Parallel reducts based on attribute significance. In: Ju, J., Greco, S., Lingras, P., et al. (eds.) Rough Set and Knowledge Technology, pp. 336–343. Springer, Heidelberg (2010)
7. Deng, D.Y., Chen, L.: Parallel reducts and F-rough sets. In: Wang, G., Li, D., Yao, Y.Y., et al. (eds.) Cloud Model and Granular Computer, pp. 210–228. Science Press, Beijing (2012). (in Chinese)
8. Deng, D.Y., Pei, M.H., Huang, H.K.: The F-rough sets approaches to the measures of concept drift. J. Zhejiang Norm. Univ. Nat. Sci. **36**(3), 303–308 (2013). (in Chinese)
9. Deng, D.Y., Xu, X.Y., Huang, H.K.: Concept drifting detection for categorical evolving data based on parallel reducts. J. Comput. Res. Dev. **52**(5), 1071–1079(2015) (in Chinese)
10. Deng, D.Y., Miao, D.Q., Huang, H.K.: Analysis of concept drifting and uncertainty in an information table. J. Comput. Res. Dev. **53**(11), 2607–2612 (2016). (in Chinese)
11. Deng, D.Y., Huang, H.K.: Double-level absolute reduction for multi-granulation rough sets. Pattern Recognit. Artif. Intell. **29**(11), 969–975 (2016). (in Chinese)
12. Deng, D.Y, Xue, H.H, Miao, D.Q., et al.: Study on criteria of attribute reduction and information loss of attribute reduction. Acta Eletron. Sin. **45**(2), (2017). (in Chinese)
13. Dubois, D., Prade, H.: Rough fuzzy sets and fuzzy rough sets. Int. J. Gen. Syst. **17**(2), 191–209 (1990)
14. Katzberg, J.D., Ziarko, W.: Variable precision extension of rough set. Foundamenta Inform. **27**(2–3), 155–168 (1996)
15. Li, D., Meng, H., Shi, X.: Membership clouds and membership cloud generators. J. Comput. Res. Dev. **32**(6), 15–20 (1995). (in Chinese)
16. Li, D., Du, Y.: Artificial Intelligence with Uncertainty. National Defence Industry Press, Beijing (2005)
17. Liu, J., Hu, Q., Yu, D.: A weighted rough set based method developed for class imbalance learning. Inf. Sci. **178**(4), 1235–1256 (2008)
18. Mendel, J.M., John, R.I.: Type-2 fuzzy sets made simple. IEEE Trans. Fuzzy Syst. **10**(2), 117–127 (2002)
19. Pawlak, Z.: Rough sets. Int. J. Inf. Comput. Sci. **11**(5), 341–356 (1982)
20. Pawlak, Z.: Rough Sets-Theoretical Aspect of Reasoning About Data. Kluwer Academic Publishers, Dordrecht (1991)
21. Pawlak, Z.: Vagueness and uncertainty: a rough set perspective. Comput. Intell. **11**(2), 227–232 (1995)
22. Pawlak, Z., Skowron, A.: Rudiments of rough sets. Inf. Sci. **177**(1), 3–27 (2007)
23. Pawlak, Z., Skowron, A.: Rough sets: some extensions. Inf. Sci. **177**(1), 28–40 (2007)
24. Pawlak, Z., Wong, S.K.M., Ziarko, W.: Rough sets: probabilistic versus deterministic approach. Int. J. Man Mach. Stud. **29**(1), 81–95 (1988)
25. Russell, B.: Vagueness. Aust. J. Philos. **1**, 84–92 (1923)
26. Shafer, G.: A Mathematical Theory of Evidence. Princeton University Press, Princeton, NJ (1976)
27. Slezak, D.: The rough bayesian model for distributed decision systems. In: Tsumoto, S., Slowinski, R., Komorowski, H.J. (eds.) Rough Sets and Current Trends in Computing (RSCTC2004), pp. 384–393. Springer, Berlin (2004)
28. Slezak, D.: Rough sets and Bayes factor. Transaction on Rough Sets III, LNAI 3400, pp. 202–229 (2005)
29. Slezak, D., Ziarko, W.: Attribute reduction in the bayesian version of variable precision rough set model. Electron. Notes Theor. Comput. Sci. **82**(4), 1–11 (2003)
30. Slezak, D., Ziarko, W.: The investigation of the bayesian rough set model. Int. J. Approx. Reason. **40**(1–2), 81–91 (2005)
31. Wong, S.K.M., Ziarko, W., Wu, J.: Comparison of the probabilistic approximate classification and the fuzzy set model. Fuzzy Sets Syst. **21**(3), 357–362 (1987)
32. Yao, Y.Y.: Probabilistic approaches to rough sets. Expert Syst. **20**(5), 287–297 (2003)

33. Yao, Y.Y., Wong, S.K.M.: A decision theoretic framework for approximating concepts. Int. J. Man Mach. Stud. **37**(6), 793–809 (1992)
34. Yao, Y.Y., Wong, S.K.M., Lingras, P.: A decision-theoretic rough set model. In: Ras, Z.W., Zemankova, M., Emrich, M.L. (eds.) Methodologies for Intelligent Systems, vol. 5, pp. 17–24. North-Holland, New York (1990)
35. Yao, Y.Y., Zhao, Y.: Attribute reduction in decision-theoretic rough set models. Inf. Sci. **178**(17), 3356–3373 (2013)
36. Zadeh, L.: Fuzzy sets. Inf. Control **8**(3), 338–353 (1965)
37. Zhu, W., Wang, F.: Reduction and axiomization of covering generalized rough sets. Inf. Sci. **152**(1), 217–230 (2003)
38. Zhu, W.: Topological approaches to covering rough sets. Inf. Sci. **177**(6), 1499–1508 (2007)
39. Zhu, W.: Generalized rough Sets based on relations. Inf. Sci. **177**(22), 4997–5011 (2007)
40. Zhu, W., Wang, F.: On three types of covering rough sets. IEEE Trans. Knowl. Data Eng. **19**(8), 1131–1144 (2007)
41. Ziarko, W.: Variable precision rough sets model. J. Comput. Syst. Sci. **46**(1), 39–59 (1993)
42. Ziarko, W.: Acquisition of hierarchy-structured probabilistic decision tables and rules from data. Expert Syst. **1**(20), 305–310 (2003)
43. Ziarko, W.: Probabilistic rough sets. In: Slezak, D., Wang, G.Y., Szczuka, M.S., et al. (eds.) Rough Sets, Fuzzy Sets, Data Mining and Granular Computing (RSFDGrC2005), pp. 283–293. Springer, Heidelberg (2005)

Directions of Use of the Pawlak's Approach to Conflict Analysis

Małgorzata Przybyła-Kasperek

Abstract The chapter briefly discusses selected applications of the rough set theory. The main aim of the chapter is to describe a model of conflict proposed by Professor Pawlak as well as its extension. In the chapter the Pawlak's model was applied to analyze the conflicts that arise between classifiers when making decisions. The example that is included in the chapter allows for comparison of the results generated by the Pawlak's model with the results generated by other approaches. Some properties in generated coalitions have been noticed.

1 Introduction

Nowadays, the ability to process data very quickly, is extremely important. Approximate reasoning provides the ability to process big data. Rough set theory, which was proposed by Professor Pawlak, gives us the opportunity to use approximate reasoning based on vague concepts.

The ability to analyze conflict situations that arise for example between classifiers is also crucial. Many mathematical models of conflict situations were proposed [17, 18, 21, 55]. The approach that was proposed by Professor Pawlak is still another approach to conflict analysis. The model is simple enough for easy computer implementation and allows to determine coalitions, the course and outcome of conflict, and takes into account the strength and strategies of agents.

This chapter describes selected applications of the rough set theory but mainly it focuses on the Pawlak's conflict model. The chapter presents the original Pawlak's model. Provides basic definitions and shows an example of the use of the original model. In the chapter three approaches, which are an extension of the Pawlak's model and were proposed by the author, are considered in more detail. In addition, some other extensions of the Pawlak's model are mentioned in the chapter. The main novelty of this chapter is to use the Pawlak's model to analyze conflicts that arise

M. Przybyła-Kasperek (✉)
Institute of Computer Science, University of Silesia,
Będzińska 39, 41-200 Sosnowiec, Poland
e-mail: malgorzata.przybyla-kasperek@us.edu.pl

G. Wang et al. (eds.), *Thriving Rough Sets*, Studies in Computational Intelligence 708, DOI 10.1007/978-3-319-54966-8_16

between classifiers when making decisions. Three different approaches to application of the Pawlak's model to analyze the conflicts between classifiers are discussed in this chapter. These approaches rely on different methods of defining an information system in a conflict model and different functions for determining the intensity of conflict between agents. Based on an example, results obtained by these approaches were compared. A measure for determining the distance between pairs of agents belonging to one coalition is proposed. Analysis of the obtained results allowed to identify the advantages and disadvantages of the discussed approaches.

The chapter is organized as follows. In the second part of this chapter a review of selected applications of the rough set theory is included. In the third section, the original Pawlak's model and its extensions are discussed. The fourth chapter describes three approaches to application of the Pawlak's model to analyze the conflicts between classifiers. In this section, an example is presented on the basis of which the results obtained using the considered approaches are compared. The chapter concludes with a short summary in the fifth section.

2 About the Rough Sets—A Review of Applications at the University of Silesia

Theory of rough set is one of the major contributions of the Professor Pawlak's work in the development of computer science. Rough sets were defined in 1982 in the paper [29]. Very quickly, this theory has found many applications [9, 13, 22, 27, 38, 51, 53]. Of course, the papers mentioned here are only a small part of applications—it is not possible to mention all, because there are so many of them. In the Institute of Computer Science at the University of Silesia rough sets were used, inter alia, in the following approaches.

The theory of rough set was used in the research conducted under the supervision of A. Wakulicz-Deja that were presented in a doctoral dissertation of P. Paszek. This approach was considered in the papers [28, 57, 58]. The study analyzed the real set of medical data that was gathered in the II Department of the Medical University of Silesia in Katowice. The set of data concerned diagnosing of patients with suspected mitochondrial diseases. In this study, the need to create, in real cases, local knowledge bases that contain different sets of conditional attributes was highlighted. The rough set theory was used in the inference process.

The rough set theory was applied to the analysis of complex medical data in the doctoral dissertation of G. Ilczuk where the supervisor was A. Wakulicz-Deja [14–16]. The study analyzed the real sets of medical data that were provided by the Electrocardiography Clinic of the Medical University of Silesia in Katowice. These data were grouped in separate tables, which were linked by relations to ensure data consistency. The study noted that the typical medical data contain dozens of tables that are related by different types of relations. These tables contain thousands of objects that are described by dozens of attributes. The method of construction of decision table containing attributes appropriate to the considered research problem, based on separate tables, was proposed.

M. Moshkov and B. Zielosko analyzed the use of partial reducts and partial decision rules in the papers [23, 24]. More precise classifiers are built on the basis of partial (approximate) reducts and decision rules than on the basis of exact reducts. In the monograph [25] concepts such as decision trees, decision rules and reducts have been widely discussed by M. Moshkov and B. Zielosko.

The rough set theory was also used in studies of many other researchers from the Institute of Computer Science at the University of Silesia [10, 26, 47, 48].

Analysis of conflicts, which is another important issue that Professor Pawlak dealt with, is also analyzed at the University of Silesia. Several publications in which this issue was considered, were prepared at this faculty [39–42]. This chapter presents yet another approach to this issue.

The aim of the review presented in this section is to show that the Institute of Computer Science at the University of Silesia is one of the centers that are engaged in the development of the Professor Pawlak's work.

3 About Conflict Analysis

Very important contribution of the Professor Pawlak's work in the development of computer science and in general to the theory of conflict and negotiation is a model of conflict analysis that was proposed in 1984 in the article [30]. This model was then developed in the papers [31–36]. This model provides a simple way to determine the relations between individuals involved in the conflict. It enables the analysis of strength of units and allows the modeling of the conflict. Basic concepts of the Pawlak's model will be presented.

It is assumed that the set Ag is a set of agents that are involved in the conflict. Opinion about certain discussed issues is expressed by each agent separately. This opinion is given by one of three values: -1 means that an agent is against, 0 neutral toward the issue 1 means favorable. This knowledge can be written in the form of an information system $S = (U, A)$, where the universe U are agents, A is a set of issues, and the set of values of $a \in A$ is equal to $V^a = \{-1, 0, 1\}$. The value $a(x)$, where $x \in U, a \in A$ is opinion of agent x about issue a. In order to better understand the conflict analysis problem and model, an example will be presented.

Example 1 We consider an example that is related to politics and elections. In the example there are six voters (agents) $U = \{1, 2, 3, 4, 5, 6\}$ and four parties (issues) $A = \{a, b, c, d\}$

a Democratic Party,
b Republican Party,
c Libertarian Party,
d Green Party.

Table 1 Information system for Example 1

U	a	b	c	d
1	−1	+1	+1	+1
2	+1	−1	0	−1
3	+1	−1	−1	0
4	0	−1	−1	0
5	+1	−1	−1	−1
6	0	+1	−1	0

The relationship of each voter to a specific issue is presented in Table 1.

In the first step of conflict analysis the relationships between agents are determined. For this purpose, a function $\phi_a : U \times U \to \{-1, 0, 1\}$ is defined for each $a \in A$:

$$\phi_a(x, y) = \begin{cases} 1 & \text{if } a(x)a(y) = 1 \text{ or } x = y, \\ 0 & \text{if } a(x)a(y) = 0 \text{ and } x \neq y, \\ -1 & \text{if } a(x)a(y) = -1. \end{cases}$$

Three relations are defined: R_a^+ alliance, R_a^0 neutrality, R_a^- conflict over $U \times U$. These relationships are expressed as follows

$$\begin{array}{l} R_a^+(x, y) \text{ if and only if } \phi_a(x, y) = 1, \\ R_a^0(x, y) \text{ if and only if } \phi_a(x, y) = 0, \\ R_a^-(x, y) \text{ if and only if } \phi_a(x, y) = -1. \end{array}$$

Each equivalence class of alliance relation R_a^+ is called coalition on a.

As can be seen, in Example 1, voters 2, 3 and 5 are allied on issue a—Democratic Party, voters 4 and 6 are neutral to this issue whereas, voter 1 and voter 2, voter 1 and voter 3, and voter 1 and voter 5 are in conflict about this issue. Coalition on issue a is a set of voters $\{2, 3, 5\}$.

Of course, we want to determine the relations between agents not only due to one attribute, but also due to a set of attributes. A function of distance between agents $\rho_B^* : U \times U \to [0, 1]$ for the set of issues $B \subseteq A$ is defined

$$\rho_B^*(x, y) = \frac{\sum_{a \in B} \phi_a^*(x, y)}{card\{B\}},$$

where

$$\phi_a^*(x, y) = \frac{1 - \phi_a(x, y)}{2} = \begin{cases} 0 & \text{if } a(x)a(y) = 1 \text{ or } x = y, \\ 0.5 & \text{if } a(x)a(y) = 0 \text{ and } x \neq y, \\ 1 & \text{if } a(x)a(y) = -1. \end{cases}$$

In the definition it was assumed that the distance between agents being in conflict is greater than distance between agents which are neutral. The function of distance between agents for the set of all issues $B = A$ is written in short as ρ^*.

There is also another way of determining the strength of the conflict between agents. For this purpose, we can use a concept of a discernibility matrix [49]. The discernibility matrix of $B \subseteq A$ has a dimension $card\{U\} \times card\{U\}$ and the element with index $x, y \in U$ is equal to

$$\delta_B(x, y) = \{a \in B : a(x) \neq a(y)\}.$$

Then a conflict function $\rho_B : U \times U \to [0, 1]$ for the set of issues $B \subseteq A$ is defined

$$\rho_B(x, y) = \frac{card\{\delta_B(x, y)\}}{card\{B\}}.$$

In this function, the distance between agents being in conflict is equal to distance between agents which are neutral.

Applying one of the two functions mentioned above, we can define the relations between agents more generally by taking into account a set of attributes. A pair $x, y \in U$ is said to be:

- allied $R^+(x, y)$, if $\rho(x, y) < 0.5$,
- in conflict $R^-(x, y)$, if $\rho(x, y) > 0.5$,
- neutral $R^0(x, y)$, if $\rho(x, y) = 0.5$.

Set $X \subseteq U$ is a coalition if for every $x, y \in X$, $R^+(x, y)$ and $x \neq y$.

The value of the function of distance between agents and the value of the conflict function for Example 1 was calculated for each pair of agents. These values are given in Table 2.

This conflict situation can be easily illustrated by a graph. A graphical representation of the conflict situation, that takes into account all issues and the function ρ^*, is presented in Fig. 1. Figure 2 shows a graphical representation of the conflict situation for the function ρ. Agents are represented by circles in the figures. When agents are allied the circles representing the agents are linked. In order to find coalitions, all cliques should be identified in the graph. So the subset of vertices such that every two vertices are linked is determined. When we consider the function ρ^* there is one coalition $\{2, 3, 4, 5\}$. When we consider the function ρ there are four coalitions $\{2, 5\}$, $\{3, 4\}$, $\{3, 5\}$ and $\{4, 6\}$. As can be seen, in the case of the use of the function ρ one big coalition $\{2, 3, 4, 5\}$ was split into three smaller $\{2, 5\}$, $\{3, 4\}$, $\{3, 5\}$. An additional coalition was also created $\{4, 6\}$, which previously has not been generated. This situation stems from the fact that both agents 4 and 6 are neutral towards the issues a and d, and then the function ρ has a smaller value than the function ρ^*.

The approach to conflict analysis proposed by Professor Pawlak was generalized and developed by other authors. In the papers [43–46, 52] of A. Skowron, S. Ramanna and J.F. Peters, the Pawlak's conflict model was developed by applying approximation spaces. In this extended model, a level of conflict that is acceptable

Table 2 Values of functions for Example 1

Function of distance between agents ρ^*

	1	2	3	4	5	6
1						
2	0.875					
3	0.875	0.25				
4	0.75	0.375	0.25			
5	1	0.125	0.125	0.25		
6	0.5	0.625	0.5	0.5	0.5	

Conflict function ρ

	1	2	3	4	5	6
1						
2	1					
3	1	0.5				
4	1	0.75	0.25			
5	1	0.25	0.25	0.5		
6	0.75	1	0.5	0.25	0.75	

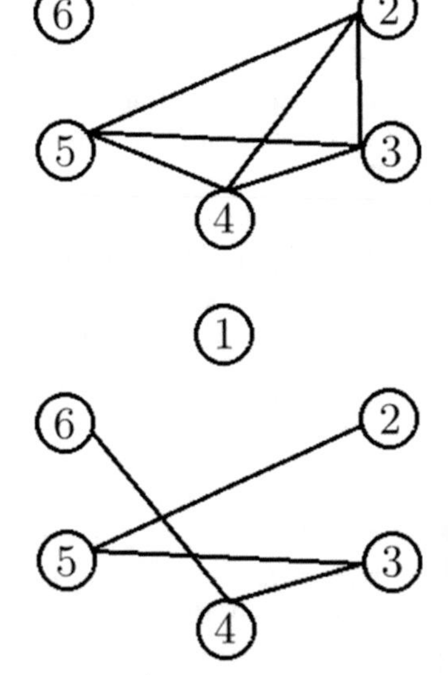

Fig. 1 A graphical representation of Example 1, the function ρ^*

Fig. 2 A graphical representation of Example 1, the function ρ

may be defined. It is also possible to specify the equivalent set of requirements based on a specified set of negotiation parameters in this model.

Another important extension of the Pawlak's model was presented in the papers [3–8, 50] by A. Skowron and R. Deja. This model can be applied in a situation in which each agent has a separate set of issues that he is interested in. The limitation of available resources is the reason that the agents get into conflicts. The main extension of the Pawlak's model is the ability to search for the causes of conflict and seek consensus on various levels.

The approach proposed by the author of the paper is yet another extension of the Pawlak's model. Conflict analysis was used to examine relations between classifiers that generate vectors of ranks or vectors of probabilities. In the cases described in the papers [39–42], classifiers make decisions on the basis of separate—local decision tables. These tables can be in various forms—sets of conditional attributes and sets of objects of different tables do not have to be equal or disjoint. However, there are no contraindications to apply the conflict model, in cases where the probability vectors or vectors of ranks are generated by various classifiers on the basis of one decision table.

The proposed development of the Pawlak's model provides three basic approaches to the creation of coalitions of classifiers. In the first two approaches coalitions are formed as a result of a one-step process that uses relations between classifiers. In the first approach (proposed in the paper [41]) disjoint coalitions are created, in the second approach (proposed in the paper [39]) non-disjoint coalitions are created. In the third approach (proposed in the papers [40, 42]) a two-step process of coalitions creation is used—the second step is a negotiations stage. Now we will describe the basic concepts of these three approaches.

Each classifier is called an agent, ag is an agent and Ag is a set of agents. It is assumed that for a classified object x and for each classifier a vector of ranks or a vector of probabilities is generated. In this second case, based on the vector of probabilities a vector of ranks is defined. Thus, for each agent ag_i the vector of ranks $[r_{i,1}(x), \ldots, r_{i,c}(x)]$, where $c = card\{V^d\}$ and V^d is a set of values of decision attribute, is generated. We define the function $\phi^x_{v_j}$ for the classified object x and each value of the decision attribute $v_j \in V^d$; $\phi^x_{v_j} : Ag \times Ag \to \{0, 1\}$

$$\phi^x_{v_j}(ag_i, ag_k) = \begin{cases} 0 & \text{if } r_{i,j}(x) = r_{k,j}(x) \\ 1 & \text{if } r_{i,j}(x) \neq r_{k,j}(x) \end{cases}$$

where $ag_i, ag_k \in Ag$.

We also define the intensity of conflict between agents using a function of the distance between agents. We define the distance between agents ρ^x for the test object x: $\rho^x : Ag \times Ag \to [0, 1]$

$$\rho^x(ag_i, ag_k) = \frac{\sum_{v_j \in V^d} \phi^x_{v_j}(ag_i, ag_k)}{card\{V^d\}},$$

where $ag_i, ag_k \in Ag$.

We say that agents $ag_i, ag_k \in Ag$ are in a friendship relation due to the object x, which is written $R^+(ag_i, ag_k)$, if and only if $\rho^x(ag_i, ag_k) < 0.5$. Agents $ag_i, ag_k \in Ag$ are in a conflict relation due to the object x, which is written $R^-(ag_i, ag_k)$, if and only if $\rho^x(ag_i, ag_k) \geq 0.5$.

In the first approach [41], disjoint groups of agents remaining in friendship relation are created. Formation of coalitions in this approach is very similar to the hierarchical agglomerate clustering method and proceeds as follows. Initially, each resource agent is treated as a separate cluster. These two steps are performed until the stop condition, which is given in the first step, is met.

1. One pair of different clusters is selected (in the very first step a pair of different resource agents) for which the distance reaches a minimum value. If the selected value of the distance is less than 0.5, then agents from the selected pair of clusters are combined into one new cluster. Otherwise, the clustering process is terminated.
2. After defining a new cluster, the value of the distance between the clusters are recalculated. The following method for recalculating the value of the distance is used. Let $\rho^x : 2^{Ag} \times 2^{Ag} \to [0, 1]$, let D_i be a cluster formed from the merging of two clusters $D_i = D_{i,1} \cup D_{i,2}$ and let it be given a cluster D_j then

$$\rho^x(D_i, D_j) = \begin{cases} \frac{\rho^x(D_{i,1}, D_j) + \rho^x(D_{i,2}, D_j)}{2} & \text{if } \rho^x(D_{i,1}, D_j) < 0.5 \\ & \text{and } \rho^x(D_{i,2}, D_j) < 0.5 \\ \max\{\rho^x(D_{i,1}, D_j), \rho^x(D_{i,2}, D_j)\} & \text{if } \rho^x(D_{i,1}, D_j) \geq 0.5 \\ & \text{or } \rho^x(D_{i,2}, D_j) \geq 0.5 \end{cases}$$

In the second approach [39], non-disjoint groups of agents who are in friendship relation are created. In this approach, the coalition is the maximum, due to the inclusion relation, subset of agents $X \subseteq Ag$ such that

$$\forall_{ag_i, ag_k \in X} \; R^+(ag_i, ag_k).$$

In the third approach [40] modified definitions of relations between agents are used. Let p be a real number that belongs to the interval $[0, 0.5)$. We say that agents $ag_i, ag_k \in Ag$ are in a friendship relation due to the object x, which is written $R^+(ag_i, ag_k)$, if and only if $\rho^x(ag_i, ag_k) < 0.5 - p$. Agents $ag_i, ag_k \in Ag$ are in a conflict relation due to the object x, which is written $R^-(ag_i, ag_k)$, if and only if $\rho^x(ag_i, ag_k) > 0.5 + p$. Agents $ag_i, ag_k \in Ag$ are in a neutrality relation due to the object x, which is written $R^0(ag_i, ag_k)$, if and only if $0.5 - p \leq \rho^x(ag_i, ag_k) \leq 0.5 + p$. The process of coalition creating consists of two stages. In the first stage, agents remaining in the friendship relations are combining into initial groups. In the second stage, neutral agents are connected to the created initial groups. However, in order to determine the opportunity to join the neutral agents, a generalized distance function between agents must be calculated. In the definition of this distance, disagreements of agents for only the highest ranks are considered. We define the function ϕ^x_G for test object x; $\phi^x_G : Ag \times Ag \rightarrow [0, \infty)$

$$\phi^x_G(ag_i, ag_j) = \frac{\sum_{v_l \in Sign_{i,j}} |r_{i,l}(x) - r_{j,l}(x)|}{card\{Sign_{i,j}\}}$$

where $ag_i, ag_j \in Ag$ and $Sign_{i,j} \subseteq V^d$ is the set of significant decision values for the pair of agents ag_i, ag_j. We also define the generalized distance between agents ρ^x_G for the test object x; $\rho^x_G : 2^{Ag} \times 2^{Ag} \rightarrow [0, \infty)$

$$\rho^x_G(X, Y) = \begin{cases} 0 & \text{if} \quad card\{X \cup Y\} \leq 1 \\ \dfrac{\sum\limits_{ag, ag' \in X \cup Y} \phi^x_G(ag, ag')}{card\{X \cup Y\} \cdot (card\{X \cup Y\} - 1)} & \text{else} \end{cases}$$

where $X, Y \subseteq Ag$. As can be easily seen, the value of the generalized distance function for two sets of agents X and Y is equal to the average value of the function ϕ^x_G for each pair of agents ag, ag' that belong to the set $X \cup Y$. In order to determine the final coalitions, for each agent, that does not belong to any initial group, we calculate the value of generalized distance function for this agent and every initial cluster. Then the agent is included to all initial clusters, for which the generalized distance does not exceed a certain threshold, which is set by the system's user. Also agents without coalition, for which the value does not exceed the threshold, are combined into a new cluster. Of course, the agents, who are in a conflict relation can not be connected into one coalition. In this way we obtain the final form of coalitions.

In the literature, other approaches to conflicts analysis can be found. Fusion methods are often used when several classifiers make decisions [20, 54]. There are also methods that are proposed in the multiple model approach [11, 19, 37] and in distributed data mining approach [1, 2]. All of these approaches are different from those considered in this chapter, because the relations between classifiers are not as thoroughly analyzed in them. In addition, assumptions relating to local tables are also completely different in these approaches. In the literature, a general framework for conflict analysis can also be found, e.g., a conflict model theory of decision making [56, 59]. It uses the expectations and values. Expectations include the anticipated consequences of the decision, and values are what the decision maker finds desirable. The basic questions about risks, consequences and alternatives are considered by the decision maker. The decision making process is presented in the form of a graph. In a situation which is considered in the paper, only vectors of ranks are available and it would be difficult to apply such a general framework, since many aspects occurring in it is not determined.

4 An Example

In this section, an example of using four different approaches to the analysis of conflicts arising between the classifiers when generating a global decision will be presented. In the first part, for the analysis of the conflict situation the Pawlak's model will be used. Although this model is not dedicated to deal with such situations, it can be used when assuming certain arrangements. Two cases of the use of the Pawlak's model will be considered. Then the three models proposed by the author of the paper will be considered. These models are extensions of the Pawlak's model. In the first approach disjoint coalitions of classifiers are created. In the second approach, coalitions are inseparable and in the third approach a two-step process of creating coalitions is used.

Consider an example in which the set of agents consists of seven classifiers $Ag = \{ag_1, ag_2, ag_3, ag_4, ag_5, ag_6, ag_7\}$. We assume that $V^d = \{v_1, v_2, v_3, v_4, v_5\}$. Each agent has generated a vector of probabilities that indicates the level of certainty with which the decisions were taken. These vectors are a reflection of the classification of a certain object x and are shown in Table 3. Based on these vectors, the vectors of ranks were designated, which are also shown in Table 3.

The conflict situation, to which we wanted to apply the Pawlak's model must be presented in the form of an information system that has the sets of attribute values equal to $\{-1, 0, 1\}$. Therefore, the vectors of ranks must be converted to this form. The universe U will be the set of agents Ag and the set of issues that being considered A will be the set of values of decision attribute V^d. Two approaches to determine the opinion of agent $ag \in U$ about issue $a \in A$ in the Pawlak's model will be considered.

4.1 *Pawlak's Model—Case Study 1*

In the first approach it was assumed, that if the classifier ag assigns the rank 1 to the value of the decision attribute $a \in V^d$ then the classifier is favorable to this decision value and $a(ag) = 1$. In other cases, we assume that the classifier is against, which

Table 3 Vectors of probabilities and vectors of ranks

Agent	Vector of probabilities	Vector of ranks
ag_1	[0.13, 0.2, 0.13, 0.2, 0.34]	[3, 2, 3, 2, 1]
ag_2	[0.15, 0.08, 0.23, 0.31, 0.23]	[3, 4, 2, 1, 2]
ag_3	[0.08, 0.17, 0.17, 0.25, 0.33]	[4, 3, 3, 2, 1]
ag_4	[0.1, 0.3, 0.2, 0.2, 0.2]	[3, 1, 2, 2, 2]
ag_5	[0.29, 0.14, 0.14, 0.29, 0.14]	[1, 2, 2, 1, 2]
ag_6	[0, 0.14, 0.14, 0.29, 0.43]	[4, 3, 3, 2, 1]
ag_7	[0.08, 0.15, 0.15, 0.38, 0.24]	[4, 3, 3, 1, 2]

Table 4 Information system—case study 1

U	v_1	v_2	v_3	v_4	v_5
ag_1	–1	–1	–1	–1	+1
ag_2	–1	–1	–1	+1	–1
ag_3	–1	–1	–1	–1	+1
ag_4	–1	+1	–1	–1	–1
ag_5	+1	–1	–1	+1	–1
ag_6	–1	–1	–1	–1	+1
ag_7	–1	–1	–1	+1	–1

Table 5 Values of the distance function between agents ρ^* and the conflict function ρ—case study 1

	ag_1	ag_2	ag_3	ag_4	ag_5	ag_6	ag_7
ag_1							
ag_2	0.4						
ag_3	0	0.4					
ag_4	0.4	0.4	0.4				
ag_5	0.6	0.2	0.6	0.6			
ag_6	0	0.4	0	0.4	0.6		
ag_7	0.4	0	0.4	0.4	0.2	0.4	

means $a(ag) = -1$. Thus, the information system in our example has the form as is shown in Table 4.

The value of the function of the distance between agents is calculated for each pair of agents. In this case, the values of the conflict function are the same as the values of the function of the distance between agents, as we do not have the neutral agents. These values are given in Table 5.

Figure 3 shows a graphical representation of the conflict situation. Agents are represented by circles in the figure. When agents are allied ($\rho(ag, ag') < 0.5$), the circles representing the agents are linked. In order to find coalitions, all cliques should be identified in the graph. So the subset of vertices such that every two vertices are linked is determined. There are two coalitions in the example $\{ag_1, ag_2, ag_3, ag_4, ag_6, ag_7\}$ and $\{ag_2, ag_5, ag_7\}$.

4.2 *Pawlak's model—Case study 2*

In the second approach the opinions of the agents are a bit softened by applying neutrality. It was assumed, that if the classifier ag assigns the rank 1 to the value of the decision attribute $a \in V^d$ then the classifier is favorable to this decision value and

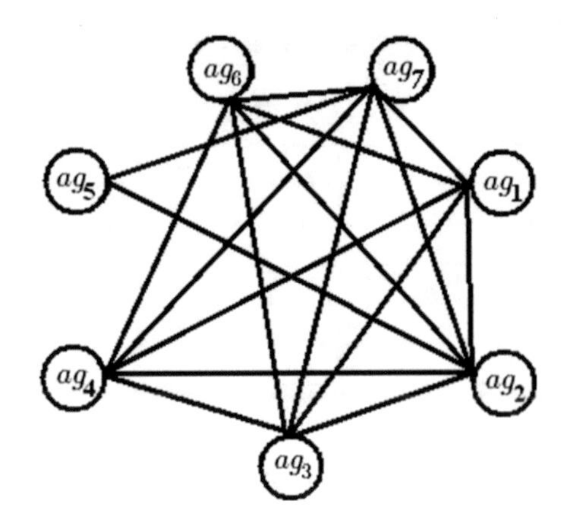

Fig. 3 A graphical representation of the example, the Pawlak's model—case study 1

Table 6 Information system—case study 2

U	v_1	v_2	v_3	v_4	v_5
ag_1	–1	0	–1	0	+1
ag_2	–1	–1	0	+1	0
ag_3	–1	–1	–1	0	+1
ag_4	–1	+1	0	0	0
ag_5	+1	0	0	+1	0
ag_6	–1	–1	–1	0	+1
ag_7	–1	–1	–1	+1	0

Table 7 Values of the distance function between agents ρ^*—case study 2

	ag_1	ag_2	ag_3	ag_4	ag_5	ag_6	ag_7
ag_1							
ag_2	0.4						
ag_3	0.2	0.3					
ag_4	0.4	0.5	0.5				
ag_5	0.6	0.5	0.6	0.6			
ag_6	0.2	0.3	0.1	0.5	0.6		
ag_7	0.3	0.2	0.2	0.5	0.5	0.2	

$a(ag) = 1$. If the classifier ag assigns the rank 2 to the value of the decision attribute $a \in V^d$ then the classifier is neutral toward this decision value and $a(ag) = 0$. In other cases, we assume that the classifier is against, which means $a(ag) = -1$. Thus, the information system in this case has the form as is shown in Table 6.

The values of the function of the distance between agents are given in Table 7. The values of the conflict function are given in Table 8. As can be easily seen in most cases the value of the function ρ^* is less than or equal to the value of the function

Table 8 Values of the conflict function ρ—case study 2

	ag_1	ag_2	ag_3	ag_4	ag_5	ag_6	ag_7
ag_1							
ag_2	0.8						
ag_3	0.2	0.6					
ag_4	0.6	0.4	0.6				
ag_5	0.8	0.4	1	0.6			
ag_6	0.2	0.6	0	0.6	1		
ag_7	0.6	0.2	0.4	0.6	0.6	0.4	

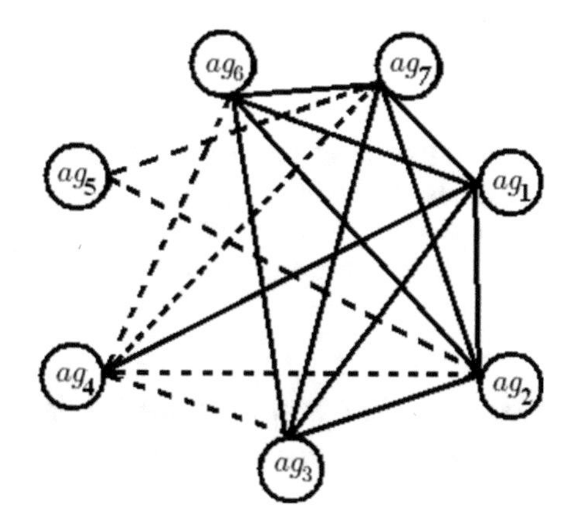

Fig. 4 A graphical representation of the conflict with the function ρ^*, the Pawlak's model—case study 2

ρ. However, in the case of pairs of agents (ag_2, ag_4), (ag_2, ag_5), and (ag_3, ag_6) the inequality is inverted. This is due to the fact that if $a(ag) = 0$ and $a(ag') = 0$, for $a \in V^d$ and $ag, ag' \in Ag$ then $\rho^*_{\{a\}}(ag, ag') > \rho_{\{a\}}(ag, ag')$.

Figure 4 shows a graphical representation of the conflict situation that is examined by the function ρ^*. As before when agents are allied $(\rho^*(ag, ag') < 0.5)$, the circles representing the agents are linked. When agents are neutral $(\rho^*(ag, ag') = 0.5)$, the circles representing the agents are connected by dotted line. In order to find coalitions, all cliques should be identified in the graph. So the subset of vertices such that every two vertices are linked is determined. There are three coalitions in the example $\{ag_1, ag_2, ag_3, ag_6, ag_7\}$, $\{ag_1, ag_4\}$ and $\{ag_5\}$.

Figure 5 shows a graphical representation of the conflict situation that is examined by the function ρ. As before when agents are allied $(\rho(ag, ag') < 0.5)$, the circles representing the agents are linked. In order to find coalitions, all cliques should be identified in the graph. There are five coalitions in the example $\{ag_1, ag_3, ag_6\}$, $\{ag_3, ag_6, ag_7\}$, $\{ag_2, ag_5\}$, $\{ag_2, ag_4\}$ and $\{ag_2, ag_7\}$.

To summarize the considered above two cases it can be seen that the coalitions that have been defined for the case study 2 (both with using the function ρ^* and the function ρ) are included in the coalitions defined in the case study 1. In the case

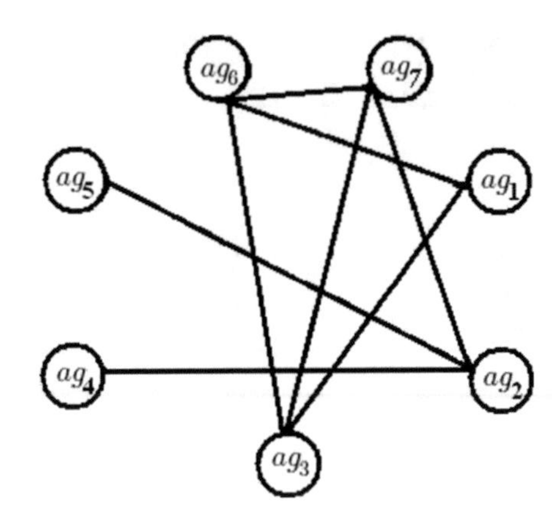

Fig. 5 A graphical representation of the conflict with the function ρ, the Pawlak's model—case study 2

study 2, the use of function ρ^* leads to the definition of fewer but more numerous coalitions, than with using the function ρ. However, these two sets of coalitions are incomparable, i.e. the set of coalitions defined with using ρ is not included in the set of coalitions defined with using ρ^*.

4.3 *The Extension of the Pawlak's Model—The Approach with Disjoint Coalitions*

As was mentioned earlier, this approach is described in detail in the paper [41]. In this approach, the value of the function of the distance between agents is calculated for each pair of agents on the basis of the vectors of rank that were presented in Table 3. In Table 9 the values of the function of the distance between agents are shown.

Then, in order to define disjoint coalitions algorithm that is presented in the previous section of the chapter is implemented. This algorithm is very similar to the hierarchical agglomerate clustering method and its subsequent steps are presented

Table 9 Values of the distance function between agents ρ^x

Agent	ag_1	ag_2	ag_3	ag_4	ag_5	ag_6	ag_7
ag_1	0						
ag_2	0.8	0					
ag_3	0.4	1	0				
ag_4	0.6	0.4	0.8	0			
ag_5	0.8	0.4	1	0.6	0		
ag_6	0.4	1	0	0.8	1	0	
ag_7	0.8	0.6	0.4	0.8	0.6	0.4	0

Table 10 Process of coalitions creating in the approach with disjoint coalitions of classifiers

Step 1

Agent	ag_1	ag_2	ag_3	ag_4	ag_5	ag_6	ag_7
ag_1	0						
ag_2	0.8	0					
ag_3	0.4	1	0				
ag_4	0.6	0.4	0.8	0			
ag_5	0.8	0.4	1	0.6	0		
ag_6	0.4	1	**0**	0.8	1	0	
ag_7	0.8	0.6	0.4	0.8	0.6	0.4	0

Step 2

Agent	ag_1	ag_2	$\{ag_3, ag_6\}$	ag_4	ag_5	ag_7
ag_1	0					
ag_2	0.8	0				
$\{ag_3, ag_6\}$	**0.4**	1	0			
ag_4	0.6	0.4	0.8	0		
ag_5	0.8	0.4	1	0.6	0	
ag_7	0.8	0.6	0.4	0.8	0.6	0

Step 3

Agent	ag_2	$\{ag_1, ag_3, ag_6\}$	ag_4	ag_5	ag_7
ag_2	0				
$\{ag_1, ag_3, ag_6\}$	1	0			
ag_4	**0.4**	0.8	0		
ag_5	0.4	1	0.6	0	
ag_7	0.6	0.8	0.8	0.6	0

Step 4

Agent	$\{ag_2, ag_4\}$	$\{ag_1, ag_3, ag_6\}$	ag_5	ag_7
$\{ag_2, ag_4\}$	0			
$\{ag_1, ag_3, ag_6\}$	1	0		
ag_5	0.6	1	0	
ag_7	0.8	0.8	0.6	0

in Table 10. In each of the steps, the selected minimum value of the distance function, for agents who will be merged into the next step, is bold in the matrix.

As a result of the algorithm's implementation we get four coalitions $\{ag_2, ag_4\}$, $\{ag_1, ag_3, ag_6\}$, $\{ag_5\}$ and $\{ag_7\}$.

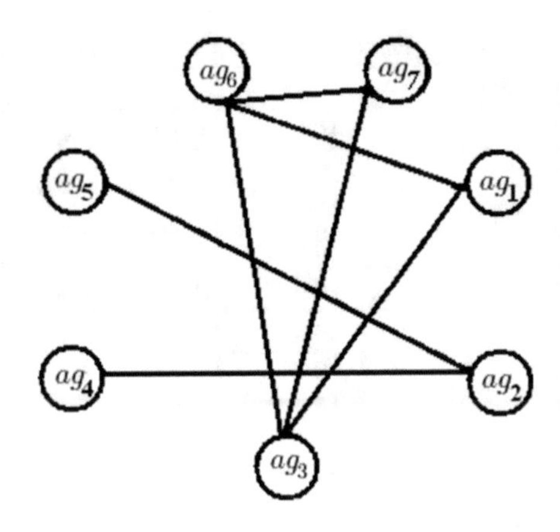

Fig. 6 A graphical representation of the example, the approach with non-disjoint coalitions

4.4 The Extension of the Pawlak's Model—The Approach with Non-disjoint Coalitions

As was mentioned earlier, this approach is described in detail in the paper [39]. In this approach, coalition is the maximum, due to inclusion relation, set of agents that remain in the friendship relation. A pair of agents (ag, ag') is in the friendship relation if and only if $\rho^x(ag, ag') < 0.5$. Based on the values of the distance function ρ^x, given in Table 9, a graphical representation of the conflict situation was made and shown in Fig. 6. When agents are in a friendship relation, the circles representing the agents are linked. In order to find coalitions, all cliques should be identified in the graph. There are four coalitions in the example $\{ag_1, ag_3, ag_6\}$, $\{ag_3, ag_6, ag_7\}$, $\{ag_2, ag_5\}$ and $\{ag_2, ag_4\}$.

4.5 The Extension of the Pawlak's Model—The Approach with Two-Step Process of Coalitions Creating

As was mentioned earlier, this approach is described in detail in the paper [40]. In this approach, the definitions of relations between agents were modified. In these definitions, parameter p occurs. Let us assume that this parameter is equal to $p = 0.1$. In the first step of the method some initial groups are created. Agents in the friendship relation are combined in the initial group. In the example only one initial group $\{ag_3, ag_6\}$ is generated, as only $\rho^x(ag_3, ag_6) < 0.5 - p = 0.4$.

In the next step of the method for each agent that has not been included in any initial groups, the values of generalized distance function are calculated. These values are given in Table 11. If a conflict occurs between the two sets of agents, then the corresponding cell in the table contains X.

Figure 7 shows a graphical representation of the neutrality relations ($0.4 \leq \rho^x(ag, ag') \leq 0.6$) in the considered conflict situation. When a pair of agents is in a

Table 11 Values of the generalized distance function between agents

	ag_1	ag_2	$\{ag_3, ag_6\}$	ag_4	ag_5	ag_7
ag_1	0					
ag_2	X	0				
$\{ag_3, ag_6\}$	0	X	0			
ag_4	1	2	X	0		
ag_5	X	1	X	$\frac{4}{3}$	0	
ag_7	X	0	$\frac{2}{3}$	X	$\frac{3}{2}$	0

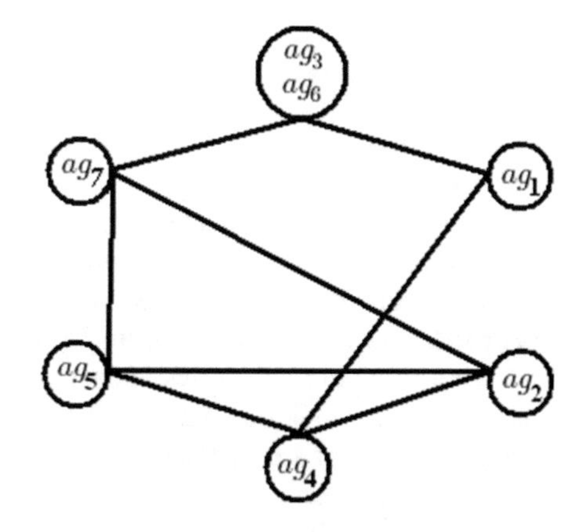

Fig. 7 A graphical representation of the example, the approach with two-step process of coalitions creating

neutrality relation, the circles representing the agents are linked. In order to find potential coalitions, all cliques should be identified in the graph. Then, for each of cliques the function value ρ^x_G is calculated and if this value meets the threshold condition, a coalition is created. Let us assume that the threshold value is equal to 2. From the graphical representation and the following calculations

$$\rho^x_G(\{ag_1\}, \{ag_4\}) = 1 \leq 2$$

$$\rho^x_G(\{ag_2, ag_5\}, \{ag_7\}) = \frac{0 + 1 + \frac{3}{2}}{3} \leq 2$$

$$\rho^x_G(\{ag_2, ag_4\}, \{ag_5\}) = \frac{1 + 2 + \frac{4}{3}}{3} \leq 2$$

$$\rho^x_G(\{ag_3, ag_6\}, \{ag_1\}) = 0 \leq 2$$

$$\rho^x_G(\{ag_3, ag_6\}, \{ag_7\}) = \frac{2}{3} \leq 2$$

Table 12 Coalitions received in the considered approaches

Approach	Coalitions
Pawlak's model—Case study 1	$\{ag_1, ag_2, ag_3, ag_4, ag_6, ag_7\}$, $\{ag_2, ag_5, ag_7\}$
Pawlak's model—Case study 2 with ρ^*	$\{ag_1, ag_2, ag_3, ag_6, ag_7\}$, $\{ag_1, ag_4\}$, $\{ag_5\}$
Pawlak's model—Case study 2 with ρ	$\{ag_1, ag_3, ag_6\}$, $\{ag_3, ag_6, ag_7\}$, $\{ag_2, ag_5\}$, $\{ag_2, ag_4\}$, $\{ag_2, ag_7\}$
Disjoint coalitions	$\{ag_2, ag_4\}$, $\{ag_1, ag_3, ag_6\}$, $\{ag_5\}$, $\{ag_7\}$
Non-disjoint coalitions	$\{ag_1, ag_3, ag_6\}$, $\{ag_3, ag_6, ag_7\}$, $\{ag_2, ag_5\}$, $\{ag_2, ag_4\}$
Two-step process of coalitions creating	$\{ag_1, ag_4\}$, $\{ag_2, ag_5, ag_7\}$, $\{ag_2, ag_4, ag_5\}$, $\{ag_1, ag_3, ag_6\}$, $\{ag_3, ag_6, ag_7\}$

we obtain the following coalitions $\{ag_1, ag_4\}$, $\{ag_2, ag_5, ag_7\}$, $\{ag_2, ag_4, ag_5\}$, $\{ag_1, ag_3, ag_6\}$, $\{ag_3, ag_6, ag_7\}$.

To summarize the considered above three approaches, that have been proposed by the author, it can be noted that the third approach with two-step process of coalitions creating generates the most comprehensive and complex coalitions. More precisely, the coalitions that have been defined using the first approach (disjoint coalitions) are included in the coalitions defined using the second approach (non-disjoint coalitions). And the coalitions that have been defined using the second approach (non-disjoint coalitions) are included in the coalitions defined using the third approach (two-step process of coalitions creating). In the papers [40, 42] it has been observed that in general, in the approach with negotiation, the clusters are more complex and better reconstruct and illustrate the views of the agents on the classification.

Table 12 shows a comparison of coalitions received in each of the considered approaches.

When comparing all considered approaches it can be seen that the Pawlak's model—Case study 2 with ρ and the approaches non-disjoint coalitions and two-step process of coalitions creating generate the most similar coalitions. The coalitions that have been defined using the Pawlak's model—Case study 2 with ρ are included in the coalitions defined using the approach with two-step process of coalitions creating. The coalitions generated using approaches the Pawlak's model: Case study 1 and Case study 2 with ρ^* are significantly more numerous compared to the coalitions generated using other approaches.

4.6 Comparative Study

In order to better evaluate and compare the coalitions that were obtained using different approaches, a measure to evaluate coalitions should be used. In the social theory, there are many measures of coalition's evaluation [12]. However, they take into account aspects such as—commitment, participation benefits and costs, stages of coalition development, member characteristics and so on—that do not occur in our considerations. Therefore, it was decided to use measure that is derived from the cluster analysis. One of the simplest and the most commonly used criteria for clusters evaluation is to minimize the sum of the distances between pairs of points belonging to the same cluster.

In our approach in order to calculate the distance between a pair of agents the vectors of ranks generated for these agents will be used. For each set of coalitions the sum of the distances between pairs of agents belonging to the same coalition is calculated according to the formula

$$W = \frac{1}{2} \sum_{k=1}^{K} \sum_{ag,ag' \in C_k} \sum_{i=1}^{card\{V^d\}} (r_{ag,i}(x) - r_{ag',i}(x))^2,$$

where K is the number of coalitions and C_k is the k-th coalition.

The distances between pairs of agents are given in Table 13 and the sum of the distances between pairs of agents belonging to the same coalition for each of the considered approaches is presented in Table 14.

As can be seen, in the approaches: Pawlak's model—Case study 1 and Case study 2 with ρ^* coalitions with the largest distance between agents within the coalition are

Table 13 Distances between pairs of agents

Agent	ag_1	ag_2	ag_3	ag_4	ag_5	ag_6	ag_7
ag_1	0						
ag_2	7	0					
ag_3	2	5	0				
ag_4	3	10	7	0			
ag_5	7	8	13	6	0		
ag_6	2	5	0	7	13	0	
ag_7	4	3	2	7	11	2	0

Table 14 Sum of the distances between pairs of agents belonging to the same coalition

Approach	Sum of the distances
Pawlak's model—Case study 1	$66 + 22 = 88$
Pawlak's model—Case study 2 with ρ^*	$32 + 3 + 0 = 35$
Pawlak's model—Case study 2 with ρ	$4 + 4 + 8 + 10 + 3 = 29$
Disjoint coalitions	$10 + 4 + 0 + 0 = 14$
Non-disjoint coalitions	$4 + 4 + 8 + 10 = 26$
Two-step process of coalitions creating	$3 + 22 + 24 + 4 + 4 = 57$

generated. That is, in the approach with Pawlak's model—Case study 1 the coalition $\{ag_1, ag_2, ag_3, ag_4, ag_6, ag_7\}$ with the sum of the distances equal to 66 was generated, and in the approach with Pawlak's model—Case study 2 with ρ^* the coalition $\{ag_1, ag_2, ag_3, ag_6, ag_7\}$ with the sum of the distances equal to 32 was generated. When the sums of the distances of all coalitions are compared, the approach with two-step process of coalitions creating is in the second place in terms of this value. However, after the precise analysis of the coalitions generated in this approach, it can be noticed that the distance within each coalition separately is not so large. Simply in this approach, we have the maximum number of coalitions and they are the most complex, and that is why the sum of the distances is relatively large. The smallest value of the sum of the distances was obtained for the approach with disjoint coalitions. This is due to the fact that in this approach, some agents remained isolated—they were not included in any coalition. Such property can be considered as a shortcoming of the approach. Because otherwise the best approach would be the approach, in which no coalition would be created, each agent was separately considered (then the sum of the distances is equal to zero).

To summarize, the measure that is used gives us some view of the form of coalitions that were created. It shows that the drawback of the approaches with Pawlak's model—Case study 1 and Case study 2 with ρ^* is that they generate coalitions, which consist of agents that are very different in their views. The disadvantage of the approach with disjoint coalitions is that it leaves agents who are isolated. Among other approaches, the approach with two-step process of coalitions creating generates the most complex and extensive coalitions.

5 Summary

Contribution of the Professor Pawlak's work in the development of computer science is significant and indisputable. The Pawlak's model of conflict is a simple and transparent tool for modeling conflict situations. This model has many extensions. In the paper application of this model to analyze conflicts that arise between classifiers was proposed. The obtained results were compared with three other approaches proposed by the author. It has been noted that in some cases the use of the Pawlak's model may generate coalitions that consist of agents that are very different in their views. In one of the case study, coalitions that were generated by the Pawlak's model are similar to coalitions which were generated by the approach with negotiations.

References

1. Caragea, D.: Learning Classifiers from Distributed, Semantically Heterogeneous, Autonomous Data Sources, Ph.D. Thesis, Iowa State University (2004)
2. Chawla, N., Eschrich, S., Hall, L.: Creating ensembles of classifiers. In: IEEE International Conference on Data Mining, pp. 580–581 (2001)

3. Deja, R.: Conflict analysis. In: Tsumoto, S., Kobayashi, S., Yokomori, T., Tanaka, H., Nakamura, A. (eds.) Proceedings of the Fourth International Workshop on Rough Sets, Fuzzy Sets and Machine Discovery, The University of Tokyo, Nov 6–8, pp. 118–124 (1996)
4. Deja, R.: Conflict model with negotiation. Bull. Pol. Acad. Sci. Tech. Sci. **44**(4), 475–498 (1996)
5. Deja, R.: Conflict analysis. In: Proceedings of the 7th European Congress on Intelligent Techniques and Soft Computing, Aachen, Germany, Sept 13–16 (1999)
6. Deja, R.: Conflict analysis. Rough set methods and applications; new developments. In: Polkowski, L. et al. (eds.) Studies in Fuzziness and Soft Computing, Physica-Verlag, pp. 491–520 (2000)
7. Deja, R.: Application of rough set theory in conflict analysis. Instytut Podstaw Informatyki Polskiej Akademii Nauk, Dissertation, supervisor: A. Skowron (2000)
8. Deja, R.: Conflict analysis. Int. J. Intell. Syst. **17**(2), 235–253 (2002)
9. Demri, S., Orłowska, E.: Incomplete Information: Structure, Inference. Complexity. Monographs in Theoretical Computer Science. Springer, Heidelberg (2002)
10. Froelich, W., Wakulicz-Deja, A.: Probabilistic Similarity-Based Reduct, Rough Sets and Knowledge Technology—6th International Conference, RSKT 2011, Banff, Canada, 9–12 Oct 2011. Proceedings, pp. 610–615 (2011)
11. Gatnar, E.: Multiple-model approach to classification and regression. PWN, Warsaw (in Polish) (2008)
12. Granner, M.L., Sharpe, P.A.: Evaluating community coalition characteristics and functioning: a summary of measurement tools. Health Educ. Res. Theory Pract. **19**(5), 514–532 (2004)
13. Grzymała-Busse, J.W.: Managing Uncertainty in Expert Systems. Kluwer Academic Publishers, Norwell (1990)
14. Ilczuk, G., Wakulicz-Deja, A.: Rough sets approach to medical diagnosis system. In: Advances in Web Intelligence Third International Atlantic Web Intelligence Conference, AWIC 2005, Lodz, Poland, 6–9 June 2005, Proceedings, pp. 204–210 (2005)
15. Ilczuk, G., Wakulicz-Deja, A.: Visualization of rough set decision rules for medical diagnosis systems. In: Rough Sets, Fuzzy Sets, Data Mining and Granular Computing, 11th International Conference, RSFDGrC 2007, Toronto, Canada, 14–16 May 2007, Proceedings, pp. 371–378 (2007)
16. Ilczuk, G., Wakulicz-Deja, A.: Selection of important attributes for medical diagnosis systems. Trans. Rough Sets **7**, 70–84 (2007)
17. Kowalski, R.: A logic-based approach to conflict resolution. Report, Department of Computing, Imperial College, pp. 1–28 (2003)
18. Kraus, S.: Strategic Negotiations in Multiagent Environments. The MIT Press, Cambridge (2001)
19. Kuncheva, L.: Combining Pattern Classifiers Methods and Algorithms. Wiley (2004)
20. Kuncheva, L., Bezdek, J.C., Duin, R.P.W.: Decision templates for multiple classifier fusion: an experimental comparison. Pattern Recognit. **34**(2), 299–314 (2001)
21. Lai, G., Li, C., Sycara, K., Giampapa, J.A.: Literature review on multi-attribute negotiations. Technical Report CMU-RI-TR-04-66, pp. 1–35 (2004)
22. Lin, T.Y., Yao, Y.Y., Zadeh, L.A. (eds.): Rough Sets, Granular Computing and Data Mining. Studies in Fuzziness and Soft Computing. Physica-Verlag, Heidelberg (2001)
23. Moshkov, M., Piliszczuk, M., Zielosko, B.: On partial covers, reducts and decision rules. Trans. Rough Sets **8**, 251–288 (2008)
24. Moshkov, M., Piliszczuk, M., Zielosko, B.: Greedy algorithm for construction of partial association rules. Fundam. Inform. **92**(3), 259–277 (2009)
25. Moshkov, M., Zielosko, B.: Combinatorial Machine Learning—A Rough Set Approach, Studies in Computational Intelligence, vol. 360. Springer (2011)
26. Nowak-Brzezińska, A., Wakulicz-Deja, A.: Exploration of knowledge bases inspired by rough set theory. In: Proceedings of the 24th International Workshop on Concurrency, Specification and Programming, Rzeszow, Poland, Sept 28–30, pp. 64–75 (2015)

27. Pal, S.K., Polkowski, L., Skowron, A. (eds.): Rough-Neural Computing: Techniques for Computing with Words. Cognitive Technologies. Springer, Heidelberg (2004)
28. Paszek, P., Wakulicz-Deja, A.: The application of support diagnose in mitochondrial encephalomyopathies. In: Rough Sets and Current Trends in Computing, Third International Conference, RSCTC 2002, Malvern, PA, USA, 14–16 Oct 2002, Proceedings, pp. 586–593 (2002)
29. Pawlak, Z.: Rough sets. Int. J. Comput. Inf. Sci. **11**, 341–356 (1982)
30. Pawlak, Z.: On conflicts. Int. J. Man-Mach. Stud. **21**, 127–134 (1984)
31. Pawlak, Z.: About Conflicts (in Polish), pp. 1–72 . Polish Scientific Publishers, Warsaw (1987)
32. Pawlak, Z.: Anatomy of conflict. Bull. Eur. Assoc. Theor. Comput. Sci. **50**, 234–247 (1993)
33. Pawlak, Z.: On some issues connected with conflict analysis. Institute of Computer Science Reports, 37/93. Warsaw University of Technology (1993)
34. Pawlak, Z.: An inquiry anatomy of conflicts. J. Inf. Sci. **109**, 65–78 (1998)
35. Pawlak, Z.: Some remarks on conflict analysis. Eur. J. Oper. Res. **166**, 649–654 (2005)
36. Pawlak, Z.: Conflicts and negotiations. In: Wang, G.-Y., Peters, J.F., Skowron, A., Yao, Y. (eds.) RSKT 2006. LNCS (LNAI), vol. 4062, pp. 12–27. Springer, Heidelberg (2006)
37. Polikar, R.: Ensemble based systems in decision making. IEEE Circuits Syst. Mag. **6**, 21–45 (2006)
38. Polkowski, L.: Rough Sets: Mathematical Foundations. Advances in Soft Computing. Physica-Verlag, Heidelberg (2002)
39. Przybyła-Kasperek, M., Wakulicz-Deja, A.: Global decision-making system with dynamically generated clusters. Inf. Sci. **270**, 172–191 (2014)
40. Przybyła-Kasperek, M., Wakulicz-Deja, A.: A dispersed decision-making system—the use of negotiations during the dynamic generation of a system's structure. Inf. Sci. **288**, 194–219 (2014)
41. Przybyła-Kasperek, M., Wakulicz-Deja, A.: Global decision-making in multi-agent decision-making system with dynamically generated disjoint clusters. Appl. Soft Comput. **40**, 603–615 (2016)
42. Przybyła-Kasperek, M., Wakulicz-Deja, A.: The strength of coalition in a dispersed decision support system with negotiations. Eur. J. Oper. Res. **252**, 947–968 (2016)
43. Ramanna, S., Peters, J.F., Skowron, A.: Generalized conflict and resolution model with approximation spaces. In: Rough Sets and Current Trends in Computing, 5th International Conference, RSCTC 2006, Kobe, Japan, 6–8 Nov 2006. Proceedings, vol. 274–283 (2006)
44. Ramanna, S., Peters, J.F., Skowron, A.: Approaches to conflict dynamics based on rough sets. Fundam. Inform. **75**(1–4), 453–468 (2007)
45. Ramanna, S., Skowron, A.: Requirements interaction an conflicts a rough set approach. In: Proceedings of the IEEE Symposium on Foundations of Computational Intelligence, FOCI 2007, part of the IEEE Symposium Series on Computational Intelligence 2007, Honolulu, Hawaii, USA, 1–5 April 2007, pp. 308–313 (2007)
46. Ramanna, S., Skowron, A., Peters, J.F.: Approximation space-based socio-technical conflict model. In: Rough Sets and Knowledge Technology, Second International Conference, RSKT 2007, Toronto, Canada, 14–16 May 2007, Proceedings, pp. 476–483 (2007)
47. Simiń ski, R.: Extraction of Rules Dependencies for Optimization of Backward Inference Algorithm, Communications in Computer and Information Science, vol. 424, pp. 191–200. Springer International Publishing (2014)
48. Simiński, R., Wakulicz-Deja, A.: Rough sets inspired extension of forward inference algorithm. In: Proceedings of the 24th International Workshop on Concurrency, Specification and Programming, Rzeszow, Poland, Sept 28–30, pp. 161–172 (2015)
49. Skowron, A., Rauszer, C.: The discernibility matrices and functions in information system. In: Słowiński, R. (ed.) Intelligent Decision Support. Handbook of Applications and Advances of the Rough Set Theory, pp. 331–362. Kluwer Academic Publishers, Dordrecht (1991)
50. Skowron, A., Deja, R.: On some conflict models and conflict resolutions. Rom. J. Inf. Sci. Technol. **3**(1–2), 69–82 (2002)

51. Skowron, A., Pal, S.K. (eds.): Special volume: rough sets, pattern recognition and data mining. Pattern Recognit. Lett. **24**(6) (2003)
52. Skowron, A., Ramanna, S., Peters, J.F.: Conflict analysis and information systems: a rough set approach. In: Rough Sets and Knowledge Technology, First International Conference, RSKT 2006, Chongqing, China, 24–26 July 2006. Proceedings, pp. 233–240 (2006)
53. Słowiński, R., Stefanowski, J. (eds.): Special issue: Proceedings of the First International Workshop on Rough Sets: State of the Art and Perspectives, Kiekrz, Poznań, Poland, 2–4 Sept 1992, Foundations of Computing and Decision Sciences, vol. 18(3–4) (1993)
54. Sosnowski, Ł., Pietruszka, A., Łazowy, S.: Election algorithms applied to the global aggregation in networks of comparators. In: Proceedings of the 2014 FedCSIS, pp. 135–144 (2014)
55. Sycara, K.: Multiagent systems. AI Mag. 79–92 (1998)
56. Tariman, J.D., Doorenbos, A., Schepp, K.G., Singhal, S., Berry, D.L.: Older adults newly diagnosed with symptomatic myeloma and treatment decision making. Oncol. Nurs. Forum **41**(4) (2014)
57. Wakulicz-Deja, A., Boryczka, M., Paszek, P.: Discretization of continuous attributes on decision system in mitochondrial encephalomyopathies. In: Rough Sets and Current Trends in Computing, First International Conference, RSCTC'98, Warsaw, Poland, 22–26 June 1998, Proceedings, pp. 483–490 (1998)
58. Wakulicz-Deja, A., Paszek, P.: Applying rough set theory to multi stage medical diagnosing. Fundam. Inform. **54**(4), 387–408 (2003)
59. Yadav, S., Kohli, N., Tiwari, V.: J. Indian Acad. Appl. Psychol. Supl. Spec. Issue **41**(3), 112–119 (2015)

Lattice Structure of Variable Precision Rough Sets

Sumita Basu

Abstract The main purpose of this chapter is to study the lattice structure of variable precision rough sets. The notion of variation in precision of rough sets have been further extended to variable precision rough set with variable classification error and its algebraic properties are also studied.

1 Introduction

Classical rough set theory as introduced by Pawlak [1, 2] is a tool for computation with data which give imprecise or vague information in terms of three valued logic. When the data set is granular in nature we are unable to observe individual objects but are forced to reason with accessible granules of knowledge. The elements of each granule can not be distinguished from the available knowledge. Due to such indiscernibility of the elements very often a subset of the entire data set (the Universal set) cannot be precisely defined. Pawlak represented the granularity by equivalence relation and defined such a set S as a suitable pair of sets $(\underline{S}, \overline{S})$ based on equivalence classes and called it a *Rough Set*. It is widely used for knowledge classification and rule learning. Finite state machine with rough transition have been reported in [3]. On the one hand, owing to the restrictions of equivalence relations, many researchers have presented various extensions [4–8], specially, covering-based rough sets [9–12] are investigated as the extensions of classical rough set theory by extending partitions to coverings. On the other hand, in classical rough set model, the approximation using the equivalence relation admits no error though may be somewhat imprecise. It is implied that $\underline{S} \subset S \subset \overline{S}$. The cardinality of the set $BN(S) = \overline{S} - \underline{S}$ will determine the precision of the representation. If $BN(S) = \phi$ the representation is exact. Increase in cardinality of $BN(S)$ will increase the imprecision of the solution.

A generalization of rough set model was proposed by Ziarko [13]. He introduced a measure of relative degree of misclassification (error) and chose to decrease the imprecision thereby increasing the error in approximation. This is an extension of

S. Basu (✉)
Bethune College, 181, Bidhan Sarani, Kolkata 700006, India
e-mail: sumi_basu05@yahoo.co.in

G. Wang et al. (eds.), *Thriving Rough Sets*, Studies in Computational Intelligence 708, DOI 10.1007/978-3-319-54966-8_17

classical rough set where the granules of knowledge are equivalence classes and called it *variable precision rough set*. Some researchers extended this concept to variable precision covering based rough set model [14].

Algebraic properties of rough set have been widely studied by researchers [15–19]. Lattice structure of rough set have been discussed in [20–22]. Algebraic properties of variable precision rough set is discussed in this chapter and it could be shown that for different classification error the set of variable precision rough sets have a lattice structure. Katzberg and Ziarko [23] introduced variable precision rough set with asymmetric bounds. Using the concept of asymmetric bounds a variable measure of degree of error have been introduced in this chapter and we call such a set *variable precision rough set with variable error*. Properties of such sets are compared with variable precision rough sets.

The chapter is organized as follows. In Sect. 2 basic concepts of rough set, variable precision rough set and lattice are introduced. In Sect. 3 structure of variable precision rough set for different classification error is explored. Section 4 is devoted to study of variable precision rough set where classification error for lower and upper approximations are not same. An example is included to explain the computation of different variable precision rough set.

2 Preliminaries

In this section some basic concepts on **Rough Sets**, **Variable Precision Rough Sets** and **Lattice** are discussed.

2.1 *Rough set*

Definition 2.1 An approximation space is defined as a pair ⟨U, R⟩, U being a non-empty set (the domain of discourse) and R an equivalence relation on it, representing indiscernibility at the object level. For $x \in R[x], R[x]$ is the set of elements of U indiscernible from x. $E = \{R[x]/x \in U\}$ is the set of elementary blocks or defining blocks of the approximation space.

Definition 2.2 A rough set X in the approximation space ⟨U, R⟩ is a pair $(\underline{X_R}, \overline{X_R})$ such that $\underline{X_R}$ (called lower approximation) & $\overline{X_R}$ (called upper approximation) are definable sets in U defined as follows:

$$\underline{X}_R = \{R[y]/y \in U \wedge R[y] \subseteq X\}$$

$$\overline{X}_R = \{R[y]/y \in U \wedge X \cap R[y] \neq \phi\}$$

The region definitely included in X is denoted by $D(X)$ and defined by $D(X) = \underline{X}_R$. $\overline{X}_R$ will be the smallest region including X. The regions defined above are defined by definable sets of U. The boundary region $BN(X)$ of the rough set X is $\underline{X}_R - \overline{X}_R$. The region not included in X is denoted by $N(X)$ and defined by $U - \overline{X}_R$.

Definition 2.3 The accuracy of approximation by the rough set X is given by

$$\alpha = \frac{card(\underline{X})}{card(\overline{X})}$$

Remark 1 If for a rough set X, $\underline{X}_R = \overline{X}_R$, i.e. $B_R = \phi$ then the rough set is precisely defined and the accuracy of approximation is 1. In general the accuracy of approximation is $\alpha \in [0, 1]$.

2.2 *Variable Precision Rough Set*

In this rough set model a set $X \subseteq U$ is approximately defined using three exactly definable sets: $D(X), BN(X)$ *and* $N(X)$. However, it may so happen that an elementary set $R[y]$ where $y \in U$ is such that although $R[y] \cap D(X) = \phi, card(R[y] \cap X)$ is quite high relative to $card(R[y])$. So inclusion of $R[y]$ in $D(X)$ will incur a small amount of error. However, if we agree to accept this error we will be able to increase the precision of the rough set so obtained. With this idea Ziarko formulated Variable Precision Rough Set (VPRS) which is defined below.

Definition 2.4 A measure of the degree of overlap between two sets X and Y with respect to X is denoted by d(X,Y) and defined by,

$$d(X, Y) = 1 - \frac{card(X \cap Y)}{card(X)}$$

Definition 2.5 A variable precision rough set (VPRS) $X(\beta)$ in the approximation space ⟨U, R⟩, is a pair $(\underline{X}_R(\beta), \overline{X}_R(\beta))$ such that $\underline{X}_R(\beta)$ & $\overline{X}_R(\beta)$ are definable sets in U defined as follows:

$$\underline{X}_R(\beta) = \{R[y]/y \in U \wedge R[y] \subset X \wedge d(R[y], X) \leq \beta\}$$

$$\overline{X}_R(\beta) = \{R[y]/y \in U \wedge X \cap R[y] \neq \phi \wedge d(R[y], X) < 1 - \beta\}$$

For the variable precision rough set model with β error a set $X \subseteq U$ is approximately defined using three sets of definable sets: $DX(\beta), BNX(\beta)$ *and* $NX(\beta)$ as follows:

$$DX(\beta) = \{R[y]/y \in U \wedge R[y] \subset X \wedge d(R[y], X) \leq \beta\}$$

$$BNX(\beta) = \overline{X}_R(\beta) - \underline{X}_R(\beta)$$

$$NX(\beta) = \{R[y]/y \in U \wedge X \cap R[y] \neq \phi \wedge d(R[y], X) \geq 1 - \beta\}$$

In general, β is chosen so that $\beta \in [0, .5)$. For given X, β, $DX(\beta)$ is the region *included* in X, $NX(\beta)$ is the region *not included* in X and $BNX(\beta)$ is the boundary region *possibly included* in X. If $BNX(\beta) = \phi$ then X is β discernible.

Definition 2.6 The accuracy of approximation by the rough set $X(\beta)$ is given by

$$\alpha X(\beta) = \frac{card(\underline{X}_R(\beta))}{card(\overline{X}_R(\beta))}$$

Proposition 2.1 *Let X be an arbitrary subset of the universe U in the approximation space $\langle U, R \rangle$, and β be the error specified then,*

1. $DX(\beta) \cup BNX(\beta) \cup NX(\beta) = U$
2. $DX(\beta) \cap BNX(\beta) = BNX(\beta) \cap NX(\beta) = DX(\beta) \cap NX(\beta) = \phi$

Proposition 2.2 *Let X be an arbitrary subset of the universe U in the approximation space $\langle U, R \rangle$, and $\beta_1 < \beta_2$ then,*

1. $\underline{X}_R(\beta_1) \subseteq \underline{X}_R(\beta_2)$
2. $\overline{X}_R(\beta_2) \subseteq \overline{X}_R(\beta_1)$
3. $DX(\beta_1) \subseteq DX(\beta_2)$
4. $NX(\beta_1) \subseteq NX(\beta_2)$
5. $BNX(\beta_2) \subseteq BNX(\beta_1)$
6. $\alpha X(\beta_1) \leq \alpha X(\beta_2)$

Proof Results 1–5 follows from the Definition 2.5. Result 6 follows from Result 1 and 2.

2.3 Lattice

Definition 2.7 Let L be a set of elements in which the binary operations $\bigcap, \bigcup$ (meet and joint respectively) and =(equality) are defined. An algebra $L = \langle L, \bigcap, \bigcup \rangle$ is a lattice if the following identities are true in L. Let $x, y, z \in L$

1. Idempotence: $x \bigcup x = x$; $x \bigcap x = x$
2. Commutativity: $x \bigcup y = y \bigcup x$; $x \bigcap y = y \bigcap x$
3. Associativity: $x \bigcup (y \bigcup z) = (x \bigcup y) \bigcup z$; $x \bigcap (y \bigcap z) = (x \bigcap y) \bigcap z$
4. Absorption: $x \bigcup (x \bigcap y) = x$; $x \bigcap (x \bigcup y) = x$

3 Order in VPRS with Respect to Classification Error β

We will follow the VPRS model as introduced by Ziarko with classification error $\beta \in [0, .5]$. For any set X and a classification error β the VPRS will be $X(\beta) = (\underline{X}_R(\beta), \overline{X}_R(\beta))$. Given X there may be a set of VPRS for different values of β as defined below:

Definition 3.1 Let $B = \{\beta_i / \beta_i \in [0, .5] \text{ and } (i \leq j \rightarrow (\beta_i \leq \beta_j)\}$. Then B is a totally ordered set.

Henceforth we will assume that $\forall i, \beta_i \in B$.

Definition 3.2 Let $\tilde{X}$ be the set of all VPRS for $X(\beta) \subset U$ where the classification error $\beta \in B$. So, $\tilde{X} = \{X(\beta)/\beta \in B\}$ so that $X(\beta) = (\underline{X}_R(\beta), \overline{X}_R(\beta))$

Proposition 3.1 *For an arbitrary subset X of the universe U, let us define* $\mathbf{DX} = \{DX(\beta)/\beta \in B\}$, *then* $\mathbf{DX}$ *is a totally ordered set with* $DX(0) = \underline{X}_R$ *as the least element and* $DX(0.5)$ *as the greatest element.*

This result follows from 3 of Proposition 2.2. Similarly we have the following propositions:

Proposition 3.2 *For an arbitrary subset X of the universe U, let us define* $\mathbf{NX} = \{NX(\beta)/\beta \in B\}$, *then* $\mathbf{NX}$ *is a totally ordered set with* $NX(0) = U - \overline{X}_R$ *as the least element and* $NX(0.5)$ *as the greatest element.*

Proposition 3.3 *For an arbitrary subset X of the universe U, let us define* $\mathbf{BNX} = \{BNX(\beta)/\beta \in B\}$, *then* $\mathbf{BNX}$ *is a totally ordered set with* $BNX(0) = \overline{X}_R - \underline{X}_R$ *as the greatest element and* $BNX(0.5)$ *as the least element.*

Definition 3.3 Let $X(\beta_1), X(\beta_2) \in \tilde{X}$ then $X(\beta_1) \subseteq X(\beta_2)$ iff $\underline{X}_R(\beta_1) \subseteq \underline{X}_R(\beta_2)$ and $\overline{X}_R(\beta_1) \subseteq \overline{X}_R(\beta_2)$.

Proposition 3.4 *For an arbitrary subset X of the universe U,*

1. $\{\underline{X}_R(\beta_i)/\beta_i \in B\}$ *is a totally ordered set with* $lub\{\underline{X}_R(\beta_i)\} = \underline{X}_R(\beta_{0.5})$ *and* $glb\{\underline{X}_R(\beta_i)\} = \underline{X}_R(0) = \underline{X}_R$
2. $\{\overline{X}_R(\beta_i)/\beta_i \in B\}$ *is a totally ordered set with* $glb\{\overline{X}_R(\beta_i)\} = \overline{X}_R(\beta_{0.5})$ *and* $lub\{\overline{X}_R(\beta_i)\} = \overline{X}_R(0) = \overline{X}_R$
3. $\underline{X}_R(0.5) \subseteq \overline{X}_R(0.5)$

Remark 2 Though DX, BNX, NX are totally ordered set $\tilde{X}$ is not necessarily so. However, we will show below that $\tilde{X}$ has a lattice structure. So, $\tilde{X}$ will be a partially ordered set.

Proposition 3.5 *If,* $\bigcup$ *and* $\bigcap$ *represent the union and intersection operation of two sets then we have the following:*

1. $\underline{X}_R(\beta_i) \bigcup \underline{X}_R(\beta_j) = \underline{X}_R(\beta_j)$ *if* $\beta_i \leq \beta_j$
2. $\underline{X}_R(\beta_i) \bigcap \underline{X}_R(\beta_j) = \underline{X}_R(\beta_i)$ *if* $\beta_i \leq \beta_j$
3. $\overline{X}_R(\beta_i) \bigcup \overline{X}_R(\beta_j) = \underline{X}_R(\beta_i)$ *if* $\beta_i \leq \beta_j$
4. $\overline{X}_R(\beta_i) \bigcap \overline{X}_R(\beta_j) = \overline{X}_R(\beta_j)$ *if* $\beta_i \leq \beta_j$

Definition 3.4 Two binary operations join ($\bigcup$) and meet ($\bigcap$) are defined on $\tilde{X}$ as follows:

$$X(\beta_1) \bigcup X(\beta_2) = ((\underline{X_{R(\beta_1)}} \bigcup \underline{X_{R(\beta_2)}}), (\overline{X_{R(\beta_1)}} \bigcup \overline{X_{R(\beta_2)}}))$$

$$X(\beta_1) \bigcap X(\beta_2) = ((\underline{X_{R(\beta_1)}} \bigcap \underline{X_{R(\beta_2)}}), (\overline{X_{R(\beta_1)}} \bigcap \overline{X_{R(\beta_2)}}))$$

Definition 3.5 Two VPRS $X(\beta_i)$ *and* $X(\beta_j)$ are said to be equal if $\underline{X}_R(\beta_i) = \underline{X}_R(\beta_j)$ and $\overline{X}_R(\beta_i) = \overline{X}_R(\beta_j)$

The approximation space remaining the same the equivalence relation R will remain the same and henceforth R will not be mentioned explicitly.

Proposition 3.6 *If,* $X(\beta_i), X(\beta_j) \in \tilde{X}$ *then*

1. $X(\beta_i) \bigcup X(\beta_j) = (\underline{X}(\beta_j), \overline{X}(\beta_i))$ *if* $\beta_i \leq \beta_j$
2. $X(\beta_i) \bigcap X(\beta_j) = (\underline{X}(\beta_i), \overline{X}(\beta_j))$ *if* $\beta_i \leq \beta_j$

Proposition 3.7 *Binary operations* $\bigcap$ *and* $\bigcup$ *are idempotent and commutative in* $\tilde{X}$

Proof From 1 of Proposition 3.6,

$$X(\beta_i) \bigcup X(\beta_i) = (\underline{X}(\beta_i), \overline{X}(\beta_i)) = X(\beta_i)$$

Also,

$$X(\beta_i) \bigcup X(\beta_j) = (\underline{X}(\beta_j), \overline{X}(\beta_i)) = X(\beta_j) \bigcup X(\beta_i) \text{ if } \beta_i \leq \beta_j$$

The result for $\bigcap$ may be proved similarly.

Proposition 3.8 *Binary operations* $\bigcap$ *and* $\bigcup$ *are associative in* $\tilde{X}$.

Proof

$$\begin{aligned} X(\beta_i) \bigcap (X(\beta_j) \bigcap X(\beta_k)) &= (X(\beta_i) \bigcap X(\beta_j)) \bigcap X(\beta_k) \\ &= (\underline{X}(\beta_i), \overline{X}(\beta_k)) \text{ if } \beta_i \leq \beta_j \leq \beta_k \\ &= (\underline{X}(\beta_i), \overline{X}(\beta_j)) \text{ if } \beta_i \leq \beta_k \leq \beta_j \\ &= (\underline{X}(\beta_k), \overline{X}(\beta_j)) \text{ if } \beta_k \leq \beta_i \leq \beta_j \\ &= (\underline{X}(\beta_j), \overline{X}(\beta_k)) \text{ if } \beta_j \leq \beta_i \leq \beta_k \\ &= (\underline{X}(\beta_j), \overline{X}(\beta_i)) \text{ if } \beta_j \leq \beta_k \leq \beta_i \\ &= (\underline{X}(\beta_k), \overline{X}(\beta_i)) \text{ if } \beta_k \leq \beta_j \leq \beta_i \end{aligned} \tag{1}$$

Hence the $\bigcap$ operation is associative. Similarly it can be shown that the $\bigcup$ operation is associative.

Proposition 3.9 *For the binary operations* $\bigcap$ *and* $\bigcup$ *absorption rule hold in* $\tilde{X}$. *So,*

$$X(\beta_i)\bigcap(X(\beta_i)\bigcup X(\beta_j)) = X(\beta_i); \quad X(\beta_i)\bigcup(X(\beta_i)\bigcap X(\beta_j)) = X(\beta_i); \ \ if\ \beta_i, \beta_j \in B$$

Proof Case I: $\beta_i \leq \beta_j$

$$X(\beta_i)\bigcap(X(\beta_i)\bigcup X(\beta_j)) = (\underline{X}(\beta_i), \overline{X}(\beta_i))\bigcap(\underline{X}(\beta_j), \overline{X}(\beta_i)) = (\underline{X}(\beta_i), \overline{X}(\beta_i)) = X(\beta_i)$$

Case II: $\beta_j \leq \beta_i$

$$X(\beta_i)\bigcap(X(\beta_i)\bigcup X(\beta_j)) = (\underline{X}(\beta_i), \overline{X}(\beta_i))\bigcap(\underline{X}(\beta_i), \overline{X}(\beta_j)) = (\underline{X}(\beta_i), \overline{X}(\beta_i)) = X(\beta_i)$$

The other part may be similarly proved.

Using Propositions 3.7, 3.8 and 3.9 we get the final result.

Proposition 3.10 $(\tilde{X}, \bigcup, \bigcap)$ *form a lattice.*

4 VPRS with Variable Classification Error (β, γ)

Discussions of VPRS show that both lower and upper approximations vary with classification error. It may so happen that for a particular problem the error admissible for the lower approximation and the error admissible for the upper approximation are different. The variable precision rough set with variable error is defined below.

Definition 4.1 A variable precision rough set with variable error (VPRSVE) X(β, γ) in the approximation space $\langle$U, R$\rangle$, is a pair $(\underline{X}_R(\beta, \gamma), \overline{X}_R(\beta, \gamma))$ such that $\underline{X}_R(\beta, \gamma)$ & $\overline{X}_R(\beta, \gamma)$ are definable sets in U defined as follows:

$$\underline{X}_R(\beta, \gamma) = \underline{X}_R(\beta) = \{R[y]/y \in U \wedge R[y] \subset X \wedge d(R[y], X) \leq \beta\}$$

$$\overline{X}_R(\beta, \gamma) = \overline{X}_R(\gamma) = \{R[y]/y \in U \wedge X \cap R[y] \neq \phi \wedge d(R[y], X) \leq (1 - \gamma)\}$$

For the VPRSVE with (β, γ) error a set $X \subseteq U$ is approximately defined using three sets of definable sets: $DX(\beta, \gamma), BNX(\beta, \gamma)$ *and* $NX(\beta, \gamma)$ as follows:

$$DX(\beta, \gamma) = \{R[y]/y \in U \wedge R[y] \subset X \wedge d(R[y], X) \leq \beta\}$$

$$BNX(\beta, \gamma) = \overline{X}_R(\gamma) - \underline{X}_R(\beta)$$

$$NX(\beta,\gamma) = \{R[y]/y \in U \wedge X \cap R[y] \neq \phi \wedge d(R[y],X) > (1-\gamma)\}$$

Remark 3 According to the requirement of the situation the boundary region of the VPRSVE $X(\beta,\gamma)$ (denoted by $BNX(\beta,\gamma) = \overline{X}_R(\gamma) - \underline{X}_R(\beta)$) is increased or decreased.

Proposition 2.1 will be modified in this case as

Proposition 4.1 *Let X be an arbitrary subset of the universe U in the approximation space* $\langle U,R\rangle$*, and* $\beta,\gamma \in [0,0.5]$ *be the error specified then,*

1. $DX(\beta,\gamma) = DX(\beta)$
2. $NX(\beta,\gamma) = NX(\gamma)$
3. $DX(\beta,\gamma) \cup BNX(\beta,\gamma) \cup NX(\beta,\gamma) = U$
4. $DX(\beta,\gamma) \cap BNX(\beta,\gamma) = BNX(\beta,\gamma) \cap NX(\beta,\gamma) = DX(\beta,\gamma) \cap NX(\beta,\gamma) = \phi$

Example 4.1 Let $U = \{x_i/i = 1,2,3.....25\}$ and R is an equivalence relation on U such that $U/R = \{[x_1],[x_2,x_3],[x_4,x_5,x_6],[x_7,x_8],[x_9],[x_{10},x_{11}],[x_{12},x_{13},x_{14},x_{15}],[x_{16}],[x_{17}],[x_{18},x_{19},x_{20}],[x_{21},x_{22},x_{23},x_{24}],[x_{25}]\}$. Let $A = \{x_3,x_4,x_5,x_{10},x_{11},x_{13},x_{14},x_{15},x_{19},x_{21}\}$.

Problem Define A with respect to the equivalence classes of U/R

Pawlakian rough set $A = (\underline{A},\overline{A})$ where $\underline{A} = \{[x_{10},x_{11}]\}$ and
$\overline{A} = \{[x_{10},x_{11}],[x_2,x_3],[x_4,x_5,x_6],[x_{12},x_{13},x_{14},x_{15}],[x_{18},x_{19},x_{20}],[x_{21},x_{22},x_{23},x_{24}]\}$, so that

$$DA = \{[x_{10},x_{11}]\}$$

$$BNA = \{[x_2,x_3],[x_4,x_5,x_6],[x_{12},x_{13},x_{14},x_{15}],[x_{18},x_{19},x_{20}],[x_{21},x_{22},x_{23},x_{24}]\}$$

$$NA = \{[x_1],[x_7,x_8],[x_9],[x_{16}],[x_{17}],[x_{25}]\}$$

For VPRS A, β can have values $0.25, 0.33, 0.5$. So there can be three possible VPRS $A(0.25), A(0.33), A(0.5)$. Thus,

$$\underline{A}(0.25) = \{[x_{10},x_{11}],[x_{12},x_{13},x_{14},x_{15}]\}$$

$$\overline{A}(0.25) = \{[x_{10},x_{11}],[x_2,x_3],[x_{18},x_{19},x_{20}],[x_{21},x_{22},x_{23},x_{24}],[x_4,x_5,x_6],[x_{12},x_{13},x_{14},x_{15}]\}$$

$$DA(0.25) = \{[x_{10},x_{11}],[x_{12},x_{13},x_{14},x_{15}]\}$$

$$BNA(0.25) = \{[x_2,x_3],[x_4,x_5,x_6],[x_{18},x_{19},x_{20}],[x_{21},x_{22},x_{23},x_{24}]\}$$

$$NA(0.25) = \{[x_1],[x_7,x_8],[x_9],[x_{16}],[x_{17}],[x_{25}]\}$$

Also,

$$\underline{A}(0.33) = \{[x_{10}, x_{11}], [x_4, x_5, x_6], [x_{12}, x_{13}, x_{14}, x_{15}]\}$$

$$\overline{A}(0.33) = \{[x_{10}, x_{11}], [x_2, x_3], [x_{18}, x_{19}, x_{20}], [x_4, x_5, x_6], [x_{12}, x_{13}, x_{14}, x_{15}]\}$$

$$DA(0.33) = \{[x_{10}, x_{11}], [x_4, x_5, x_6], [x_{12}, x_{13}, x_{14}, x_{15}]\}$$

$$BNA(0.33) = \{[x_2, x_3], [x_{18}, x_{19}, x_{20}]\}$$

$$NA(0.33) = \{[x_1], [x_7, x_8], [x_9], [x_{16}], [x_{17}], [x_{21}, x_{22}, x_{23}, x_{24}], [x_{25}]\}$$

and,

$$\underline{A}(0.5) = \{[x_{10}, x_{11}], [x_2, x_3], [x_4, x_5, x_6], [x_{12}, x_{13}, x_{14}, x_{15}]\} = \overline{A}(0.5)$$

$$DA(0.5) = \{[x_{10}, x_{11}], [x_2, x_3], [x_4, x_5, x_6], [x_{12}, x_{13}, x_{14}, x_{15}]\}$$

$$BNA(0.5) = \phi$$

$$NA(0.5) = \{[x_1], [x_7, x_8], [x_9], [x_{16}], [x_{17}], [x_{21}, x_{22}, x_{23}, x_{24}], [x_{25}], [x_{18}, x_{19}, x_{20}]\}$$

Six VPRSVE are possible for A defined with respect to given approximation space of which $A(0.25, 0.33)$ is given below:

$$\underline{A}(0.25, 0.33) = \underline{A}(0.25) = \{[x_{10}, x_{11}], [x_{12}, x_{13}, x_{14}, x_{15}]\}$$

$$\overline{A}(0.25, 0.33) = \overline{A}(0.33) = \{[x_{10}, x_{11}], [x_2, x_3], [x_{18}, x_{19}, x_{20}], [x_4, x_5, x_6], [x_{12}, x_{13}, x_{14}, x_{15}]\}$$

$$DA(0.25, 0.33) = \{[x_{10}, x_{11}], [x_{12}, x_{13}, x_{14}, x_{15}]\}$$

$$BNA(0.25, 0.33) = \{[x_2, x_3], [x_4, x_5, x_6], [x_{18}, x_{19}, x_{20}]\}$$

$$NA(0.25, 0.33) = \{[x_1], [x_7, x_8], [x_9], [x_{16}], [x_{17}], [x_{21}, x_{22}, x_{23}, x_{24}], [x_{25}]\}$$

5 Conclusion

In this chapter algebraic properties of set of VPRS for a particular imprecise set X have been studied. In order to define such an imprecise set the approximation space is partitioned into three regions, the included region ($DX(\beta)$), the boundary region ($BNX(\beta)$) and the rejection region ($NX(\beta)$). For a particular X with variations of β

the regions vary. It could also be shown that the set of all VPRS for the set X forms a lattice. We extended the classification error β to a pair (β, γ) and explained its use with an example. The included region, boundary region and rejection region for a VPRSVE is defined and it is shown that these three regions partition the approximation space. Study of the algebraic properties of VPRSVE is an open area of research.

References

1. Pawłak, Z.: Rough sets. Int. J. Comput. Inf. Sci. **11**(5), 341–356 (1982)
2. Pawłak, Z.: Rough classification. Int. J. Man-Mach. Stud. **20**(5), 469–483 (1984)
3. Basu, S.: Rough finite-state machine. Cybern. Syst. **36**, 107–124 (2005)
4. Skowron, A., Stepaniuk, J.: Tolerance approximation spaces. Fundamenta Informaticae **27**(2–3), 245–253 (1996)
5. Slowinski, R., Vanderpooten, D.: A generalized definition of rough approximations based on similarity. IEEE Trans. Knowl. Data Eng. **12**(2), 331–336 (2000)
6. Yao, Y.Y.: On generalizing Pawłak approximation operators. In: Rough Sets and Current Trends in Computing. Lecture Notes in Computer Science, vol. 1424, pp. 298–307. Springer, Berlin, Germany (1998)
7. Yao, Y.Y.: Relational interpretations of neighborhood operators and rough set approximation operators. Inf. Sci. **111**(14), 239–259 (1998)
8. Yao, Y.Y.: Constructive and algebraic methods of the theory of rough sets. Inf. Sci. **109**(14), 21–47 (1998)
9. Zhu, W., Wang, F.: A new type of covering rough sets. In: Proceedings of the IEEE International Conference on Intelligent Systems, pp. 444–449. London, UK, Sept 2006
10. Zhu, W.: Basic concepts in covering-based rough sets. In: Proceedings of the 3rd International Conference on Natural Computation (ICNC 07), pp. 283–286, Aug 2007
11. Wang, S., Zhu, P., Zhu, W.: Structure of covering-based rough sets. Int. J. Math. Comput. Sci. **6**, 147–150 (2010)
12. Zhu, W.: Relationship among basic concepts in covering-based rough sets. Inf. Sci. **179**(14), 2478–2486 (2009)
13. Ziarko, W.: Variable precision rough set model. J. Comput. Syst. Sci. **46**(1), 39–59 (1993)
14. Zhu, Y., Zhu, W.: A variable precision covering-based rough set model based on functions. Hindawi Publishing Corporation. Sci. World J. **2014**, Article ID 210129, 5 pages
15. Zhu, W.: Relationship between generalized rough sets based on binary relation and covering. Inf. Sci. **179**(3), 210–225 (2009)
16. Wang, C., Chen, D., Sun, B., Hu, Q.: Communication between information systems with covering based rough sets. Inf. Sci. **216**, 17–33 (2012)
17. Banerjee, M., Chakraborty, M.K.: Rough sets through algebraic logic. Fundam. Inform. **28**(3–4), 211–221 (1996)
18. Pomykala, J., Pomykala, J.A.: The stone algebra of rough sets. Bull. Polish Acad. Sci. Math. **36**(7–8), 495–508 (1988)
19. Banerjee, M., Chakraborty, M.K.: Algebras from rough sets. In: Pal, S.K., Polkowski, L., Skowron, A. (eds.) Rough-Neural Computing—Techniques for Computing with Words. Springer, Heidelberg (preprint)
20. Cattaneo, G., Ciucci, D.: Lattices with interior and closure operators and abstract approximation spaces. In: Peters, J.F., Skowron, A., Wolski, M., Chakraborty, M.K., Wu, W.-Z. (eds.) Transactions on Rough Sets X. LNCS, vol. 5656, pp. 67–116. Springer, Heidelberg (2009)
21. Jarvinen, J.: Lattice theory for rough sets. In: Transactions on Rough Sets VI. Series Lecture Notes in Computer Science, vol. 4374, pp. 400–498

22. Samanta, P., Chakraborty, M.K.: Generalized rough sets and implication lattice. In: Transactions on Rough Sets XIV, pp. 183–201
23. Katzberg, J., Ziarko, W.: Variable precision rough sets with asymmetric bounds. In: Rough Sets, Fuzzy Sets and Knowledge Discovery, pp. 167–177. Springer (1994)

Part IV
Rough Set Based Data Mining

Mining for Actionable Knowledge in Tinnitus Datasets

Katarzyna A. Tarnowska, Zbigniew W. Ras and Pawel J. Jastreboff

Abstract This chapter describes the application of decision and action rules mining to the problem area of tinnitus treatment and characterization. The chapter presents the process of a Tinnitus Retraining Therapy treatment protocol, which is to be automatized with classification and action rules. The tinnitus dataset collected at Emory University School of Medicine in Atlanta, as well as preprocessing steps performed on the data are described. Next, a series of experiments on association and action rule extraction are presented. Selected outcome rules are listed in a form of medical hypotheses. An analysis and interpretation of sample rules are provided together with their validation in accordance with expert medical knowledge.

1 Introduction

Recently, there has been an increasing interest in business analytics and big data tools to understand and drive industries evolution. The healthcare industry is also interested in new methods to analyze data and provide better care. Given the wealth of data that various institutions are accumulating, it is natural to take advantage of data driven decision-making solutions. Modern computing techniques, including machine learning, intelligent data analysis and decision support systems technologies, provide a new promising way to better understand, further improve and support the treatment. The main motivation for researching this topic is to study and analyze the possibilities of applying modern information technologies and machine learning methods in the area of medicine. Machine learning and data exploration

---

K.A. Tarnowska (✉) · Z.W. Ras
University of North Carolina, Charlotte, USA
e-mail: ktarnows@uncc.edu

Z.W. Ras
Warsaw University of Technology, Polish-Japanese Academy of IT,
Warsaw, Poland
e-mail: ras@uncc.edu

P.J. Jastreboff
Emory University School of Medicine, Atlanta, USA
e-mail: pjastre@emory.edu

G. Wang et al. (eds.), *Thriving Rough Sets*, Studies in Computational Intelligence 708, DOI 10.1007/978-3-319-54966-8_18

methods should help in understanding relationships among the treatment factors and audiological measurements, in order to better understand tinnitus treatment. Understanding the relationships between patterns among treatment factors would help to optimize the treatment process. Additionally, different preprocessing techniques will be used so that to transform the tinnitus dataset into more suitable for machine understanding.

2 Background

Tinnitus, popularly known as "ringing in the ears", nowadays, affects a significant portion of the population—according to some estimations about 10–20% general population. Causes of tinnitus are often not clear—it is associated with hearing loss, ear infections, acoustic neuroma, Menere's syndrome, aging and side-effect of some drugs. There is no cure for it and treatment methodologies prove ineffective in many cases and some methods of treatment work well for some patients but not necessary for the others (must be highly personalized).

Tinnitus Retraining Therapy is a highly successful method of treatment proposed and developed by Dr. Jastreboff. The patients are categorized into one of four groups of tinnitus based on interview, audiological and medical evaluation (see Table 1). The therapy consists of a series of counseling sessions accompanied by use of devices called sound generators. Treatment progress and results were historically collected by Dr. Jastreboff resulting in a database of demographic and medical data of patients, as well as a series of metrics measuring treatment progress for each visit.

Our motivation was to further study factors behind therapy's effectiveness in order to collect actionable knowledge gathered by Dr. Jastreboff over several years of treatment (1999–2005). This would allow for further proliferation of therapy, introducing objectivity and standardization of the therapy in places lacking expertise in the field.

Table 1 Determining categories of tinnitus patients [1]

Category	Hyperacusis	Prolonged sound-induced exacerbation	Subjective hearing loss	Impact on life	Treatment
0	–	–	–	Low	Counseling only
1	–	–	–	High	Sound generator set at mixing point
2	–	–	Present	High	Hearing aid with stress on enrichment of the auditory background
3	Present	–	Not relevant	High	Sound generators set above threshold of hearing
4	Present	Present	Not relevant	High	Sound generators set at threshold; very slow increase of sound level

3 Approach

The approach based on the action rules presents a new method of machine learning, which solves problems that traditional methods, such as classification or association rules, cannot handle. The purpose here is to analyze data in order to improve the understanding of the data and seek specific actions to enhance the decision-making process. In contrast to learning the association rules, the action rule approach mines actionable patterns that can be employed to reach a desired goal (such as to increase treatment progress) instead of merely extracting passive relations between variables. Since its introduction in 2000 [2], action rules have been successfully applied in many domain areas including business [2], medical diagnosis and treatment [3], and music automatic indexing and retrieval [4].

Action rules seem to be especially promising in the field of medical data, as a doctor can examine the effect of treatment decisions on a patient's improved state. For example, in the tinnitus dataset, such an indicator for tracking improvement progress would be a Total score attribute, calculated by the sum of the responses from the interview form.

3.1 Origins in Rough Set Theory

Concepts of *Action Rule, Reducts, Decision Table* and *Information System* have their origins in the theory of Rough Sets, developed by Professor Zdzisław Pawlak at the beginning of 1980s [5]. The theory proposed a novel approach to the formal representation of knowledge description, and since its introduction was developing extensively all around the word, confirming its usefulness in practical settings.

3.2 Decision Rules

The decision rule, for a given decision table, is a rule in the form: $(\phi \rightarrow \delta)$, where ϕ is called *premise* (or *assumption*) and δ is called *conclusion* (or *thesis*) of the rule. The premise for an atomic rule can be a single term or a conjunction of k elementary conditions: $\phi = p_1 \wedge p_2 \wedge ... \wedge p_n$, and δ is a decision attribute. Decision rule describing a class K_j means that objects, which satisfy (match) the rule's premise, belong to K_j.

Each rule can be characterized by the following features:

- length(r) = number of descriptors in the premise of the rule,
- $[r]$ = a set of objects from U matching the rule's premise,
- support(r) = number of objects from U matching the rule's premise: $|[r]|$ (relative support is further divided by number of objects N),

- confidence(r) = reliability of the rule: $\frac{|[r] \cap DEC_k|}{|[r]|}$ number of objects matching both rule's premise and conclusion, divided by absolute support.

3.2.1 Classification Rules

In the context of prediction problem, decision rules generated from training dataset, are used for classifying new objects (for example classifying a new patient for tinnitus category). New objects are understood as objects that were not used for the rules induction (new patients coming to the doctor). The new objects are described by attribute values (for instance a patient with conducted audiological evaluation and form responses). The goal of classification is to assign a new object to one of the decision classes. Prediction is performed by matching the object description with the rule antecedents.

3.3 Action Rules

An *action* is understood as a way of controlling or changing some of attribute values in an information system to achieve desired results [6]. An *action rule* is defined [2] as a rule extracted from an information system, that describes a transition that may occur within objects from one state to another, with respect to decision attribute, as defined by the user. In nomenclature, action rule is defined as a term: $[(\omega) \wedge (\alpha \rightarrow \beta) \rightarrow (\Phi \rightarrow \Psi)]$, where ω denotes conjunction of fixed condition attributes, $(\alpha \rightarrow \beta)$ are proposed changes in values of flexible features, and $(\Phi \rightarrow \Psi)$ is a desired change of decision attribute (action effect). Action rule discovery applied to tinnitus dataset could, for example, suggest a change in a flexible attribute, such as type of sound generator instrument, to help "reclassify" or "transit" an object (patient) to a different category and consequently, attain better treatment effectiveness.

An action rule is built from *atomic action sets*.

Definition 1 *Atomic action term* is an expression $(a, a_1 \rightarrow a_2)$, where a is attribute, and $a_1, a_2 \in V_a$, where V_a is a domain of attribute a.

If $a_1 = a_2$ then a is called stable on a_1.

Definition 2 By *action sets* we mean the smallest collection of sets such that:

1. If t is an atomic action term, then t is an action set.
2. If t_1, t_2 are action sets, then $t_1 \wedge t_2$ is a candidate action set.
3. If t is a candidate action set and for any two atomic actions $(a, a_1 \rightarrow a_2)$, $(b, b_1 \rightarrow b_2)$ contained in t we have $a \neq b$, then t is an action set. Here b is another attribute $(b \in A)$, and $b_1, b_2 \in V_b$.

Definition 3 By an *action rule* we mean any expression $r = [t_1 \Rightarrow t_2]$, where t_1 and t_2 are action sets.

The interpretation of the action rule r is, that by applying the action set t_1, we would get, as a result, the changes of states in action set t_2.

Example 1 Assuming that a, b and d are stable attribute, flexible attribute and decision attribute respectively in S, expressions (a, a_2), $(b, b_1 \rightarrow b_2)$, $(d, d_1 \rightarrow d_2)$ are examples of atomic action sets. Expression (a, a_2) means that the value a_2 of attribute a remains unchanged, $(b, b_1 \rightarrow b_2)$ that value of attribute b is changed from b_1 to b_2. Expression $r = [\{(a, a_2) \wedge (b, b_1 \rightarrow b_2)\} \Rightarrow \{(d, d_1 \rightarrow d_2)\}]$ is an example of an action rule meaning that if value a_2 of a remains unchanged and value of b will change from b_1 to b_2, then the value of d will be expected to transition from d_1 to d_2. Rule r can be also perceived as the composition of two association rules r_1 and r_2, where $r_1 = [\{(a, a_2) \wedge (b, b_1)\} \Rightarrow (d, d_1)]$ and $r_2 = [\{(a, a_2) \wedge (b, b_2)\} \Rightarrow (d, d_2)]$.

In other words, if we apply action rule r on a patient satisfying rule r_1, then it is also expected that this patient will satisfy rule r_2. The confidence of action rule r is defined as (confidence of r_1) x (confidence of r_2).

3.4 Meta Actions

Action rules are mined on the entire set of objects in S. Meta-actions, on the other hand, are chosen based on the action rules. They are formally defined as higher level concepts used to model a generalization of action rules in an information system [2]. They trigger actions that cause transitions in values of some flexible attributes in the information system. These changes, in turn, result in a change of decision attributes' values.

Definition 4 Let $M(S)$ be a set of meta-actions associated with an information system S. Let $a \in A$, $x \in X$, and $M \subset M(S)$. Applying the meta-actions in the set M on object x will result in $M(a(x)) = a(y)$, where object x is converted to object y by applying all meta-actions in M to x.

Example 2 Let $M(S)$, where $S = (X, A)$, be a set of meta-actions associated with an information system S. In addition let $T = \{v_{i,j} : j \in J_i, x_i \in X\}$ be the set of ordered transactions, patient visits, such that $v_{i,j} = [(x_i, A(x_i)_j)]$, where $A(x_i)_j$ is a set of attribute values $\{a(x_i) : a \in A\}$ of the object x_i for the visit represented uniquely by the visit identifier j. Each visit represents the current state of the object (patient) and current diagnosis. For each patient's two consecutive visits $(v_{i,j}, v_{i,j+1})$, where meta-actions were applied at visit j, it is possible to extract an *action set*. In this example, an *action set* is understood as an expression that defines a change of state for a distinct attribute that takes several values (multivalued attribute) at any object state. For example $\{a_1, a_2, a_3\} \rightarrow \{a_1, a_4\}$ is an action set that defines a change of values for attribute $a \in A$ from the set $\{a_1, a_2, a_3\}$ to $\{a_1, a_4\}$, where $\{a_1, a_2, a_3, a_4\} \subseteq V_a$ [7].

These action sets resulting from the application of meta-actions represent the actionable knowledge needed by practitioners. However, not every patient reacts the same way to the same meta-actions, because patients may have different preconditions. In other words, some patients can be partially affected by the meta actions and may have other side-effects. So, there is a need to introduce personalization on meta actions when executing action rules. The problem of personalized meta-actions is a fairly new topic that creates room for new improvements. There is a minor work on the personalization of meta-actions done so far [8]. Action sets have to be additionally mined for the historical patterns. To evaluate these action set patterns some frequency measure for all patients has to be used (for example support or confidence). There is a room for improvements in personalized meta action mining, as well. In healthcare for instance, meta actions representing patient's treatments can be mined from doctor's prescription. In addition to action rule mining in healthcare, meta actions present an interesting area for personalized treatments mining.

4 Experiments

4.1 Dataset

The progress of treatment with Tinnitus Retraining Therapy (habituation of tinnitus) was monitored and collected in Tinnitus and Hyperacusis Center at Emory University School of Medicine. Original sample of 555 patients, described by forms during initial or follow-up visits, collected by Dr. Jastreboff, was used. Additionally, the Tinnitus Handicap Inventory was administered to individuals during their visits to the Center. The database consists of tuples identified with patient and visit numbers and have been developed over years by inserting patients' information from paper forms (devised by doctor Jastreboff).

The raw dataset was organized into 11 tables including data on:

- *Demographics*—includes al the demographics information such as address, age, gender, occupation, work status.
- *Pharmacology*—information on medications taken by a patient.
- *Visits*—the main inventory of visits and their outcomes, timestamped.
- *Audiological measurements*—carried out by physician at visits.
- *Initial and follow-up forms' questions on tinnitus*, sound tolerance and hearing problem—the answers are mostly Likert scale.
- *Newman form questions*—contain patient's subjective opinion on impact of tinnitus on three areas of their lives: emotional, functional and catastrophical, along with summary values and total score of all of them.
- *Instruments*—sound generators used within the therapy with the details such as type, model, etc.
- *REM*—settings used for sound generators as a part of the therapy.

4.2 Preprocessing and Feature Extraction

The raw dataset was preprocessed: tables were merged into one dataset of visits. The dataset was found to be incomplete and inconsistent in terms of visits' numbering and timestamps, which had to be fixed manually. Some columns contained too many missing values, so they had to be discarded.

New features were introduced as described below.

4.2.1 Tinnitus Background

Binary features related to Tinnitus background, such as: STI—*Stress Tinnitus Induced*, NTI—*Noise Tinnitus Induced*, etc., developed based on the textual descriptions in *T Induced* and *H Induced* columns in the *Demographics* table—for example STI was identified with keywords, such as 'divorce', 'excessive work', etc. NTI—with 'noise exposure', 'shooting guns'. Other binary attributes developed to indicate a tinnitus/hyperacusis cause were related to specific medical conditions:

- *HLTI—Hearing Loss Tinnitus Induced*—covers patients who associated their tinnitus with a hearing loss.
- *DETI—Depression Tinnitus Induced*—relates tinnitus symptoms to depression.
- *AATI—Auto Accident Tinnitus Induced*—whether tinnitus emerged as a result of auto accident, which involved head injuries.
- *OTI—Operation Tinnitus Induced*—patients after surgeries.
- *OMTI—Other Medical*—patients, whose tinnitus was related to medical conditions other than a hearing loss, depression or an operation—patients with acoustic neuroma, Lyme's disease, ear infections, obsessive compulsive disorder and others.

4.2.2 Temporal Features

Having information about patient's date of birth, as well as date of the first visit, a column, informing what was the age of the patient when they started treatment, can be derived. Temporal information could be also extracted from *T induced* column (or *H induced*), which often contains data about how long ago or the date the tinnitus (or hyperacusis) appeared. This way *DTI*/*DHI* columns were developed-by checking each tuple of a patient and calculating it manually. Having this information it was possible to derive a number of new features: the age of a patient when tinnitus started, as well as the time elapse between the tinnitus onset and the initial visit to doctor. It can potentially lead to discovering the knowledge on an impact of patient's age at the start of the treatment, the age when tinnitus began, and time elapse from the tinnitus symptoms onset to the treatment start, on the effectiveness of particular treatment methods in TRT.

To summarize the work on temporal features development, following new columns were added to the original database:

- *DTI—Date Tinnitus Induced*—date column derived from text columns,
- *DHI—Date Hyperacusis Induced*—analogous to the above, but derived from *H induced* column—these both new attributes convey general information about when "the problem" started and both were developed manually,
- *AgeInd*—patient's age when the problem (tinnitus or hyperacusis) was induced—derived from *DOB* and *DTI*/*DHI* columns,
- *AgeBeg*—patient's age when they started TRT treatment (first visit to doctor Jastreboff)—derived from *DOB* and *Date* (of visit 0) columns,
- numerical columns *DAgo*, *WAgo*, *MAgo*, *YAgo*—informing how many days, weeks, months, and years ago the problem started,
- binary columns calculated on the basis of columns above: *Y30*, *Y20*, *Y10*, *Y5*, *Y3*, *Y1*, *M6*, *M3*, *M1*, *W2*, *W1*, *D1* informing to which group of time elapse, between the tinnitus onset to the treatment start, a patient belongs (Y—years, M—months, W—weeks, D-days, and numerical value). For example, having "True" value in *Y5* column for the given patient, means that the problem was induced between 5 to 10 years before starting TRT treatment.

4.2.3 Binary Features for Medications Taken by Patients

Instead of maintaining a list of medications for each patient, they were altered into pivotal features. By pivoting the data values on the medication column, the resulting set contains a single row per patient. This single row lists all the medication taken by a patient, with the medication names shown as column names, and a binary value (True/False) for the columns. Pivot transformation was deployed with PL/SQL procedures. Each distinct value in Medication column of *Pharmacology* table was developed into additional column. Bit values in the column indicate, for each patient-visit tuple, whether the medication denoted by a column name was taken. As a result, 311 additional features were developed, each for distinct medication. Similar approach was taken to *Application* column in *Pharmacology* table. Values in this column describe patients' medical problems that are associated with the taken medications. As a result, additional 161 columns were developed for each separate medical state (for example "anxiety", "asthma", "insomnia", "ulcers", etc.).

4.3 Feature Selection

Feature selection experiments were performed along with classification (in WEKA) to choose the most relevant subset of features. The most important features for the diagnosis classification purposes proved to be audiological measurements.

5 Diagnostic Rule Extraction

5.1 Methodology

The associations of interest to mine for are factors affecting patient's category of tinnitus, such as audiological measurements, demographics data, forms' answers and pharmacology. The simplified process of diagnosis is presented in Fig. 1. The treatment approach varies according to category; thus, accurate placement of patients into these categories is critical to provide proper treatment.

Experiments on diagnostic rule discovery (association rules) were carried out with LISp-Miner system, which offers exploratory data analysis, implemented by its own procedures, called GUHA—highly optimized algorithm for rules generation [9]. The GUHA method, an original Czech data mining method with strong theoretical background, uses definition of the set of association rules (or G-rules in Ac4ft-Miner) to generate and verify particular rules on the data provided to the system. Algorithm does not use Apriori-like, but bit-string approach to mine rules. Premise and conclusion of the GUHA rule (relevant pattern) are defined in terms of boolean attributes, which are, in turn, defined as conjunction or disjunction of boolean attributes or literals.

5.2 Results

The results from the system are printed as general hypotheses (all factors were used for an algorithm). The following printings show example of generated comprehensive rules along with confidence ad support values for each rule (the explanation for the attribute's abbreviations is provided in Appendix in Table 4). These obtained from experiments targeting the best confidence can be interpreted as being more

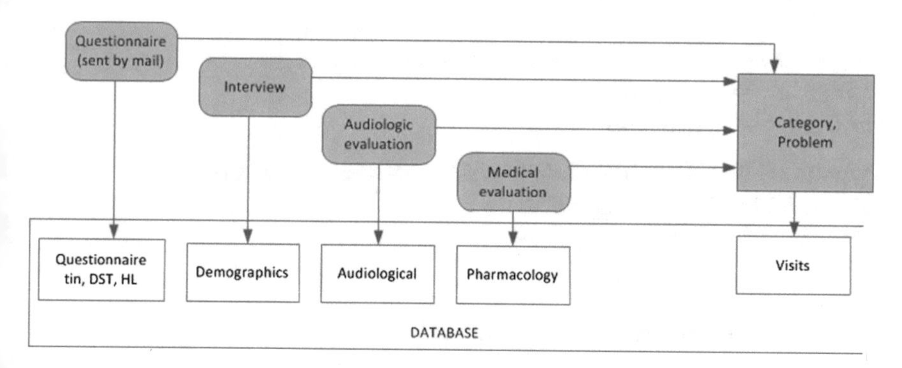

Fig. 1 Factors and data flow in the process of determining patient's category and problem

accurate, but less general. On the other hand, rules extracted with such settings, so that to obtain best support, hold true more generally (in greater population).

Besides, a series of experiments was run for each area of interest separately in relation to patient's category (to obtain more detailed results), which are discussed in the subsequent subsections:

- Interview $\Longrightarrow$ Category
- Audiology $\Longrightarrow$ Category
- Demographics $\Longrightarrow$ Category
- Pharmacology $\Longrightarrow$ Category
- Pharmacology $\Longrightarrow$ Tinnitus

Some interesting findings were obtained. We show only the most interesting examples of rules with best support and confidence, among many rules generated. In the listings below, confidence and support values are provided for each rule, where support is defined as percentage of objects in the whole dataset satisfying that rule.

5.2.1 Comprehensive—Most General rules (names of attributes are explained in Table 4)

Examples of rules for each category, mined from all the relevant attributes, with the highest support:

Hypotheses 1 $H\,EL < 1 \Longrightarrow_{0.52;0.04} C(0)$
$L\,SD \geq 100 \wedge LL4 \geq 999 \wedge LR8 \geq 999 \wedge R\,SD \geq 100 \Longrightarrow_{0.5;0.04} C(0)$
$L\,SD \geq 100 \wedge LL4 \geq 999 \wedge LL8 \geq 999 \wedge LR8 \geq 999 \wedge R\,SD \geq 100 \Longrightarrow_{0.5;0.04} C(0)$

Hypotheses 2 $LL12 \geq 999 \wedge LR12 \geq 999 \wedge R\,SD \geq 100 \Longrightarrow_{0.58;0.11} C(1)$
$LR12 \geq 999 \wedge T\,EL \geq 8 \Longrightarrow_{0.57;0.09} C(1)$
$T\,An \geq 8 \wedge R\,SD \geq 100 \Longrightarrow_{0.56;0.09} C(1)$

Hypotheses 3 $L4 \geq 65 \Longrightarrow_{0.62;0.1} C(2)$
$HL\,pr \geq 5 \Longrightarrow_{0.54;0.14} C(2)$

Hypotheses 4 $H\,An \geq 8 \wedge H\,Sv \geq 7.5 \Longrightarrow_{0.55;0.1} C(3)$
$H\,Sv \geq 7.5 \Longrightarrow_{0.5;0.11} C(3)$
$H\,EL \geq 8 \Longrightarrow_{0.5;0.11} C(3)$

Hypotheses 5 $L\,SD \geq 100 \wedge L4<10 \wedge LL3<75 \Longrightarrow_{0.67;0.02} C(4)$
$L3<5 \wedge LL3<75 \Longrightarrow_{0.59;0.02} C(4)$
$L4<10 \wedge LL3<75 \Longrightarrow_{0.53;0.02} C(4)$

5.2.2 Comprehensive—Most Accurate

Examples of rules for each category, mined from all the relevant attributes, with the highest confidence:

Hypotheses 6 $DST(N) \Longrightarrow_{0.78;0.01} C(0)$
$Concert(0) \Longrightarrow_{0.75;0.01} C(0)$
$Rest(0) \Longrightarrow_{0.75;0.01} C(0)$

Hypotheses 7 $R3(<15;20)) \wedge T\,An \geq 8 \Longrightarrow_{0.94;0.03} C(1)$
$LL2 \geq 999 \wedge LR12 \geq 999 \wedge R4(<15;20)) \wedge T\,EL \geq 8 \Longrightarrow_{0.94;0.03} C(1)$
$LR12 \geq 999 \wedge R4(<15;20)) \wedge T\,EL \geq 8 \Longrightarrow_{0.94;0.03} C(1)$
$R4(<15;20)) \wedge T\,Sv \geq 8 \Longrightarrow_{0.94;0.03} C(1)$

Hypotheses 8 $LR8 \geq 999 \wedge R6 \geq 75 \wedge T\,Sv \geq 8 \Longrightarrow_{0.96;0.04} C(2)$
$LL8 \geq 999 \wedge LR8 \geq 999 \wedge R6 \geq 75 \wedge T\,Sv \geq 8 \Longrightarrow_{0.96;0.04} C(2)$
$LR6 \geq 999 \wedge LR8 \geq 999 \wedge R2 \geq 45 \wedge R3 \geq 60 \wedge R6 \geq 75 \Longrightarrow_{0.95;0.03} C(2)$
$LR6 \geq 999 \wedge LR8 \geq 999 \wedge R2 \geq 45 \wedge R4 \geq 65 \wedge R6 \geq 75 \Longrightarrow_{0.95;0.03} C(2)$
$LR6 \geq 999 \wedge LR8 \geq 999 \wedge R2 \geq 45 \wedge R4 \geq 65 \wedge R8 \geq 75 \Longrightarrow_{0.95;0.03} C(2)$
$L2 \geq 50 \wedge L3 \geq 60 \wedge LR8 \geq 999 \wedge R6 \geq 75 \Longrightarrow_{0.95;0.03} C(2)$

Hypotheses 9 $LL3(<85;91)) \wedge H\,pr \geq 7 \wedge H\,Sv \geq 7.5 \Longrightarrow_{1;0.03} C(3)$
$LL3(<85;91)) \wedge H\,An \geq 8 \wedge H\,EL \geq 8 \wedge H\,Sv \geq 7.5 \Longrightarrow_{1;0.03} C(3)$
$LL3(<85;91)) \wedge H\,EL \geq 8 \wedge H\,Sv \geq 7.5 \Longrightarrow_{1;0.03} C(3)$
$LR1 < 74 \wedge LR2 < 74 \wedge LR6 < 78 \wedge H\,pr \geq 7 \wedge H\,An \geq 8 \Longrightarrow_{0.94;0.03} C(3)$

5.2.3 Interview $\Longrightarrow$ Category

Hypotheses 10 $H\,EL < 1 \Longrightarrow_{0.52;0.04} C(0)$
$H\,An < 1.5 \wedge H\,EL < 1 \wedge H\,Sv < 1.5 \Longrightarrow_{0.51;0.03} C(0)$

Hypotheses 11 $HL\,pr < 0.5 \wedge T\,EL \geq 8 \Longrightarrow_{0.55;0.06} C(1)$
$H\,pr < 0.5 \wedge HL\,pr < 0.5 \Longrightarrow_{0.58;0.04} C(1)$

Hypotheses 12 $HL\,pr \geq 5 \wedge T\,EL \geq 8 \Longrightarrow_{0.57;0.07} C(2)$
$HL\,pr \geq 5 \wedge T\,Sv \geq 8 \Longrightarrow_{0.57;0.06} C(2)$
$HL\,pr \geq 5 \wedge T\,An \geq 8 \Longrightarrow_{0.55;0.07} C(2)$

Hypotheses 13 $H\,An \geq 8 \wedge H\,EL \geq 8 \wedge H\,Sv \geq 7.5 \Longrightarrow_{0.58;0.09} C(3)$
$H\,pr \geq 7 \wedge H\,An \geq 8 \wedge H\,EL \geq 8 \wedge H\,Sv \geq 7.5 \Longrightarrow_{0.58;0.08} C(3)$

The obtained rules seem to confirm the expert's (medical) knowledge:

- patients categorized into 0 group have a problem a low impact on life (*H EL* is low),
- category-1 patients have significant tinnitus problem, but without hyperacusis (*H pr* is low) and there is no significant hearing loss (*HL pr* is low),
- category 2 is characterized on the other hand with significant hearing loss ($HL\,pr \geq 5$),
- category 3 is associated by the expert with significant hyperacusis problem—obtained hypotheses show association of high values of *H An*, *H Sv* and *H EL* with this category.

5.2.4 Audiology $\Longrightarrow$ Category

Hypotheses 14 $L\ SD \geq 100 \wedge LL4 \geq 999 \wedge LR8 \geq 999 \wedge R\ SD \geq 100 \Longrightarrow_{0.5;0.04} C(0)$
$L\ SD \geq 100 \wedge LL4 \geq 999 \wedge LL8 \geq 999 \wedge LR8 \geq 999 \wedge R\ SD \geq 100 \Longrightarrow_{0.5;0.04} C(0)$

Hypotheses 15 $LL12 \geq 999 \wedge LR12 \geq 999 \wedge R\ SD \geq 100 \Longrightarrow_{0.58;0.11} C(1)$
$LL12 \geq 999 \wedge R\ SD \geq 100 \Longrightarrow_{0.55;0.12} C(1)$

Hypotheses 16 $LR8 \geq 999 \wedge R4 \geq 65 \Longrightarrow_{0.78;0.08} C(2)$
$L2 \geq 50 \Longrightarrow_{0.7;0.1} C(2)$

Hypotheses 17 $LR6 < 78 \Longrightarrow_{0.63;0.07} C(3)$
$LR2 < 74 \Longrightarrow_{0.62;0.07} C(3)$

Hypotheses 18 $L\ SD \geq 100 \wedge L4 < 10\ AND\ LL3 < 75 \Longrightarrow_{0.67;0.02} C(4)$
$L3 < 5 \wedge LL3 < 75 \Longrightarrow_{0.59;0.02} C(4)$

For the second tested area generated hypotheses inform that a basic audiogram with LDLs is the crucial test for diagnosis. Based on obtained rules, it can be concluded that the lower the tolerance, the more severe category of tinnitus should be assigned to a patient. According to our medical expertise, the found results were interesting and also in theory it is expected to have strong correlation of THI with LDL.

5.2.5 Demographics $\Longrightarrow$ Category

Hypotheses 19 $Country(USA) \wedge MedNr(<3;4)) \wedge State(GA) \Longrightarrow_{0.56;0.02} C(0)$

Hypotheses 20 $AgeBeg(<50;55)) \wedge Country(USA) \wedge G(m) \Longrightarrow_{0.58;0.02} C(1)$
$AgeBeg(<50;55)) \wedge G(m) \Longrightarrow_{0.56;0.02} C(1)$
$Country(USA) \wedge G(m) \wedge M6(yes) \Longrightarrow_{0.5;0.02} C(1)$

Hypotheses 21 $AgeBeg \geq 68 \Longrightarrow_{0.58;0.03} C(2)$
$G(m) \wedge MedNr \geq 5 \wedge T\ side(yes) \Longrightarrow_{0.53;0.03} C(2)$

Hypotheses 22 $Work(h) \Longrightarrow_{0.69;0.02} C(3)$
$Country(USA) \wedge G(f) \wedge M1(yes) \Longrightarrow_{0.83;0.01} C(3)$
$AgeBeg \geq 40 \wedge Country(USA) \wedge AgeInd(<30;38)) \Longrightarrow_{0.71;0.01} C(3)$
$Occup(homemaker) \Longrightarrow_{0.71;0.01} C(3)$
$G(f) \wedge STI(yes) \Longrightarrow_{0.5;0.01} C(3)$

Hypotheses 23 $Country(USA) \wedge G(m) \wedge MedNr(3) \wedge Y10(yes) \Longrightarrow_{0.8;0.01} C(4)$

Some relevant patterns of patients' demographics in particular categories were also found out. For example, as a rule, patients with tinnitus of low effect on their lives (that is, category-0) came from the state of Georgia in the USA (that is nearby the clinic) and were affected with 3 other afflictions (were taking three types of medications for treating them). According to our medical expertise, that probably just reflects the fact that long distance patients with low level of severity did not bother to come as it would involve cost and effort; coming was much easier for people from Georgia. Another common pattern for patients in category 1 was: a male aged 50–55 from the USA, whose tinnitus had started 6–12 months before he began TRT.

It could also be observed that category-2 patients are typically older (age when they began treatment typically higher than 68 years old, as a rule), they had been taking more medications (5 and more) and their tinnitus was associated with taking these medications (*T side(yes)*). According to our medical knowledge, we can confirm that older patients are taking more medications. Also, hearing loss, which has to be present for Category 2, is strongly correlated with the age.

Relevant patterns for Category 3 included:

- the patients who worked at home (and also their tinnitus was induced by medications),
- the patients occupied with homemaking,
- the female patients with the tinnitus induced 1–3 months before they went to a doctor,
- females whose tinnitus was associated with stressful situations,
- patients relatively young (younger than 40 years old, whose problem started at 30–38 years old), living in the USA.

Pattern found for patients with the fourth category, included males curing three other afflictions with the corresponding medications, whose tinnitus is 10–20 years old.

It should be noted that the demographic-based rules must not be primarily used in diagnosis, and the medical knowledge confirms it. Patient's category should not be based on their age, place of residence, occupation, etc., but rather on more objective medical factors, such as, audiological measures or interview. Nevertheless, they reveal some common demographic patterns in categories of patients treated in the past, which may bring additional knowledge, used as heuristics or hints in the rule-based decision support system.

5.2.6 Pharmacology $\Longrightarrow$ Category

Another series of experiments were focused on discovering patterns relating additional patients' afflictions and medication taken in order to cure them, to the category of tinnitus treatment.

Hypotheses 24 *Ativan(yes)* $\wedge$ *Anxiety disorder(yes)* $\Longrightarrow_{0.58;0.01}$ *C(1)*
Klonopin(yes) $\wedge$ *Panic disorder(yes)* $\wedge$ *Seizures(yes)* $\Longrightarrow_{0.53;0.01}$ *C(1)*
Depression disorder(yes) $\wedge$ *Panic disorder(yes)* $\wedge$ *Seizures(yes)* $\Longrightarrow_{0.5;0.02}$ *C(1)*

Preliminary results have shown that patients with accompanying depression, anxiety or panic disorders were assigned to Category 1, while patients with hypertension, for example, belonged to category 2. Relevant group of patients treated for anxiety, panic/seizures or depression disorders (with Ativan/Klonopin) was diagnosed with the first category of tinnitus. According to our medical expertise, these drugs are routinely prescribed by physicians for treating tinnitus, in order to decrease anxiety or depression.

Hypotheses 25 $Angin(yes) \wedge Hypertension(yes) \Longrightarrow_{0.69;0.02} C(2)$

Patients with hypertension and angina can be hypothetically classified into the second category of tinnitus (with 69% confidence). According to our medical expertise, typically these conditions are associated with aging which in turn is strongly associated with hearing loss.

5.2.7 Pharmacology $\Longrightarrow$ Tinnitus

The last group of experiments aimed at finding out which drugs might cause side-effect of tinnitus:

Hypotheses 26 $Norvasc(yes) \wedge T\ side(yes) \Longrightarrow_{0.67;0.01} C(2)$
$Prozac(yes) \wedge T\ side(yes) \Longrightarrow_{0.6;0.01} C(1)$
$Synthroid(yes) \wedge T\ side(yes) \Longrightarrow_{0.6;0.01} C(2)$
$Atenolol(yes) \wedge T\ side(yes) \Longrightarrow_{0.56;0.01} C(2)$
$Celebrex(yes) \wedge T\ side(yes) \Longrightarrow_{0.56;0.01} C(2)$
$Klonopin(yes) \wedge T\ side(yes) \Longrightarrow_{0.56;0.01} C(1)$

The first medication is applied for hypertension and angina, the second for depression, bulimia nervosa, OCD. Synthroid is used in thyroid hormone therapy, Atenolol reduces blood pressure (treats hypertension). Celebrex acts anti -inflammatory and Klonopin—anti-panic and anti-seizure.

The conclusion from the experiment is that these medications should be further investigated on their side-effects. Patients taking them and seeking help for their tinnitus might recover simply after stop taking them or switching to other complementary pharmaceuticals, with no such side-effects. It might also save time on complex tinnitus therapy, avoiding unnecessary actions. As for depression, however, it is not clear, whether this disorder is a cause or an effect of tinnitus. According to our medical expertise it can be both.

6 Treatment Rule Extraction

6.1 Methodology

As stated earlier, action rules should help in choosing the right course of treatment within Tinnitus Retraining Therapy. The treatment process and a data flow within it is shown in Fig. 2.

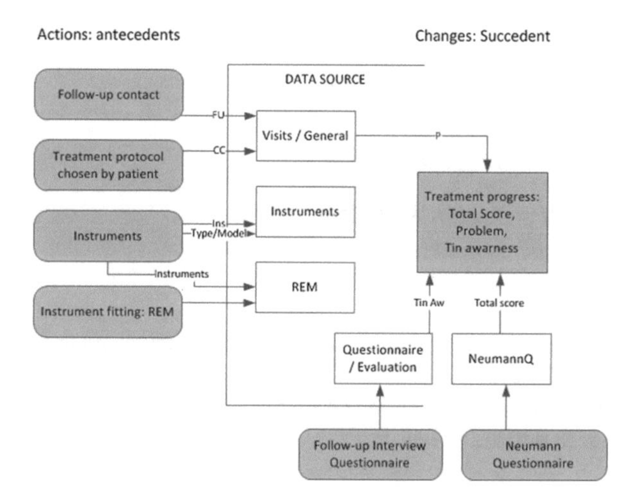

Fig. 2 The process and data flow of treatment actions and TRT tracking

Appropriate tasks were defined in LISp-Miner [9] based on analysis of the process. Treatment actions include: treatment protocol (relevant for each category), applying a sound generator and setting the generator (REM). An attribute chosen to track treatment and improvement is a Total score, which indicates severity of tinnitus according to the following scale: 0–16 -slight, 18–36 -mild, 38–56 -moderate, 58–76 -severe, and 78–100 -catastrophic handicap. The aim of extracting the action rules is to find treatment actions that lead to changes in severity of a patient's tinnitus from higher to lower. *A Total Score* attribute was missing for about half of visits registered in a tinnitus database and the same for *A Tinnitus Awareness* attribute. Even when considering case of both *Tsc* or *Taw* still about 40% values were missing. To handle this problem and retain all the tuples for visits which may contain useful information about treatment actions, an algorithm for imputation of missing values was developed and applied.

6.2 Decision Attribute Development

In order to find out action rules that indicate improvement, the decision attribute was further preprocessed and new derived features were developed:

- *ChTsc*—Change in Total Score
- *ChTaw*—Change in Tinnitus Awareness
- *PerChTsc*—Percentage Change in Tinnitus Score
- *PerChTaw*—Percentage Change in Tinnitus Awareness

On the top of these, one decision attribute was developed: a new change attribute for X indicator at visit v of a given patient, in a following way:

Definition 5 $CH_{X,v} =$

- $NULL$, for $X_n = NULL$ or $X_{n+1} = NULL$
- 0, for $X_n = 0$ and $X_{n+1} = 0$
- $\frac{-100}{dist_{n+1,n}}$, for $X_n = 0$ and $X_{n+1} > 0$
- $(100\% * \frac{X_{n+1}-X_n}{X_n})/(dist_{n+1,n})$, for $X_n \neq 0$

where:

- $X = Tsc$ or $X = Taw$
- X_n is a measurement of X at v, or the closest previous measurement of X from v: $DATE(v) \geq DATE(X_n)$
- X_{n+1} is the closest next measurement of X since the visit v: $DATE(v) < DATE(X_{n+1})$
- *dist* is a distance defined as below:

Definition 6 $dist_{n+1,n} =$

- $NULL$, for $CH_{X_n,v} = NULL$
- $DATEDIFF(weeks, DATE(X_{n+1}), DATE(X_n))$, for $DATE(X_{n+1}) > DATE(X_n)$

Algorithm calculates the change and distance values for each visit based on the definitions presented above.

One final change attribute is a combined change attribute defined as follows:

Definition 7 $CH = ChTsc$ and $distCh = distTsc$
in the following cases (in order of priority):

- Ch_{Tsc} is not *NULL* and Ch_{Taw} is *NULL*—this is the most obvious case—we choose a change in indicator that is available,
- *ScT* is not *NULL* and *AwT* is *NULL*—the case when both change features are available for the tuple, but change for *Sc t* is accurate, while *ChTaw* is approximated by "neighboring" previous and next measurements,
- Ch_{Tsc} is not *NULL* and Ch_{Taw} is not *NULL* and $distTsc < distTaw$—there are values for change attributes for both indicators, as well as current values of indicators themselves (*Sc t* and *Aw T*)—a change value associated with lower distance is chosen (it is assumed that treatment effectiveness measured in shorter time distance is more accurate).

Analogously:
$CH = ChTaw$ and $distCh = distTaw$

when:

- Ch_{Tsc} is *NULL* and Ch_{Taw} is not *NULL*,
- *ScT* is *NULL* and *AwT* is not *NULL*
- Ch_{Tsc} is not *NULL* and Ch_{Taw} is not *NULL* and $distTsc > distTaw$.

Table 2 *Ch* and *treat len* attributes definition in LISp-Miner

Group	Att name	Attribute meaning	Type	Cat	Sample
Temporal	*Ch*	Percentage change per week	Inter	5	Better
	$treat_{len}$	Length of treatment in weeks	Inter	10	<1;4)

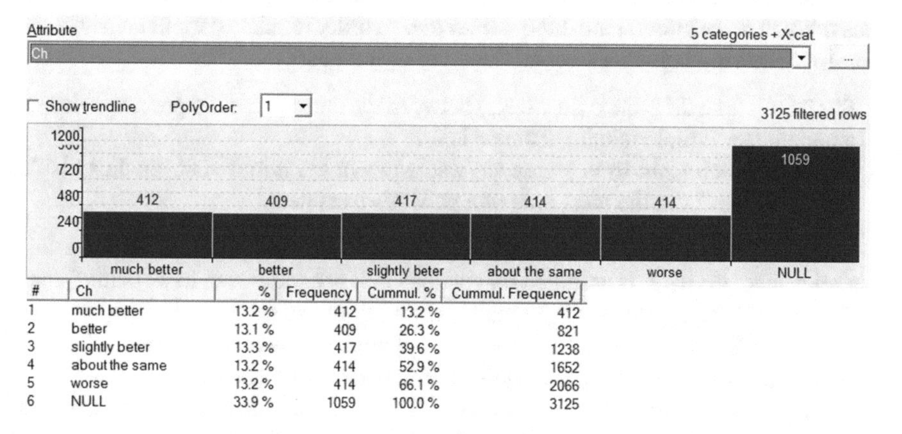

#	Ch	%	Frequency	Cummul. %	Cummul. Frequency
1	much better	13.2 %	412	13.2 %	412
2	better	13.1 %	409	26.3 %	821
3	slightly beter	13.3 %	417	39.6 %	1238
4	about the same	13.2 %	414	52.9 %	1652
5	worse	13.2 %	414	66.1 %	2066
6	NULL	33.9 %	1059	100.0 %	3125

Fig. 3 Frequency distribution of categories for *Ch* attribute in tinnitus dataset

Table 3 Category names and corresponding intervals for *Ch* attribute

Category name	Ch
Much better	$< -99; -4.9107)$
Better	$< -4.9107; -2)$
Slightly better	$< -2; 0.4348)$
About the same	$< -0.4348; 0.8418)$
Worse	$< 0.8418; 1944.7369)$

The last case, not resolved by the two above, is when Ch_{Tsc} is not *NULL* and Ch_{Taw} is not *NULL* and $distTsc = distTaw$. Then a "combined" change is calculated as an average of both:

$$CH = \frac{Ch_{Tsc} + Ch_{Taw}}{2} \text{ and } distCh = distTaw = distTsc.$$

A new attribute *Ch* (with corresponding *distCh* attribute) was introduced to LISp-Miner environment under *Temporal* group of attributes (see Table 2).

Figure 3 shows categories defined for a *Ch* attribute, as intervals, along with their balanced frequency (absolute, relative and cumulated).

There are 5 categories for a change value: "worse" for positive values of *Ch*, "about the same" for no change, and three categories for different magnitudes of negative values: “slightly better”, “better” and “much better”. Corresponding intervals for each category are shown in Table 3.

6.3 Distance Features

An additional column, indicating length of treatment of a given measure (*distCh*), was defined as an interval attribute—*treat len*. In order to relate patient’s visits temporally, following columns were additionally developed:

- *distPrev*—time difference (in weeks) between the current and the previous visit of a patient (for initial visit the distance is 0),
- *dist0*—for each visit: time elapse (in weeks) from the initial visit, the last visit’s *dist0* informs about the total time of a patient’s treatment,

After defining the additional attributes in LISp-Miner, they were used for defining relevant patterns. It is assumed that actions that generally lead to a “better” condition are interesting (for now, no matter if an improvement is slight, moderate or significant). The procedure is enforced to generate only interesting action rules (for treatment purposes) and generates only effective treatment actions.

With a new, accurate change attribute *Ch* for the succedent part, developed as described above, final choice of the most reliable rules can be made.

Besides considering a *Ch* attribute in the experimental setup, also temporal dependencies between actions and their effects were considered, suggesting a change in the length of treatment with a particular method (*treat* attribute).

6.4 Instrument Fitting

(names of all attributes are explained in the Table 4):

Hypotheses 27 *Instr(GHI):Freq LE(<3000;3150))* $\rightarrow$ *Freq LE* $\geq$ *3775)* $\Longrightarrow_{0.32;37;8}$ *Ch(better/much better/slightly better)*
Instr(SG): Mix R SL(<9;10)) $\rightarrow$ *Mix R SL(<11;12))* $\Longrightarrow_{0.27;8;11}$ *Ch(better/much better/slightly better)*

Instr(SG): Mix R SL(<9;10)) $\rightarrow$ *Mix R SL(<15;17))* $\Longrightarrow_{0.27;8;8}$ *Ch(better/much better/slightly better)*

Instr(GHI): Mix L SL(<7;8)) $\rightarrow$ *Mix L SL<2* $\Longrightarrow_{0.27;8;8}$ *Ch(better/much better/slightly better)*

Instr(GHI): Mix L SL(<7;8)) $\rightarrow$ *Mix L SL(<11;12))* $\Longrightarrow_{0.27;8;8}$ *Ch(better/much better/slightly better)*

Instr(SG): Freq LE(<2670;2800)) ∧ Freq RE(<2670;2800)) → Freq LE(<2500; 2670)) ∧ Freq RE(<2500;2670)) $\Longrightarrow_{0.23;6;7}$ *Ch(better/much better/slightly better)*

Instr(GHI): Th L SPL(<36;37)) → Th L SPL(<37;38)) $\Longrightarrow_{0.17;8;9}$ *Ch(better/ much better/slightly better)*

Instr(GHI): Mix R SL(<6;7)) → Mix R SL(<9;10)) $\Longrightarrow_{0.17;9;8}$ *Ch(better/much better/slightly better)*

Instr(SG): Freq RE(<3000;3150)) → Freq RE(<2500;2670)) $\Longrightarrow_{0.11;9;12}$ *Ch(better/much better)*

Instr(SG): Freq RE(<3000;3150)) → Freq RE(<2500;2670)) $\Longrightarrow_{0.03;5;6}$ *Ch(slightly better) → Ch(better/much better)*

Instr(SG): Freq LE(<2670;2800)) → Freq LE(<2500;2670)) $\Longrightarrow_{0.1;12;9}$ *Ch(better/much better/slightly better)*

Instr(SG): Freq LE(<2670;2800)) → Freq LE(<3000;3150)) $\Longrightarrow_{0.1;8;11}$ *Ch(better/much better)*

Instr(SG) ∧ Model(TR COE): Freq RE(<2500;2670)) → Freq RE(<2670;2800)) $\Longrightarrow_{0.09;10;10}$ *Ch(better/much better/slightly better)*

Instr(SG) ∧ Model(TR COE): Freq RE(<2500;2670)) → Freq RE(<3000;3150)) $\Longrightarrow_{0.08;10;12}$ *Ch(better/much better/slightly better)*

Instr(SG): Freq RE(<2670;2800)) → Freq RE(<2500;2670)) $\Longrightarrow_{0.08;11;12}$ *Ch(better/much better/slightly better)*

Instr(GHS): Freq RE(<2800;3000)) → Freq RE(<2670;2800)) $\Longrightarrow_{0.07;11;12}$ *Ch(better/much better/slightly better)*

Instr(SG): Th R SPL(<33;34)) → Th R SPL(<36;37)) $\Longrightarrow_{0.02;8;9}$ *Ch(better/much better/slightly better)*

Type(GHH): Freq RE(<2670;2800)) → Freq RE(<3000;3150)) $\Longrightarrow_{0.02;8;13}$ *Ch(better/much better/slightly better)*

FU(A) ∧ Instr(GHI) ∧ Freq RE(<3000;3150)): treat(<6;8)) → treat(<5;6)) $\Longrightarrow_{0.1;9;8}$ *Ch(better/much better/slightly better)*

The above action rules, related to the instruments' fitting with REM, include rules for the following types of instruments: "SG" (sound generators generally), "GHI" (general type of sound generator that includes both GHI hard and GHI soft models), particular types: “GHS” (GHI soft) and “GHH” (GHI hard), up to specific model, such as “TRI-COE”. The following settings of the instruments were considered in the variable antecedent parts of rules: *Freq RE, Freq LE, Mix R SL, Mix L SL, Th R SPL, Th L SPL*. These constitute quite a significant subset of settings for fitting the instruments.

For example, the last action rule from Hypotheses 27 informs that the probability of a successful treatment increases by 10 percentage points, when the “Audiological/counseling” treatment combined with the setting of "GHI" instrumentation to "Freq RE" at in <3000;3150) shortens from 6–8 weeks to 5–6 weeks.

6.5 Treatment Protocol

The second rule from the listing below informs that changing the treatment of a patient in Category-1 from the treatment protocol “0” lasting 12–16 weeks to the treatment protocol “1” for more than 32 weeks, should increase improvement by 61 percentage points.

Hypotheses 28 $Cat(0){:}\ CC(0) \rightarrow CC(1) \Longrightarrow_{0.33;42;9} Ch(better/slightly\ better)$

$Cat(0){:}\ CC(0) \wedge treat(<12;16)) \rightarrow CC(1) \wedge treat \geq 32 \Longrightarrow_{0.61;42;9} Ch(better/slightly\ better)$

$Cat(3){:}\ Instr(GHH) \wedge FU(T) \wedge CC(3) \rightarrow Instr(TCI\text{-}C) \wedge FU(A) \wedge CC(2) \Longrightarrow_{0.33;8;9} Ch(better/much\ better/slightly\ better)$

$Cat(3){:}\ Instr(Viennatone) \wedge CC(3) \rightarrow Instr(TCI\text{-}C) \wedge CC(2) \Longrightarrow_{0.25;8;8} Ch(slightly\ better/better/much\ better)$

$Cat(3){:}\ CC(0) \rightarrow CC(2) \Longrightarrow_{0.18;8;22} Ch(slightly\ better/better/much\ better)$

$Cat(1){:}\ CC(0) \rightarrow CC(1) \Longrightarrow_{0.14;9;491} Ch(slightly\ better/better/much\ better)$

$Cat(3){:}\ CC(0) \rightarrow CC(3) \Longrightarrow_{0.08;8;239} Ch(slightly\ better/better/much\ better)$

6.6 Treatment Personalized for Demographics

The following hypotheses were generated for the tasks that were defined in order to maximize treatment personalization.

Hypotheses 29 $AgeBeg(<50;55)) \wedge G(m) \wedge Cat(1) \wedge T\ side(yes){:}\ MedNr \geq 5 \rightarrow MedNr(<2;3)) \Longrightarrow_{0.55;14;8} Ch(slightly\ better/better)$

$G(m) \wedge Cat(1) \wedge OMTI(yes) \wedge T\ side(yes){:}\ MedNr(<3;4)) \rightarrow MedNr(<4;5)) \Longrightarrow_{0.41;9;18} Ch(slightly\ better/better/much\ better)$

$AgeBeg(<50;55)) \wedge G(m) \wedge AgeInd(<50;56)) \wedge T\ side(yes){:}\ CC(2) \rightarrow CC(1) \Longrightarrow_{0.55;14;11} Ch(slightly\ better/better/much\ better)$

$G(m) \wedge Cat(1) \wedge OMTI(yes) \wedge T\ side(yes){:}\ F(T) \rightarrow F(A) \Longrightarrow_{0.23;9;16} Ch(slightly\ better/much\ better)$

$AgeBeg(<55;60)) \wedge G(m) \wedge Cat(1) \wedge T\ side(yes){:}\ Instr(GHS) \rightarrow Instr(GHH) \Longrightarrow_{0.19;8;8} Ch(better/much\ better)$

6.7 Treatment Personalized for Tinnitus Background

Hypotheses 30 $OMTI(yes) \wedge T\ side(yes){:}\ Instr(Viennatone) \wedge FU(T) \rightarrow Instr(GHH) \wedge FU(A) \Longrightarrow_{0.56;8;8} Ch(slightly\ better/better)$

$NTI(yes) \wedge G(m){:}\ Instr(GHS) \rightarrow Instr(GHH) \Longrightarrow_{0.33;28;8} Ch(slightly\ better/better/much\ better)$

$G(m) \wedge OMTI(yes) \wedge M6(yes) \wedge Cat(1){:}\ FU(T) \rightarrow FU(A) \Longrightarrow_{0.3;5;10} Ch(slightly\ better/better/much\ better)$

$OMTI(yes) \wedge T\ side(yes){:}\ Work(h) \rightarrow Work(w) \Longrightarrow_{0.3;13;11} Ch(slightly\ better/better)$

$OMTI(yes) \wedge G(f){:}\ Instr(GHS) \rightarrow Instr(GHH) \Longrightarrow_{0.28;10;8} Ch(slightly\ better/better)$

$OMTI(yes) \wedge T\ side(yes) \wedge Cat(1){:}\ Instr(GHS) \rightarrow Instr(GHH) \Longrightarrow_{0.3;13;11} Ch(slightly\ better/better)$

$OMTI(yes) \wedge G(f){:}\ Instr(Viennatone) \rightarrow Instr(GHH) \Longrightarrow_{0.25;8;8} Ch(slightly\ better/better/much\ better)$

$OMTI(yes) \wedge T\ side(yes) \wedge Cat(1){:}\ Instr(Viennatone) \wedge FU(T) \rightarrow Instr(GHS) \wedge FU(A) \Longrightarrow_{0.24;8;8} Ch(slightly\ better/better)$

$G(m) \wedge NTI(yes) \wedge M3(yes) \wedge Cat(3){:}\ FU(A) \rightarrow FU(T) \Longrightarrow_{0.18;6;8} Ch(slightly\ better/better/much\ better)$

OMTI(yes) $\wedge$ *Instr(GHS): treat* $\geq$ *32* $\rightarrow$ *treat(<5;6))* $\Longrightarrow_{0.06;9;6}$ *Ch(slightly better/better/much better)*

OMTI(yes) $\wedge$ *FU(T): treat(<21;32))* $\rightarrow$ *treat(<8;10))* $\Longrightarrow_{0.01;11;14}$ *Ch(slightly better/better/much better)*

The two last rules hypothesize that in case of medical-induced tinnitus (*OMTI(yes)*), it should be advantageous to shorten the treatment with "GHS" instrumentation from "above 32 weeks" to 5–6 weeks, as well as shorten the telephone-based treatment from 21–32 weeks to 8–10 weeks.

6.8 Treatment Personalized for Medical Condition

The relevant action rules, which consider other diseases in a patient, include: patients with ulcers, hypertension, seizures, depression/anxiety disorders. The treatment actions include: reducing the number of medications (which are also associated with tinnitus as a side-effect), changing instrumentation (for example, from "GHS" to "HA", or from "Viennatone" to "GHS"), but also changing place of residence (for example, state "NY" to "WI", "GA" to "IL").

Hypotheses 31 *G(m)* $\wedge$ *T side(yes)* $\wedge$ *Ulcers(yes): Med(*$\geq$*5)* $\wedge$ *State(GA)* $\rightarrow$ *Med(<2;3))* $\wedge$ *State(IL)* $\Longrightarrow_{0.73;10;8}$ *Ch(slightly better/better)*

G(m) $\wedge$ *T side(yes)* $\wedge$ *Ulcers(yes)* $\wedge$ *Erosive arthritis(yes)* $\wedge$ *GERD(yes): Med(*$\geq$*5)* $\wedge$ *State(GA)* $\rightarrow$ *Med(<2;3))* $\wedge$ *State(IL)* $\Longrightarrow_{0.71;10;8}$ *Ch(slightly better/better)*

Cat(1) $\wedge$ *T side(yes)* $\wedge$ *Hypertension(yes): Med(*$\geq$*5)* $\wedge$ *FU(T)* $\rightarrow$ *Med(<4;5))* $\wedge$ *FU(A)* $\Longrightarrow_{0.56;12;11}$ *Ch(slightly better/better/much better)*

G(m) $\wedge$ *T side(yes)* $\wedge$ *Seizures(yes): Instr(GHS)* $\wedge$ *State(NY)* $\rightarrow$ *Instr(HA)* $\wedge$ *State(WI)* $\Longrightarrow_{0.53;9;8}$ *Ch(slightly better/much better)*

G(m) $\wedge$ *T side(yes)* $\wedge$ *Depression(yes)* $\wedge$ *Panic disorder (yes)* $\wedge$ *Seizures(yes): Instr(GHS)* $\wedge$ *State(NY)* $\rightarrow$ *Instr(HA)* $\wedge$ *State(WI)* $\Longrightarrow_{0.53;9;8}$ *Ch(slightly better/much better)*

Cat(1) $\wedge$ *T side(yes)* $\wedge$ *Depression(yes)* $\wedge$ *Anxiety disorder (yes): Instr(GHH)* $\wedge$ *Med(*$\geq$*5)* $\rightarrow$ *Instr(GHS)* $\wedge$ *Med(<4;5))* $\Longrightarrow_{0.48;9;12}$ *Ch(slightly better/better/much better)*

G(m) $\wedge$ *T side(yes)* $\wedge$ *Depression(yes): Med(*$\geq$*5)* $\wedge$ *State(GA)* $\rightarrow$ *Med(<2;3)))* $\wedge$ *State(WI)* $\Longrightarrow_{0.47;14;10}$ *Ch(slightly better/much better)*

G(m) ∧ T side(yes) ∧ Seizures(yes): Med(≥5) ∧ FU(A) → Med(<2;3)) ∧ FU(T) $\Longrightarrow_{0.47;9;8}$ *Ch(slightly better/better/much better)*

OMTI(yes) ∧ T side(yes) ∧ Depression(yes): Instr(Viennatone) ∧ Med(≥5) → Instr(GHS) ∧ Med(<4;5)) $\Longrightarrow_{0.42;21;12}$ *Ch(slightly better/much better)*

6.9 Meta Actions

Following hypotheses show examples of meta actions generated for the patient with an ID 01054 (that is, a set of effective actions, for this particular patient).

Hypotheses 32 *THC(01054) ∧ Cat(1): Instr(VSS) ∧ F(T) → Instr(V - AMTI) ∧ F(A)* $\Longrightarrow_{0.67;1;4}$ *Ch(slightly better/better/much better)*

THC(01054) ∧ Cat(1): Freg LE(<2500;2670)) ∧ Freg RE(<2500;2670)) ∧ Mix R SPL(<51;52)) → Freg LE(<2120;2380)) ∧ Freg RE(<2380;2500)) ∧ Mix R SPL(<53;55)) $\Longrightarrow_{0.5;1;1}$ *Ch(better/much better)*

THC(01054) ∧ Cat(1): Freg LE(<3000;3150)) ∧ Freg RE(<3000;3150)) ∧ Mix R SL(<9;10)) → Freg LE(<2500;2670)) ∧ Freg RE(<2500;2670)) ∧ Mix R SL(<14;15)) $\Longrightarrow_{0.5;1;1}$ *Ch(slightly better/much better)*

THC(01054) ∧ Cat(1): Freg LE(<3000;3150)) ∧ Freg RE(<3000;3150)) ∧ Mix R SL(<9;10)) → Freg LE(<2120;2380)) ∧ Freg RE(<2380;2500)) ∧ Mix R SL(<11;12)) $\Longrightarrow_{0.5;1;1}$ *Ch(slightly better/better/much better)*

THC(01054) ∧ Cat(1): Mix R SL(<13;14)) ∧ Mix R SPL(<51;52)) ∧ Th R SPL(<38;39)) → Mix R SL(<11;12)) ∧ Mix R SPL(<53;54)) ∧ Th R SPL(<42;43)) $\Longrightarrow_{0;1;1}$ *Ch(better/much better)*

The above sets of actions for the patient "01054" (meta-actions) are examples of an effective treatment undertaken for this case (profile) of a patient. It can be also observed that the last set of actions (the last hypothesis) brought no results.

7 Discussion

7.1 Summary of Experiments on Rule Extraction

Experiments described in the two previous sections were conducted in order to extract knowledge on tinnitus diagnosis and treatment, in the form of rules—decision rules and action rules, whose theoretical background was presented in Sect. 3. While the former should help in understanding the relations between different diagnosis factors, the latter suggest a course of treatment (action) leading to improvement in a patient's condition. The experiments on finding association rules can also help in analyzing the collected data in terms of patient's characteristics and discover patterns that are not obvious from a medical point of view.

However, the main advantage of the proposed approach based on rule extraction, is a possibility to automatically retrieve knowledge in the form of rules, without engaging time of a medical expert. It seems promising, as experts are usually not widely available. Additionally, knowledge engineering based on interviewing experts is quite time- consuming. Often, it is also cumbersome for experts to formulate their knowledge in the form of specific rules, as they often make decisions intuitively, based on experience. This knowledge, on the other hand, is hidden in large databases which can be extracted in the form of rules that imitate human behavior. This methodology is particularly interesting and useful for building a rule-based decision support system.

The discovered rules can be either exploited in a qualitative way by an expert, or used to perform classification (scoring) of incoming objects. Ultimately, automatically extracted rules should be built into the rule engine of decision support system. An appropriate mechanism of automatic rule execution (or alternatively inference engine) should be implemented. The relevant rules could be then evoked by matching new data with the rules' premises (antecedents) and their conclusions (succedents) can be presented to the system user.

Rule extraction, in contrary to the method based on building a classifier provides a better insight into different diagnostic and treatment factors. It also enables customizing the associations which are supposed to be discovered. Also, when implemented into a knowledge base of a decision support system, they can potentially provide an explanatory mechanism. It means, the decision the system arrives at, can be explained by means of antecedent parts of the rules that were triggered. It can also potentially serve for educational purposes. The personnel untrained in tinnitus treatment can learn tinnitus diagnosis and treatment by using the system and its explanatory functionalities, which imitate the behavior and decision making processes of a human expert.

To sum up the experiments, the discovered rules confronted with expert knowledge confirm the correctness of the approach and methodology. The discovered new, unknown patterns provide additional knowledge to an expert that otherwise could not be easily noticed from a large and complex dataset. It is important to note that the discovered knowledge should be treated as hypotheses, which nevertheless, have to be either confirmed by an expert or by a controlled study, designed to validate the hypothetical claims. In particular, the rules generated and presented in this work should not be used for any diagnosis or treatment decision, or suggest any particular course of treatment. The work within this research is experimental and aimed at presenting potential application of action rules and meta actions in the area of medical diagnosis and treatment. The final validity check of the presented approach can be done by comparing clinical results with the extracted knowledge.

7.2 Conclusions

The work within this chapter verifies a possibility of applying theory of traditional machine learning techniques, such as classification and association rules, as well as novel data mining methods, including action rules and meta actions, to a practical decision problem in the area of medicine.

The work included a series of data preprocessing steps, building new features and testing the proposed approach with the use of chosen methodologies and tools. The tests on knowledge discovery approach were divided into: testing the classification model first (not presented in this chapter), extracting the decision rules and generating action rules/ meta actions. New temporal features were introduced to describe the sparse records of patients' visits. Next, they were used in building a classification model and extracting the rules. Interesting and potentially novel rules relating treatment factors to symptoms were revealed.

Appendix: Attribute Definition

Table 4 The definition of attributes related to tinnitus patients and their visits in LISp-Miner: an attribute's group, name (short), meaning, type, number of categories, sample value(s)

Group	Att name	Attribute meaning	Type	Cat	Sample
General	THC	Patient identifier	Nom	583	00001
	V	Visit number	Nom	36	0, 1, 2, 3, 4
	P	Problems in order	Nom	15	H, HLT, TL
	Miso	Misophonia	Nom	2	Yes/no
	Miso treat	Miso treatment protocol	Nom	4	1, 2, 3, 4
	FU	Follow-up contact	Nom	5	A, C, T, E
	DP	Dependency of H presence	Nom	2	Yes/no
	REM	Real-ear measurements	Nom	2	Yes/no
	C	Category assigned by doctor	Nom	6	0, 1, 2, 3, 4
	CC	Category chosen by patient	Nom	6	0, 1, 2, 3, 4
Audiological	R25	RE pure-tone thresh 0.25 kHz	Inter	8	<−10;0)
	R50	RE pure-tone thresh 0.50 kHz	Inter	8	<−5;0)
	...	...	...	...	...
	L25	LE pure-tone thresh 0.25 kHz	Inter	8	<−5;0)
	...	...	...	...	...
	LR50	Loudness Discomfort Level R	Inter	8	<12;75)

(continued)

Table 4 (continued)

Group	Att name	Attribute meaning	Type	Cat	Sample
	LL50	Loudness Discomfort Level L	Inter	8	<11;77)
	...	...	...	...	...
	T PR	T pitch match	Inter	40	<0.35;1)
	T Rm	RE match type	Nom	4	NB,NBN
	T LR	T loudness match dB	Inter	4	<4;22)
	Th R	RE threshold of hearing	Inter	50	< − 10;−2)
	MRR	RE minimal masking level	Inter	8	<0;26)
	...	...	...	...	...
Demographics	AgeBeg	Age when treatment began	Inter	7	<8;40)
	G	Gender	Nom	2	f/m
	Occup	Occupation	Nom	54	Engineer
	Work	Work status	Nom	4	h, r, s, w
	MedNr	Number of medications taken	Inter	5	<1;2)
	Country	Country of residence	Nom	9	USA, Chile
	State	State of residence	Nom	31	AL, GA
	Zip	Zip code of residence	Nom	181	01742
Tinnitus	AgeInd	Age problem started	Inter	7	<7;30)
	AATI	Tin induced by auto accident	Nom	2	Yes/no
	DETI	Tin induced by depression	Nom	2	Yes/no
	HLTI	Tin assoc with hearing loss	Nom	2	Yes/no
	NTI	Tin induced by noise	Nom	2	Yes/no
	STI	Tin induced by stress	Nom	2	Yes/no
	OTI	Tin induced by operation	Nom	2	Yes/no
	OMTI	Tin induced by other medical	Nom	2	Yes/no
	T side	Tin as side-effect of pharm	Nom	2	Yes/no
	Gradual	Gradual onset of tinnitus	Nom	2	Yes/no
	Sudden	Sudden onset of tinnitus	Nom	2	Yes/no
Condition	Aches	Aches present?	Nom	2	Yes/no
	...		...	...	...
	Menieres	Menieres disease present?	Nom	2	Yes/no
	...		...	...	...
	Vertigo	Vertigo present?	Nom	2	Yes/no
Medication	Accupiril	Accupiril taken?	Nom	2	Yes/no
	...		...	...	...
	Zyrtec	Zyrtec taken?	Nom	2	Yes/no

(continued)

Table 4 (continued)

Group	Att name	Attribute meaning	Type	Cat	Sample
Instruments	Ins	Instrument category	Nom	3	SG, HA
	Type	Type of instrument	Nom	6	GHH
	Model	Model of instrument	Nom	16	BTE
	Ins vis	Instrument (Visits table)	Nom	43	BTE, GHS
	Instr	Instrument (Question table)	Nom	32	SG, GHS
REM	Freg RE	Right-ear measurements	Inter	12	<39;2000)
	Th R SPL		Inter	25	<22;26)
	Mix R SPL		Inter	25	<4;31)
	Mix R SL		Inter	15	<0;2)
	Tol R SPL		Inter	25	<7;29)
	Tol R SL		Inter	15	<4;8)
	Max R SPL		Inter	25	<43;48)
	Max R SL		Inter	25	<6;11)
	Freg LE	Left-ear measurements	Inter	12	<50;2000)
	...		...	...	...
Interview	An t	% of time when annoyed	Inter	10	<0;2)
	Aw t	% of time when aware	Inter	10	<0;7.5)
	Out	Outcome	Nom	4	B,N,S,W
	T sv	Severity of tinnitus	Inter	5	<0;3)
	T an	Annoyance of tinnitus	Inter	5	<0;3)
	T EL	Tinnitus effect on life	Inter	5	<0;2)
	T pr	Tinnitus as a problem	Inter	5	<0;2.5)
	H pr	Hyperacusis as a problem	Inter	5	<0;0.5)
	HL pr	Hearing loss as a problem	Inter	5	<0;0.5)
	DST	Oversensitivity y/n	Nom	2	Y, N
	Phys	Physical discomfort y/n	Nom	2	Y, N
	Descr	Descr of troublesome sound	Nom	43	Sirens
	Concert	Activity prevented	Nom	5	0, 2, 4
	Shopp	Shopping prevented	Nom	5	0, 2, 4
	Mov	Movies prevented	Nom	5	0, 2, 4
	Wrk	Work prevented	Nom	5	0, 2, 4
	Rest	Restaurants prevented	Nom	5	0, 2, 4
	Drv	Driving prevented	Nom	5	0, 2, 4
	Sport	Sports prevented	Nom	5	0, 2, 4
	Church	Church prevented	Nom	5	0, 2, 4
	House	Housekeeping prevented	Nom	5	0, 2, 4

(continued)

Table 4 (continued)

Group	Att name	Attribute meaning	Type	Cat	Sample
	Child	Childcare prevented	Nom	5	0, 2, 4
	Soc	Social activities prevented	Nom	5	0, 2, 4
	Oth	Other activities prevented	Nom	5	0, 2, 4
	H sv	Severity of DST	Inter	5	<0;1.5)
	H an	Annoyance of DST	Inter	5	<0;1.5)
	H EL	DST effect on life	Inter	5	<0;1)
	Pr	Program assessment	Nom	3	Y, N, U
	Ret	Returning Instruments	Nom	2	Y, N
NewmanQ	F1	Difficult to concentrate?	Nom	3	0, 2, 4
	F2	Difficult to hear people?	Nom	3	0, 2, 4
	E3	Tin makes you angry?	Nom	3	0, 2, 4
	F4	Tin makes you confused?	Nom	3	0, 2, 4
	C5	Feel desperate?	Nom	3	0,2,4
	E6	Complain about your tin?	Nom	3	0, 2, 4
	F7	Sleeping problems?	Nom	3	0,2,4
	C8	Feel cannot escape your tin?	Nom	3	0, 2, 4
	F9	Tin interfere social activities?	Nom	3	0, 2, 4
	E10	Feel frustrated?	Nom	3	0,2,4
	C11	Feel have a terrible disease?	Nom	3	0, 2, 4
	F12	Difficult for you to enjoy life?	Nom	3	0, 2, 4
	F13	Job /house responsibilities?	Nom	3	0, 2, 4
	E14	Tin make you often irritable?	Nom	3	0, 2, 4
	F15	Difficult for you to read?	Nom	3	0, 2, 4
	E16	Tinnitus make you upset?	Nom	3	0, 2, 4
	E17	Stress on your relationships?	Nom	3	0, 2, 4
	F18	Difficult to focus attention?	Nom	3	0, 2, 4
	C19	No control over your tinnitus?	Nom	3	0, 2, 4
	F20	Tin makes you often tired?	Nom	3	0, 2, 4
	E21	Tin makes you depressed?	Nom	3	0, 2, 4
	E22	Tinnitus makes you anxious?	Nom	3	0, 2, 4
	C23	Cannot cope with your tin?	Nom	3	0, 2, 4
	F24	Tin worse when under stress?	Nom	3	0, 2, 4
	E25	Tin makes you feel insecure?	Nom	3	0, 2, 4
	Sc F	Total score: Function	Inter	6	<0;6)
	Sc E	Total score: Emotion	Inter	6	<0;4)
	Sc C	Total score: Catastrophic	Inter	6	<0;2)
	Sc T	Total score: sum of above	Inter	5	Mild

References

1. Jastreboff, P.J., Jastreboff, M.M.: Tinnitus retraining therapy (TRT) as a method for treatment of tinnitus and hyperacusis patients. J.-Am. Acad. Audiol. **11**(3), 162–177 (2000)
2. Ras, Z.W., Wieczorkowska, A.: Action-rules: How to increase profit of a company. In: Principles of Data Mining and Knowledge Discovery, pp. 587–592. Springer (2000)
3. Wasyluk, H., Ras, Z.W., Wyrzykowska, E.: Application of action rules to HEPAR clinical decision support system. Experimental and Clinical Hepatology **4**, 46–48 (2008)
4. Ras, Z.W., Wieczorkowska, A.: Advances in Music Information Retrieval. Studies in Computational Intelligence, vol. 274. Springer (2010)
5. Pawlak, Z., Marek, W.: Rough sets and information systems. ICS. PAS. Reports **441**, 481–485 (1981)
6. Im, S., Ras, Z., Tsay, L.-S.: Action reducts. In: Foundations of Intelligent Systems, pp. 62–69. Springer (2011)
7. Touati, H., Ras, Z.W., Studnicki, J., Wieczorkowska, A.: Mining surgical Meta-actions effects with variable diagnoses' number. In: Foundations of Intelligent Systems, pp. 254–263. Springer (2014)
8. Wang, J.: Encyclopedia of Business Analytics and Optimization. IGI Global (2014)
9. Simunek, M.: LISp-Miner Control Language description of scripting language implementation. J. Syst. Integration **5**(2), 28–44 (2014)

Rough-Granular Computing for Relational Data

Piotr Hońko

Abstract Rough set theory and granular computing have widely been applied in data mining. They have been used separately as well as a combined approach called rough-granular computing. The usefulness of this approach in data mining is the driving force for employing it to improve processing of relational data. This chapter introduces three rough-granular approaches dedicated to handle complex data such as relational one. Each of them processes relational data as granules and use the tolerance rough set model to deal with possible uncertainty in data. The chapter also compares the three approaches in terms of construction of information system, information granules, and approximation spaces.

1 Introduction

Handling uncertainty in data is a challenging task in the field of data mining. A powerful framework intended for this issue is provided by rough set theory [17]. It was proposed by Professor Zdzisław Pawlak in early 1980s as a mathematical tool to deal with uncertainty in data. Although being a standalone field, rough set theory is considered as one of the main granular computing tools.

Granular computing is a relatively new, rapidly growing field of research (see, e.g. [1, 4, 5, 12, 19, 24, 28]). It can be viewed as a label of theories, methodologies, techniques, and tools that make use of granules in the process of problem solving [29].

A granule is a collection of entities drawn together by indistinguishability, similarity, proximity or functionality [30]. Therefore, a granule can be defined as any object, subset, class, or cluster of a given universe. The process of the formation of granules is called granulation. To clearly differentiate granulation from clustering, the semantic aspect of GC is taken into account. Therefore, we treat information

P. Hońko (✉)
Faculty of Computer Science, Bialystok University of Technology,
Wiejska 45A, 15-351 Białystok, Poland
e-mail: p.honko@pb.edu.pl

G. Wang et al. (eds.), *Thriving Rough Sets*, Studies in Computational
Intelligence 708, DOI 10.1007/978-3-319-54966-8_19

granulation as a semantically meaningful grouping of elements based on a given criterion [3].

In recent years, one can observe a trend in data mining towards the application of granular computing based on the rough set approach. This newly emerging approach is called rough-granular computing [23, 26].

Techniques of granular computing, especially rough sets, have widely been applied in the field of data mining (see, e.g. [2, 18, 23]). Methods of rough sets have also found application in mining data stored in multiple tables, i.e. relational data mining. Namely, it has found application in tasks such as eliminating unimportant data (see, e.g. [25]); the analysis of invalid, missing, and indistinguishable data (see, e.g. [13, 15]); reducing data size (see, e.g. [14]); relational classification rules generation (see, e.g. [14, 16, 27]). A rough set model in those approaches was used as a separate tool, i.e. it was not embedded in the granular computing framework.

Constructing a rough set model for processing data stored in a relational structure is not a trivial task. A relational database considered in the context of data mining tasks (e.g. classification) has a specified table (target table) that includes objects to be analyzed and it can be treated as the counterpart of the single table database. The remaining relational database tables (background tables) include additional data that is directly or indirectly associated with the target table. For that reason, a lot of, or even most, essential information about target objects can be hidden in the background tables.

The crucial problem when applying rough sets to relational data is, therefore, to construct an approximation space. Such a space should include essential information about target objects, background objects, as well as relationships among them.

The goal of this chapter is to present three approaches for processing relational data using rough set tools. They all are defined in the paradigm of granular computing and are constructed based on the tolerance rough set model, which was originally developed for single table databases. The underlying idea of building a rough-granular approach to relational data is to use the benefits of both: granular computing to define a given problem at a proper level of granularity, and rough set theory to deal with uncertainty in relational data.

The remaining of the chapter is organized as follows. Section 2 introduces a rough-granular computing defined for single table databases. Sections 3, 4, and 5 present three rough-granular approaches constructed, respectively, based on sum of approximation spaces, relational approximation space, and compound approximation spaces. Section 6 compares the three approaches in terms of essential steps such as construction of information system, information granules, and approximation spaces. Section 7 provides concluding remarks.

2 Rough-Granular Computing for Single Table Data

This section provide a rough-granular computing based approach for processing data stored in single table databases.

Rough-granular computing can be viewed as rough set theory interpreted in the framework of granular computing and applied to discovering knowledge from databases. Elementary granules in this approach are represented by indiscernibility or similarity classes. Higher level granules, which correspond to rough approximations, are constructed based on elementary granules that totally (lower approximation) or partially (upper approximation) belong to the concept under consideration. These granules are the basis for discovering relevant patterns (e.g. classification rules) describing the concept.

Depending on the data type and the task to be performed, different rough set models can be used as the core of rough-granular computing. The tolerance rough set model [20, 25] can be taken due to its flexibility in tuning parameters. This model is defined using the notion of information system, and is in fact a generalization of the standard rough set model introduced by Professor Zdzisław Pawlak.

Definition 1 (*information system*) [17] An information system is a pair $IS = (U, A)$, where U is a non-empty finite set of objects, called the universe, and A is a non-empty finite set of attributes. Each attribute $a \in A$ is treated as a function $a : U \rightarrow V_a$, where V_a is the value set of a.

Granules in a information system are defined using the following language. Let $\Sigma(IS)$ denote the set of formulas, i.e. Boolean combinations of descriptors over $IS = (U, A)$. Descriptors are of the form (a in V) where $a \in A$ and $V \subseteq V_a$.

Definition 2 [21] (*set of formulas*) The set $\Sigma(IS)$ formulas is defined recursively by

1. (a in V) $\in \Sigma(IS)$ for any $a \in A$ and $V \subseteq V_a$,
2. if $\alpha \in \Sigma(IS)$, then $\neg\alpha \in \Sigma(IS)$,
3. if $\alpha, \beta \in \Sigma(IS)$, then $\alpha \wedge \beta \in \Sigma(IS)$,
4. if $\alpha, \beta \in \Sigma(IS)$, then $\alpha \vee \beta \in \Sigma(IS)$.

Let $||\alpha||_{IS} \subseteq U$ denote the semantics of α in IS.

Definition 3 [21] (*semantics of formulas*) The semantics of formulas from $\Sigma(IS)$ with respect to an information system $IS = (U, A)$ is defined recursively by

1. $||a \text{ in } V||_{IS} = \{x \in U : a(x) \in V\}$,
2. $||\neg\alpha||_{IS} = U \setminus ||\neg\alpha||_{IS}$,
3. $||\alpha \wedge \beta||_{IS} = ||\alpha||_{IS} \cap ||\beta||_{IS}$,
4. $||\alpha \vee \beta||_{IS} = ||\alpha||_{IS} \cup ||\beta||_{IS}$.

An approximation space is defined based on an information system as follows.

Definition 4 [20] (*approximation space*) A parameterized approximation space $AS_{\#,\$}$ for an information system $IS = (U, A)$ is defined by $AS_{\#,\$} = \left(U, I_{\#}, \nu_{\$}\right)$, where

1. U is a non-empty set of objects,
2. $I_{\#} : U \rightarrow P(U)$ is an uncertainty function,
3. $\nu_{\$} : P(U) \times P(U) \rightarrow [0, 1]$ is a rough inclusion function.

For every object, the uncertainty function defines a set of similarly described objects (elementary granule). The function can be defined as follows.

Definition 5 (cf. [20]) (uncertainty function) Let $IS = (U, A)$ be an information system. An uncertainty function $I_{B,\varepsilon}$ is defined by

$$I_{B,\varepsilon}(x) = \bigcap_{a \in B} I_{a,\varepsilon_a}(x)$$

where $x \in U, B \subseteq A, \varepsilon = (\varepsilon_a : a \in B)$ is a vector of thresholds such that $\varepsilon_a \geq 0$ for $a \in B$, $I_{a,\varepsilon_a}(x) = \{y \in U : d_a(x, y) \leq \varepsilon_a\}$, and $d_a : U \times U \to [0, \infty)$ is a distance measure.

The rough inclusion function defines the degree of inclusion of a set X in a set Y, where $X, Y \subseteq U$. Depending on its definition, the rough inclusion function can satisfy different properties. The following properties can be considered.

1. $\forall_{A,B \subseteq U} \; A \subseteq B \Rightarrow \nu_\$(A, B) = 1 \; (p_1)$,
2. $\forall_{A,B \subseteq U} \; \nu_\$(A, B) = 1 \Leftrightarrow A \subseteq B \; (p_2)$,
3. $\forall_{A,B,C \subseteq U} \; \nu_\$(B, C) = 1 \Rightarrow \nu_\$(A, B) \leq \nu_\$(A, C) \; (p_3)$,
4. $\forall_{A,B,C \subseteq U} \; B \subseteq C \Rightarrow \nu_\$(A, B) \leq \nu_\$(A, C) \; (p_4)$,
5. $\forall_{A,B \subseteq U} \; \nu_\$(A, B) = 0 \Leftrightarrow A \cap B = \emptyset \; (p_5)$.

We call $\nu_\$$ rough inclusion function (RIF), quasi-rough inclusion function (q-RIF), or weak quasi-rough inclusion function (weak q-RIF) if it satisfies properties p_2 and p_3, p_1 and p_3, or p_1 and p_4, respectively [6]. Property p_5 is optional.

The following rough inclusion functions will be used.

Definition 6 [20] (rough inclusion functions) The rough inclusion $\nu_{l,u}(X, Y)$ of a set X in a set Y is defined by

$$\nu_{l,u}(X, Y) = f_{l,u}\left(\nu_{SRI}(X, Y)\right), \text{where } f_{l,u}(t) = \begin{cases} 0 & \text{if } 0 \leq t \leq l \\ \frac{t-l}{u-l} & \text{if } l < t < u \\ 1 & \text{if } \quad t \geq u \end{cases},$$

$0 \leq l < u \leq 1$and $\nu_{SRI}(X, Y) = \begin{cases} \frac{card(X \cap Y)}{card(X)} & \text{if } X \neq \emptyset \\ 1 & \text{if } X = \emptyset \end{cases}$ is the standard rough inclusion.

Note that if $l = 0$ and $u = 1$, then the rough inclusion $\nu_{l,u}$ is equivalent to the standard rough inclusion ν_{SRI}.

The lower and upper approximations (higher level granules) of a concept are defined as follows.

Definition 7 [20] (approximations of a subset in $AS_{\#,\$}$) For an approximation space $AS_{\#,\$} = \left(U, I_\#, \nu_\$\right)$ and any subset $X \subseteq U$, the lower and the upper approximations are defined respectively by

$$LOW\left(AS_{\#,\$},X\right)=\left\{x\in U:\nu_{\$}\left(I_{\#}(x),X\right)=1\right\},$$

$$UPP\left(AS_{\#,\$},X\right)=\left\{x\in U:\nu_{\$}\left(I_{\#}(x),X\right)\rangle 0\right\}.$$

Symbols #, $ denote vectors of parameters which can be tuned in the process of concept approximation.

Granules in an approximation space are constructed while computing similarity classes.

Definition 8 (*similarity class as a granule*) Given an approximation space $AS_{(B,\varepsilon),\$} = \left(U, I_{(B,\varepsilon)}, \nu_{\$}\right)$ constructed based on an information system $IS = (U, A)$, where $B \subseteq A$, and $(x, y) \in I_{(B,\varepsilon)} \Leftrightarrow \forall_{a\in B}|a(x) - a(y)| \leq \varepsilon_a, \varepsilon_a \in \varepsilon$. Let $\alpha_i = \bigwedge_{a\in B}(a, [a(x_i) - \varepsilon_a, a(x_i) + \varepsilon_a])$ where $x_i \in U$.
A similarity class defined by uncertainty function $I_{B,\varepsilon}(x_i)$ can be expressed by the granule $(\alpha_i, ||\alpha_i||_{IS})$.

Example 1 Consider the following database.

Customer					
id	Name	Age	Gender	Income	Class
1	Adam Smith	36	Male	1500	Yes
2	Tina Jackson	33	Female	2500	Yes
3	Ann Thompson	30	Female	1800	No
4	Susan Clark	30	Female	1800	Yes
5	Eve Smith	26	Female	2500	Yes
6	John Clark	29	Male	3000	Yes
7	Jack Thompson	33	Male	1800	No

Let $AS_{(B,\varepsilon),(l,u)} = \left(U, I_{B,\varepsilon}, \nu_{l,u}\right)$ be an approximation space, where $U = \{o_i : 1 \leq i \leq 7\}$, o_i correspond to i-th object from table *customer*, $B = \{age, income\}, \varepsilon = (\varepsilon_{age}, \varepsilon_{income}) = (3, 500)$, the distance measure is $d(x, y) = |a(x) - a(y)|, l = 0.33, u = 0.67$. The similarity class $I_{B,\varepsilon}(x_i)$ of an object $x_i \in U$ can be expressed be the granule $(\alpha_i, ||\alpha_i||_{IS})$ where $\alpha_i = (age, [age(x_i) - 3, age(x_i) + 3]) \wedge (income, [income(x_i) - 500, income(x_i) + 500])$.

Let $X_1 = \{1, 2, 4, 5, 6\}$ be the set (i.e. concept) to be approximated.
The table below shows the similarity classes and their rough inclusion degrees in X.

$o_i \in U$	$I_{B,\varepsilon}(o_i)$	$\nu_{SRI}(I_{B,\varepsilon}(o_i), X)$
1	$\{1,7\}$	0.5
2	$\{2\}$	1
3	$\{3,4,7\}$	0.33
4	$\{3,4,7\}$	0.33
5	$\{5,6\}$	1
6	$\{5,6\}$	1
7	$\{1,3,4,7\}$	0.5

We obtain the following approximations (higher level granules)
$LOW(AS_{(B,\varepsilon),(l,u)}, X) = \{2, 5, 6\}$, $UPP(AS_{(B,\varepsilon),(l,u)}, X) = \{1, 2, 5, 6, 7\}$.

3 Sum of Approximation Spaces

This section introduces an approach that is based on a sum of approximation spaces, each constructed based on one information system [22].

3.1 Sum of Information Systems and Their Granules

The structure used to store data is a sum of information systems.

Definition 9 (*sum of information systems*) Let $IS_i = (U_i, A_i)$ for $i = 1, \ldots, k$ be information systems. The sum of IS_i ($i = 1, \ldots, k$), denoted by $+(IS_1, \ldots, IS_k)$, is defined by

1. The objects of $+(IS_1, \ldots, IS_k)$ consist of tuples $(x_1, \ldots, x_k)$ of objects from IS_i, i.e. $U = U_1 \times \cdots \times U_k$.
2. The attributes of $+(IS_1, \ldots, IS_k)$ consist of the attributes of IS_i-copy with disjoint attribute sets.

To define dependencies among particular systems the constrained sum of information systems is introduced.

Definition 10 (*constrained sum of information systems*) Let $IS_i = (U_i, A_i)$ for $i = 1, \ldots, k$ be information systems and let $R \subseteq U_1 \times \cdots \times U_k$ be a constraint relation. The constrained sum of IS_i ($i = 1, \ldots, k$), denoted by $+_R(IS_1, \ldots, IS_k)$, is defined by

1. The objects of $+_R(IS_1, \ldots, IS_k)$ consist of k-tuples $(x_1, \ldots, x_k)$ of objects from R, i.e. all objects from $U_1 \times \cdots \times U_k$ satisfying the constraint R.
2. The attributes of $+_R(IS_1, \ldots, IS_k)$ consist of the attributes of $A_1, \ldots, A_k$ where distinct copies are made for attributes in common.

Constraints in a constrained sum of information system can be defined internally, i.e. by Boolean combination of attribute-value descriptors where attributes come from particular information systems, or externally, i.e. by measurable attributes different than those from particular information systems. The information system $+_R(IS_1, \ldots, IS_k)$ can also be defined as a subsystem of $+(IS_1, \ldots, IS_k)$ by imposing on it a constraint being the characteristic function of the relation R.

Granules in sums of information systems are defined using the language defined in Sect. 2.

Definition 11 (*granules in: information system, sum of information systems, constrained sum of information systems*)

1. A granule in IS constructed over a formula $\alpha \in \Sigma(IS)$ is defined by $(\alpha, ||\alpha||_{IS})$.
2. A granule in $+(IS_1, IS_2)$ constructed over formulas $\alpha \in \Sigma(IS_1)$ and $\beta \in \Sigma(IS_2)$ is defined by $(\alpha \wedge \beta, ||\alpha||_{IS_1} \times ||\beta||_{IS_2})$.
3. A granule in $+_R(IS_1, IS_2)$ constructed over formulas $\alpha \in \Sigma(IS_1)$ and $\beta \in \Sigma(IS_2)$ is defined by $(\alpha \wedge \beta, (||\alpha||_{IS_1} \times ||\beta||_{IS_2}) \cap R)$.

Granules in $+(IS_1, \ldots, IS_k)$ and $+_R(IS_1, \ldots, IS_k)$ can be defined analogously to those from $+(IS_1, IS_2)$ and $+_R(IS_1, IS_2)$, respectively.

3.2 Approximation Spaces Constructed Based on Sum of Information Systems

The sum of information system is used to define an approximation space.

Definition 12 (*sum of approximation spaces*) Let $AS_{\#_i} = (U_i, I_{\#_i}, \nu_{SRI})$ be an approximation space for the information system IS_i, where $i = 1, \ldots, k$. The sum of approximation spaces $+(AS_{\#_1}, \ldots, AS_{\#_k})$ for the sum of information systems $+(IS_1, \ldots, IS_k)$ is defined by

1. the universe $U = U_1 \times \cdots \times U_k$,
2. the uncertainty function $I_{\#_1,\ldots,\#_k}((x_1, \ldots, x_k)) = I_{\#_1}(x_1) \times \cdots \times I_{\#_k}(x_k)$,
3. the inclusion relation $\nu_{SRI}(X_1 \times \cdots \times X_k, Y_1 \times \cdots \times Y_k) = \nu_{SRI}(X_1, Y_1) \cdot \cdots \cdot \nu_{SRI}(X_k, Y_k)$.

Definition 13 (*approximations of a subset in* $+(AS_{\#_1}, \ldots, AS_{\#_k})$) For an approximation space $+(AS_{\#_1}, \ldots, AS_{\#_k}) = \left(U, I_{\#_1,\ldots,\#_k}, \nu_{SRI}\right)$ and any subset $X_1 \times \cdots \times X_k \subseteq U$, the lower and the upper approximations are defined respectively by

$$LOW\left(+(AS_{\#_1}, \ldots, AS_{\#_k}), X_1 \times \cdots \times X_k\right) =$$

$$\left\{(x_1, \ldots, x_k) \in U : \nu_{SRI}\left(I_{\#_1,\ldots,\#_k}\left((x_1, \ldots, x_k)\right), X_1 \times \cdots \times X_k\right) = 1\right\},$$

$$UPP\left(+(AS_{\#_1}, \ldots, AS_{\#_k}), X_1 \times \cdots \times X_k\right) =$$

$$\left\{(x_1, \ldots, x_k) \in U : \nu_{SRI}\left(I_{\#_1,\ldots,\#_k}\left((x_1, \ldots, x_k)\right), X_1 \times \cdots \times X_k\right) > 0\right\}.$$

An approximation of a compound concept (i.e. $X_1 \times \cdots \times X_k \subseteq U$) can be computed based on approximations of its particular components (i.e. X_i, $1 \leq i \leq k$).

Proposition 1 *For an approximation space* $+(AS_{\#_1}, \ldots, AS_{\#_k})$ *and any subset* $X_1 \times \cdots \times X_k \subseteq U$ *we obtain*

$$LOW\left(+(AS_{\#_1}, \ldots, AS_{\#_k}), X_1 \times \cdots \times X_k\right) = LOW(AS_{\#_1}, X_1) \times \cdots \times LOW(AS_{\#_k}, X_k),$$

$$UPP\left(+(AS_{\#_1}, \ldots, AS_{\#_k}), X_1 \times \cdots \times X_k\right) = UPP(AS_{\#_1}, X_1) \times \cdots \times UPP(AS_{\#_k}, X_k).$$

A proof of this proposition can be found in [22].

Definition 14 (*constrained sum of approximation spaces*) Let $AS_{\#_i} = (U_i, I_{\#_i}, \nu_{SRI})$ be an approximation space for the information system IS_i, where $i = 1, \ldots, k$, and let $R \subseteq U_1 \times \cdots \times U_k$ be a constraint relation. The constrained sum of approximation spaces $+_R(AS_{\#_1}, \ldots, AS_{\#_k})$ for the sum of information systems $+_R(IS_1, \ldots, IS_k)$ is defined by

1. the universe $(U_1 \times \cdots \times U_k) \cap R = R$,
2. the uncertainty function $I_{\#_1,\ldots,\#_k}((x_1, \ldots, x_k)) = (I_{\#_1}(x_1) \times \cdots \times I_{\#_k}(x_k)) \cap R$,
3. the inclusion relation $\nu_{SRI}(X_1 \times \cdots \times X_k, Y_1 \times \cdots \times Y_k) = \nu_{SRI}(X_1, Y_1) \cdot \cdots \cdot \nu_{SRI}(X_k, Y_k)$.

Definition 15 (*approximations of a subset in* $+_R(AS_{\#_1}, \ldots, AS_{\#_k})$) For an approximation space $+_R(AS_{\#_1}, \ldots, AS_{\#_k}) = \left(U, I_{\#_1,\ldots,\#_k}, \nu_{SRI}\right)$ and any subset $X_1 \times \cdots \times X_k \subseteq U$, the lower and the upper approximations are defined respectively by

$$LOW\left(+_R(AS_{\#_1}, \ldots, AS_{\#_k}), X_1 \times \cdots \times X_k\right) =$$

$$\left\{(x_1, \ldots, x_k) \in R : \nu_{SRI}\left(I_{\#_1,\ldots,\#_k}\left((x_1, \ldots, x_k)\right), X_1 \times \cdots \times X_k\right) = 1\right\},$$

$$UPP\left(+_R(AS_{\#_1}, \ldots, AS_{\#_k}), X_1 \times \cdots \times X_k\right) =$$

$$\left\{(x_1, \ldots, x_k) \in R : \nu_{SRI}\left(I_{\#_1,\ldots,\#_k}\left((x_1, \ldots, x_k)\right), X_1 \times \cdots \times X_k\right) > 0\right\}.$$

An approximation of a compound concept in $+_R(AS_{\#_1}, \ldots, AS_{\#_k})$ can be computed based on its approximation obtained in $+(AS_{\#_1}, \ldots, AS_{\#_k})$.

Proposition 2 *For an approximation space* $+_R(AS_{\#_1}, \ldots, AS_{\#_k})$ *and any subset* $X_1 \times \cdots \times X_k \subseteq U$ *we obtain*

$$LOW\left(+_R(AS_{\#_1}, \ldots, AS_{\#_k}), X_1 \times \cdots \times X_k\right) =$$
$$R \cap LOW\left(+(AS_{\#_1}, \ldots, AS_{\#_k}), X_1 \times \cdots \times X_k\right),$$

$$UPP\left(+_R(AS_{\#_1}, \ldots, AS_{\#_k}), X_1 \times \cdots \times X_k\right) =$$
$$R \cap UPP\left(+(AS_{\#_1}, \ldots, AS_{\#_k}), X_1 \times \cdots \times X_k\right).$$

A proof of this proposition can be found in [22].

Example 2 Given an extended version of the database from Example 1.

Customer

id	Age	Gender	Income	Class
1	36	Male	1500	Yes
2	33	Female	2500	Yes
3	30	Female	1800	No
4	30	Female	1800	Yes
5	26	Female	2500	Yes
6	29	Male	3000	Yes
7	30	Male	1800	No

Product

id	Name	Price
1	Bread	2.00
2	Butter	3.50
3	Milk	2.50
4	Tea	5.00
5	Coffee	6.00
6	Cigarettes	6.50

Purchase

id	cust_id	prod_id	Amount	Date
1	1	1	1	24/06
2	1	3	2	24/06
3	2	1	1	25/06
4	2	3	1	26/06
5	4	6	1	26/06
6	4	2	3	27/06
7	5	5	2	27/06
8	6	4	1	27/06

1. Let $IS_1 = (U_1, A_1)$ and $IS_2 = (U_2, A_2)$ be information systems corresponding to relations *customer* and *product*, respectively. Consider the approximation space $+(AS_{B_1}, AS_{B_2})$ where $AS_{B_1} = (U_1, I_{(B_1,\varepsilon_1)}, \nu_{SRI})$ and $AS_{B_2} = (U_2, I_{(B_2,\varepsilon_2)}, \nu_{SRI})$ are constructed respectively based on IS_1 and IS_2, and $B_1 = \{age, income\}, B_2 = \{price\}, \varepsilon_1 = (\varepsilon_{age}, \varepsilon_{income}) = (3, 500), \varepsilon_2 = (\varepsilon_{price}) = (2.00)$.
   The similarity class $I_{(B_1,\varepsilon_1),(B_2,\varepsilon_2)}((x_i, y_j))$ of an object $(x_i, y_j) \in U_1 \times U_2$ can be expressed be the granule $(\alpha_i \wedge \beta_j, ||\alpha_i||_{IS_1} \times ||\beta_j||_{IS_2})$ where $\alpha_i = (age, [age(x_i) - 3, age(x_i) + 3]) \wedge (income, [income(x_i) - 500, income(x_i) + 500])$ and $\beta_j = (price, [price(y_i) - 2.00, price(y_i) + 2.00])$.

$x_1 \in U_1$	$I_{B_1}(x_1)$	$\nu_{SRI}(I_{B_1}(x_1), X_1)$	$x_2 \in U_2$	$I_{B_2}(x_2)$	$\nu_{SRI}(I_{B_2}(x_2), X_2)$
1	$\{1,7\}$	0.5	1	$\{1,2,3\}$	0
2	$\{2\}$	1	2	$\{1,2,3,4\}$	0.25
3	$\{3,4,7\}$	0.33	3	$\{1,2,3\}$	0
4	$\{3,4,7\}$	0.33	4	$\{2,4,5\}$	0.67
5	$\{5,6\}$	1	5	$\{4,5\}$	1
6	$\{5,6\}$	1	6	$\{6\}$	1
7	$\{1,3,4,7\}$	0.5	–	–	–

   Let $X_1 = \{1,2,4,5,6\} \subset U_1$, $X_2 = \{4,5,6\} \subset U_2$, and $X_1 \times X_2$ be the set to be approximated.
   The table below shows the similarity classes and their rough inclusion degrees in the respective sets.

   We obtain the following approximations $LOW(+(AS_{B_1}, AS_{B_2}), X_1 \times X_2) = \{2,5,6\} \times \{5,6\}$ and $UPP(+(AS_{B_1}, AS_{B_2}), X_1 \times X_2) = \{1, \ldots, 7\} \times \{2,4,5,6\}$.
2. Consider an approximation space $+_R(AS_{B_1}, AS_{B_2})$, where AS_{B_1} and AS_{B_2} are defined as in the previous point, and where $R \subset U_1 \times U_2$ is defined by the condition: *A customer bought a product*. In fact, R is defined by the *purchase* table, i.e. $R = \pi_{cust\_id,prod\_id}(purchase)$.

$(x_1, x_2) \in R$	$I_{(B_1,\varepsilon_1),(B_2,\varepsilon_2)}((x_1, x_2))$	$\nu_{SRI}(I_{(B_1,\varepsilon_1),(B_2,\varepsilon_2)}((x_1, x_2)), X_1 \times X_2)$
$(1,1)$	$\{(1,1),(1,3)\}$	0
$(1,3)$	$\{(1,1),(1,3)\}$	0
$(2,1)$	$\{(2,1),(2,3)\}$	0
$(2,3)$	$\{(2,1),(2,3)\}$	0
$(4,6)$	$\{(4,6)\}$	1
$(4,2)$	$\{(4,2),(6,4)\}$	0.5
$(5,5)$	$\{(5,5)\}$	1
$(6,4)$	$\{(4,2),(6,4)\}$	0.5

   The table below shows the similarity classes and their rough inclusion degrees in the concept.

We obtain the following approximations $LOW(+_R(AS_{A_1}, AS_{A_2}), X_1 \times X_2) = \{(4,6),(5,5)\}$ and $UPP(+_R(AS_{A_1}, AS_{A_2}), X_1 \times X_2) = \{(4,6),(4,2),(5,5),(6,4)\}$.

Except for the possibility to deal with uncertain data, the approach based on the sum of approximation spaces also enables to express patterns (e.g. classification rules generated for approximations) in a simple manner than it is possible using a relational language. Namely, patterns can be expressed using an attribute-value language and constraints showing dependencies among particular information systems. The construction of patterns can also be simplified compared with that of relational patterns since operations such as pattern refinement and pattern satisfiability are less complex when using an attribute-value language.

4 Relational Approximation Spaces

This section introduce an approach that is based on an information system dedicated to processing relational data [8, 9].

4.1 Relational Information System and Its Granules

To consider objects apart from the tables they belong to, the notion of relational object is used.

Definition 16 (*relational object*) Given a database relation with the schema $R(a_1, a_2, \ldots, a_n)$. An expression of the form $R(v_1, v_2, \ldots, v_n)$ is an object of R if and only if $(v_1, v_2, \ldots, v_n)$ is a tuple of R.

In this approach a relational database is represented by an information system that is constructed based on a standard information system.

We will use D_T and D_B to denote, respectively, the sets of target and background relations of database $D = T \cup B$.

Let $U_{D_T} = \bigcup_{R \in D_T} R$ and $U_{D_B} = \bigcup_{R \in D_B} R$ be, respectively, the set of all target and background objects of database D. Subsequently, let $A_{D_T} = \bigcup_{R \in D_T} A_R$[1] and $A_{D_B} = \bigcup_{R \in D_B} A_R$ be, respectively, the set of all attributes of the target and background relations of database D.

The following representation of a relational database is introduced.

Definition 17 (*information system for a relational database*) A relational database $D = T \cup B$ is represented by an information system $IS_D = (U_D, A_D)$, where

[1] A_R denotes here the set of all attributes of relation R.

1. $U_D = U_{D_T} \cup U_{D_B}$ is a non-empty finite set of objects, called the universe,
2. $A_D = A_{D_T} \cup A_{D_B}$ is a non-empty finite set of attributes.

The information system defined above includes objects together with the names of tables they belong to. Information on table joins is not directly stored int the system. They can be reconstructed based on metadata on primary and foreign keys.

In this approach essential information acquired from the relational data are descriptions of target objects. These descriptions are used, in a sense, to identify the objects, i.e., based on their descriptions the objects are compared to one another or to relational patterns. For each target object its description based on the background relations is constructed. To construct such descriptions the notion of related set is introduced [7, 9].

Definition 18 (*related objects*) Object o is related to object o' if and only if there exists a key attribute joining o with o'.[2]

In this approach, the key attribute is, in general, understood as an important attribute for joining tables. It is usually a primary or foreign key. However, in some cases, it can also be another attribute by which one table can be joined with another table or with itself.

A target object's description is expressed by a set of background objects joined with the target object.

Definition 19 (*related set*) A related set of a target object o, denoted by $rlt(o)$, is a set of background objects directly or indirectly related to the target object.

Each target object in the approach is processed along with its related set.

For a given target object one can usually obtain more than one description, each of which describes the object with different precision. The objective is to choose an appropriate description of the target object with respect to a given data mining task. The precision of the target object's description (i.e., the related set) can be tuned by its depth level. To define a related set of a given depth level, Definitions 18 and 19 are generalized.

Definition 20 (*n-related objects*) Object o_0 is n-related to object o_n if and only if there exist $n-1$ objects such that o_i is related to o_{i+1}, where $n > 0$ and $0 \leq i \leq n-1$.

Definition 21 (*n-related set*) The n-th depth level related set of a target object o, denoted by $rlt^n(o)$, is a set of background objects, each of which are m-related to object o and $m \leq n$.

A related set of a given target object can be viewed as its specific description. In order to derive relational patterns, the target object's description is generalized. To obtain a general description of a target object itself and its related set, they are both generalized.

---

[2]The tables the objects belong to are not assumed to be different.

Definition 22 (*generalized object*) Let $o = R(v_1, v_2, \dots, v_n)$ be an object where $(v_1, v_2, \dots, v_n)$ is a tuple of a relation R. A generalized object o, denoted by o_{gen}, is defined by

$$o_{gen} = o\sigma$$

where $\sigma = \{v_{i_1}/t_1, v_{i_2}/t_2, \dots v_{i_m}/t_m\}$ is a substitution such that $v_{i_j} \in \{v_1, v_2, \dots, v_n\}$ $(j = 1, \dots, m,\ m \leq n)$, and t_i is either a variable, a list of constants, or symbol "_" if the component is not important for the consideration.

Definition 23 (*generalized related set*) Let $rlt(o) = \{o_1, \dots, o_n\}$ be the related set of a target object o. A generalized related set of a target object o, denoted by $rlt_{gen}(o)$, is defined by

$$rlt_{gen}(o) = rlt(o)\sigma = \{o_1\sigma_1, \dots, o_n\sigma_n\}$$

where σ is a substitution, there exists $\sigma_0 \subseteq \sigma$ such that $o_{gen} = o\sigma_0$, and $\sigma_i \subseteq \sigma$ $(i = 1, \dots, n)$.

A generalized n-related set is defined in an analogous way.

Related sets can be generalized in a variety of ways (for more, details see [7]). A method for generalization can be developed taking into consideration language bias.

Generalized target objects and their related sets are used to define information granules. For this purpose the method for constructing information granules [21] is expanded to a relational case.

In this approach an elementary granule is defined by a conjunction of relational descriptors, i.e., expressions of the form $R(t_1, t_2, \dots, t_n)$, where R is a relation name, and t_i $(1 \leq i \leq n)$ are the terms (constants or variables).
Given information system $IS_D = (U_D, A_D)$.

1. A generalized target object o_{gen} of object o from IS_D is a trivial elementary granule, i.e., a single relational descriptor. The meaning (i.e., semantics) of the granule, denoted by $SEM_{IS_D}(o_{gen})$, is the set of target objects that satisfy the descriptor.
2. A generalized related set $rlt_{gen}(o)$ of target object o from IS_D is an elementary granule where each descriptor is constructed based on a background relation. The meaning of the granule, denoted by $SEM_{IS_D}(rlt_{gen}(o))$, is the set of target objects for each of which there exists a substitution such that each descriptor under the substitution is satisfied.
3. A generalized target object o_{gen} with its generalized related set $rlt_{gen}(o)$ is represented by the granule $(o_{gen}, rlt_{gen}(o))$. The meaning of the granule is $SEM_{IS_D}\left((o_{gen}, rlt_{gen}(o))\right) = (SEM_{IS_D}(o_{gen}), SEM_{IS_D}(rlt_{gen}(o)))$.

The information granules as defined above can be viewed as an abstract representation of relational data. The accuracy level of the representation can easily be changed by taking another depth level of the related sets.

4.2 Approximation Spaces Constructed Based on Relational Information Systems

The information system defined for relational database is used to construct an approximation space.

Definition 24 (*approximation space* $AS^i_{\#,\$}$) An approximation space $AS^i_{\#,\$}$ for a database $D = T \cup B$ represented by the information system $IS_D = (U_D, A_D)$ is defined by $AS^i_{\#,\$} = \left(U^i, I_\#, \nu_\$\right)$, where

1. $U^i = \left\{(o, rlt^i(o)) : o \in U_T\right\}$ is a non-empty set of granules,
2. $I_\# : U^i \to P\left(U^i\right)$ is an uncertainty function,
3. $\nu_\$: P\left(U^i\right) \times P\left(U^i\right) \to [0, 1]$ is a rough inclusion function.

Definition 25 (*similarity of objects*) Let o and o' be relational objects constructed over a relation R. The similarity of objects o and o' for attribute subset $B \subseteq R.A$ is computed as follows

$$sim_B(o, o') = \begin{cases} \frac{\sum_{a\in B} sim_a(a(o), a(o'))}{|B|} & \text{if } B \neq \emptyset, \\ 0 & \text{if } B = \emptyset, \end{cases}$$

where $sim_a(v, v')$ is any measure that returns the similarity of values v and v' of an attribute a.

Since a target object may be joined with more than one object of the same relation, the measure that operates on sets of objects is introduced.

Definition 26 (*similarity of sets*) Let S_R and S'_R be sets of relational objects constructed over a relation R such that $|S_R| \leq |S'_R|$. The similarity of sets S_R and S'_R is computed as follows

$$R\_sim_B(S_R, S'_R) = \frac{max\left\{\sum_{i=1}^{|S_R|} sim_B(P[i], P'[i]) : P' \in perm(S'_R)\right\}}{|S'_R|}$$

where $perm(S)$ is the set of all permutations of a set S, and P is a certain permutation of S_R.

Due to operating on permutations the measure is suitable for relatively small sets. The next measure operate on sets including objects of different relations.

Let $\mathcal{B} = \{B_R : R \in S \cap S', B \subseteq R.A\}$ where S and S' are sets of relational objects.

Definition 27 (*similarity of sets of relational objects*) The similarity of sets S and S' of relational objects is computed as follows

$$S\_sim_{\mathcal{B}}(S, S') = \frac{\sum_{R\in rel(S)\cap rel(S')} R\_sim_{B_R}(S_R, S'_R)}{|rel(S) \cup rel(S')|}$$

Definition 28 (*similarity of granules*) Let $g = (o, rlt(o))$ and $g' = (o', rlt(o'))$ be granules such that $o, o' \in U_{D_T}$. The similarity of granules g and g' is computed as follows

$$g\_sim_{\mathcal{B}}(g, g') = S\_sim_{\mathcal{B}}(\{o\} \cup rlt(o), \{o'\} \cup rlt(o')).$$

Definition 29 (*uncertainty function*) The uncertainty function $I_{\mathcal{B},\varepsilon}$ is defined as follows

$$I_{\mathcal{B},\varepsilon}(g) = \left\{g' \in U^i : g\_sim_{\mathcal{B}}\left(g, g'\right) \geq \varepsilon\right\},$$

where $\varepsilon \in (0, 1]$ is a similarity threshold.

When the uncertainty function is defined one can compute approximations of any subset of U^i using measures analogous to those from Definition 7.

If the standard rough inclusion and the full similarity of granules are considered (i.e. the uncertainty function is $I_{\mathcal{B},1}$), then the quality of approximations increases together with the increase of the depth level.

Proposition 3 *Let* $AS^i_{(\mathcal{B},1),SRI} = (U^i, I_{\mathcal{B},1}, \nu_{SRI})$ *and* $AS^j_{(\mathcal{B},1),SRI} = (U^j, I_{\mathcal{B},1}, \nu_{SRI})$ *be approximation spaces such that* $i \leq j$ *and let* $X^i \subseteq U^i$ *and* $X^j \subseteq U^j$ *be subsets constructed based on the same subset of* U_T. *The following hold*

$$LOW(AS^i_{(\mathcal{B},1),SRI}, X^i) \subseteq LOW(AS^j_{(\mathcal{B},1),SRI}, X^j),$$

$$UPP(AS^j_{(\mathcal{B},1),SRI}, X^j) \subseteq UPP(AS^i_{(\mathcal{B},1),SRI}, X^i).$$

A proof can be constructed based on the fact that $I_{\mathcal{B},1}(g) \subseteq I_{\mathcal{B},1}(g')$ where $g \in U^j$ and $g' \in U^i$ are constructed based on the same object from U_T.

Definition 30 (*approximation space* $genAS^i_{\#,\$}$) An approximation space $genAS^i_{\#,\$}$ for a database $D = T \cup B$ represented by the information system $IS_D = (U_D, A_D)$ is defined by $genAS^i_{\#,\$} = \left(U^i_{gen}, I_{\#}, \nu_{\$}\right)$, where

1. $U^i_{gen} = \left\{(o_{gen}, rlt^i_{gen}(o)) : o \in U_T\right\}$ is a non-empty set of granules,
2. $I_{\#} : U^i_{gen} \to P\left(U^i_{gen}\right)$ is an uncertainty function,
3. $\nu_{\$} : P\left(U^i_{gen}\right) \times P\left(U^i_{gen}\right) \to [0, 1]$ is a rough inclusion function.

To compute the similarity of generalized objects of the same relation, the measure from Definition 19 can be used. Here, attributes that are replaced in a generalized object with variables are treated as nominal.

The mentioned measures can be used if a syntactic comparison of generalized objects is sufficient. Otherwise the following measure can be applied.

Definition 31 (*semantic similarity of objects*) Let o_{gen} and o'_{gen} be relational objects constructed over the same relation. The semantic similarity of objects o_{gen} and o'_{gen} for attribute subset B is computed as follows

$$sim'_B(o_{gen}, o'_{gen}) = \begin{cases} 1 & \text{if} \quad \exists_\sigma o_{gen}\sigma = o'_{gen} \wedge o'_{gen}\sigma^{-1} = o_{gen}; \\ 0 & \text{otherwise.} \end{cases}$$

For generalized approximation spaces we have analogous properties to those from Proposition 3.

Example 3

1. Consider the following approximation space for the database from Example 2: $AS^2_{(\mathscr{B},\varepsilon),SRI} = (U^2, I_{\mathscr{B},\varepsilon}, \nu_{SRI})$, where $\mathscr{B} = \{B_0, B_1\}, B_0 = \{age, income\}, B_1 = \{price\}, \varepsilon = 0.5$.

   Universe U^2 consists of the following granules $g_1 = (c_1, \{p_1, p_2, p'_1, p'_3\}), g_2 = (c_2, \{p_3, p_4, p'_1, p'_3\}), g_3 = (c_3, \emptyset), g_4 = (c_4, \{p_5, p_6, p'_6, p'_2\}), g_5 = (c_5, \{p_7, p'_5\}), g_6 = (c_6, \{p_8, p'_4\}), g_7 = (c_7, \emptyset)$.[3]

   The similarity of granules for attributes from $\mathscr{B}$ is computed using the function $sim_a(v, v') = 1 - \frac{|v-v'|}{maxV_a - minV_a}$. To compare the whole granules, the function from Definition 28 is used.

   Let $X = \{1, 2, 4, 5, 6\}$ be the set to be approximated.

   The table below shows the similarity classes and their rough inclusion degrees in X.

$g \in U^2$	$I_{\mathscr{B},\varepsilon}(g)$	$\nu_{SRI}(I_{\mathscr{B},\varepsilon}(g), X)$
1	$\{1,2,4\}$	1
2	$\{1,2,4,6\}$	1
3	$\{3,7\}$	0
4	$\{1,2,4,5,6\}$	1
5	$\{4,5,6\}$	1
6	$\{2,4,5,6\}$	1
7	$\{3,7\}$	0

   We obtain that $LOW(AS^2_{(\mathscr{B},\varepsilon),SRI}, X) = UPP(AS^2_{(\mathscr{B},\varepsilon),SRI}, X) = X$.

2. Consider the approximation spaces $AS^2_{(\mathscr{B},\varepsilon),SRI} = (U^2_{gen}, I_{\mathscr{B},\varepsilon}, \nu_{SRI})$, where $\mathscr{B}$ and ε are defined as in the previous point.
   Universe U^2_{gen} can have the following form after generalization:

   $g_1 = (c(A, \_, 30, \_, 1500, yes), \{p(B, A, C, \_, \_), p'(C, \_, \{2.00, 2.50\})\}$,
   $g_2 = (c(A, \_, 33, \_, 2500, yes), \{p(B, A, C, \_, \_), p'(C, \_, \{2.00, 2.50\})\}$,
   $g_3 = (c(A, \_, 30, \_, 1800, no), \emptyset)$,

[3] Symbols c_i, p_i, p'_i denote the i-th object of tables *customer*, *purchase*, *product*, respectively.

$g_4 = (c(A, \_, 30, \_, 1800, yes), \{p(B, A, C, \_, \_), p'(C, \_, \{12.00, 3.50\})\}$
$g_5 = (c(A, \_, 26, \_, 2500, yes), \{p(B, A, C, \_, \_), p'(C, \_, 6.00)\},$
$g_6 = (c(A, \_, 29, \_, 3000, yes), \{p(B, A, C, \_, \_), p'(C, \_, 5.00)\},$
$g_7 = (c(A, \_, 30, \_, 1800, no), \emptyset).$

The same results regarding approximations of X can be obtained using functions from the previous point and the following one to measure the similarity of sets of prices: $sim(V, V') = 1 - |avg(V) - avg(V')| \frac{min\{|V|,|V'|\}}{max\{|V|,|V'|\}}$ where $avg(V)$ returns the average of values from V, and $\frac{min\{|V|,|V'|\}}{max\{|V|,|V'|\}}$ is used to take into account a possible difference in the cardinalities of V and V'.

This approach enables to construct a granular representation of relational objects. Such a representation can speed up the process of pattern generation. Namely, the information (i.e. dependencies between target objects and their related background ones) to be used in the pattern construction is included in the granular representation. Depending on the depth level one can obtain a more or less general granular representation of relational objects. When a more general representation is taken, the descriptions of some target objects may become indistinguishable. To deal with this issue the above described relational approximation space can be used.

5 Compound Approximation Spaces

This section introduces an approach that is a relational extension of the granular data mining approach dedicated to dealing with uncertainty in data [10, 11].

5.1 Compound Information Systems and Their Granules

Each table of a database is represented by an information system.

Definition 32 (*information system for a database table*) An information system for a database table with the schema $R_i(id, a_1, \ldots, a_m)$ is a pair $IS_i = (U_i, A_i)$, where $U_i = \{x : x \in R_i\}$ and $A_i = \{id, a_1, a_2, \ldots, a_m\}$.[4]

The compound information system corresponding to m database tables is defined as follows.

Definition 33 (*compound information system $IS_{(1,2,\ldots,m)}$*) Let $IS_i = (U_i, A_i)$ be information systems, where $1 \leq i \leq m$ and $m > 1$ is a fixed number. A compound information system $IS_{(1,2,\ldots,m)}$ is defined by

[4]The index (i.e. the relation identifier) is omitted if this does not lead to a confusion.

$$IS_{(1,2,\ldots,m)} = \times(IS_1, IS_2, \ldots, IS_m) = \left(\prod_{i=1}^{m} U_i, \bigcup_{i=1}^{m} A_i\right). \quad (1)$$

To allow the connections between tables that occur in the original database or are defined by an expert a constrained version of the compound information system is introduced.

A constraint, denoted by $\bowtie_{\Theta}$, is defined by the theta join on disjunction of the formulas from Θ.

The constrained compound information system corresponding to m database tables is defined as follows.

Definition 34 (*constrained compound information system* $IS^{\Theta}_{(1,2,\ldots,m)}$) A constrained compound information system $IS^{\Theta}_{(1,2,\ldots,m)}$ is defined by

$$IS^{\Theta}_{(1,2,\ldots,m)} = \bowtie_{\Theta} (IS_1, IS_2, \ldots, IS_m) = (U_1 \bowtie_{\Theta} U_2 \bowtie_{\Theta} \cdots \bowtie_{\Theta} U_m, \bigcup_{i=1}^{m} A_i). \quad (2)$$

For each information system that corresponds to a database table a description language is defined. The language enables to define formulas that are used for constructing information granules.

Let $A = A_{des} \cup A_{key}$, where $IS = (U, A)$ is an information system and A_{des} (A_{key}) is the set of descriptive (key) attributes. The descriptive language for IS is denoted by $L_{IS} = L_{IS_{des}} \cup L_{IS_{key}}$. An atomic formula and its negation are defined in L_{IS} by their syntax and semantics.

Definition 35 (*syntax and semantics of an atomic formula in* $L_{IS} = L_{IS_{des}} \cup L_{IS_{key}}$) The syntax and semantics of atomic formulas in a language L_{IS} are defined by

1. $a \in A_{des}, v \in V_a \Rightarrow (a, v) \in L_{IS_{des}}$ and $SEM_{IS_{des}}(a, v) = \{x \in U : a(x) = v\}$,
2. $a \in A_{des}, V \subseteq V_a \Rightarrow (a, V) \in L_{IS_{des}}$ and $SEM_{IS_{des}}(a, V) = \{x \in U : a(x) \in V\}$,
3. $\alpha \in L_{IS_{des}} \Rightarrow \neg\alpha \in L_{IS_{des}}$ and $SEM_{IS_{des}}(\neg\alpha) = U \backslash SEM_{IS_{des}}(\alpha)$,
4. $a, a' \in A_{key} \Rightarrow (a, a') \in L_{IS_{key}}$ and $SEM_{IS_{key}}(a, a') = \{x \in U : a(x) = a'(x)\}$,
5. $\alpha \in L_{IS_{key}} \Rightarrow \neg\alpha \in L_{IS_{key}}$ and $SEM_{IS_{key}}(\neg\alpha) = U \backslash SEM_{IS_{key}}(\alpha)$.

More advanced formulas are constructed recursively using logical operators such as conjunction and disjunction. For more details, see [10].

The above defined language facilitates the construction of formulas that express not only simple features of objects (i.e. formulas with descriptors of the form $(a, v) \in L_{IS_{des}}$) but also relationships between the features (i.e. formulas with descriptors of the form $(a, a') \in L_{IS_{key}}$).

Using any granule description language L, one can defined granules of the form $(\alpha, SEM(\alpha))$, where $\alpha \in L$.

A description language corresponding to two database tables is constructed as follows.

Let $L_{IS_{(i,j)}} = L_{IS_{i\vee j}} \cup L_{IS_{i\wedge j}}$, where $L_{IS_{i\vee j}}$ consists of formulas from L_{IS_i} and L_{IS_j}, and $L_{IS_{i\wedge j}}$ consists of formulas constructed over both IS_i and IS_j.

Definition 36 (*syntax and semantics of an atomic formula in* $L_{IS_{(i,j)}}$) The syntax and semantics of atomic formulas in a language $L_{IS_{(i,j)}}$ are defined by

1. $\alpha \in L_{IS_i} \Rightarrow \alpha \in L_{IS_{i\vee j}}$ and $SEM_{IS_{i\vee j}}(\alpha) = SEM_{IS_i}(\alpha) \times U_j$,
2. $\alpha \in L_{IS_j} \Rightarrow \alpha \in L_{IS_{i\vee j}}$ and $SEM_{IS_{i\vee j}}(\alpha) = U_i \times SEM_{IS_j}(\alpha)$,
3. $\alpha \in L_{IS_{i\vee j}} \Rightarrow \neg\alpha \in L_{IS_{i\vee j}}$ and $SEM_{IS_{i\vee j}}(\neg\alpha) = (U_i \times U_j)\backslash SEM_{IS_{i\vee j}}(\alpha)$,
4. $a \in (A_i)_{key}, a' \in (A_j)_{key} \Rightarrow (a, a') \in L_{IS_{i\wedge j}}$ and $SEM_{IS_{i\wedge j}}(a, a') = \{(x, y) \in U_i \times U_j : a(x) = a'(y)\}$,
5. $\alpha \in L_{IS_{i\wedge j}} \Rightarrow \neg\alpha \in L_{IS_{i\wedge j}}$ and $SEM_{IS_{i\wedge j}}(\neg\alpha) = (U_i \times U_j)\backslash SEM_{IS_{i\wedge j}}(\alpha)$.

The above defined language makes it possible to construct formulas that show features of pairs of objects from different universes. Furthermore, the formulas can also show the relationship between the objects themselves (i.e. formulas with a descriptor of the form $(a, a') \in L_{IS_{i\wedge j}}$).

A description language can be extended to $L_{IS_{(m)}}$ defined for a compound information system $IS_{(m)}$.

Definition 37 (*syntax and semantics of an atomic formula in* $L_{IS_{(m)}}$) The syntax and semantics of atomic formulas in a language $L_{IS_{(m)}}$ are defined by

1. $\alpha \in L_{IS_i} \Rightarrow \alpha \in L_{IS_{(m)}}$ and $SEM_{IS_{(m)}}(\alpha) = U_1 \times \cdots \times U_{i-1} \times SEM_{IS_i}(\alpha) \times U_{i+1} \times \cdots \times U_m$,
2. $\alpha \in L_{IS_{(i,j)}} \Rightarrow \alpha \in L_{IS_{(m)}}$ and $SEM_{IS_{(m)}}(\alpha) = \{(x_1, \ldots, x_i, \ldots, x_j, \ldots, x_m) \in \prod_{k=1}^{m} U_k : (x_i, x_j) \in SEM_{IS_{(i,j)}}(\alpha)\}$,
3. $\alpha \in L_{IS_{(m)}} \Rightarrow \neg\alpha \in L_{IS_{(m)}}$ and $SEM_{IS_{(m)}}(\neg\alpha) = (U_1 \times \cdots \times U_m)\backslash SEM_{IS_{(m)}}(\alpha)$.

Since knowledge discovery is focused on selected database tables only, usually one table (i.e. the target table), the semantics of $L_{IS_{(m)}}$ is extended by the following

1. $\alpha \in L_{IS_{(m)}} \Rightarrow SEM^{\pi_i}_{IS_{(m)}}(\alpha) = \pi_{A_i}(SEM_{IS_{(m)}}(\alpha))$, where $1 \le i \le m$,[5]
2. $\alpha \in L_{IS_{(m)}} \Rightarrow SEM^{\pi_{i_1,i_2,\ldots,i_k}}_{IS_{(m)}}(\alpha) = \pi_{A_{i_1},A_{i_2},\ldots,A_{i_k}}(SEM_{IS_{(m)}}(\alpha))$, where

   $1 \le i_1, i_2, \ldots, i_k \le m$ and $k < m$.

The syntax and semantics of $L_{IS^{\Theta}_{(m)}}$ are defined in the same way as in Definition 37. It is enough to replace $IS_{(i,j)}$, $IS_{(m)}$, and the $\times$ operation with $IS^{\Theta}_{(i,j)}$, $IS^{\Theta}_{(m)}$, and the $\bowtie_{\Theta}$ operation, respectively.

[5] $\pi_A(\bullet)$ is understood as a projection over the attributes from A.

5.2 Approximation Spaces Constructed Based on Compound Information Systems

Firstly, the notion of approximation space is slightly redefined.

Definition 38 (*approximation space AS_ω*) An approximation space AS_ω for an information system $IS = (U, A)$ is defined by

$$AS_\omega = (U, I_\omega, \nu_\omega) \tag{3}$$

where $\omega = (\#, \$), I_\omega = I_\#, \nu_\omega = \nu_\$$.

Definition 39 (*compound approximation space $AS_{\omega_{(m)}}$*) Let $AS_{\omega_i} = (U_i, I_{\omega_i}, \nu_{\omega_i})$, where $1 \leq i \leq m$ and $m > 1$, be approximation spaces for information systems $IS_i = (U_i, A_i)$. A compound approximation space $AS_{\omega_{(m)}}$ for a compound information system $IS_{(m)} = \times(IS_i, \ldots, IS_m)$ is defined by

$$AS_{\omega_{(m)}} = \times(AS_{\omega_1}, \ldots, AS_{\omega_m}) = (U_{\omega_{(m)}}, I_{\omega_{(m)}}, \nu_{\omega_{(m)}}) \tag{4}$$

where

1. $U_{\omega_{(m)}} = \prod_{i=1}^{m} U_i$,
2. $\forall_{(x_1,\ldots,x_m)\in U_{\omega_{(m)}}} \; I_{\omega_{(m)}}((x_1, \ldots, x_m)) = \prod_{i=1}^{m} I_{\omega_i}(x_i)$,
3. $\forall_{X_i, Y_i \in U_i, 1\leq i\leq m} \; \nu_{\omega_{(m)}}(\prod_{i=1}^{m} X_i, \prod_{i=1}^{m} Y_i) = \prod_{i=1}^{m} \nu_{\omega_i}(X_i, Y_i)$.

Approximations of a set in a compound approximation space are defined as follows.

Definition 40 (*approximations of a set in $AS_{\omega_{(m)}}$*) Let $AS_{\omega_{(m)}} = (U_{\omega_{(m)}}, I_{\omega_{(m)}}, \nu_{\omega_{(m)}})$ be a CAS and $X_i \subseteq U_i$, where $1 \leq i \leq m$. The lower and upper approximations of the set $\prod_{i=1}^{m} X_i$ in $AS_{\omega_{(m)}}$ are defined, respectively, by

$$LOW(AS_{\omega_{(m)}}, \prod_{i=1}^{m} X_i) = \{(x_1, \ldots, x_m) \in U_{\omega_{(m)}} : \nu_{\omega_{(m)}}(I_{\omega_{(m)}}((x_1, \ldots, x_m)), \prod_{i=1}^{m} X_i) = 1\},$$

$$UPP(AS_{\omega_{(m)}}, \prod_{i=1}^{m} X_i) = \{(x_1, \ldots, x_m) \in U_{\omega_{(m)}} : \nu_{\omega_{(m)}}(I_{\omega_{(m)}}((x_1, \ldots, x_m)), \prod_{i=1}^{m} X_i) > 0\}.$$

Approximations of a set in $AS_{\omega_{(m)}}$ have properties analogous to those defined in $+(AS_{\#_1}, \ldots, AS_{\#_k})$ (see Proposition 1).

Definition 41 (*constrained compound approximation space $AS^{\Theta}_{\omega_{(m)}}$*) Let $AS_{\omega_i} = (U_i, I_{\omega_i}, \nu_{\omega_i})$ be approximation spaces for information systems $IS_i = (U_i, A_i)$, and $\nu_{\$} = \nu_{\omega_i}$ for $1 \leq i \leq m$. Let also $\Theta = \{\theta_1, \theta_2, \ldots, \theta_n\}$ be a set of joins of AS_{ω_i} such that $\underset{1<j\leq m}{\forall} \; \underset{i<j}{\exists} \; U_i \bowtie_{\Theta} U_j \neq \emptyset$ (each approximation space joins with some earlier considered space).
A constrained compound approximation space $AS^{\Theta}_{\omega_{(m)}}$ for a constrained compound information system $IS^{\Theta}_{(m)}$ is defined by

$$AS^{\Theta}_{\omega_{(m)}} = \bowtie_{\Theta} (AS_{\omega_1}, \cdots, AS_{\omega_m}) = (U^{\Theta}_{\omega_{(m)}}, I^{\Theta}_{\omega_{(m)}}, \nu^{\Theta}_{\omega_{(m)}}) \tag{5}$$

where

1. $U^{\Theta}_{\omega_{(m)}} = U_1 \bowtie_{\Theta} \cdots \bowtie_{\Theta} U_m$,
2. $\underset{(x_1,\ldots,x_m)\in U_1 \bowtie_{\Theta} \cdots \bowtie_{\Theta} U_m}{\forall} \quad I^{\Theta}_{\omega_{(m)}}((x_1, \ldots, x_m)) = I_{\omega_1}(x_1) \bowtie_{\Theta} \cdots \bowtie_{\Theta} I_{\omega_m}(x_m)$,
3. $\underset{X_i, Y_i \in U_i, 1\leq i\leq m}{\forall} \quad \nu^{\Theta}_{\omega_{(m)}}(X_1 \bowtie_{\Theta} \cdots \bowtie_{\Theta} X_m, Y_1 \bowtie_{\Theta} \cdots \bowtie_{\Theta} Y_m) = \nu_{\$}(X_1 \bowtie_{\Theta} \cdots \bowtie_{\Theta} X_m,$ $Y_1 \bowtie_{\Theta} \cdots \bowtie_{\Theta} Y_m)$.

Approximations of a set in a constrained compound approximation space are defined as follows.

Definition 42 (*approximations of a set in $AS^{\Theta}_{\omega_{(m)}}$*) Let $AS^{\Theta}_{\omega_{(m)}} = (U^{\Theta}_{\omega_{(m)}}, I^{\Theta}_{\omega_{(m)}}, \nu^{\Theta}_{\omega_{(m)}})$ be a CCAS and $X_i \subseteq U_i$, where $1 \leq i \leq m$. The lower and upper approximations of the set $X_1 \bowtie_{\Theta} \cdots \bowtie_{\Theta} X_m$ in $AS^{\Theta}_{\omega_{(m)}}$ are defined, respectively, by

$$LOW(AS^{\Theta}_{\omega_{(m)}}, X_1 \bowtie_{\Theta} \cdots \bowtie_{\Theta} X_m) =$$

$$\{(x_1, \ldots, x_m) \in U^{\Theta}_{\omega_{(m)}} : \nu^{\Theta}_{\omega_{(m)}}(I^{\Theta}_{\omega_{(m)}}((x_1, \ldots, x_m)), X_1 \bowtie_{\Theta} \cdots \bowtie_{\Theta} X_m) = 1\},$$

$$UPP(AS^{\Theta}_{\omega_{(m)}}, X_1 \bowtie_{\Theta} \cdots \bowtie_{\Theta} X_m) =$$

$$\{(x_1, \ldots, x_m) \in U^{\Theta}_{\omega_{(m)}} : \nu^{\Theta}_{\omega_{(m)}}(I^{\Theta}_{\omega_{(m)}}((x_1, \ldots, x_m)), X_1 \bowtie_{\Theta} \cdots \bowtie_{\Theta} X_m) > 0\}.$$

The lower approximation of a compound concept (i.e. $X_1 \bowtie_{\Theta} \cdots \bowtie_{\Theta} X_m \subseteq U^{\Theta}_{\omega_{(m)}}$) may differ from that computed based on the lower approximations of its particular components (i.e. X_i, $1 \leq i \leq m$). Thanks to this, additional knowledge can be provided by the lower approximation of a compound concept.

Proposition 4 *Let $AS_{\omega_{(m)}}$ be a compound approximation space and $AS^{\Theta}_{\omega_{(m)}}$ its constrained version such that ν_{ω_i} $(1 \leq i \leq m)$ and $\nu^{\Theta}_{\omega_{(m)}}$ are RIFs. For any subset $X_1 \bowtie_{\Theta} \cdots \bowtie_{\Theta} X_m \subseteq U^{\Theta}_{\omega_{(m)}}$ we have*

$$LOW(AS_{\omega_1}, X_1) \bowtie_{\Theta} \cdots \bowtie_{\Theta} LOW(AS_{\omega_m}, X_m) \subseteq LOW(AS^{\Theta}_{\omega_{(m)}}, X_1 \bowtie_{\Theta} \cdots \bowtie_{\Theta} X_m),$$

$UPP(AS_{\omega_1}, X_1) \bowtie_\Theta \cdots \bowtie_\Theta LOW(AS_{\omega_m}, X_m) = UPP(AS^\Theta_{\omega_{(m)}}, X_1 \bowtie_\Theta \cdots \bowtie_\Theta X_m).$

A proof of this proposition can be found in [11].

Example 4 1. The approximation space that corresponds to $+(AS_{B_1}, AS_{B_2})$ from Example 2 is $AS_{\omega_{(2)}} = \times(AS_{\omega_1}, AS_{\omega_2}) = (U_{\omega_{(2)}}, I_{\omega_{(2)}}, \nu_{\omega_{(2)}})$, where AS_{ω_1} and AS_{ω_2} are constructed respectively based on $IS_1 = (U_1, A_1)$ and $IS_2 = (U_2, A_2)$, i.e. $U_{\omega_{(2)}} = U_1 \times U_2$, $I_{\omega_{(2)}}$ is constructed with $I_{\omega_1} = I_{\{age,income\},\varepsilon_1}$ and $I_{\omega_2} = I_{\{price\},\varepsilon_2}$ $(\varepsilon_1 = (\varepsilon_{age}, \varepsilon_{income}) = (3, 500), \varepsilon_2 = (\varepsilon_{price}) = (2.00))$, $\nu_{\omega_{(2)}}$ is constructed with $\nu_{\omega_1} = \nu_{SRI}$ and $\nu_{\omega_1} = \nu_{SRI}$.

The similarity class $I_{\omega_{(2)}}((x_i, y_j))$ of an object $(x_i, y_j) \in U_{\omega_{(2)}}$ can be expressed be the granule $(\alpha_{ij}, SEM_{IS_{(2)}}(\alpha_{ij})$ where $\alpha_i = (age, [age(x_i) - 3, age(x_i) + 3]) \wedge (income, [income(x_i) - 500, income(x_i) + 500]) \wedge (price, [price(y_i) - 2.00, price(y_i) + 2.00])$.

The approximations of $X_1 \times X_2 = \{1, 2, 4, 5, 6\} \times \{4, 5, 6\} \subset U_{\omega_{(2)}}$ are as those from Example 2, i.e. $LOW(AS_{\omega_{(1,2)}}, X_1 \times X_2) = \{2, 5, 6\} \times \{5, 6\}$ and $UPP(AS_{\omega_{(1,2)}}, X_1 \times X_2) = \{1, \dots, 7\} \times \{2, 4, 5, 6\}$. Moreover, we obtained $\pi_1(LOW(AS_{\omega_{(2)}}, X_1 \times X_2)) = \{2, 5, 6\}, \pi_1(UPP(AS_{\omega_{(2)}}, X_1 \times X_2)) = \{1, \dots, 7\}$, $\pi_2(LOW(AS_{\omega_{(2)}}, X_1 \times X_2)) = \{5, 6\}, \pi_2(UPP(AS_{\omega_{(2)}}, X_1 \times X_2)) = \{2, 4, 5, 6\}$.

2. The approximation space that corresponds to $+_R(AS_{B_1}, AS_{B_2})$ from Example 2 is $AS^\Theta_{\omega_{(3)}} = \bowtie(AS_{\omega_1}, AS_{\omega_2}, AS_{\omega_3}) = (U^\Theta_{\omega_{(3)}}, I^\Theta_{\omega_{(3)}}, \nu^\Theta_{\omega_{(3)}})$, where $\Theta = \{\theta_1, \theta_2\}, \theta_1 = (customer.id, purchase.cust\_id), \theta_2 = (purchase.prod\_id, product.id)$. Furthermore, $AS_{\omega_1}, AS_{\omega_2}$, and AS_{ω_3} are constructed respectively based on $IS_1 = (U_1, A_1)$ (*customer*), $IS_2 = (U_2, A_2)$ (*purchase*), and $IS_2 = (U_2, A_2)$ (*product*). Namely, $U^\Theta_{\omega_{(3)}} = U_1 \bowtie_\Theta U_2 \bowtie_\Theta U_3$, $I_{\omega_1} = I_{\{age,income\},\varepsilon_1}, I_{\omega_2} = I_{\emptyset,null}$,[6] and $I_{\omega_3} = I_{\{price\},\varepsilon_2}$ $(\varepsilon_1 = (\varepsilon_{age}, \varepsilon_{income}) = (3, 300), \varepsilon_2 = (\varepsilon_{price}) = (2.00))$.

The similarity class $I_{\omega_{(3)}}((x_i, x_j, x_k))$ of an object $(x_i, x_j, x_k) \in U^\Theta_{\omega_{(3)}}$ can be expressed be the granule $(\alpha_{ijk}, SEM_{IS^\Theta_{(2)}}(\alpha_{ij}))$ where $\alpha_i = (age, [age(x_i) - 3, age(x_i) + 3]) \wedge (income, [income(x_i) - 500, income(x_i) + 500]) \wedge (price, [price(x_k) - 2.00, price(x_k) + 2.00])$.

Let $X = X_1 \bowtie_\Theta U_2 \bowtie_\Theta X_3 = \{1, 2, 4, 5, 6\} \bowtie_\Theta U_2 \bowtie_\Theta \{4, 5, 6\} = \{(4, 5, 6), (5, 7, 5), (6, 8, 4)\} \subset U_{\omega_{(3)}}$ be the concept to be approximated.

The table below shows the similarity classes and their rough inclusion degrees in X.

We obtain the following approximations $LOW(AS^\Theta_{\omega_{(2)}}, X) = \{(4, 5, 6), (5, 7, 5)\}$ and $UPP(AS^\Theta_{\omega_{(2)}}, X) = \{4, 5, 6), (4, 6, 2), (5, 7, 5), (6, 8, 4)\}$. We also have $\pi_1(LOW(AS^\Theta_{\omega_{(2)}}, X)) = \{4, 5\}, \pi_1(UPP(AS^\Theta_{\omega_{(2)}}, X)) = \{4, 5, 6\}$, $\pi_3(LOW(AS^\Theta_{\omega_{(2)}}, X)) = \{5, 6\}$ and $\pi_3(UPP(AS^\Theta_{\omega_{(2)}}, X)) = \{2, 4, 5, 6\}$.

The compound approximation spaces enable to approximate not only a combination of concepts defined for particular relational database tables, but also the relationships that take place for these concepts. Thanks to this, additional knowledge

[6] The uncertainty function $I_{\emptyset,null}$ is used when no condition is defined.

$(x_1, x_2) \in U_1 \bowtie_\Theta U_2$	$I^\Theta_{\omega_{(1,2)}}((x_1, x_2))$	$\nu_{\omega_{(1,2)}}(I^\Theta_{\omega_{(1,2)}}((x_1, x_2)), X_1 \bowtie_\Theta X_2)$
$(1, 1, 1)$	$\{(1, 1, 1), (1, 2, 3)\}$	0
$(1, 2, 3)$	$\{(1, 1, 1), (1, 2, 3)\}$	0
$(2, 3, 1)$	$\{(2, 3, 1), (2, 4, 3)\}$	0
$(2, 4, 3)$	$\{(2, 3, 1), (2, 4, 3)\}$	0
$(4, 5, 6)$	$\{(4, 5, 6)\}$	1
$(4, 6, 2)$	$\{(4, 6, 2), (6, 8, 4)\}$	0.5
$(5, 7, 5)$	$\{(5, 7, 5)\}$	1
$(6, 8, 4)$	$\{(4, 6, 2), (6, 8, 4)\}$	0.5

based on obtained approximation can be acquired compared with that obtained by direct joining approximations of the concepts defined in particular tables.

Furthermore, patterns in this approach can be constructed and expressed in a simpler way since an extended attribute-value language is used, which has a simpler syntax than a relational one.

6 Comparative Study

This section compares the three approaches in terms of construction of information system, information granules, and approximation spaces. For simplicity's sake the approaches from Sects. 3, 4, and 5 will be referred to as $A1$, $A2$, and $A3$, respectively.

1. Information system for relational data.
   The data structure in $A1$ enables to join different information systems into one system called sum of information systems. In the basic version all particular systems are independent of one another. The extended version, called constrained sum of information systems, makes it possible to define a relation between any two particular systems. This structure can be used to store relational data in such a way that each particular system corresponds to a basic database table, and the relation between two systems corresponds to the join table. In spite of the fact that a constraint used in a constrained sum of information systems is more general than the theta join of two tables, it can only be used for a limited group of join tables. Namely, those tables which include only key attributes (see Example 2).
   Approach $A2$ uses an information system that can collect relational objects from all database tables. The universe in this data structure is divided into two subsets, each of which includes either target or background objects. The system enables to store data from any relational database that is devoted to mining data, i.e. at least one target table is specified. It is also assumed that the target table is fixed before transforming the database into its corresponding information system. The change of the target table, which may take place in e.g. association discovery, requires rebuilding the information system.
   Approach $A3$ provides two versions of a compound information system. The basic version joins particular information systems, each of which corresponds to

one database table. The constrained version enables to impose constraints on two or more particular information systems. Such a constraint may reflect a relationship that occurs in the original relational database, but it also enables to introduce another limitation defined by an expert.
The data structures in this approach are similar to those from $A1$. Namely, the compound information system is coincident with the sum of information systems, whereas its constrained version corresponds in a sense to the constrained sum of information systems. On the one hand, the constrained compound information system can be treated as a special case of the constrained sum of information systems since the constraint, which is generally defined in the latter, can be formed by using formulas that join tables. On the other hand, the system in $A3$ is more general since it enables to transform any join tables, also those that join more than two tables as well as those that include not only key attributes, but also descriptive ones.

2. Granules in information systems.
In $A1$ granules in particular information systems are defined using the granule description language defined in the approach for single table databases (see Sect. 2). In the constrained sum of information system, the syntax of a granule is the same as in the non-constrained version, whereas the semantic of the granule is that from the non-constrained version limited to those objects that belong to the constraint defined by relation R. Such an approach enables to use directly the standard granule description language for defining granules in a system consisting of any number of particular information systems.
In $A2$ a basic granule is formed based on one target object and the set of background objects related to the target one. The relationship between a target object and a background one can be defined based on the relationship between the target and background tables. These objects can also be joined using a relationship introduced by an expert. The approach enables also to define information granules, each of which is constructed based on one basic granule. The construction relies on a generalization of a basic granule by replacing its constants with appropriate variables. The set of such information granules can be seen as an abstract representation of relational data, and it facilitates the process of discovering important regularities in relational data.
In $A3$ information granules are defined in particular information systems using the description language adapted to defining granules in a single table of a relational database. To define granules that take into account the relationship between tables, the description language is extended by introducing formulas that express the join of two tables. Therefore, the expressiveness of the description language is increased since it can use formulas constructed based not only on an attribute-value condition, but also on an attribute-attribute condition where the two attributes can be either descriptive ones or key ones.
3. Approximation space.
Each of the approaches adapts the tolerance rough set model and constructs it based on the information system defined in the approach.

In $A1$ an approximation space is defined for each particular system of the sum of information systems in the same way as for a single table database (see Sect. 2). An approximation space for the whole sum of information systems is defined by combining the universes, the uncertainty functions, and the rough inclusion functions from particular approximation spaces. Such an approach facilitates the process of computing approximations of a concept defined in the whole space since it can be done by merging partial results obtained in particular spaces. The approximations in the constrained version are computed based on those from the non-constrained version by limiting them using the constraint defined by relation R.

In $A2$ the standard approximation space (see Sect. 2) is adapted to work on the universe that consists of granules. Since each of granules includes a target object and its related set the main effort to adapt the tolerance rough set model is to adjust the uncertainty function. To this end, a similarity measure used by the uncertainty function is defined to compare granules. The comparison is done at each level of granules, i.e. additional functions are used to measure similarity of two attributes, two objects of the same relation, two sets of objects of the same relation, two sets of objects of any relations, and finally two target objects together with their related sets.

Like the sum of information systems, the compound approximation space in $A3$ makes it possible to define approximations of a compound concept based on the partial results obtained for particular approximation spaces. The constrained version, in turn, enables to approximate a more advanced concept than that being a combination of concepts defined in particular spaces. Namely, such a concept can also include an information how database tables are joined. Thank to this, it is possible to approximate not only a combination of concepts, but also the relationship that occurs among the tables. Furthermore, the approach enables to consider approximations of a compound concept in the context of an approximation subspace, in particular the subspace corresponding to one table.

It is worth mentioning that the three approaches can be used to process relational data that is stored in a typical relational database or in any database consisting of multiple tables. All the approaches provide tools to express any constraints or relationships that occur in the database. Thanks to this, it is possible to deal with e.g. functional and multi-valued dependencies.

7 Conclusions

This chapter has presented and compared three rough-granular computing based approaches to processing relational data. They all use information granules to describe information hidden in data. The core of the three approaches is the tolerance rough set model.

The following main advantages of using the approaches can be pointed out.

1. Patterns are generated from relational data in a simpler way and they are expressed using a simple language than a relational one (Approaches $A1$ and $A3$).
2. Pattern can be generated faster when they are constructed based on a granular representation of relational data (Approach $A2$).
3. Additional knowledge can be obtained when approximating not only concepts from particular relational database tables, but also the relationships that take place for these concepts (Approach $A3$).

The choice of the approach for a given task can be done based on the characteristics given below.

1. Sum of approximation spaces.
   The approach is dedicated to complex data that can be expressed as a combination of information systems that can alternatively interact with one another, where the way of interaction is defined by a constraint. The data structure enables to store a relational database where each join table includes only key attributes. A concept in a sum of approximation spaces can be approximated using a combination of partial results obtained in particular spaces.
2. Relational approximation space.
   A relational database used in this approach is required to have at least one specified target table that includes objects to be analyzed. The remaining tables, i.e. background ones, should include objects that can be used to build descriptions of the target objects that are joined with them. The approach enables to approximate a concept of target objects in the context of their descriptions. They can be adjusted to a given problem by e.g. changing the depth level used during construction of the descriptions.
3. Compound approximation space.
   The approach enables to store tables of a relational database and relationships that occur in the original database. It is also possible to define by an expert additional constraints that are relevant for a given task. The concept to be approximated can be a combination of concepts defined in particular database tables. It can also include the information on relationships among tables. Taking into account a relationship enables to produce additional approximations that may be different from those for the concept not including the relationship. The result obtained for a compound concept can be limited to any table in order to analyze the result for a particular table in the context of the remaining ones.

The approaches have been presented in this chapter show that the notion of rough set, originally defined by Professor Zdzisław Pawlak for single table databases, can also be useful for processing relation data. Namely, rough set theory enables to provide an approximated description of a compound concept defined over many tables as well as to analyze a concept of one table in the context of information hidden in the remaining part of the database.

Acknowledgements The project was partially funded by the National Science Center awarded on the basis of the decision number DEC-2012/07/B/ST6/01504.

References

1. Apolloni, B., Bassis, S., Rota, J., Galliani, G.L., Gioia, M., Ferrarim L.: A neurofuzzy algorithm for learning from complex granules. Granul. Comput. 1–22 (2016)
2. Bargiela, A., Pedrycz, W.: Granular Computing: An Introduction. Kluwer Academic Publishers, Boston (2003)
3. Bargiela, A., Pedrycz, W.: Toward a theory of granular computing for human-centered information processing. IEEE Trans. Fuzzy Syst. **16**(2), 320–330 (2008)
4. Ciucci, D.: Orthopairs and granular computing. Granul. Comput. **1**(3), 159–170 (2016)
5. Dubois, D., Prade, H.: Bridging gaps between several forms of granular computing. Granul. Comput. **1**(2), 115–126 (2016)
6. Gomolińska, A.: Rough approximation based on weak q-RIFs. Trans. Rough Sets **10**, 117–135 (2009)
7. Hońko, P.: Simialrity-based classification in relational databases. Fundam. Inform. **101**(3), 187–213 (2010)
8. Hońko, P.: Rough-granular computing based relational data mining. In: Greco, S., Bouchon-Meunier, B., Coletti, G., Fedrizzi, M., Matarazzo, B., Yager, R.R. (eds.) Advances on Computational Intelligence. Communications in Computer and Information Science, vol. 297, pp. 290–299. Springer, Berlin (2012)
9. Hońko, P.: Association discovery from relational data via granular computing. Inform. Sci. **234**, 136–149 (2013)
10. Hońko, P.: Description languages for relational information granules. Fundam. Inform. **137**(3), 323–340 (2015)
11. Hońko, P.: Compound approximation spaces for relational data. Int J Approx Reasoning **71**, 89–111 (2016)
12. Li, J., Mei, C., Xu, W., Qian, Y.: Concept learning via granular computing: a cognitive viewpoint. Inform. Sci. **298**, 447–467 (2015)
13. Liu, C., Zhong, N.: Rough problem settings for ILP dealing with imperfect data. Comput. Intell. **17**(3), 446–459 (2001)
14. Martienne, E., Quafafou, M.: Learning logical descriptions for document understanding: a rough sets-based approach. In: Polkowski, L., Skowron, A. (eds.) Rough Sets and Current Trends in Computing, pp. 202–209. Springer, LNCS, Berlin (1998)
15. Midelfart, H., Komorowski, H.J.: A rough set approach to inductive logic programming. In: Ziarko, W., Yao, Y.Y. (eds.) Rough Sets and Current Trends in Computing, vol. 2005, pp. 190–198. Springer, LNCS (2000)
16. Milton, R.S., Maheswari, V.U., Siromoney, A.: Rough sets and relational learning. In: Transactions on Rough Sets I, LNCS, vol. 3100, pp. 321–337. Springer (2004)
17. Pawlak, Z.: Rough Sets. Theoretical Aspects of Reasoning about Data. Kluwer Academic, Dordrecht (1991)
18. Pedrycz, W., Skowron, A., Kreinovich, V.: Handbook of Granular Computing. Wiley, New York (2008)
19. Peters, G., Weber, R.: DCC: a framework for dynamic granular clustering. Granul. Comput. **1**(1), 1–11 (2016)
20. Skowron, A., Stepaniuk, J.: Tolerance approximation spaces. Fundam. Inform. 245–253 (1996)
21. Skowron, A., Stepaniuk, J.: Information granules: towards foundations of granular computing. Int. J. Intell. Syst. **16**(1), 57–85 (2001)
22. Skowron, A., Stepaniuk, J.: Constrained sums of information systems. In: Tsumoto, S., Slowinski, R., Komorowski, H.J., Grzymala-Busse, J.W. (eds.) Rough Sets and Current Trends in Computing. Lecture Notes in Computer Science, vol. 3066, pp. 300–309. Springer (2004)
23. Skowron, A., Stepaniuk, J., Swiniarski, R.: Modeling rough granular computing based on approximation spaces. Inform. Sci. **184**(1), 20–43 (2012)
24. Skowron, A., Jankowski, A., Dutta, S.: Interactive granular computing. Granul. Comput. **1**(2), 95–113 (2016)

25. Stepaniuk, J.: Knowledge discovery by application of rough set models. In: Polkowski, S.T., Lin, T. (eds.) Rough Set Methods and Applications: New Developments in Knowledge Discovery in Information Systems, pp. 137–233. Physica, Heidelberg, Germany (2000)
26. Stepaniuk, J.: Rough-Granular Computing in Knowledge Discovery and Data Mining. Studies in Computational Intelligence, vol. 152. Springer, Berlin (2008)
27. Stepaniuk, J., Hońko, P.: Learning first-order rules: a rough set approach. Fundam. Inform. **61**, 139–157 (2004)
28. Xu, Z., Wang, H.: Managing multi-granularity linguistic information in qualitative group decision making: an overview. Granul. Comput. **1**(1), 21–35 (2016)
29. Yao, Y.Y.: Granular computing: basic issues and possible solutions. In: Wang, P. (ed.) Proceedings of the 5th Joint Conference on Information Sciences (JCIS), Association for Intelligent Machinery, pp. 186–189 (2000)
30. Zadeh, L.A.: Towards a theory of fuzzy information granulation and its centrality in human reasoning and fuzzy logic. Fuzzy Set Syst. **90**(2), 111–127 (1997)

The Boosting and Bootstrap Ensembles for the Pair Classifier Based on the Dual Indiscernibility Matrix

Piotr Artiemjew, Bartosz Nowak, Lech Polkowski and Przemyslaw Gorecki

Abstract In search for subsets of data sets relevant in the classification of data tasks, we have exploited the betweenness relation adopted from axiomatic elementary geometry. It has turned out that this relation can partition data into two subsets, one of them dense in the sense that each object in this set is a convex combination of a finite number of objects in it, the other, to the contrary, consisting of objects endowed with outliers, i.e., pairs of (attribute, value) not possessed by any other object. A technical tool for singling out those subsets is the Dual Indiscernibility Matrix defined as a counterpart to a well-known Discernibility Matrix of Skowron-Rauszer. On the basis of those ideas, the pair Classifier has been introduced. It is its main feature that test objects are approximated to a certain degree by pairs of training objects which are not required to cover the object completely. In this chapter, we examine selected methods for stabilization of the Pair Classifier like Bootstrap Ensemble, Arcing based Bootstrap, Ada–Boost with Monte Carlo split. We present results of experiments with some standard data sets. Consecutive sections are dedicated to basics of the method: Dual Indiscernibility Matrix, Kernel and residuum in data sets, Pair Classifier, Experiments, Discussion of results, Conclusion and Perspectives.

P. Artiemjew (✉) · B. Nowak · L. Polkowski · P. Gorecki
Faculty of Mathematics and Computer Science, University of Warmia and Mazury, Olsztyn, Poland
e-mail: artem@matman.uwm.edu.pl

B. Nowak
e-mail: bnowak@matman.uwm.edu.pl

L. Polkowski
e-mail: polkow@pjwstk.edu.pl

P. Gorecki
e-mail: pgorecki@matman.uwm.edu.pl

L. Polkowski
Polish-Japanese Academy IT, Warszawa, Poland

G. Wang et al. (eds.), *Thriving Rough Sets*, Studies in Computational Intelligence 708, DOI 10.1007/978-3-319-54966-8_20

1 Introduction: Rough Set Theory

Rough set theory was proposed by Pawlak [13] as a tool for reasoning about *uncertain concepts* by means of a precise language. The basic notion of that theory, the *approximation space*, meant a pair of the form (U, R), where U denoted a set of *objects* and R was an *equivalence relation* on the set U. In this scenario, concepts defined as subsets of the set U were stratified into *exact* and *inexact (rough)* by means of the relation R and its equivalence classes, denoted $[x]_R$ for each object $x \in U$. A concept $X \subseteq U$ was termed *exact* if and only if X could be represented as a union of a family of equivalence classes of R, in the contrary case the concept X is rough. For a rough concept X, its description in terms of R was possible only by means of approximations from below and from above.

The *lower approximation* $\underline{R}X$ of the concept X is defined as the union $\bigcup\{[x]_R : [x]_R \subseteq X\}$ and the *upper approximation* $\overline{R}X$ is defined as the union $\bigcup\{[x]_R : [x]_R \cap X \neq \emptyset\}$. Clearly, $\underline{R}X \subseteq X \subseteq \overline{R}X$ and the three are equal if and only if X is exact. The reader will find a detailed algebraic analysis of approximations in Pawlak [14].

This approach to uncertain notions has paid in cases of raw data assembled in data tables. This tables were formalized in Pawlak [14] as *information systems*. An information system is a pair $IS = (U, A)$ where U is a set of objects and A is a set of *attributes* of objects. Each attribute $a \in A$ is formalized as a mapping on the set U into a set V of *possible values*; the symbol $a(u)$ denotes the value of the attribute a on the object u. To render in this setting the basic notion of an approximation space, one invokes the *Leibniz Principle* of *identitas indiscernibilium*, see Leibniz [10], which states that objects which by no available means can be discerned should be regarded as identical. Application of the Leibniz Principle leads to the hierarchy of *indiscernibility relations* $\{IND_B : B \subseteq A\}$ where each relation IND_B is defined as the set of pairs:

$$\{(x, y) : \bigwedge a \in B.a(x) = a(y)\}. \tag{1}$$

The equivalence relation IND_B defines the family of $B-exact$ concepts and the family of $B-rough$ concepts. An information system, reflecting the user choice of attributes, can be confronted with the expert in a given process whose evaluation of objects is reflected in the *decision attribute* d. The *decision system* is a triple (U, A, d) where (U, A) is an information system and $d \notin A$ is a decision attribute, see Pawlak [14].

To describe objects in an information/decision system $(U, A)/(U, A, d)$, one uses the language of *descriptors*; a descriptor is a formula of the form $(a, a(u))$, where $a \in A \cup \{d\}$ is an attribute and $u \in U$ is an object. Descriptors are interpreted in the set U of objects and the meaning of the descriptor $(a, a(u))$ is

$$[(a, a(u))] = \{v \in U : a(v) = a(u)\}. \tag{2}$$

Descriptors relevant for an object $u \in U$, are collected in the *information set of u*, denoted $Inf(u)$ and defined as:

$$Inf(u) = \{(a, a(u)) : a \in A\}. \tag{3}$$

The relation between the set $Inf = \{Inf(u) : u \in U\}$ and the decision is expressed by *decision rules* of the form:

$$\bigwedge Inf(u) \Rightarrow (d, d(u)). \tag{4}$$

2 Mereology and Rough Mereology

Mereology in its classical form was created by Leśniewski [11], see Polkowski [16, 17]. Its basic notion is the notion of a *part* relation, denoted π. Parts are required to satisfy the following conditions on a set of objects U:

$$1. \bigwedge_{u \in U} \neg\pi(u, u) \; 2. \; \pi(u, v) \wedge \pi(v, w) \Rightarrow \pi(u, w). \tag{5}$$

The relation between mereology and rough set theory becomes evident when one realizes that the subset relation in the set theory satisfies the requirements to be a part relation, hence, a characterization of a rough concept X is: *X is rough if and only if* $\pi(\underline{R}X, X)$ *if and only if* $\pi(X, \overline{R}X)$. It follows that the substantial relational-algebraic fragment of rough set theory can as well be formalized in the language of mereology.

2.1 Rough Mereology: Basic Notions

It was realized early in the development of rough set theory that the indiscernibility relations do not give satisfactory flexibility in dealing with complex objects and data and the need was felt was more parameterized theory, see Polkowski et al. [21] as the first work in which tolerance relations were introduced into rough set theory, and from that time many papers were produced on similarity relations in rough sets.

One way of introducing similarity and tolerance relations into rough sets was proposed in Polkowski and Skowron [19, 20], as *rough mereology* which Achille Varzi termed ‘fuzzified mereology’, see Varzi [26]. Rough mereology aimed at capturing the essential features of partial containment and its basic notion has been proposed as a ternary relation $\mu(u, v, r)$ read ‘the object u is a part to a degree of r in the object v’, where r is a real number in the interval [0,1] and u, v are objects in a set of objects U. The relation μ was called in [19] a *rough inclusion* and the requirements it has been subjected to have been as follows

- $\mu(u, v, 1)$ if and only if there exists a weak tolerance relation τ satisfying the conditions: 1. $tau(u, u)$; 2. $\tau(u, v) \wedge \tau(v, u) \Rightarrow u = v$; 3. $\tau(u, v) \wedge \tau(v, w) \Rightarrow \tau(u, w)$.
- $\mu(u, v, 1) \wedge \mu(w, u, r) \Rightarrow \mu(, v, r)$.
- $\mu(u, v, r) \wedge s < r \Rightarrow \mu(u, v, s)$.

The choice of a particular rough inclusion depends on the context; for instance, in the problem of intelligent control of a formation of autonomous mobile robots, the rough inclusion was defined on closed convex subsets of the plane as:

$$\mu(C, D, r) \Leftrightarrow \frac{||C \cap D||}{||C||} \geq r, \tag{6}$$

where $||X||$ denotes the area of X.

2.2 Rough Mereology in Information and Decision Systems. Meregeometry in Data

For an information system $IS = (U, A)$ (a similar analysis holds for any decision system, so we omit this case for now), we define as standard the *Łukasiewicz rough inclusion* in whose definition the Jan Łukasiewicz idea of partial truth value is exploited, see Łukasiewicz [12]

$$\mu_L(u, v, r) \Leftrightarrow \frac{|Inf(u) \cap Inf(v)|}{|Inf(u)|} \geq r. \tag{7}$$

We recall that $|X|$ denoted the cardinality of X.

For a rough inclusion $\mu(u, v, r)$, one defines the mereological distance function $\kappa_\mu$as follows:

$$kappa(u, v) = min\{argmax_r \mu(u, v, r), argmax_s \mu(v, u, s)\}. \tag{8}$$

The distance function κ allows for introduction of geometry into the information system. First, we define the mereological counterpart of the relation of *nearness* due to Johan van Benthem in [2], denoted $N(u, v, w)$ and read 'u is nearer to than w', see Polkowski [16]:

$$N(u, v, w) \Leftrightarrow \kappa(u, v) \geq \kappa(w, v). \tag{9}$$

The relation of nearness is a prelude to the relation of betweenness. The relation of betweenness was introduced by Alfred Tarski in his lectures on elementary geometry in Warsaw University in 1926-7, and it was along with the relation of equidistance the basis for axiomatization of elementary Euclidean geometry, see Tarski and Givant [25]. In Tarski axiomatization the meaning of the relation $Btw(u, v, w)$ was that 'u is between v and w', i.e., u is a point in the segment with endpoints v, w.

The abstract version of the betweenness relation due to van Benthem [2] is the relation $B(u, v, w)$ defined as follows,

$$B(u, v, w) \Leftrightarrow \bigwedge_t [(t = u) \vee N(u, v, t) \vee N(u, w, t)], \tag{10}$$

which means that for each object t distinct from u, the object u is nearer to one at least of endpoints v, w than t.

3 Dual Indiscernibility Matrix (DIM), Kernel, Residuum

We consider an information system $IS = (U, A)$ with the value set V of possible attribute values denoted by the generic symbol q. For the system IS, we define, see Polkowski [15], the matrix DIM as the matrix of the form $[c_{a,q}]_{A \times V}$, where

$$c_{a,q} = \{u \in U : a(u) = q\}. \tag{11}$$

We generalize the betweenness relation B of (10) to the relation GB of *generalized betweenness* in the context of the information system IS, to express betweenness of an object among a finite number of other objects,

$$GB(u, v_1, v_2, ..., v_k) \Leftrightarrow \bigwedge t.(t = u) \vee \bigvee j \leq k.\kappa(u, v_j) \geq \kappa(t, v_j). \tag{12}$$

We interpret this notion of betweenness in case when κ is induced by the Łukasiewicz rough inclusion (7).

Theorem 1 (Polkowski [15]) *$GB(u, v_1, v_2, ..., v_k)$ is true if and only if the information set $Inf(u)$ can be partitioned into sets $C_1, C_2, ..., C_k$, where $C_i \subset Inf(v_i)$ for $i = 1, 2, ..., k$. Letting $\alpha_i = \frac{card(C_i)}{card(A)}$ for $i = 1, 2, ..., k$, we represent u as the convex combination of $v_1, v_2, ..., v_k$ with $\sum_{i=1}^{k} \alpha_i = 1$, each α_i positive.*

For the information system IS, applying DIM to it, we single out the set

$$Res(IS) = \{u \in U : \bigvee (a, q).c_{a,q} = \{u\}, \tag{13}$$

i.e., u is an outlier in the sense that the value q of a is taken only on u. We call the set $Res(IS)$ the *residuum of IS*. The complement $U \backslash Res(IS)$ is called the *kernel of IS* denoted $Ker(IS)$.

The idea behind the kernel is that objects in it are approximated by some other objects in it in the sense of the generalized betweenness GB, close to it in the sense of (12), hence this idea is similar to the idea of nearest neighbors NN; the difference is in the fact that u in (12) is built from fragments of $v_1, v_2, ..., v_k$ making the idea of closeness expressed stronger than in case of NN.

Table 1 Classification results

Database	Set tested	Accuracy of C4.5	Accuracy of k-NN	Number of samples
Adult	Whole set	0.857 ± 0.003	0.837 ± 0.003	39074.0
	Ker	0.853 ± 0.004	0.835 ± 0.003	22366.0
	Res	0.849 ± 0.003	0.833 ± 0.003	16708.0
PID	whole set	0.733 ± 0.027	0.723 ± 0.021	614.4
	Ker	0.704 ± 0.037	0.711 ± 0.032	212.9
	Res	0.724 ± 0.035	0.745 ± 0.030	401.5
Fertility	Whole set	0.852 ± 0.073	0.866 ± 0.060	80.0
Diagnosis	*Ker*	0.846 ± 0.075	0.880 ± 0.064	71.6
	Res	0.852 ± 0.068	0.880 ± 0.064	8.4
German	whole set	0.713 ± 0.023	0.732 ± 0.025	800.0
Credit	*Ker*	0.671 ± 0.045	0.714 ± 0.038	98.9
	Res	0.712 ± 0.023	0.726 ± 0.030	701.1
Heart	Whole set	0.750 ± 0.054	0.825 ± 0.048	216.0
Disease	*Ker*	0.742 ± 0.061	0.822 ± 0.051	109.2
	Res	0.767 ± 0.054	0.827 ± 0.041	106.8

Both subsets, the kernel and the residuum yield very satisfactory results when standard well-known classifiers are applied to them. Table 1 shows a comparison among accuracy of C4.5 and kNN classifiers applied, respectively, to the whole set, the kernel and the residuum, see [18].

4 The Pair Classifier

The success of the kernel in representing the whole system, as shown in Table 1, suggests a further simplification of the procedure for approximation of test objets by training objects and here we propose a study of approximation of test objects by pairs of training objects successively approximating test objects best. We consider anew a decision system (U, A, d) in the task of classification, and we split the set U into the training set U_{trn} and the test set U_{tst}.

Given a test object $u_{tst} \in U_{tst}$, we define the *best training cover of level 0 for* u_{tst} denoted $best - 00 - u_{tst}$, as follows,

$$best - 00 - u_{tst} = argmax_{v_{trn}} |Inf(v_{trn}) \cap Inf(u_{tst}|. \tag{14}$$

Next, we define the *second best training cover of level 0 for* u_{tst} denoted $best - 01 - u_{tst}$,

$$best - 01 - u_{tst} = argmax_{v_{trn} \neq best-00-u_{tst}} |Inf(v_{trn}) \cap Inf(u_{tst}|. \quad (15)$$

This two training objects constitute the level 0 of approximation: $L0 = \{best - 00 - u_{tst}, best - 01 - u_{tst}\}$. Assuming training objects of levels up to the level k, Lk, have been defined for the test object u_{tst}, we define training objects of the level (k+1), L(k+1), as:

$$best - (k+1)0 - u_{tst} = argmax_{v_{trn} \in U \setminus \bigcup \{Li: i=0,1,\ldots,k\}} |Inf(v_{trn}) \cap Inf(u_{tst}|, \quad (16)$$

and $best - (k+1)1 - u_{tst} =$

$$argmax_{v_{trn} \in U \setminus [\bigcup \{Li: i=1,2,\ldots,k|\} \cup \{best-(k+1)0-u_{tst}} |Inf(v_{trn}) \cap Inf(u_{tst}|. \quad (17)$$

The variable *max* stores the largest number *m* such that Lm is defined.

Having levels defined pairs in them are pooled according to a chosen scheme and their pooled set of decision values is subjected to Majority Voting with the random tie resolution. We can have the following schemes,

- Only pairs of level 0;
- Pairs of the specified best level;
- All pairs up to a specified level;
- All pairs selected up to Lmax.

Table 2 shows results of a comparison among Pair Classifier and k-NN, and, Bayes classifiers. The symbol Lx denotes the whole system, as shown in Table 1, level of covering, Pair-0 is the simple pair classifier with approximations by pairs and Pair-best denotes the best result over levels studied.

Table 2 Pair classifier

Database	kNN	Bayes	Pair-best	Pair-0
Adult	0.841	0.864	0.853 L1	0.823
Australian	0.855	0.843	0.859 L4,5	0.859
Diabetes	0.631	0.652	0.721 L0	0.710
German credit	0.730	0.704	0.722 L1	0.721
Heart disease	0.837	0.829	0.822 L1	0.800
Hepatitis	0.890	0.845	0.892 L0	0.831
Congressional voting	0.938	0.927	0.928L0	0.928
Mushroom	1.0	0.910	1.0 L0	1.0
Nursery	0.578	0.869	0.845 L0	0.845
Soybean large	0.928	0.690	0.910 L0	0.910

5 Stabilization of Pair Classifier by Some Selected Methods

As the last leg in this study, we propose to apply to the Pair classifier some methods of stabilization. We propose to stabilize results of classification by Pair Classifier by Bootstrap Aggregating, Boosting based on Arcing, Boosting based Ada–Boost with Monte carlo split and Random Forests. We recall below for the benefit of the reader the principles of all of them. Ensemble methods own their resurgence to the question:'can a set of weak learners create a strong learner?', see Kearns [8] and Kearns and Valiant [9]. Methods for making this goal possible are known under the common name of boosting, see Freund and Schapire [5], Shapire [23, 24], Zhou [27, 28]. We briefly recapitulate the approaches to boosting selected by us.

5.1 *Bootstrap Aggregating*

This method proposed by Breiman [3] known also as *bagging*, produces from the training set *Trn* of size n, a number of copies of size n by random sampling with replacement; the number m of copies is a parameter. Each copy is used to produce a classification on the test part of data and in case of classification, the final result is produced by Majority Voting. Bootstrap aggregating is also used in statistical analysis of data in order to reduce variance in high-variance classifiers like regresion or decision trees.

5.2 *Bosting Based on Arcing*

The idea of bagging has been strengthened, see Freund-Shapire [5, 6] and Shapire [24], cf. also Breiman [4] by coupling it with the idea of 'arcing' ('adaptive resampling and combining'). As with bagging, the user specifies the number of steps in the procedure. In the first step, the copy of the training set is i.i.d. drawn with uniform probability. In subsequent steps, probabilities with which copies are re-sampled change. Assuming the probability distribution at the kth step to be P^k, and the classifier induced from the kth copy be C^k, its error is

$$d(u) = 1 \text{ if u is classified incorrectly else } 0. \tag{18}$$

One lets

$$\varepsilon_k = \sum_x P^k(x) \times d(x) \tag{19}$$

as the error of classification by C^k averaged over all objects in the kth copy. Then the parameter

$$\beta_k = \frac{1-\varepsilon_k}{\varepsilon_k} \tag{20}$$

is computed, and, the probability distribution for the (k+1)st copy is defined as

$$P^{k+1}(u) = \frac{P^k(u) \times \beta_k^{d^k(u)}}{\sum_v P^k(v) \times \beta_k^{d^k(v)}}, \tag{21}$$

the sum extended over all v's in the (k+1)st copy. It follows that arcing assigns a greater weight to worse classifiers. Finally, after the assumed number of steps, all classifiers vote, the classifier C^k with the weight equal to $log\beta_k$.

5.3 *AdaBoost with Monte Carlo Split*

AdaBoost, see Freund-Shapire [7], see Rojas [22], meaning 'adaptive boosting' is a proposition on the Kearns question 'how weak learners may combine into a strong learner?'. AdaBoost consists in adaptively adding consecutive weak classifiers in a judicious way. The rational choice of the next weak classifier provided by AdaBoost is based on the following consideration. For the training sample of the form $\{(x_i, y_i) : i = 1, 2, ..., n\}$, where each $y_i \in \{+1, -1\}$, $+1$ meaning correct classification, -1 misclassification, consider ensemble of classifiers $C \in \mathscr{C}$ of finite cardinality, and assume that after (k-1)-steps the boosted classifier is of the form $C^{(k-1)} = a_1C_1 + a_2C_2 + \cdots + a_{m-1}C_{m-1}$ and at the mth step we want to choose C_m and its weight a_m in order to make a better classifier.

Introducing the exponential loss $\sum_{i=1}^{k} exp(-y_iC_m(x_i))$, and the weight $w_i^{(m)} = exp(-y_iC_{m-1}(x_i)$, we can represent the error by the mth classifier as $\sum_{i=1}^{k} w_i^{(m)} exp(-y_ia_mC_m(x_i))$. It follows (see works quoted) that the best mth classifier C_m is the one that minimizes the error $\sum_{i=1}^{k} w_i^{(m)}$ and the weight a_m is given as $\frac{1}{2}ln\frac{1-\varepsilon_m}{\varepsilon_m}$ where $\varepsilon_m = \frac{\sum_{y_i \neq C_m(x_i)} w_i^{(m)}}{\sum_{i=1}^{k} w_i^{(m)}}$. In AdaBoost with Monte Carlo split the *Trn* set is split according to a fixed ratio, usually about 0.6. Except for this the algorithm is as the previous one.

6 Experimental Session

In the experimental part we have carried out experiments on the real data from the UCI Repository [29]. We have carried out experiments with use of multiple Cross Validation 5 (CV-5) [3]. We use Australian, Pima Indians Diabetes and Heart Disease data sets. The boosting, stabilisation methods are performed with pair classifier five times and the average results are presented on the plots. Considering the results from [1] we have chosen for pair classifier the parameters max = 1 and descriptors indiscernibility ratio $\varepsilon = 0.01$. The experimental session was extensive but due to lack of space we show only a few selected results.

7 Discussion of Results

In this section we have the results of our experimental session. In Figs. 1, 2 and 3, we have the result for Australian credit, Heart disease and Diabetes, respectively. In each figure, on the top, we have the result for Ensemble of Bootstraps, in the middle the result for Boosting based on Arcing and the bottom plots are for Boosting based on the Monte Carlo split. On the plots we have the average result from five series of 50 iterations of learning and the quality of classification is presented in the sense of global accuracy of classification. When looking at all results, it turns out that the stabilisation of classification for the committee starts from about 20 iterations of learning, and after that the standard deviation of results was less than 0.005. In all cases there is high stabilisation of classification for committee of classifiers in comparison with the result for each separate weak classifier. It is difficult to point to one ensemble method which could work best for all data, it is evident that the quality is highly dependent on data internal logic. For instance, the Ensemble of Bootstraps works best for Australian credit, in case of Heart disease the Boosting based on Arcing seems to be the best, but in case of Diabetes, Ensemble of Bootstraps and Boosting based on Monte carlo split works best with a similar accuracy.

8 Conclusions

In this work we have checked the effectiveness of selected ensemble models for our novel dual matrix based Pair Classifier. We use three more popular methods of boosting, Ensemble of Bootstraps, Arcing and Boosting by Monte Carlo split. We have investigated the version of the pair classifier introduced after proposal by Polkowski in [18]. We have carried out experiments on the real data from the UCI repository. The result show the classification quality improvement for pair based classification with use of all boosting variants. There is no one rule for choice of the proper ensemble model for any data, the effectiveness depends on the type of data set. For instance, in case of Australian credit data set the Ensemble of Bootstraps works best. In case of Heart disease data set the Boosting based on Arcing seems to be the best. And in case of Diabetes data set the Ensemble of Bootstraps and Boosting based on Monte Carlo split wins. In all cases there is significant improvement of classification accuracy for committee of classifiers in comparison with the single weak classifier. The classifier stabilizes after a few iterations of learning.

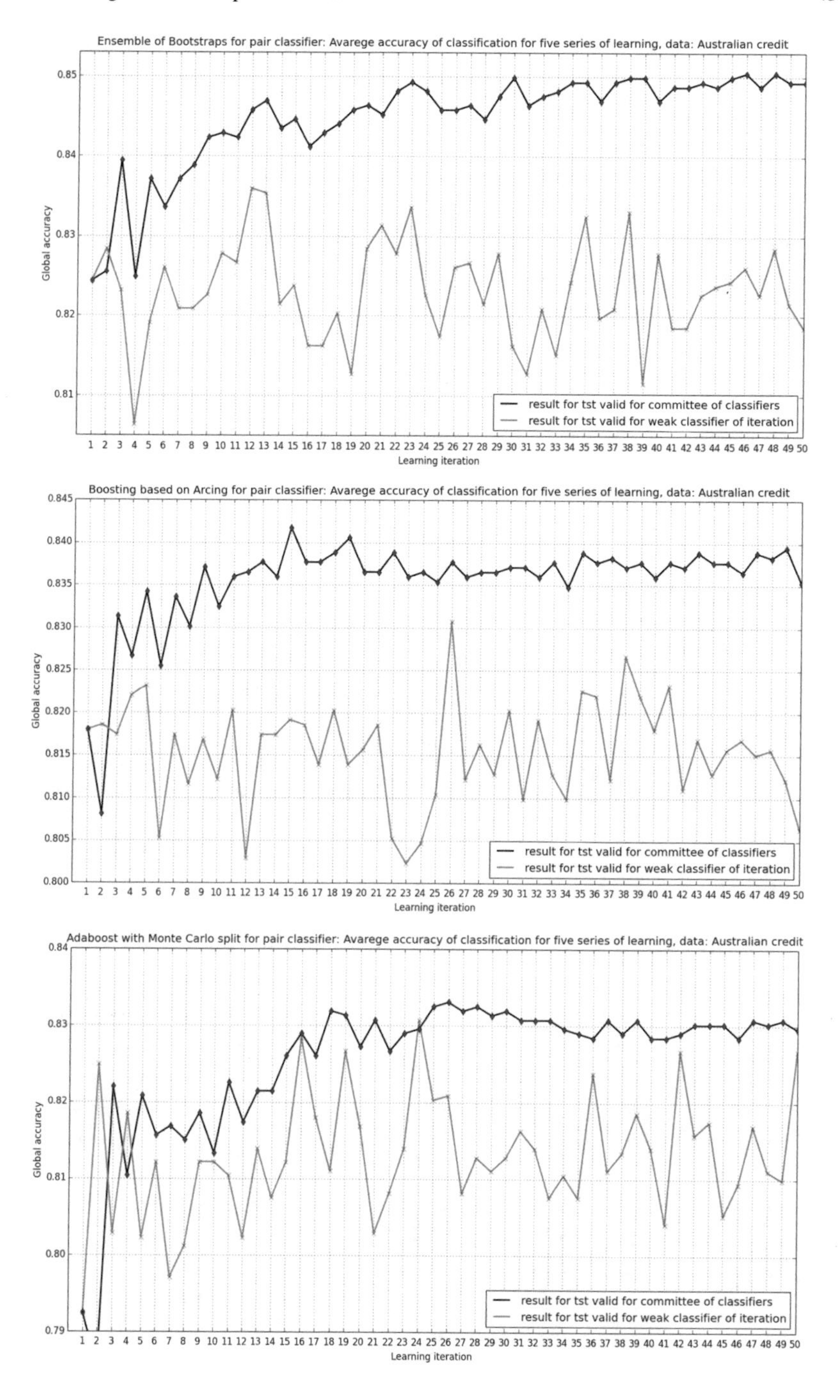

Fig. 1 The result for 5×50 iterations of learning for Australian credit for pair classifier. The first picture show the result for Ensemble of Bootstraps, the second for Arcing, and the last one for Ada-Boost

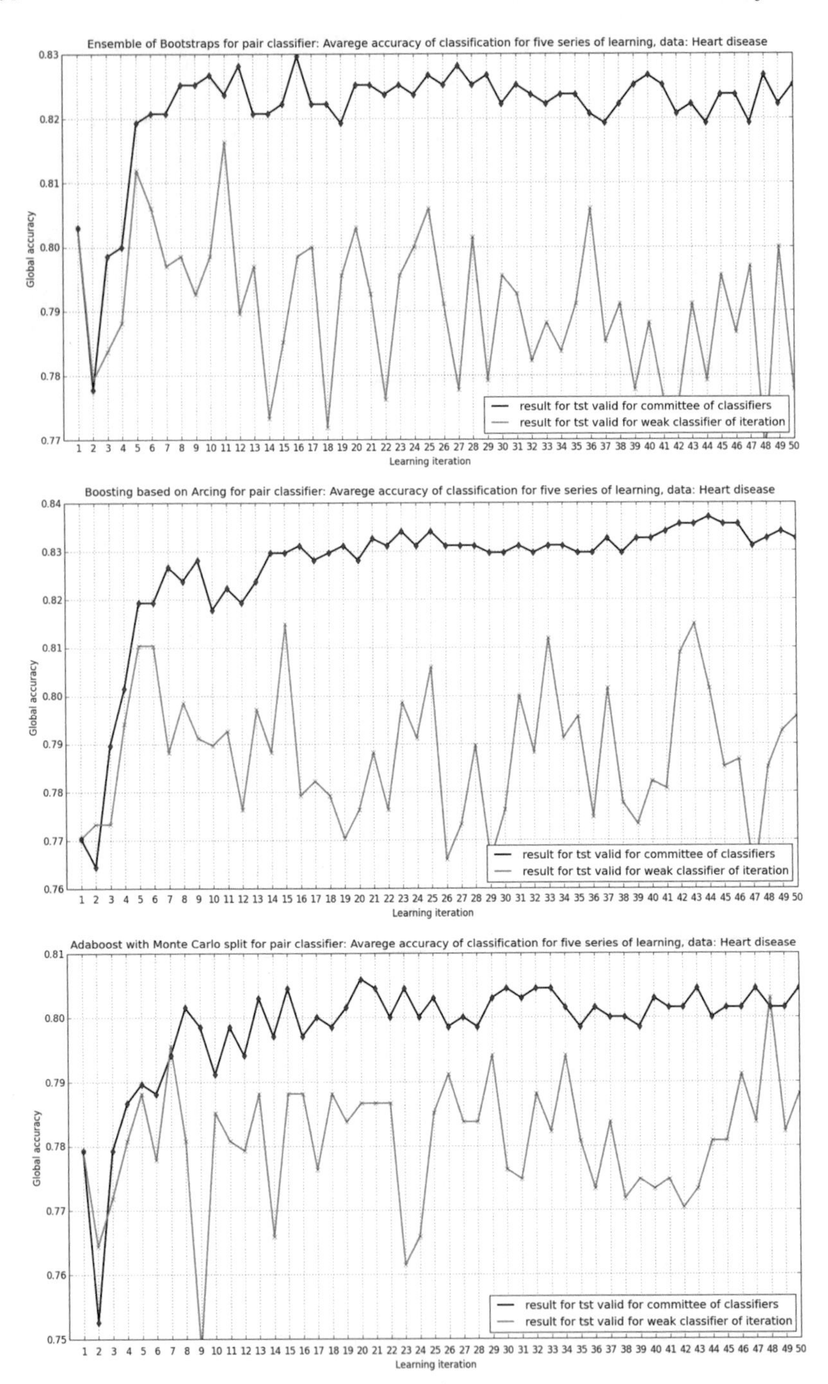

Fig. 2 The result for 5×50 iterations of learning for Heart disease for pair classifier. The first picture show the result for Ensemble of Bootstraps, the second for Arcing, and the last one for Ada-Boost

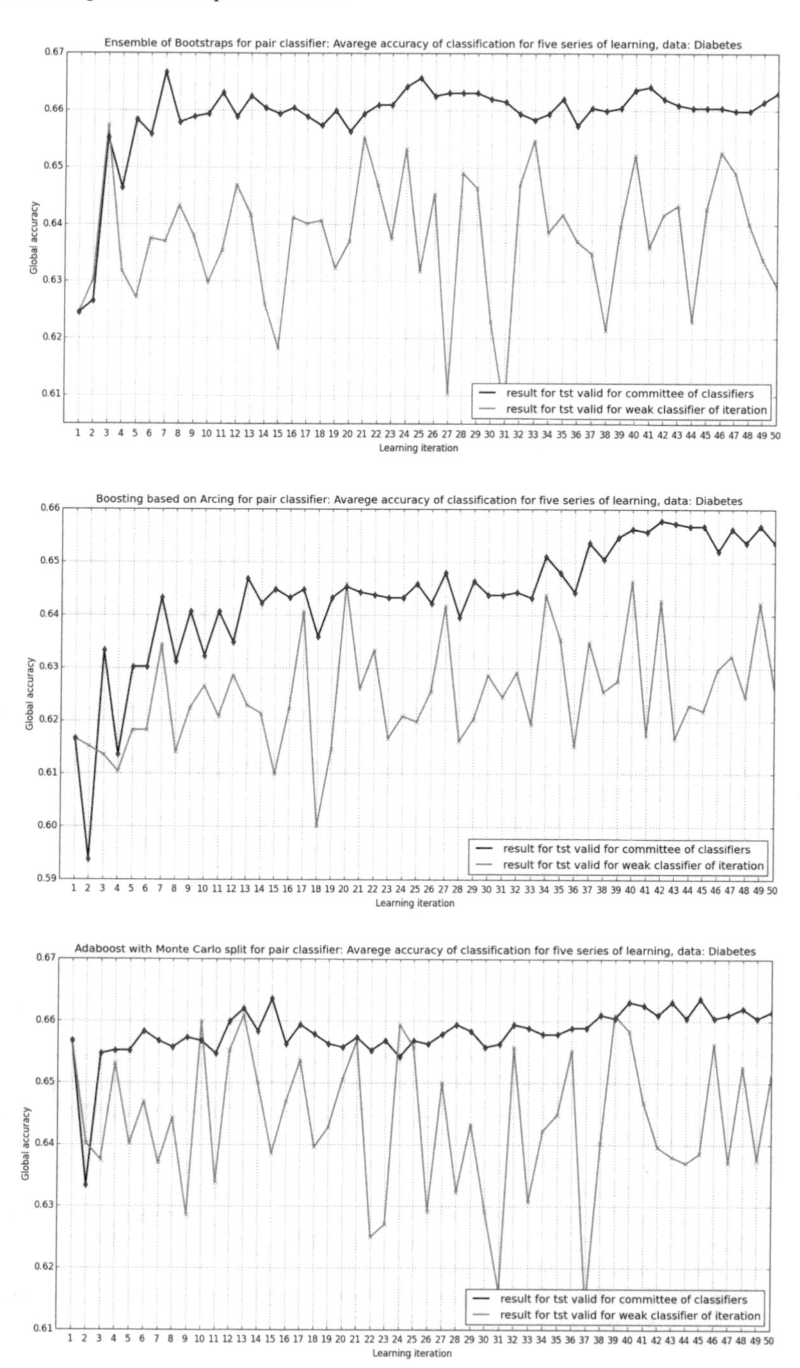

Fig. 3 The result for 5×50 iterations of learning for Diabetes for pair classifier. The first picture show the result for Ensemble of Bootstraps, the second for Arcing, and the last one for Ada-Boost

In the future works, we would like to check the effectiveness of pair classifier on the approximated data sets. Additionally we have plan to extend the basic version of proposed classifier focusing on the decision classes of the training data and to check the version based on the fuzzy sets.

Acknowledgements The research has been supported by grant 1309–802 from Ministry of Science and Higher Education of the Republic of Poland.

References

1. Artiemjew, P., Nowak, B., Polkowski, L.: A new classifier based on dual indiscernibility matrix. In: Proceedings of ICIST 2016 to Appear in Communications in Computer and Information Sciences. Springer (2016)
2. van Benthem, J.: The Logic of Time. Springer Synthese Library (1991)
3. Breiman, L.: Bagging predictors. Mach. Learn. **24**(2), 23–40 (1996)
4. Breiman, L.: Arcing classifier. Ann. Stat. **26**(3), 801–849 (1998)
5. Freund, Y., Schapire, R.E.: Experiments with a new boosting algorithm. In: Proceedings of the 13th International Conference Machine Learning, pp. 148–156. Morgan Kaufmann, San Francisco (1996)
6. Freund, Y., Schapire, R.E.: A decision-theoretic generalization of online learning and an application to boosting. J. Comput. Syst. Sci. **55**, 119–139 (1997)
7. Freund, Y., Schapire, R.E.: A short introduction to boosting. J. Jpn. Soc. Artif. Intell. **14**(5), 771–780 (1999)
8. Kearns, M.: Thoughts on hypothesis boosting. Manuscript (unpublished)(Machine Learning class project, December 1988) (1988)
9. Kearns, M., Valiant, L.: Crytographic limitations on learning Boolean formulae and finite automata. Symposium on Theory of computing (ACM) **21**, 433–444 (1989)
10. Leibniz, G.W.: Philosophical papers and letters. In Loemker (1969)
11. Leśniewski, S.: Foundations of the General Theory of Sets (in Polish: Podstawy Oglnej Teoryi Zbiorw). The Polish Scienific Circle, Moscow (1916)
12. Łukasiewicz, J.: Die Logischen Grundlagen der Wahrscheinlichtkeitsrechnung. Krakw (1913)
13. Pawłak, Z.: Rough sets. Int. J. Comput. Inf. Sci. **11**, 341–356 (1982)
14. Pawłak, Z.: Rough Sets. Theoretical Aspects of Reasoning about Data. Kluwer, Dordrecht (1991)
15. Polkowski, L.: Betweenness, Łukasiewicz rough inclusions, Euclidean representations in information systems, hyper–granules, conflict resolution. In: Proceedings of 24th Workshop CS&P (Concurrency, Specification, Programming), Vol. 1492/97–110. Rzeszow University, Poland (2015). http://ceur-ws.org/
16. Polkowski, L.: Approximate reasoning by parts. An Application of Rough Mereology. ISRL, vol. 20. Springer, Berlin (2011)
17. Polkowski, L.: Mereology in engineering and computer science. In: Calosi, C., Graziani, P. (eds.) Mereology and the Sciences. Springer Synthese Library, vol. 371, pp. 217–292. Springer International, Cham, Switzerland (2015)
18. Polkowski, L., Nowak, B.: Betweenness, Łukasiewicz rough inclusions, euclidean representations in information systems, hyper-granules, conflict resolution. Fundamenta Informaticae **147**(2–3), 337–352 (2016)
19. Polkowski, L., Skowron, A.: Rough mereology. In: Proceedings of ISMIS'94. LNAI, vol. 869, pp. 85–94 (1994)
20. Polkowski, L., Skowron, A.: Rough mereology. A new paradigm for approximate reasoning. Int. J. Approx. Reason. **15**(4), 333–365 (1997)

21. Polkowski, L., Skowron, A., Zytkow, J.: Tolerance based rough sets. In: Soft Computing. Simulation Concil Inc., San Diego (1996)
22. Rojas, R.: AdaBoost and the super bowl of classifiers, a tutorial introduction to adaptive boosting. Technical report, Freie University, Berlin (2009)
23. Schapire, R.E.: The Boosting approach to machine learning: an overwiev. Mathematical Sciences Research Institute, Workshop on Non-linear Estimation and Classification (2003)
24. Schapire, R.E.: Using output codes to boost multiclass learning problems. In: Proceedings of the 14th International Conference on Machine Learning, pp. 313–321. Morgan Kaufmann, San Francisco (1997)
25. Tarski, A., Givant, S.: Tarski's system of Geometry. Bull. Symb. Log. **5**(2), 175–214 (1999)
26. Varzi, A.: Mereology. In: Stanford Encyclopedia of Philosophy. https://plato.stanford.edu
27. Zhou, Z.-H.: Ensemble Methods: Foundations and Algorithms. Chapman and Hall/CRC (2012)
28. Zhou, Z.-H.: Boosting 25 years. CCL 2014 Keynote (2014)
29. UCI Repository. http://archive.ics.uci.edu/ml/

Printed in the United States
By Bookmasters